Prospects of Plant-Based Vaccines in Veterinary Medicine

Jacqueline MacDonald
Editor

Prospects of Plant-Based Vaccines in Veterinary Medicine

Springer

Editor
Jacqueline MacDonald
Western University
London, ON
Canada

ISBN 978-3-030-07941-3 ISBN 978-3-319-90137-4 (eBook)
https://doi.org/10.1007/978-3-319-90137-4

Printed on acid-free paper

This Springer imprint is published by the registered company Springer International Publishing AG part of Springer Nature
The registered company address is: Gewerbestrasse 11, 6330 Cham, Switzerland

Preface

With the high-profile research into recombinant plant-made therapeutics for Ebola and Zika virus, the commercialization and widespread use of such products are likely to occur in the future. Plant-made therapeutics have a variety of advantages over those made in traditional systems; yet their most fruitful application may be in veterinary medicine, due to less stringent regulations and a higher need for low-cost products. This book provides an in-depth explanation of the advantages and current limitations of recombinant plant-made vaccines for use in veterinary medicine. It discusses the background and up-to-date scientific achievements on plant-made vaccines for the most commonly targeted veterinary infections, written by top scientists in the field.

London, Canada Jacqueline MacDonald

Contents

History and Promise of Plant-Made Vaccines for Animals

Ed Rybicki

Abstract Plant-made vaccines are now a well-established and well-tested concept in veterinary medicine—yet the only product so far licenced was never produced commercially. This is puzzling, given the breadth of exploration of plant-made animal vaccines, and their immunogenicity and efficacy, over more than twenty years of research. The range of candidate vaccines that have been tested in laboratory animal models includes vaccines for *E. coli*, *Salmonella, Yersinia pestis*, foot and mouth disease virus, rabbit haemorrhagic disease virus, rabbit and canine and bovine papillomaviruses, mink enteritis and porcine circovirus, and lately also bluetongue virus, among many others. There are many proofs of efficacy of such vaccines, and regulatory pathways appear to have been explored for their licencing. This review will briefly explore the history of plant-made vaccines for use in animals, and will discuss the unique advantages of plant-made vaccines for use in a veterinary medicine setting in detail, with a proposal of their relevance within the "One Health" paradigm.

Keywords Plant-made vaccines · Therapeutic vaccine · One health
Transient expression · *Agrobacterium tumefaciens* · *Nicotiana benthamiana*
Regulatory · cGMP · FDA · EMEA

1 Introduction

Plant-produced vaccines for veterinary medicine are an exciting prospect, largely because of the possibilities of producing protein-based vaccines' including edible vaccines' at low cost, at almost any scale, and potentially locally and on demand. They have also been controversial because of the very real possibilities of contamination of the human food supply with vaccine-producing transgenic plants, and

E. Rybicki (✉)
Biopharming Research Unit, Department of Molecular and Cell Biology, University of Cape Town, Cape Town, South Africa
e-mail: ed.rybicki@uct.ac.za

J. MacDonald (ed.), *Prospects of Plant-Based Vaccines in Veterinary Medicine*,
https://doi.org/10.1007/978-3-319-90137-4_1

because of concerns around the possibility of immunological tolerance developing to oral or edible vaccines. However, one set of problems that many foresaw—regulatory and production problems—has not eventuated, and in fact the environment now seems primed very favourably for their introduction.

The main justifications for plant-made vaccines are that vaccine antigen production in plants is safe; that it is both cheap and highly scalable; that plants produce and process eukaryote-derived proteins much better than can bacteria or even yeasts; that use of plants would allow for production of vaccines in the developing world where they are needed most; and of vaccines or therapeutics that will never be produced economically by other technologies.

However, despite more than twenty years of development, there are still no plant-produced vaccines or biologics available for animals—although there are in fact products licenced for and in use in humans.

This review will explore the early history of plant-produced vaccines with an emphasis on proofs of principle and of efficacy, what the recent development of robust, stable transient plant production systems for vaccine antigens could mean for veterinary medicine, and the potential of plant-produced vaccines to advance both animal and potentially human health' under the banner of the One Health Movement.

2 Early History of Plant-Made Animal Vaccines

While viral proteins have probably been the most common vaccine candidates made in plants (reviewed in Rybicki 2014), it was expression of a bacterial protein—*Escherichia coli* heat labile enterotoxin (LT-B)—that first proved that veterinary-relevant antigens could be produced in plants, and provided the first proof of principle for edible vaccines. LT-B produced in transgenic tobacco or potatoes (Haq et al. 1995) was functionally equivalent to *E coli*-produced protein in specific assays, and immunisation of mice by oral gavage with plant material elicited systemic and mucosal toxin neutralising antibodies. Moreover, fresh potato containing LT-B was immunogenic in mice when eaten.

An early virus vaccine candidate was one against mink enteritis virus (MEV) disease: this was novel in that it comprised chimaeric Cowpea mosaic virus (CPMV) virions incorporating a short linear epitope from MEV VP2 capsid protein and displaying it on the surface of virions, produced by inoculation of bean plants with an infectious cDNA clone of rCPMV (Dalsgaard et al. 1997). This conferred protection against clinical disease and virtually abolished virus shedding—and given that the epitope sequence used is found in MEV, canine parvovirus, and feline panleukopenia virus, the same vaccine could potentially also protect against these viruses.

Another early virus vaccine candidate was against rabbit haemorrhagic disease virus (RHDV): this was made by expressing the whole RHDV VP60 capsid protein in transgenic potatoes; parenteral immunisation with plant extracts was protective in rabbits (Castanon et al. 1999). Subsequently, another study demonstrated that an edible vaccine consisting of leaves of transgenic plants containing presumably partially-assembled VP60 subunits, was an effective priming vaccine for later baculovirus-derived parenterally-delivered vaccine (Gil et al. 2006).

The first report of a Foot and mouth disease virus (FMDV) plant-made antigen was of expression in plant protoplasts of a VP1-derived peptide of FMDV as an insertion into the minor coat protein of a replicating CPMV as a demonstration of antigen presentation (Usha et al. 1993). However, the first proof of efficacy was done using transgenic *Arabidopsis thaliana* expressing whole VP1: parenteral immunisation of mice with leaf extracts elicited antibodies that bound to VP1 and to intact FMDV particles, and all immunised mice were protected against virulent FMDV challenge (Carrillo et al. 1998). The Wigdorovitz group went on to demonstrate that mice could be protected against FMDV challenge after oral or parenteral vaccination with extracts of transgenic alfalfa expressing VP1 (Wigdorovitz et al. 1999), or immunisation with leaf extracts of tobacco plants expressing VP1 via a recombinant Tobacco mosaic virus vector (Wigdorovitz et al. 1999). A refinement of these achievements included transgenic expression in alfalfa of amino acid residues 135–160 of VP1 (VP135-160) fused to glucuronidase (GUS), which both allowed selection of strongest expressers by assay of enzyme activity, and was protective in mice (Dus Santos et al. 2002). Another novel application of carrier technology was the insertion of VP1 amino acids 140–160 (G-H loop) in an interior region of the hepatitis B virus core antigen gene (HBcAg), and expression of the chimaera in transgenic *Nicotiana tabacum*. The chimaeric protein formed virus-like particles (VLPs) in the tobacco leaves, and mice immunised intraperitoneally with a soluble extract were protected against viral challenge (Huang et al. 2005).

An early attempt at showing the feasibility of making an anthrax vaccine was the expression in transgenic *N. tabacum* of the protective antigen (PA) protein of *Bacillus anthracis*, possibly the best target for a subunit vaccine because it alone is protective (Aziz et al. 2002), although it went no further than showing cytolytic activity of the protein. Soon after, the same group went on to express PA in transplastomic *N. tabacum,* with significant yield increases but still no efficacy trial (Aziz et al. 2005). Another investigation of transplastomic tobacco by Henry Daniell's group was more thorough: yields were high (2.5 g/kg in fresh leaf tissue), the protein was protected in chloroplasts from protease cleavage and was stable when stored in leaves or as crude extracts, and was biologically active (Watson et al. 2004). While they did not show immunogenicity or protection, the authors speculated that "With an average yield of 172 mg of PA per plant using an experimental transgenic cultivar grown in a greenhouse, 400 million doses of vaccine (free of contaminants) could be produced per acre". The Daniell group subsequently showed that chloroplast-derived PA was equal in potency to the natural product from *B anthracis*, and that mice immunised subcutaneously with partially purified chloroplast-derived PA with adjuvant produced high IgG titres and survived challenge with lethal doses of toxin (Koya et al. 2005).

A different sort of approach to anthrax, and one of the first attempts at making a therapeutic antibody in plants, was taken by Vidadi Yusibov's group, who used the technique of transient *Agrobacterium* infiltration-mediated expression in

N. benthamiana to produce a human-derived PA-specific monoclonal antibody (Hull et al. 2005). The antibody neutralised toxin activity both in vitro and in vivo at a comparable level to hybridoma-produced antibodies. The Yusibov group at what became Fraunhofer USA Center for Molecular Biotechnology later used the same transient expression technology to separately express artificial antigens comprising domain 4 of PA or domain 1 of *B anthracis* lethal factor (LF), fused in-frame with lichenase (LicKM), a thermostable enzyme from *Clostridium thermocellum* (Chichester et al. 2007). Mice immunised with a combination of the two antigens produced high titres of mainly IgG1, and sera could neutralise the effects of anthrax lethal toxin (LeTx) in vitro.

Rabies vaccines made in plants included an early yet highly sophisticated candidate that was composed of the Alfalfa mosaic virus (AMV) CP fused to an artificial polypeptide containing rabies virus G protein amino acids 253–275, and N protein amino acids 404–418, and expressed either in *N. tabacum* plants transgenic for AMV replicase, or via rTMV in either *N. benthamiana* or spinach (Yusibov et al. 2002). The plants made particles containing AMV-derived RNA, encapsidated with chimaeric CP: raw spinach leaves were orally immunogenic in mice and in human volunteers. A simpler candidate was the G protein alone, with plant signal peptide and ER retention signal, made in transgenic *N. tabacum* (Ashraf et al. 2005). While yields were relatively low (0.38% of total soluble leaf protein), purified protein injected peritoneally in mice elicited protective immunity against lethal intracerebral challenge with live rabies virus—an excellent proof of both principle and efficacy.

Plant-made animal rotavirus vaccines were an early target, with a stand-out study by Yu and Langridge (2001) providing evidence that transgenic potato could produce fusion proteins consisting of cholera toxin (CT) B and A2 subunits fused with murine rotavirus enterotoxin and enterotoxigenic *E coli* fimbrial antigen, respectively. Fusion antigens assembled in potato tubers into cholera holotoxin-like structures that bound enterocytes, and elicited serum and intestinal antibodies after oral immunisation in mice. Moreover, passively immunised mouse neonates were partially protected against diarrhoea after rotavirus challenge, demonstrating that combination vaccines for viral and bacterial pathogens may be made in plants. A simpler approach to rotavirus prevention was expression of a His-tagged VP8* fragment of bovine rotavirus (BRV) VP4 in *N. benthamiana* via recombinant TMV, purification of the antigen by Ni^{2+} chromatography, and intraperitoneal immunisation of adult female mice (Perez Filgueira et al. 2004). Eighty-five percent of suckling mice born from these mothers were protected from BRV challenge, compared to 35% immunised with an irrelevant control antigen. The same group also showed that a fusion protein made in transgenic alfalfa consisting of a short peptide derived from BRV VP4 fused to GUS was immunogenic both when given intraperitoneally and orally to adult female mice, and their sucklings were protected against challenge (Wigdorovitz et al. 2004). Another group used transgenic alfalfa to produce human rotavirus VP6, and showed that female mice gavaged with alfalfa extract containing oligoCpG as an adjuvant developed high titres of antibodies both systemically and mucosally, and their pups were partially protected against simian rotavirus challenge.

The same animal model first used to show the efficacy of insect cell-made papillomavirus virus-like particle (VLP)-based vaccines (Breitburd et al. 1995) was also used to demonstrate the efficacy of two very different plant-made papillomavirus vaccines, a few years after the demonstration that Human papillomavirus L1 major capsid protein virus-like particles could be produced in transgenic tobacco or potato (Biemelt et al. 2003; Varsani et al. 2003; Warzecha et al. 2003). Cottontail rabbit papillomavirus (CRPV), the cause of the famous "jackalope" sightings in the USA, provides an excellent model system in domestic rabbits for investigation of prophylactic and therapeutic papillomavirus vaccines (Breitburd et al. 1997). Accordingly, in the first study CRPV L1 major capsid protein-containing extracts were prepared either from transgenic *N. tabacum* or *N. benthamiana* infected with recombinant TMV, and used with Freund's incomplete adjuvant to immunise rabbits that were subsequently challenged with live virus (Kohl et al. 2006). Although the vaccines appeared to contain small aggregates of CRPV L1 rather than intact VLPs, and immune rabbit sera failed to neutralise CRPV infectivity in an in vitro assay, the rabbits were protected from wart development (Kohl et al. 2006). In the second study, infectious recombinant TMV was used to surface display, via fusion to the capsid protein, a peptide consisting of amino acids 94–122 of the L2 minor capsid protein from either CRPV or the Rabbit oral papillomavirus (ROPV) (Palmer et al. 2006). Groups of rabbits received either or both vaccines, and were challenged with live CRPV or ROPV. Immune rabbit sera reacted with whole L2 protein, and CRPV-specific sera neutralised CRPV pseudovirion infectivity. Rabbits receiving the CRPV or CRPV + ROPV vaccines were completely protected against CRPV infection, and those receiving ROPV alone were weakly protected against CRPV. These studies demonstrated for the first time that plant-made papillomavirus vaccines based on L1 protein or L2-derived peptide had real potential as prophylactic vaccines, for use in animals as well as in humans. Strangely, given that Bovine papillomaviruses (BPV) had been used for many years as model systems for anti-wart vaccination, it was not until 2012, with transient agroinfiltration-mediated expression of BPV-1 VLPs in *N. benthamiana*, that a candidate plant-made BPV L1 VLP-based vaccine was successfully made, although no efficacy trials were done (Love et al. 2012).

Expression of animal vaccine components in seeds of transgenic plants was attempted quite early on, with Lamphear et al. (2002) in 2002 reviewing their own earlier work on maize seed expression of the B subunit of *E coli* heat-labile enterotoxin and the TGEV S protein, with data on the potency, efficacy, and stability of these vaccines. Another report followed in 2002 on the expression in maize seed of the S envelope protein of transmissible gastroenteritis coronavirus (TGEV) of swine, and its protective efficacy in piglets fed with the seed (Jilka 2002). This followed an earlier demonstration of oral immunogenicity of the S protein N-terminal domain in transgenic potato tubers (Gomez et al. 2000). Rabies too was a target for maize seed expression, with a report of G protein expression in transgenic maize seed to 1% of total soluble protein, and complete protection in a heterologous rabies strain challenge of mice orally immunized with one dose of ~50 μg of G protein in seed extract (Loza-Rubio et al. 2008). The same group later showed that sheep orally given a single dose of the transgenic maize seed containing ~2 mg of the G protein were protected to the same extent as those immunized with a commercial inactivated vaccine (Loza-Rubio et al. 2012). The authors claimed that "this is the first study in which an orally administered edible vaccine showed efficacy in a polygastric model", which was an important development.

Maize was a popular target for both production and storage of recombinant proteins in early molecular farming times (see Streatfield et al. 2003); however, other hosts were used too. For instance, the haemagglutinin (H) protein of Rinderpest virus was expressed in transgenic pigeon pea to 0.49% of total soluble protein (Satyavathi et al. 2003), and also in peanuts for a product that was both orally and parenterally immunogenic in mice (Khandelwal et al. 2004); so too was glycoprotein B (gB) of human cytomegalovirus in seeds of transgenic tobacco (Tackaberry et al. 2003), the fusion (F) glycoprotein of Newcastle disease virus in transgenic rice seed (Yang et al. 2007), and the serotype-specific VP2 protein of Bluetongue virus in transgenic peanuts (Athmaram et al. 2006).

Most of these efforts were negated, however, by the one big scandal to have hit molecular farming as far as the use of food plants for vaccine production is concerned. In 2002, APHIS inspectors found volunteer TGEV CP-expressing maize growing in soybean fields in two locations that were used to grow ProdiGene Inc's TGEV transgenic maize in the previous season (APHIS 2008)—and in one, the soybeans were harvested with the maize plants still standing and sent to a storage facility, where they were mixed with a large volume of other seeds. The company was fined and paid substantial cleanup costs, had to develop a new compliance implementation programme, and the US Dept of Agriculture issued new guidelines for trials of such products. This had an unfortunate knock-on effect for molecular farming, in that it resulted in an effectively voluntary moratorium on the use of food crops for recombinant protein production worldwide.

The one major success story of early work on veterinary vaccines was the approval by the US Department of Agriculture's Center for Veterinary Biologics of Dow AgroSciences' injectable Newcastle disease virus (NDV) haemagglutin-based vaccine for poultry, that had been made in a suspension cultured *N. tabacum* cell line. Sadly, the product was never sold: the company only wanted '… to demonstrate that our Concert™ Plant-Cell-Produced system is capable of producing a vaccine that is safe and effective and to demonstrate that it meets the requirements for approval under the rigorous USDA regulatory system. NDV is well known and understood by the regulatory agency, so it served as an excellent model to prove this new technology' (Rybicki 2009).

3 New Developments in Plant Expression Technology

The early historical account of molecular farming for veterinary vaccines given above gives an idea of the array of technologies available and used up to the mid-2000s: transgenic and transplastomic expression of subunit proteins; recombinant plant viruses either used to express whole vaccine candidate genes, or to display chosen peptides fused to their capsid proteins; fusion of vaccine protein genes to carrier proteins to improve immunogenicity, including by inherent adjuvant properties; candidate parenteral and oral vaccines to both viruses and bacteria; therapeutics for animals made in plants; use of plant cell cultures to make antigens. Many proofs of principle were obtained, for candidate vaccines against a wide range of viral and bacterial disease agents; and proofs of efficacy for vaccines delivered orally or parenterally, in whole plant material or as extracts.

While all of these aspects are still currently used in molecular farming, developments that have revolutionised the field were first, the widespread adoption of *Agrobacterium*-mediated transient expression (agroinfiltration) of recombinant proteins; and second, the use of "deconstructed" plant virus-derived vectors delivered via *Agrobacterium* to amplify expression (reviewed in Rybicki 2010). These innovations enabled the advent of high-throughput testing of expression constructs, coupled with very rapid and generally higher yield production of vaccine antigens once optimal construct design had been determined. For example, our group investigated, via agroinfiltration techniques, three different codon usage schemes and three different intracellular localization strategies for optimization of Human papillomavirus type 16 L1 protein expression in *N. benthamiana*, in one large experiment over only 7 days (Maclean et al. 2007).

Use of deconstructed TMV-based vectors delivered by *Agrobacterium* routinely has allowed significant increases of antigen yield, up to grams per kilogram fresh tissue weight (Gleba et al. 2014; Klimyuk et al. 2014). The so-called TMV-based "launch vectors" of Fraunhofer USA have also allowed significant yield increases and rapid production of antigens (Chichester et al. 2013; Shamloul et al. 2014). Improved non-replicating hyper-translational (HT) expression vectors derived from Cowpea mosaic virus RNA2 have also allowed significantly higher yields via agroinfiltration (Sainsbury et al. 2008, 2009) and the possibility of multiple genes from the same vector (Saxena et al. 2016); so too has the use of a ssDNA geminivirus-derived set of vectors by different groups (Huang et al. 2009; Regnard et al. 2010), and other ssDNA plant (or other host) virus-derived vectors (Rybicki and Martin 2014).

The number of peptide display vectors/chimaeric protein fusion partners has multiplied: while self-replicating rTMV was once state of the art, now one may choose between TMV- and Potato virus X (PVX)-based vectors (Lico et al. 2015), Cucumber mosaic virus (CMV) CP (Nemchinov and Natilla 2007; Zhao and Hammond 2005), Bamboo mosaic virus (Yang et al. 2007), PVX-vectored Alternanthera mosaic virus (AltMV) CP gene (Tyulkina et al. 2011), lichenase (LickM), cholera toxin B subunit (CTB), AMV CP, and GUS, as mentioned earlier. Plant virus virions in particular are now seen as easily-made nanoparticles suitable for a number of vaccine-relevant purposes (Steele et al. 2017), including as self-adjuvanting peptide-based vaccine display vehicles (Lebel et al. 2015; Leclerc 2014), and excellent inducers of cross-presentation by MHC receptors (Hanafi et al. 2010).

The use of tags or small peptide fusion partners is now also considerably more sophisticated, with a variety of specialized tags to choose from. These include the now-ubiquitous 6xHis tag, used for Ni^{2+} or other immobilised metal affinity chromatography (IMAC) protein purification technique; a new "Cysta-tag" for the same purpose (Sainsbury et al. 2016); the N-terminal proline-rich domain of maize seed gamma zein (Zera) that induces the formation of ER-located protein bodies (Torrent et al. 2009); elastin-like polypeptides (ELPs) with repeating pentapeptide 'VPGXG' sequences, or hydrophobins—small fungal proteins which alter the hydrophobicity of the fusion partner—both of which also form protein bodies (Conley et al. 2011). As examples, our group has recently successfully used ELP fusion to the CP of Beak and feather disease virus (BFDV) of parrots to aid in both accumulation and purification of the protein as a candidate vaccine (Duvenage et al. 2013). We have also used the Zera tag as a protein body display vehicle for an ectopic M2e moiety common to all influenzavirus A types, which could serve as a universal vaccine for these viruses (Mbewana et al. 2015). Another potentially veterinary use of Zera was in the enhancement of *Yersinia pestis* F1-V antigen fusion protein accumulation: this was ~3× higher than F1-V alone in three different host plant systems—namely, *N. benthamiana*, alfalfa and *N. tabacum* NT1 suspension-cultured cells (Alvarez et al. 2010).

The expression vehicles themselves have also been subject to engineering: it is now possible to precisely control glycosylation of plant-made proteins. This can be done by knock-out modification via RNA interference (RNAi) technology of the plant glycosyltransferases beta1,2-xylosyltransferase (XylT) and core alpha1,3-fucosyltransferase (FucT). These enzymes are responsible for the transfer of beta1,2-linked xylose and core alpha1,3-linked fucose residues to glycoprotein N-glycans, which are plant-specific modifications not found in mammalian glycoproteins (Strasser et al. 2008). It is also possible to use transient co-expression technologies to modify glycosylation (Castilho and Steinkellner 2016), as well as to achieve almost completely native sialylated recombinant proteins by expression of whole mammalian glycosylation pathways in plants (Castilho et al. 2010; Steinkellner and Castilho 2015). It is possible to abolish N-glycosylation entirely, by co-expression of bacterial PNGase F (Mamedov and Yusibov 2013). One can also control endogenous plant proteases that may limit recombinant protein accumulation: for example, transient co-expression of secreted A1/S1 protease inhibitor tomato cathepsin D inhibitor (SlCDI) significantly lowered A1 and S1 protease activities in the *N. benthamiana* apoplast, while increasing recombinant protein content by ~45% (Goulet et al. 2012). It was found that co-expression of tomato cystatin SlCYS8, which inhibits C1A proteases, increased the transient expression yield of a monoclonal antibody in *N. benthamiana* by nearly 40% (Robert et al. 2013). It is also possible to reduce protease activity in cell suspension cultures by expression of specific antisense RNAs, resulting in significantly increased accumulation of recombinant antibodies (Mandal et al. 2014).

While suspension-cultured plant cells have been used for many years for molecular farming—and in fact were used for the only USDA-licenced plant-produced animal vaccine, against NDV—new developments have made them an even more attractive prospect for low-cost vaccine production. Use of flow cytometry with cell sorting, formerly the province of mammalian cell culture work only, has allowed high-expressing MAb-producing tobacco BY-2 cell lines from a heterogeneous population of cells by selecting the co-expressed fluorescent marker protein DsRed (Kirchhoff et al. 2012). However, one of the most exciting recent developments with this technology is the advent of the "cell pack": this is a technique for getting highly efficient (up to 100%) *Agrobacterium*-mediated transient transformation of suspension-cultured cells that have been captured by suction onto a filter (Rademacher 2014). Cell packs can be tiny (Eppendorf tube tips) or large (e.g.: centimetres deep in a 20 cm Buchner funnel); protein expression occurs in immobilised cells in the presence of minimal liquid media, and can continue for days (https://tinyurl.com/k22da6q). The technology is ideal for rapid and high-throughput screening of expression—and the possibility exists for taking cells back into culture and selecting for permanent transfection. These are important developments, because of the acceptability of the products of plant cell cultures for production of biologics to regulatory bodies (see below). Another production host highly suited to industrial-scale production is microalgae: they are easier to establish and use than plant cell cultures, and share all the same advantages of scalability, contained growth, and consistent transgene expression levels (Specht and Mayfield 2014).

A very important development for molecular farming has been the development of protocols for increasing yields and implementing industrial-scale production and downstream processing of vaccines and biologics, without which no large-scale trials could take place, or routine manufacturing occur. A useful development was use of a transgenic *N. tabacum*/*N. glauca* hybrid that does not synthesize alkaloids, is highly vigorous, can easily be propagated by vegetative cuttings and does not produce viable pollen, which greatly aids biocontainment (Ling et al. 2012). The application of techniques more familiar to chemical engineers is also advantageous: for example, it proved possible, by sequential use of fractional factorial designs and response surface methodology, to optimize culture media for MAb production in transgenic tobacco BY-2 cells, and to increase MAb yields up to 31-fold after 10 days of culture compared to use of standard media (Vasilev et al. 2013). The Fraunhofer IME group have described generic chromatography-based strategies focusing on the binding behaviours of host cell proteins to chromatography resins under varying conditions of pH and conductivity (Buyel and Fischer 2014). Another useful technique from that group is a comprehensive description of the use of heat treatment of either intact leaves or of plant extracts to facilitate the industrial-scale removal of host cell proteins, optimised by a design-of-experiments approach that will also be familiar to engineers (Buyel et al. 2016). Many of these and other strategies used to optimise yields in molecular farming are reviewed here (Twyman et al. 2013).

The establishment by various companies and institutes of facilities suitable for manufacture of animal and clinical trial material is also a very welcome development. As examples, the long-established Kentucky BioProcessing Inc (KBP) is a contract manufacturer capable of production from transgenic plants or transiently transfected plants, using the U.S. Food and Drug Administration's current Good Manufacturing Practices (cGMP) for pharmaceuticals, at scales up to thousands of kilograms of plants per week (https://www.kentuckybioprocessing.com/). They have recently produced and stockpiled Mabs against Ebolaviruses.

Another contract manufacturing firm with large production capacity is iBio Inc: like KBP, they have a wide range of patents on their proprietary gene expression technology (Holtz et al. 2015). They are also partnering with a range of agencies and companies, including with the Brazilian Oswaldo Cruz Foundation for plant-made yellow fever vaccine, and the US Dept of Defense and the Bill & Melinda Gates Foundation for influenza vaccines (http://www.ibioinc.com/).

The Fraunhofer USA Center for Molecular Biotechnology (http://www.fhcmb.org/) is a not-for-profit research and development organisation, that offers "... plant-based protein production, purification, scale-up and GMP manufacturing to support the development of vaccines, therapeutics and diagnostics", also with proprietary expression platforms, and can take products right through to fill and finish. The Fraunhofer IME in Aachen also has a state-of-the-art mechanised plant production facility still under construction as of 2017.

4 Regulatory Approvals

The regulatory environment has changed for the better, even though it was not in truth as inimical as first supposed: this was borne out by the fact that as early as 2006, the Cuban regulatory agencies and the USDA had approved plant-made MAbs for the purification of an already-licenced yeast-made hepatitis B vaccine, and the tobacco cell-made NDV vaccines, respectively (Rybicki 2009).

As another early example, the Fraunhofer IME molecular farming group published in 2004 that use of whole plants for biologics production lacks intrinsic benefits of cell culture techniques, such as precise control over growth conditions, batch-to-batch product consistency, sterile containment, and it being much harder to be in compliance with good manufacturing practice (GMP) (Hellwig et al. 2004). They pointed out that plant cell suspension cultures have all the merits of microbial and animal cell cultures, have an established track record for secondary metabolite production, and are far cheaper to use.

These justifications notwithstanding, the same group later noted, in a review on GMP issues for plant-made proteins in whole plants, that: "When [plant-derived] recombinant proteins are intended for medical use... they fall under the same regulatory guidelines for manufacturing that cover drugs from all other sources, and when such proteins enter clinical development this includes the requirement for production according to [GMP]. In principle, the well-characterized GMP regulations that apply to pharmaceutical proteins produced in bacteria and mammalian cells are directly transferrable to plants" (Fischer et al. 2012). They subsequently were able to get GMP manufacturing authorisation from German authorities for making MAbs from transgenic *N. tabacum* for a phase I clinical trial (Ma et al. 2015).

Other entities have also scaled and regularised production to allow production of materials for animal and clinical trial—and one of the most successful has been Medicago Inc., who presently has routine large-scale production of influenzavirus A haemagglutinin (HA)-based VLPs for use in advanced human clinical trial

(D'Aoust et al. 2010). In 2012, Medicago Inc. succeeded in manufacturing 10 million doses of an H1N1 VLP-based influenza vaccine candidate in one month, by Phase 1-appropriate cGMP, as part of the US Defense Advanced Research Projects Agency (DARPA)-funded challenge (DARPA 2012).

A group in Japan has also recently developed a GMP-compliant production process for a transgenic rice seed-based cholera vaccine—MucoRice-CTB—which is simply polished, powdered seed, now in clinical trial (Kashima et al. 2016).

As evidence of the increasing maturity of veterinary molecular farming, one of the editors of this book has co-authored a recent article on regulatory and commercial hurdles hampering the advance to market of plant-produced veterinary vaccines, covering developing business plans, assessing market opportunities, manufacturing scale-up, financing, protecting and using intellectual property, and regulatory approval (MacDonald et al. 2015).

5 Future of Plant Expression to Make Biologics for Veterinary Use

At first sight, molecular farming appears the ideal way to make recombinant protein-based veterinary vaccines: production of active ingredients is markedly cheaper per unit mass than by use of any animal tissue-culture system, and generally cheaper than yeast or bacterial culture (Rybicki 2010); partially-purified or unprocessed extracts are highly unlikely to contain any animal pathogens; edible and oral vaccines appear highly feasible; the financial barrier to entry for manufacture appears far lower than for conventionally-made vaccines. It is possible to efficiently make bacterial proteins using bacterial-derived translational machinery in chloroplasts in transplastomic plants, as well as to make other proteins at very high yield; conventional transgenics have been used to make many vaccine candidates, with many proofs of efficacy; transient expression technologies have revolutionised the field in terms of providing high yields and very rapid development times from concept to product. And yet, only one product—Dow's NDV vaccine—is registered for use, and that is not sold.

It is possible that heavy investment by big industry players in conventional manufacturing technologies has stalled their uptake of molecular farming technology for veterinary vaccines and biologics: this has certainly been true for human biologics. However, perhaps developments from the human field could be used as a spur for uptake of veterinary vaccines and biologics: an example here is the licencing of Protalix Biotherapeutics' Elelyso® or glucocerebrosidase, a therapeutic for a genetic mitochondrial enzyme deficit called Gaucher disease, made using transgenic carrot cell lines in 800 litre plastic bag fermenters (http://protalix.com/about/elelyso/). A contamination of Genzyme's mammalian cell production facilities in 2009 with a mammalian calicivirus led to the FDA allowing Protalix to supply the drug to patients who needed it, and to accelerated licensure (Bethencourt 2009).

The company has also successfully tested oral administration of drugs in plant cells, which would be a highly welcome development: they claim that "Oral delivery of protein therapies [is] possible due to the unique cellulose wall of plant cells that makes them resistant to degradation when passing through the digestive tract" (Protalix 2017). Another apposite example was the fortuitous availability of a plant-made anti-Zaire ebolavirus MAb cocktail known as ZMapp™, at the height of the recent West African Ebola disease outbreak (reviewed in Rybicki 2014). This was made by transient expression in *N. benthamiana*, and only a few clinical trial doses were available: these were used under the humanitarian principle, and later the MAbs were cleared for use by the FDA in an efficacy trial just before the end of the epidemic (LeafBio 2016).

Both these examples are of niche products that were not being made at large scale or for a large market by conventional techniques, and for which there was a sudden, pressing need that could not be supplied by other means. This could provide motivation for small companies to either develop inexpensive vaccines for emerging diseases, or to target niche vaccines or niche therapeutics, in the knowledge that large established entities are unwilling to take the risk.

One example for the former possibility comes from the recent emergence of bluetongue virus (BTV) disease in sheep and small ruminants in Europe, due to northward spread of the insect vector with climate change (Purse et al. 2008): while attenuated live vaccines are available—South Africa presently uses a cocktail of 24 such viruses—concerns in Europe about reassortment of virus dsRNA genome components between vaccinated and naturally diseased animals, as well as of the safety of the vaccines in terms of possible under-attenuation which may result in disease development in certain sheep breeds (Niedbalski 2011), mean these are not being used. The irregular occurrence of outbreaks, and the limited number of strains involved, mean that stockpiling vaccines is desirable. However, killed vaccines still require growing potentially dangerous viruses, and while it is possible to make VLPs in cell cultures and these are effective (Pearson and Roy 1993; Roy et al. 1994), the technology is too expensive for farm animal use. It is fortunate, therefore, that it is also possible to make BTV-8 VLPs via transient expression in *N. benthamiana*, and these are as effective in a single injected dose as the commercial vaccine (Thuenemann et al. 2013). There are currently no plans to manufacture this or other plant-produced BTV vaccines for the European or other markets; however, this may soon change.

An example for a niche vaccine product comes from ours and others' work on beak and feather disease virus (BFDV) vaccines: psittacines are highly valued companion animals; however, there are very few vaccines for their diseases, and none yet available for BFDV. While some recent work in this area has shown that recombinant CP can be made in *E coli* and in insect cells (Heath et al. 2006; Patterson et al. 2013; Stewart et al. 2007), that it appears to be protective (Bonne et al. 2009) and that this can apparently form VLPs (Sarker et al. 2015), it still appears that the protein is too expensive to produce for use as a vaccine. While initial work with plant production of BFDV CP was disappointing due to low yields, recent work from our group (Duvenage et al. 2013) showed a significant increase in BFDV yield due to fusion with elastin-like polypeptide (ELP), and good immunogenicity in mice. This, coupled with a very simple purification protocol enabled by ELPylation (Conley et al. 2009), could allow scalable, cheap production of BFDV vaccines.

While therapeutics such as MAbs or other biologics for veterinary use are generally limited to high-value companion animals, plant production could open up a hitherto neglected market niche. One excellent example is the manufacture in Japan of canine interferon-α (Tabayashi and Matsumura 2014): this is done via transgenic strawberries in a completely enclosed GMP-compliant facility, and the product is powdered strawberry extract given orally, to combat canine periodontal disease. Another very recent example in dogs, albeit with them being used as a model for human disease, was the proof that lyophilised transplastomic lettuce leaves expressing CTB fusions of coagulation factor IX (FIX) could be used orally in feed for >300 days in haemophilia B dogs with no ill effects—and that this treatment resulted in robust suppression of IgG/inhibitor and IgE formation against intravenously-provided FIX, and a marked shortening of blood coagulation times (Herzog et al. 2017).

An example for agricultural use is the oral dosing of pigs with transgenic *Arabidopsis thaliana* seeds containing designer IgAs against enterotoxigenic *E coli* (ETEC) (Virdi et al. 2013): this product consisted of dimeric llama-derived heavy chain variable region fused to the Fc portion of a porcine IgA and the porcine IgA J chain and secretory component, which allowed production of dimeric secretory IgA-like antibodies (VHH-IgA). In a piglet feed-challenge experiment with ETEC, dosing piglets with 20 mg/d per pig VHH-IgA produced a progressive decline in bacterial shedding and a significantly higher weight gain than seen in control or other experimental pigs.

A highly novel plant-made therapeutic product was the receptor binding domain of the tailspike protein Gp9 from the P22 bacteriophage: this is known to reduce *Salmonella* colonisation in the chicken gut (Miletic et al. 2015). Purified ELP-fused Gp9 bound to *Salmonella enterica* serovar Typhimurium in vitro, and feeding lyophilized leaves containing Gp9-ELP to newly hatched chickens showed that it has the potential to control *Salmonella* contamination in commercially-raised fowl. These and other experiments are reviewed here (Juarez et al. 2016; Topp et al. 2016), in articles that make an excellent case for plant-made immunotherapeutics for veterinary use.

6 The One Health Approach and Its Relevance for Modern Veterinary Vaccines

The One Health concept has as one of its central themes the integration of opportunities for vaccine-based approaches for the prevention of zoonotic and emerging diseases across veterinary and human medicine (Monath 2013), and three different frameworks for the use of vaccines in these contexts have been formulated. Framework I vaccines are used to protect humans and economically valuable animals, where neither are central to the transmission cycle: a good example here would be West Nile virus, which is a mosquito-borne flavivirus spread around the world by birds, and which incidentally infects animals like horses as well as humans. Framework II vaccines, on the other hand, are intended for use in domesticated animals as a means of preventing disease in both animals and humans: examples of disease agents here would be *Brucella abortis, E coli* O157, and rabies, influenza, Rift Valley fever, and Hendra viruses. Framework III vaccines are for immunising non-domesticated animals in order to prevent transmission of disease agents to humans and domesticated animals: examples here are the use of oral bait rabies, and *Mycobacterium bovis* and Lyme disease vaccines.

A set of disease agents which exemplify the potential strength of the One Health approach are influenza viruses, and they have in fact been the focus of a number of international meetings and planning sessions (Chien 2013; Dwyer and Kirkland 2011; Kahn et al. 2014; Ludwig et al. 2014; Powdrill et al. 2010; Short et al. 2015). The unique mix of hosts that occurs in intensive agricultural environments that could give rise to pandemics—swine, birds and humans—is a major cause of international concern; so too is the development of suitable vaccines for the prevention of infection in domesticated birds, farmed swine, and humans. Plants have been shown to be highly useful for the production of influenza vaccines, and indeed possibly the fastest ever production at scale of an influenzavirus A strain vaccine—1 month for 10 million doses—was done by Medicago Inc. for H1N1pdm 2009 HA VLPs in 2012 (Rybicki 2014). Medicago also managed in 2013, as an exercise to demonstrate preparedness, to produce grams of cGMP-grade plant-made H7 HA-only VLPs only 19 days after accessing the H7 HA gene cDNA sequence, in response to an outbreak in China in the same year. The fact that plant-made influenza vaccines have worked very well in animal models means that they should be trialled extensively in domestic fowl and swine, to see if the maintenance of the viruses in these hosts can be curbed. As for companion animals, there is even a canine influenza vaccine candidate: following a 2004 H3N8 outbreak in the US, a group in Canada used the plant-derived filamentous Malva mosaic virus (MaMV) nanoparticles as a vaccine platform to display the highly conserved ectopic M2e peptide and to increase its immunogenicity. Together with the adjuvant OmpC derived from *Salmonella typhi,* the vaccine was protective against both the homologous virus and a heterosubtypic strain of influenza in mice, as well as eliciting antibodies reactive with M2e peptides derived from H9N2, H5N1 and H1N1 strains and being immunogenic in dogs (Leclerc et al. 2013).

Given that brucellosis is listed as a One Health priority, it is worth noting that a transgenic plant-produced *B abortus* outer membrane protein (U-Omp19) was an effective oral vaccine in mice against a systemic challenge, eliciting an adaptive IL-17 immune response (Pasquevich et al. 2011)—and that the protein has significant adjuvant activity, and oral vaccination of mice with U-Omp19 plus *Salmonella* antigens was protective against virulent challenge with *S typhimurium* (Risso et al. 2017).

It is important to realise that, while vaccines are the target of this review, One Health products can also be reagents to be used in more effective or cheaper diagnostic kits, and in particular for point-of-care devices, or for research laboratory use—and especially proteins that could be both a reagent and used as a candidate vaccine in animals and possibly humans. A few of the best potential One Health targets for plant-made dual-function proteins would be proteins from Middle Eastern respiratory syndrome (MERS) coronavirus (Wirblich et al. 2017), Nipah and Hendra viruses (Landford and Nunn 2012; Mackenzie et al. 2003), diagnostic/vaccine candidate proteins from Rift Valley Fever and Crimean-Congo haemorrhagic fever viruses (Kortekaas 2014; Monath 2013). Inexpensive and abundant proteins made from these agents could first serve as reagents in the development of cheap point-of-care diagnostics, and then as vaccine candidates in animals, if appropriate, and then possibly in humans.

A useful example here is of the expression both by agroinfiltration in *N. benthamiana* as a reagent, and in transgenic *N. tabacum* roots and leaves as a vaccine, of a fused GcGn envelope glycoprotein-encoding gene from Crimean-Congo haemorrhagic fever virus (Ghiasi et al. 2011). The protein yield was 1–2 mg/kg fresh plant weight. Transgenic material was orally immunogenic, and elicited humoral and mucosal antibody responses, and antibodies bound inactivated virus used as a vaccine booster in some experiments. Agroinfiltration-produced GnGc was used as a reagent in ELISA to detect immune responses. Another study from our group was of the production of CCHFV N protein in *N. benthamiana* by agroinfiltration specifically as a reagent for use in diagnostic tests (Atkinson et al. 2016): a plant codon-optimised and 6xHis tagged N protein gene was found to accumulate best as a soluble protein in the cytoplasm, from which it could be easily purified by ammonium sulphate fractionation and immobilised Ni^{2+} column chromatography. Purified NP was used in a validated indirect ELISA to detect anti-CCHFV IgG in sera from convalescent human patients: this was successful for 13/13 samples, with no readings for samples from patients with no history of CCHFV infection. The results were 100% concordant with those from a commercially available immunofluorescent assay. Given that soluble N protein is hard to produce and difficult to purify from insect cell cultures, the plant-made product would seem to be a desirable replacement.

7 Conclusions

While the same has been said in many venues over more than twenty years now, the field of molecular farming really does seem to be near to meeting its initial promise for veterinary use. All of the technology that is required for efficient, high-yield production of biologics is in place; downstream processing modalities have been well worked out by a number of near- and cGMP-compliant facilities; many candidate vaccines for a wide variety of pathogens have been tested; therapeutic biologics too for veterinary use are now feasible; regulatory agencies seem agreeable to considering plant-made products. The generally shorter regulatory path, the possibility of using less stringently purified products, and the very real possibility of using oral vaccines and therapeutics, should also be highly attractive for product developers. I sincerely hope, then, that realisation of the promise comes very soon.

References

Alvarez ML, Topal E, Martin F, Cardineau GA (2010) Higher accumulation of F1-V fusion recombinant protein in plants after induction of protein body formation. Plant Mol Biol 72:75–89

APHIS (2008) Noncompliance history

Ashraf S, Singh PK, Yadav DK, Shahnawaz M, Mishra S, Sawant SV, Tuli R (2005) High level expression of surface glycoprotein of rabies virus in tobacco leaves and its immunoprotective activity in mice. J Biotechnol 119:1–14

Athmaram TN, Bali G, Devaiah KM (2006) Integration and expression of Bluetongue VP2 gene in somatic embryos of peanut through particle bombardment method. Vaccine 24:2994–3000

Atkinson R, Burt F, Rybicki EP, Meyers AE (2016) Plant-produced Crimean-Congo haemorrhagic fever virus nucleoprotein for use in indirect ELISA. J Virol Methods 236:170–177

Aziz MA, Singh S, Anand Kumar P, Bhatnagar R (2002) Expression of protective antigen in transgenic plants: a step towards edible vaccine against anthrax. Biochem Biophys Res Commun 299:345–351

Aziz MA, Sikriwal D, Singh S, Jarugula S, Kumar PA, Bhatnagar R (2005) Transformation of an edible crop with the pagA gene of Bacillus anthracis. FASEB J 19:1501–1503

Bethencourt V (2009) Virus stalls Genzyme plant. Nat Biotechnol 27:681

Biemelt S, Sonnewald U, Galmbacher P, Willmitzer L, Muller M (2003) Production of human papillomavirus type 16 virus-like particles in transgenic plants. J Virol 77:9211–9220

Bonne N, Shearer P, Sharp M, Clark P, Raidal S (2009) Assessment of recombinant beak and feather disease virus capsid protein as a vaccine for psittacine beak and feather disease. J Gen Virol 90:640–647

Breitburd F, Kirnbauer R, Hubbert NL, Nonnenmacher B, Trin-Dinh-Desmarquet C, Orth G, Schiller JT, Lowy DR (1995) Immunization with viruslike particles from cottontail rabbit papillomavirus (CRPV) can protect against experimental CRPV infection. J Virol 69:3959–3963

Breitburd F, Salmon J, Orth G (1997) The rabbit viral skin papillomas and carcinomas: a model for the immunogenetics of HPV-associated carcinogenesis. Clin Dermatol 15:237–247

Buyel JF, Fischer R (2014) Generic chromatography-based purification strategies accelerate the development of downstream processes for biopharmaceutical proteins produced in plants. Biotechnol J 9:566–577

Buyel JF, Hubbuch J, Fischer R (2016) Comparison of tobacco host cell protein removal methods by blanching intact plants or by heat treatment of extracts. J Visualized Exp: JoVE

Carrillo C, Wigdorovitz A, Oliveros JC, Zamorano PI, Sadir AM, Gomez N, Salinas J, Escribano JM, Borca MV (1998) Protective immune response to foot-and-mouth disease virus with VP1 expressed in transgenic plants. J Virol 72:1688–1690

Castanon S, Marin MS, Martin-Alonso JM, Boga JA, Casais R, Humara JM, Ordas RJ, Parra F (1999) Immunization with potato plants expressing VP60 protein protects against rabbit hemorrhagic disease virus. J Virol 73:4452–4455

Castilho A, Strasser R, Stadlmann J, Grass J, Jez J, Gattinger P, Kunert R, Quendler H, Pabst M, Leonard R, Altmann F, Steinkellner H (2010) In planta protein sialylation through overexpression of the respective mammalian pathway. J Biol Chem 285:15923–15930

Castilho A, Steinkellner H (2016) Transient Expression of Mammalian Genes in N. benthamiana to Modulate N-Glycosylation. Methods Mol Biol 1385:99–113

Chichester JA, Musiychuk K, de la Rosa P, Horsey A, Stevenson N, Ugulava N, Rabindran S, Palmer GA, Mett V, Yusibov V (2007) Immunogenicity of a subunit vaccine against *Bacillus anthracis*. Vaccine 25:3111–3114

Chichester JA, Manceva SD, Rhee A, Coffin MV, Musiychuk K, Mett V, Shamloul M, Norikane J, Streatfield SJ, Yusibov V (2013) A plant-produced protective antigen vaccine confers protection in rabbits against a lethal aerosolized challenge with Bacillus anthracis Ames spores. Human Vaccines Immunotherapeutics 9:544–552

Chien YJ (2013) How did international agencies perceive the avian influenza problem? The adoption and manufacture of the 'One World, One Health' framework. Sociol Health Illn 35:213–226

Conley AJ, Joensuu JJ, Jevnikar AM, Menassa R, Brandle JE (2009) Optimization of elastin-like polypeptide fusions for expression and purification of recombinant proteins in plants. Biotechnol Bioeng 103:562–573

Conley AJ, Joensuu JJ, Richman A, Menassa R (2011) Protein body-inducing fusions for high-level production and purification of recombinant proteins in plants. Plant Biotechnol J 9:419–433

D'Aoust MA, Couture MM, Charland N, Trepanier S, Landry N, Ors F, Vezina LP (2010) The production of hemagglutinin-based virus-like particles in plants: a rapid, efficient and safe response to pandemic influenza. Plant Biotechnol J 8:607–619

Dalsgaard K, Uttenthal A, Jones TD, Xu F, Merryweather A, Hamilton WD, Langeveld JP, Boshuizen RS, Kamstrup S, Lomonossoff GP, Porta C, Vela C, Casal JI, Meloen RH, Rodgers PB (1997) Plant-derived vaccine protects target animals against a viral disease. Nat Biotechnol 15:248–252

DARPA (2012) DARPA makes 10 million strides in the race to contain a hypothetical pandemic. US Defense Advanced Research Projects Agency

Dus Santos MJ, Wigdorovitz A, Trono K, Rios RD, Franzone PM, Gil F, Moreno J, Carrillo C, Escribano JM, Borca MV (2002) A novel methodology to develop a foot and mouth disease virus (FMDV) peptide-based vaccine in transgenic plants. Vaccine 20:1141–1147

Duvenage L, Hitzeroth II, Meyers AE, Rybicki EP (2013) Expression in tobacco and purification of beak and feather disease virus capsid protein fused to elastin-like polypeptides. J Virol Methods 191:55–62

Dwyer DE, Kirkland PD (2011) Influenza: one health in action. N S W Public Health Bull 22:123–126

Fischer R, Schillberg S, Hellwig S, Twyman RM, Drossard J (2012) GMP issues for recombinant plant-derived pharmaceutical proteins. Biotechnol Adv 30:434–439

Ghiasi SM, Salmanian AH, Chinikar S, Zakeri S (2011) Mice orally immunized with a transgenic plant expressing the glycoprotein of Crimean-Congo hemorrhagic fever virus. Clin Vaccine Immunol: CVI 18:2031–2037

Gil F, Titarenko E, Terrada E, Arcalis E, Escribano JM (2006) Successful oral prime-immunization with VP60 from rabbit haemorrhagic disease virus produced in transgenic plants using different fusion strategies. Plant Biotechnol J 4:135–143

Gleba YY, Tuse D, Giritch A (2014) Plant viral vectors for delivery by Agrobacterium. Curr Top Microbiol Immunol 375:155–192

Gomez N, Wigdorovitz A, Castanon S, Gil F, Ordas R, Borca MV, Escribano JM (2000) Oral immunogenicity of the plant derived spike protein from swine-transmissible gastroenteritis coronavirus. Arch Virol 145:1725–1732

Goulet C, Khalf M, Sainsbury F, D'Aoust MA, Michaud D (2012) A protease activity-depleted environment for heterologous proteins migrating towards the leaf cell apoplast. Plant Biotechnol J 10:83–94

Hanafi LA, Bolduc M, Gagne ME, Dufour F, Langelier Y, Boulassel MR, Routy JP, Leclerc D, Lapointe R (2010) Two distinct chimeric potexviruses share antigenic cross-presentation properties of MHC class I epitopes. Vaccine 28:5617–5626

Haq TA, Mason HS, Clements JD, Arntzen CJ (1995) Oral immunization with a recombinant bacterial antigen produced in transgenic plants. Science 268:714–716

Heath L, Williamson AL, Rybicki EP (2006) The capsid protein of beak and feather disease virus binds to the viral DNA and is responsible for transporting the replication-associated protein into the nucleus. J Virol 80:7219–7225

Hellwig S, Drossard J, Twyman RM, Fischer R (2004) Plant cell cultures for the production of recombinant proteins. Nat Biotechnol 22:1415–1422

Herzog RW, Nichols TC, Su J, Zhang B, Sherman A, Merricks EP, Raymer R, Perrin GQ, Häger M, Winberg B, Daniell H (2017) Oral tolerance induction in hemophilia B dogs fed with transplastomic lettuce. Mol Ther 25:512–522

Holtz BR, Berquist BR, Bennett LD, Kommineni VJ, Munigunti RK, White EL, Wilkerson DC, Wong KY, Ly LH, Marcel S (2015) Commercial-scale biotherapeutics manufacturing facility for plant-made pharmaceuticals. Plant Biotechnol J 13:1180–1190

Huang Y, Liang W, Wang Y, Zhou Z, Pan A, Yang X, Huang C, Chen J, Zhang D (2005) Immunogenicity of the epitope of the foot-and-mouth disease virus fused with a hepatitis B core protein as expressed in transgenic tobacco. Viral Immunol 18:668–677

Huang Z, Chen Q, Hjelm B, Arntzen C, Mason H (2009) A DNA replicon system for rapid high-level production of virus-like particles in plants. Biotechnol Bioeng 103:706–714

Hull AK, Criscuolo CJ, Mett V, Groen H, Steeman W, Westra H, Chapman G, Legutki B, Baillie L, Yusibov V (2005) Human-derived, plant-produced monoclonal antibody for the treatment of anthrax. Vaccine 23:2082–2086

Jilka J (2002) An oral vaccine in maize protects against transmissible gastroenteritis virus in swine. In: Erickson L, Yu W-J, Brandle J, Rymerson R (eds) Molecular farming of plants and animals for human and veterinary medicine. Springer, Netherlands, pp 223–236

Juarez P, Virdi V, Depicker A, Orzaez D (2016) Biomanufacturing of protective antibodies and other therapeutics in edible plant tissues for oral applications. Plant Biotechnol J 14:1791–1799

Kahn RE, Ma W, Richt JA (2014) Swine and influenza: a challenge to one health research. Curr Top Microbiol Immunol 385:205–218

Kashima K, Yuki Y, Mejima M, Kurokawa S, Suzuki Y, Minakawa S, Takeyama N, Fukuyama Y, Azegami T, Tanimoto T, Kuroda M, Tamura M, Gomi Y, Kiyono H (2016) Good manufacturing practices production of a purification-free oral cholera vaccine expressed in transgenic rice plants. Plant Cell Rep 35:667–679

Khandelwal A, Renukaradhya GJ, Rajasekhar M, Sita GL, Shaila MS (2004) Systemic and oral immunogenicity of hemagglutinin protein of rinderpest virus expressed by transgenic peanut plants in a mouse model. Virology 323:284–291

Kirchhoff J, Raven N, Boes A, Roberts JL, Russell S, Treffenfeldt W, Fischer R, Schinkel H, Schiermeyer A, Schillberg S (2012) Monoclonal tobacco cell lines with enhanced recombinant protein yields can be generated from heterogeneous cell suspension cultures by flow sorting. Plant Biotechnol J 10:936–944

Klimyuk V, Pogue G, Herz S, Butler J, Haydon H (2014) Production of recombinant antigens and antibodies in Nicotiana benthamiana using 'magnifection' technology: GMP-compliant facilities for small- and large-scale manufacturing. Curr Top Microbiol Immunol 375:127–154

Kohl T, Hitzeroth II, Stewart D, Varsani A, Govan VA, Christensen ND, Williamson AL, Rybicki EP (2006) Plant-produced cottontail rabbit papillomavirus L1 protein protects against tumor challenge: a proof-of-concept study. Clin Vaccine Immunol 13:845–853

Kortekaas J (2014) One Health approach to Rift Valley fever vaccine development. Antiviral Res 106:24–32

Koya V, Moayeri M, Leppla SH, Daniell H (2005) Plant-based vaccine: mice immunized with chloroplast-derived anthrax protective antigen survive anthrax lethal toxin challenge. Infect Immun 73:8266–8274

Lamphear BJ, Streatfield SJ, Jilka JM, Brooks CA, Barker DK, Turner DD, Delaney DE, Garcia M, Wiggins B, Woodard SL, Hood EE, Tizard IR, Lawhorn B, Howard JA (2002) Delivery of subunit vaccines in maize seed. J Control Release 85:169–180

Landford J, Nunn M (2012) Good governance in 'one health' approaches. Rev Sci Tech 31:561–575

LeafBio (2016) LeafBio announces conclusion of ZMapp™ clinical trial

Lebel ME, Chartrand K, Leclerc D, Lamarre A (2015) Plant viruses as nanoparticle-based vaccines and adjuvants. Vaccines (Basel) 3:620–637

Leclerc D, Rivest M, Babin C, Lopez-Macias C, Savard P (2013) A novel M2e based flu vaccine formulation for dogs. PLoS ONE 8:e77084

Leclerc D (2014) Plant viral epitope display systems for vaccine development. Curr Top Microbiol Immunol 375:47–59

Lico C, Benvenuto E, Baschieri S (2015) The two-faced potato virus X: from plant pathogen to smart nanoparticle. Front Plant Sci 6:1009

Ling HY, Edwards AM, Gantier MP, Deboer KD, Neale AD, Hamill JD, Walmsley AM (2012) An interspecific Nicotiana hybrid as a useful and cost-effective platform for production of animal vaccines. PLoS ONE 7:e35688

Love AJ, Chapman SN, Matic S, Noris E, Lomonossoff GP, Taliansky M (2012) In planta production of a candidate vaccine against bovine papillomavirus type 1. Planta

Loza-Rubio E, Rojas E, Gomez L, Olivera MT, Gomez-Lim MA (2008) Development of an edible rabies vaccine in maize using the Vnukovo strain. Dev Biol 131:477–482

Loza-Rubio E, Rojas-Anaya E, Lopez J, Olivera-Flores MT, Gomez-Lim M, Tapia-Perez G (2012) Induction of a protective immune response to rabies virus in sheep after oral immunization with transgenic maize, expressing the rabies virus glycoprotein. Vaccine 30:5551–5556

Ludwig S, Zell R, Schwemmle M, Herold S (2014) Influenza, a One Health paradigm–novel therapeutic strategies to fight a zoonotic pathogen with pandemic potential. Int J Med Microbiol 304:894–901

Ma JK, Drossard J, Lewis D, Altmann F, Boyle J, Christou P, Cole T, Dale P, van Dolleweerd CJ, Isitt V, Katinger D, Lobedan M, Mertens H, Paul MJ, Rademacher T, Sack M, Hundleby PA, Stiegler G, Stoger E, Twyman RM, Vcelar B, Fischer R (2015) Regulatory approval and a first-in-human phase I clinical trial of a monoclonal antibody produced in transgenic tobacco plants. Plant Biotechnol J 13:1106–1120

MacDonald J, Doshi K, Dussault M, Hall JC, Holbrook L, Jones G, Kaldis A, Klima CL, Macdonald P, McAllister T, McLean MD, Potter A, Richman A, Shearer H, Yarosh O, Yoo HS, Topp E, Menassa R (2015) Bringing plant-based veterinary vaccines to market: managing regulatory and commercial hurdles. Biotechnol Adv 33:1572–1581

Mackenzie JS, Field HE, Guyatt KJ (2003) Managing emerging diseases borne by fruit bats (flying foxes), with particular reference to henipaviruses and Australian bat lyssavirus. J Appl Microbiol 94(Suppl):59S–69S

Maclean J, Koekemoer M, Olivier AJ, Stewart D, Hitzeroth II, Rademacher T, Fischer R, Williamson AL, Rybicki EP (2007) Optimization of human papillomavirus type 16 (HPV-16) L1 expression in plants: comparison of the suitability of different HPV-16 L1 gene variants and different cell-compartment localization. J Gen Virol 88:1460–1469

Mamedov T, Yusibov V (2013) In vivo deglycosylation of recombinant proteins in plants by co-expression with bacterial PNGase F. Bioengineered 4:338–342

Mandal MK, Fischer R, Schillberg S, Schiermeyer A (2014) Inhibition of protease activity by antisense RNA improves recombinant protein production in Nicotiana tabacum cv. Bright Yellow 2 (BY-2) suspension cells. Biotechnol J 9:1065–1073

Mbewana S, Mortimer E, Pera FF, Hitzeroth II, Rybicki EP (2015) Production of H5N1 influenza virus matrix protein 2 ectodomain protein bodies in tobacco plants and in insect cells as a candidate universal influenza vaccine. Front Bioeng Biotechnol 3:197

Miletic S, Simpson DJ, Szymanski CM, Deyholos MK, Menassa R (2015) A plant-produced bacteriophage tailspike protein for the control of salmonella. Front Plant Sci 6:1221

Monath TP (2013) Vaccines against diseases transmitted from animals to humans: a one health paradigm. Vaccine 31:5321–5338

Nemchinov LG, Natilla A (2007) Transient expression of the ectodomain of matrix protein 2 (M2e) of avian influenza A virus in plants. Protein Expr Purif 56:153–159

Niedbalski W (2011) Bluetongue vaccines in Europe. Pol J Vet Sci 14:299–304
Palmer KE, Benko A, Doucette SA, Cameron TI, Foster T, Hanley KM, McCormick AA, McCulloch M, Pogue GP, Smith ML, Christensen ND (2006) Protection of rabbits against cutaneous papillomavirus infection using recombinant tobacco mosaic virus containing L2 capsid epitopes. Vaccine 24:5516–5525
Pasquevich KA, Ibanez AE, Coria LM, Garcia Samartino C, Estein SM, Zwerdling A, Barrionuevo P, Oliveira FS, Seither C, Warzecha H, Oliveira SC, Giambartolomei GH, Cassataro J (2011) An oral vaccine based on U-Omp19 induces protection against B. abortus mucosal challenge by inducing an adaptive IL-17 immune response in mice. PLoS ONE 6: e16203
Patterson EI, Swarbrick CM, Roman N, Forwood JK, Raidal SR (2013) Differential expression of two isolates of beak and feather disease virus capsid protein in Escherichia coli. J Virol Methods 189:118–124
Pearson LD, Roy P (1993) Genetically engineered multi-component virus-like particles as veterinary vaccines. Immunol Cell Biol 71(Pt 5):381–389
Perez Filgueira DM, Mozgovoj M, Wigdorovitz A, Dus Santos MJ, Parreno V, Trono K, Fernandez FM, Carrillo C, Babiuk LA, Morris TJ, Borca MV (2004) Passive protection to bovine rotavirus (BRV) infection induced by a BRV VP8* produced in plants using a TMV-based vector. Arch Virol 149:2337–2348
Powdrill TF, Nipp TL, Rinderknecht JL (2010) One health approach to influenza: assessment of critical issues and options. Emerg Infect Dis 16:e1
Protalix (2017) Orally delivered proteins
Purse BV, Brown HE, Harrup L, Mertens PP, Rogers DJ (2008) Invasion of bluetongue and other orbivirus infections into Europe: the role of biological and climatic processes. Rev Sci Tech 27:427–442
Rademacher T (2014) Method for the generation and cultivation of a plant cell pack. Google Patents
Regnard GL, Halley-Stott RP, Tanzer FL, Hitzeroth II, Rybicki EP (2010) High level protein expression in plants through the use of a novel autonomously replicating geminivirus shuttle vector. Plant Biotechnol J 8:38–46
Risso GS, Carabajal MV, Bruno LA, Ibanez AE, Coria LM, Pasquevich KA, Lee SJ, McSorley SJ, Briones G, Cassataro J (2017) U-Omp19 from brucella abortus is a useful adjuvant for vaccine formulations against salmonella infection in mice. Front Immunol 8:171
Robert S, Khalf M, Goulet MC, D'Aoust MA, Sainsbury F, Michaud D (2013) Protection of recombinant mammalian antibodies from development-dependent proteolysis in leaves of Nicotiana benthamiana. PLoS ONE 8:e70203
Roy P, Bishop DH, LeBlois H, Erasmus BJ (1994) Long-lasting protection of sheep against bluetongue challenge after vaccination with virus-like particles: evidence for homologous and partial heterologous protection. Vaccine 12:805–811
Rybicki EP (2009) Plant-produced vaccines: promise and reality. Drug Discov Today 14:16–24
Rybicki EP (2010) Plant-made vaccines for humans and animals. Plant Biotechnol J 8:620–637
Rybicki EP (2014) Plant-based vaccines against viruses. Virol J 11:205
Rybicki EP, Martin DP (2014) Virus-derived ssDNA vectors for the expression of foreign proteins in plants. Curr Top Microbiol Immunol 375:19–45
Sainsbury F, Lavoie PO, D'Aoust MA, Vezina LP, Lomonossoff GP (2008) Expression of multiple proteins using full-length and deleted versions of cowpea mosaic virus RNA-2. Plant Biotechnol J 6:82–92
Sainsbury F, Thuenemann EC, Lomonossoff GP (2009) pEAQ: versatile expression vectors for easy and quick transient expression of heterologous proteins in plants. Plant Biotechnol J 7:682–693
Sainsbury F, Jutras PV, Vorster J, Goulet MC, Michaud D (2016) A chimeric affinity tag for efficient expression and chromatographic purification of heterologous proteins from plants. Front Plant Sci 7:141

Sarker S, Ghorashi SA, Swarbrick CM, Khandokar YB, Himiari Z, Forwood JK, Raidal SR (2015) An efficient approach for recombinant expression and purification of the viral capsid protein from beak and feather disease virus (BFDV) in Escherichia coli. J Virol Methods 215–216:1–8
Satyavathi VV, Prasad V, Khandelwal A, Shaila MS, Sita GL (2003) Expression of hemagglutinin protein of Rinderpest virus in transgenic pigeon pea [*Cajanus cajan* (L.) Millsp.] plants. Plant Cell Rep 21:651–658
Saxena P, Thuenemann EC, Sainsbury F, Lomonossoff GP (2016) Virus-derived vectors for the expression of multiple proteins in plants. Methods Mol Biol 1385:39–54
Shamloul M, Trusa J, Mett V, Yusibov V (2014) Optimization and utilization of agrobacterium-mediated transient protein production in nicotiana. J Vis Exp: JoVE
Short KR, Richard M, Verhagen JH, van Riel D, Schrauwen EJ, van den Brand JM, Manz B, Bodewes R, Herfst S (2015) One health, multiple challenges: the inter-species transmission of influenza A virus. One Health 1:1–13
Specht EA, Mayfield SP (2014) Algae-based oral recombinant vaccines. Front Microbiol 5:60
Steele JF, Peyret H, Saunders K, Castells-Graells R, Marsian J, Meshcheriakova Y, Lomonossoff GP (2017) Synthetic plant virology for nanobiotechnology and nanomedicine. Wiley Interdiscip Rev Nanomed Nanobiotechnol
Steinkellner H, Castilho A (2015) N-Glyco-engineering in plants: update on strategies and major achievements. Methods Mol Biol 1321:195–212
Stewart ME, Bonne N, Shearer P, Khalesi B, Sharp M, Raidal S (2007) Baculovirus expression of beak and feather disease virus (BFDV) capsid protein capable of self-assembly and haemagglutination. J Virol Methods 141:181–187
Strasser R, Stadlmann J, Schahs M, Stiegler G, Quendler H, Mach L, Glossl J, Weterings K, Pabst M, Steinkellner H (2008) Generation of glyco-engineered Nicotiana benthamiana for the production of monoclonal antibodies with a homogeneous human-like N-glycan structure. Plant Biotechnol J 6:392–402
Streatfield SJ, Lane JR, Brooks CA, Barker DK, Poage ML, Mayor JM, Lamphear BJ, Drees CF, Jilka JM, Hood EE, Howard JA (2003) Corn as a production system for human and animal vaccines. Vaccine 21:812–815
Tabayashi N, Matsumura T (2014) Forefront study of plant biotechnology for practical use: development of oral drug for animal derived from transgenic strawberry. Soc Biotechnol J Japn 92:537–539
Tackaberry ES, Prior F, Bell M, Tocchi M, Porter S, Mehic J, Ganz PR, Sardana R, Altosaar I, Dudani A (2003) Increased yield of heterologous viral glycoprotein in the seeds of homozygous transgenic tobacco plants cultivated underground. Genome 46:521–526
Thuenemann EC, Meyers AE, Verwey J, Rybicki EP, Lomonossoff GP (2013) A method for rapid production of heteromultimeric protein complexes in plants: assembly of protective bluetongue virus-like particles. Plant Biotechnol J 11:839–846
Topp E, Irwin R, McAllister T, Lessard M, Joensuu JJ, Kolotilin I, Conrad U, Stoger E, Mor T, Warzecha H, Hall JC, McLean MD, Cox E, Devriendt B, Potter A, Depicker A, Virdi V, Holbrook L, Doshi K, Dussault M, Friendship R, Yarosh O, Yoo HS, MacDonald J, Menassa R (2016) The case for plant-made veterinary immunotherapeutics. Biotechnol Adv 34:597–604
Torrent M, Llop-Tous I, Ludevid MD (2009) Protein body induction: a new tool to produce and recover recombinant proteins in plants. Methods Mol Biol 483:193–208
Twyman RM, Schillberg S, Fischer R (2013) Optimizing the yield of recombinant pharmaceutical proteins in plants. Curr Pharm Des 19:5486–5494
Tyulkina LG, Skurat EV, Frolova OY, Komarova TV, Karger EM, Atabekov IG (2011) New viral vector for superproduction of epitopes of vaccine proteins in plants. Acta Naturae 3:73–82
Usha R, Rohll JB, Spall VE, Shanks M, Maule AJ, Johnson JE, Lomonossoff GP (1993) Expression of an animal virus antigenic site on the surface of a plant virus particle. Virology 197:366–374

Varsani A, Williamson AL, Rose RC, Jaffer M, Rybicki EP (2003) Expression of Human papillomavirus type 16 major capsid protein in transgenic Nicotiana tabacum cv. Xanthi. Arch Virol 148:1771–1786

Vasilev N, Gromping U, Lipperts A, Raven N, Fischer R, Schillberg S (2013) Optimization of BY-2 cell suspension culture medium for the production of a human antibody using a combination of fractional factorial designs and the response surface method. Plant Biotechnol J 11:867–874

Virdi V, Coddens A, De Buck S, Millet S, Goddeeris BM, Cox E, De Greve H, Depicker A (2013) Orally fed seeds producing designer IgAs protect weaned piglets against enterotoxigenic Escherichia coli infection. Proc Natl Acad Sci U S A 110:11809–11814

Warzecha H, Mason HS, Lane C, Tryggvesson A, Rybicki E, Williamson AL, Clements JD, Rose RC (2003) Oral immunogenicity of human papillomavirus-like particles expressed in potato. J Virol 77:8702–8711

Watson J, Koya V, Leppla SH, Daniell H (2004) Expression of Bacillus anthracis protective antigen in transgenic chloroplasts of tobacco, a non-food/feed crop. Vaccine 22:4374–4384

Wigdorovitz A, Carrillo C, Dus Santos MJ, Trono K, Peralta A, Gomez MC, Rios RD, Franzone PM, Sadir AM, Escribano JM, Borca MV (1999a) Induction of a protective antibody response to foot and mouth disease virus in mice following oral or parenteral immunization with alfalfa transgenic plants expressing the viral structural protein VP1. Virology 255:347–353

Wigdorovitz A, Perez Filgueira DM, Robertson N, Carrillo C, Sadir AM, Morris TJ, Borca MV (1999b) Protection of mice against challenge with foot and mouth disease virus (FMDV) by immunization with foliar extracts from plants infected with recombinant tobacco mosaic virus expressing the FMDV structural protein VP1. Virology 264:85–91

Wigdorovitz A, Mozgovoj M, Santos MJ, Parreno V, Gomez C, Perez-Filgueira DM, Trono KG, Rios RD, Franzone PM, Fernandez F, Carrillo C, Babiuk LA, Escribano JM, Borca MV (2004) Protective lactogenic immunity conferred by an edible peptide vaccine to bovine rotavirus produced in transgenic plants. J Gen Virol 85:1825–1832

Wirblich C, Coleman CM, Kurup D, Abraham TS, Bernbaum JG, Jahrling PB, Hensley LE, Johnson RF, Frieman MB, Schnell MJ (2017) One-Health: a safe, efficient, dual-use vaccine for humans and animals against middle east respiratory syndrome coronavirus and rabies virus. J Virol 91

Yang C-D, Liao J-T, Lai C-Y, Jong M-H, Liang C-M, Lin Y-L, Lin N-S, Hsu Y-H, Liang S-M (2007a) Induction of protective immunity in swine by recombinant bamboo mosaic virus expressing foot-and-mouth disease virus epitopes. BMC Biotechnol 7:62

Yang ZQ, Liu QQ, Pan ZM, Yu HX, Jiao XA (2007b) Expression of the fusion glycoprotein of Newcastle disease virus in transgenic rice and its immunogenicity in mice. Vaccine 25: 591–598

Yu J, Langridge WH (2001) A plant-based multicomponent vaccine protects mice from enteric diseases. Nat Biotechnol 19:548–552

Yusibov V, Hooper DC, Spitsin SV, Fleysh N, Kean RB, Mikheeva T, Deka D, Karasev A, Cox S, Randall J, Koprowski H (2002) Expression in plants and immunogenicity of plant virus-based experimental rabies vaccine. Vaccine 20:3155–3164

Zhao Y, Hammond RW (2005) Development of a candidate vaccine for Newcastle disease virus by epitope display in the Cucumber mosaic virus capsid protein. Biotechnol Lett 27:375–382

Plant Transformation Strategies

Verónica Araceli Márquez-Escobar, Omar González-Ortega and Sergio Rosales-Mendoza

Abstract In this chapter, a general outlook on the plant transformation approaches is provided with emphasis in applications related to molecular farming. The rationale of nuclear, chloroplast, and transient expressions mediated by viral vectors are reviewed. Implications of such technologies in terms of protein yields, post-translational modifications, scalability, and production time scale are critically analyzed. New trends in plant genetic engineering are also identified and perspectives on how these technologies might influence the molecular farming field are provided.

Keywords Nuclear transformation · Chloroplast transformation
Stable transformation · Transient transformation · Transplastomic technologies

1 Introduction

Over 100 years ago a bacterium causing tumors in plants was described (Smith and Townsend 1907). The tumor formation (crown gall disease) resulted from the ability of an alphaproteobacteria, named *Agrobacterium tumefaciens*, to transfer

V. A. Márquez-Escobar · S. Rosales-Mendoza (✉)
Laboratorio de Biofarmacéuticos Recombinantes, Facultad de Ciencias Químicas, Universidad Autónoma de San Luis Potosí, Av. Dr. Manuel Nava 6, 78210 San Luis Potosí, SLP, Mexico
e-mail: rosales.s@fcq.uaslp.mx

V. A. Márquez-Escobar · S. Rosales-Mendoza
Sección de Biotecnología, Centro de Investigación En Ciencias de La Salud Y Biomedicina, Universidad Autónoma de San Luis Potosí, Avenida Sierra Leona 550, Lomas Segunda Sección, 78210 San Luis Potosí, SLP, Mexico

O. González-Ortega
Laboratorio de Bioseparaciones, Facultad de Ciencias Químicas, Universidad Autónoma de San Luis Potosí, Av. Dr. Manuel Nava 6, 78210 San Luis Potosí, SLP, Mexico

J. MacDonald (ed.), *Prospects of Plant-based Vaccines in Veterinary Medicine*,
https://doi.org/10.1007/978-3-319-90137-4_2

DNA to the host cell; leading to the stable insertion of bacterial DNA called T-DNA (Tzfira and Citovsky 2006). Following this line, the discovery of the Ti plasmid (Ti, tumor induction) and the elucidation of the mechanisms mediating the gene transfer from *Agrobacterium* into the plant cell were described mainly by the Belgium group headed by Marc Van Montagu and Jeff Schell. This knowledge opened the possibility of using this bacterium as a delivery system for genetic engineering purposes in plants, thus beginning the transgenic plants era (Herrera-Estrella et al. 1983; De Block et al. 1984).

The ability to create transgenic plants has allowed plants to be used as bioreactors to produce biopharmaceuticals (BP) (Paul and Ma 2011). The plant-made BP technology is based on the introduction of the gene of interest into a wildtype plant generating a transgenic plant, in which the particular gene of interest codes for the therapeutic agent. The approaches to transform plants for this purpose can be classified in two categories: stable and transient (Tzfira et al. 2004; Krenek et al. 2015; Fig. 1). According to the stable transformation approach, either the nucleus or the chloroplast can be targeted to insert the heterologous DNA into the corresponding genome; generating a heritable trait. Stable transformation methodologies comprise three main steps: (1) delivery of the foreign DNA to the plant tissue, (2) selection of transformed plant tissues according to the marker gene used, and (3) establishment of in vitro lines (undifferentiated cells or tissues) or whole plants (Fig. 2). On the other hand, transient transformation is based on the introduction of the foreign DNA into the nuclear plant genome regardless integration events

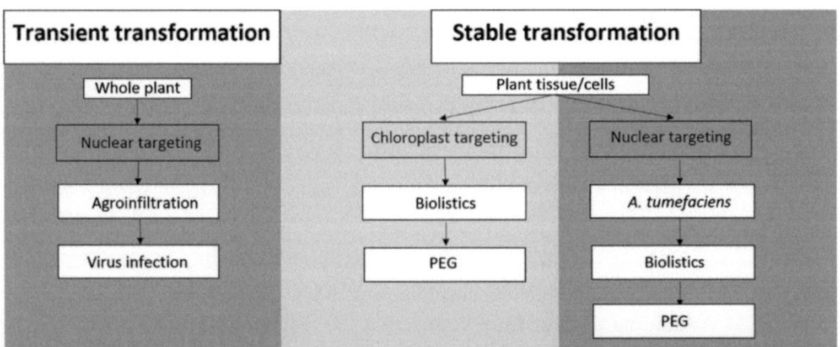

Yield	Very high	High	Modest-low
Post - translational modifications	Complex (glycosylation and disulfide bonds)	Limited to disulfide bonds	Complex (glycosylation and disulfide bonds)
Timescale	Short	Very long	Long
DNA insertion into genome	Possible, Random insertions	Site-specific	Random insertions
Gene silencing	No	No	Very possible
Transgene containment	Possible, through genetic modification	Yes, due to maternal inheritance	Possible, through genetic modification
Multiple gene transcription	Possible	Yes	No

Fig. 1 Schematic summary of the plant transformation approaches and their features

(Fig. 3). Under this approach the target BP is expressed, in the short term, in plant tissues of adult plants; with the subsequent harvest of the plant material that is typically used to purify the target BP.

2 Stable Transformation Strategies

2.1 *Agrobacterium-Mediated Transformation*

The most simple and common method to achieve nuclear transformation relies on the use of *A. tumefaciens,* which is capable of introducing large segments of DNA with minimal rearrangement, high efficiency, and low number of insertions (Gelvin 2003; Păcurar et al. 2011). Briefly, the process involves the action of vir proteins, most of them encoded by the 200 kb tumor-inducing plasmid (Ti plasmid, pTi) in the bacterium. Vir A and Vir G are constitutively expressed, whereas expression of Vir B, C, D, and E are dependent on the activation of VirG. Upon interaction with the plant cell, Agrobacterium senses the chemical exudates from the plant, including acetosyringone that is commonly secreted by wounded cells. At this stage, VirA acts as a chemoreceptor of phenolic compounds. In parallel, sugars from the plant are detected by the chvE protein, which increases VirA protein sensitivity to phenolic compounds. Afterwards VirA protein phosphorylates the VirG protein, leading to the activation of VirG that acts as a transcription factor with the subsequent activation of vir operons (Vir B, C, D, and E).

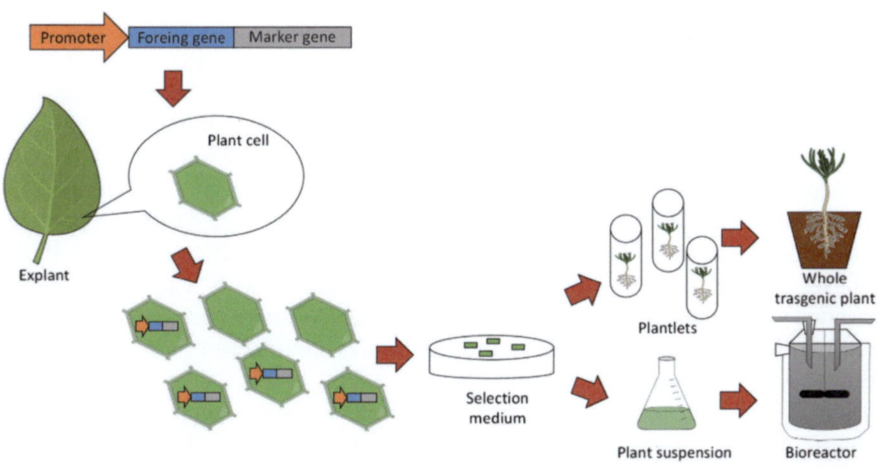

Fig. 2 Schematic methodology for stable transformation approaches. (1) The foreign DNA is delivered into the plant cell, (2) plant explants are placed onto selection medium to recover cells that stably integrated the foreign DNA, (3) transformed lines are propagated to establish undifferentiated cell lines or perform a regeneration step to yield whole plants

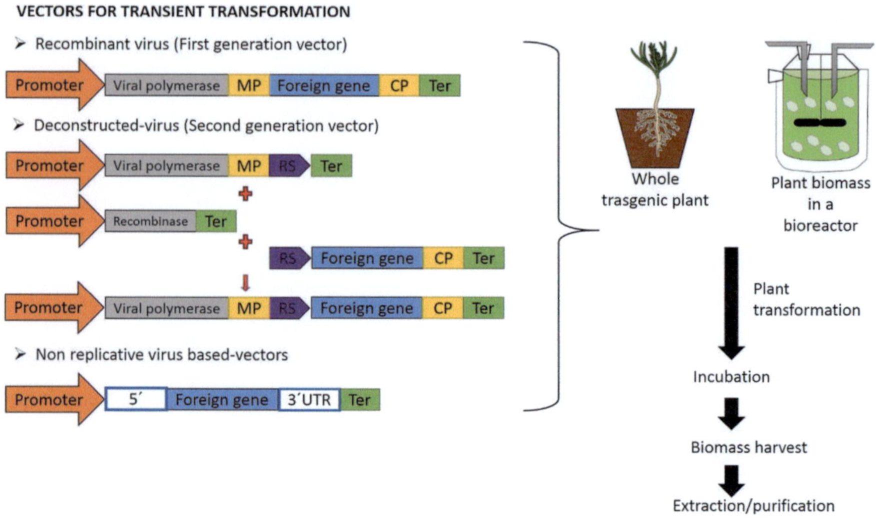

Fig. 3 Schematic methodology for transient transformation approaches. (1) The foreign DNA is delivered to the plant cell as a deconstructed viral vector or recombinant virus. (2) Plant tissues from whole plants or biomass produced in bioreactors are subjected to transformation by either mechanical inoculation or Agrobacterium-mediated DNA delivery (Agroinfiltration). (3) Plants/tissues are incubated for a short period (up to one week) for accumulation of the recombinant protein, in this step inductors are applied when expression relies on an inducible promoter. (4) Plant biomass is harvested to proceed to the extraction and purification of the plant-made recombinant protein/viruses. MP, movement protein; RS, recombination site; CP, capsid protein; Ter, terminator

The pTi plasmid also harbors the transfer DNA (T-DNA), which is the region transferred from the bacterium to the plant cell, and which carries the genes responsible for generation of crown gall tumors. Genes present in the T-DNA comprise oncogenes responsible for phytohormone production, such as auxins and cytokinins; as well as genes responsible for the biosynthesis of opines, a group of amino acid derivatives acting as carbon source for the bacterium. The T-DNA is flanked by 25–28 base-pair repeats called borders. VirC, D and E proteins coordinate the process of T-DNA excision and form a Vir proteins/T-DNA complex that is translocated into the plant cell. During this process VirD1/D2 complex nicks the T-DNA at both border sequences. VirD2 attaches to the 5′ end of the single stranded cut T-DNA, and VirE2 coats the T-DNA strand forming the T-DNA transport complex that is exported into the host cell cytoplasm through a channel formed by VirD4 and VirB proteins.

Vir D4 and Vir B constitute a type IV secretory system required for the transference of the Vir-T-DNA complex. Vir proteins not only form the channel but also works as ATPases providing energy to assemble the channel or for the exporting processes. Although it is known that during host cell attachment the bacterium synthesizes cellulose fibrils and rhicadhesin, a protein that helps sticking to the cell wall, the complex set of plant cell-bacteria interactions is not yet fully understood.

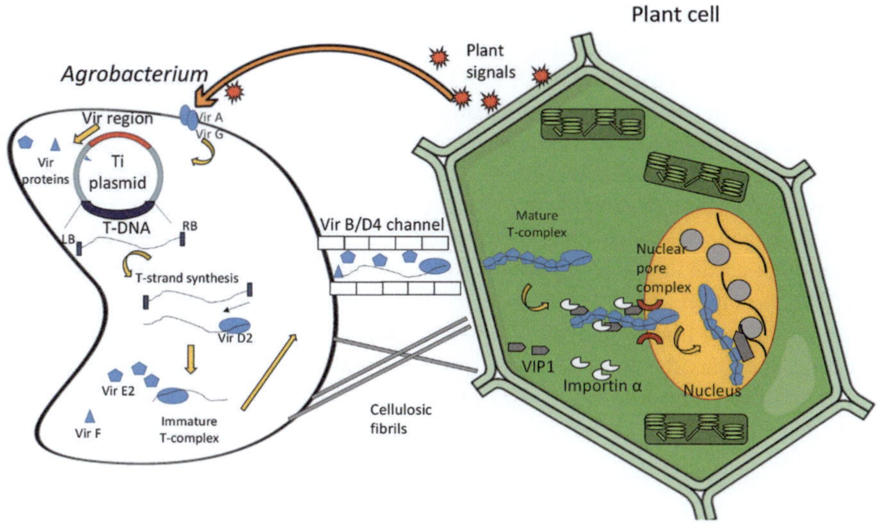

Fig. 4 Mechanisms for *Agrobacterium*-mediated DNA transfer into the plant cell genome. Wound-released phenolic plant compounds are recognized by VirA leading to the activation of VirG; with a subsequent expression of a set of vir genes located in the Ti plasmid. The T-DNA region is splitted from the Ti plasmid by the VirD1/D2 complex to yield a single-stranded DNA molecule. The attachment of Agrobacterium to the plant cell is mediated by the cellulosic fibrils and a VirB/VirD4 channel is formed. The immature T-DNA and some other Vir proteins, such as VirE2 and VirF, are transported through the VirB/VirD4 channel. Once in the cytosol of the plant cell, VirE2 and VirF adhere to the T-DNA inducing its maturation; with a subsequent nuclear translocation mediated by VIP1 and the importin α. Finally, the T-DNA reaches the chromatin and proteosomal degradation uncoats the T-DNA followed by DNA integration into the nuclear genome

Once the T-DNA is inside of the plant cell, the VirE2 and VirD2 proteins are recognized by the host's importin alpha thanks to the nuclear localization signals that exist in these vir proteins. In this process the plant protein VIP1 helps VirE2 to reach the importin. Importin alpha subsequently interacts with importin beta and the nuclear pore complex allowing the transfer of the T-DNA into the nucleus. Finally in the nucleus, the T-DNA integration to the host genome takes place when the plant protein VIP2 targets the T-DNA into areas of chromatin that are being actively transcribed (Tzfira and Citovsky 2002; Gelvin 2010; Dafny-Yelin et al. 2008). In order to insert the foreign DNA into the host plant cell, VirC selects and cuts into a strand of host DNA.

Since the T-DNA region is the mobile element upon infection by *Agrobacterium*, genetic engineers take advantage of this fine mechanism to achieve nuclear transformation of plant cells by replacing the T-DNA genes with appropriate expression cassettes for the gene of interest and a marker gene, which are flanked by the direct repeat borders (Fig. 4). Sequences from the Ti plasmid have been divided between two vectors in a binary system for nuclear genetic engineering: the shuttle vector contains an artificial T-DNA carrying the heterologous

DNA flanked by the left and right borders, while the helper plasmid encodes the *vir* genes required for plant transformation (Hellens et al. 2000; Karimi et al. 2002; Earley et al. 2006; Hoekema et al. 1983).

Agrobacterium-mediated transformation leads to a variable number of insertions that occur in random loci of the plant genome, which can lead to modest expression of the transgene (the achieved protein yield is generally below 1% of total soluble protein, TSP) (Francis and Spiker 2005), phenotypic alterations (Schnell et al. 2015), and gene silencing. Another limitation of *Agrobacterium*-mediated transformation consists in the fact that some plant species are not efficiently infected by this transformation vector. For instance some monocots such as wheat, rice and maize; which are highly relevant crops, are naturally recalcitrant to *A. tumefaciens* infection and thus the transformation protocols should be optimized to overcome this limitation (van Wordragen and Dons 1992; Gelvin 2003). These important issues led to the development of alternative transformation protocols.

3 Biolistics

The biolistics method is based on the delivery of "microcarriers", which are micro sized tungsten or gold particles coated with the DNA of interest, into the target cells by using high velocity acceleration (Daniell 1993, 1997). This technology allows inserting the target DNA in either the nuclear or chloroplast genomes.

Although biolistics represents an alternative to transform Agrobacterium-recalcitrant species at the nuclear level, some disadvantages of the technique include the following: multiple insertions of the transgene frequently occur, which could lead to silencing events; and lack of fine definition of the DNA segment that is inserted in the host, which obliges for detailed screening of the transformants to rescue those having the full length DNA of interest (Kikkert et al. 2005). Biolistics is generally applied to species that are recalcitrant to *Agrobacterium*-mediated transformation. In the case of chloroplast transformation, biolistics constitutes an attractive method since DNA vectors are efficiently delivered into the chloroplasts, whereas no reproducible success on the use of *A. tumefaciens* has been achieved for chloroplast transformation. Thanks to the efficient recombination machinery that singularly exists in the chloroplast, the desired insertion can be mediated by double homologous recombination mediated by left and right flanking sequences of about 1–2 kb each, which belong to the host plastid genome and define the insertion site (Svab et al. 1990; Staub and Maliga 1992; Svab and Maliga 1993; Maliga 1993). In this manner, site-specific insertion in the chloroplast genome is achieved, increasing transformation efficiency, avoiding position effects and minimizing the insertion of undefined DNA fragments. However, it should be considered that nuclear insertions might occur and transformants should be screened to discard such events in the selected lines.

The correct coating of the microcarriers with the DNA of interest is critical for biolistics; this step involves mixing the DNA with calcium chloride, which provides a positive charged surface for the adherence of the DNA to the microcarrier; and spermidine, a cationic polyamine that protects DNA from degradation by cellular nucleases and allows DNA adsorption onto the particles (Brune et al. 1991; Thomas et al. 1996).

4 Transient Transformation Approaches

Stable nuclear transformation strategies have some limitations for the production of heterologous proteins since the achieved yields are usually low, long periods are required to generate transformed lines, and the transformation protocols are often inefficient for many species of interest. Transient transformation is an alternative approach to overcome these limitations. While the stable transformation process takes months, transient transformation can take 4–15 days with expression levels up to 80% TSP (5 g of recombinant protein per kilogram of fresh weight) (Gleba et al. 2005). The current technologies based in transient transformation rely in the use of Agrobacterium to deliver the expression vector into the target cells. An important number of achievements in this area have been obtained by using expression vectors (replicative or non-replicative) based on viral elements delivered by Agrobacterium. Such viral vectors have led to the highest yields observed in transient expression approaches in plants.

Vector design is essential to succeed in transient transformation methods, which comprise the transfer of the heterologous DNA by Agrobacterium or viruses. Several factors should be considered to obtain efficient expression vectors, such as the length of the gene of interest, host tissue, specific organelle targeting, etc. (Gleba et al. 2004). Among the efficient expression systems, some of them take advantage of the plant RNA virus machinery; which allows for higher viral replication rates and efficient protein expression in the host cell (Gleba et al. 2007). Some of the RNA plant viruses that have been used for the production of plant viral vectors are the Tobacco mosaic virus (TMV), Cowpea mosaic virus (CPMV), Potato virus X (PVX), Alfalfa mosaic virus (AMV), Plum pox virus (PPV), and Geminivirus (Hefferon 2012).

First generation vectors are based on a “full-virus” vector strategy in which the original viral promoter is conserved to drive foreign gene replication and expression. Recombinant proteins are produced individually or as fusions with the viral coat protein (CP), and the machinery of normal viral replication cycle is preserved; including viral replication, host infection, translation, assembly of mature virions, cell-to-cell movement, movement through the whole plant, reprogramming of the host biosynthetic machinery, and suppression of gene silencing. Expression levels attained with these first generation vectors are higher than those for stable transformation, reaching up to 10% of total soluble protein (TSP). However, a drawback of these first generation vectors is the size limitation of the protein of interest; proteins larger than 30 kDa are poorly expressed or fail to express because large CP fusions compromise viral movement or because the larger genome is too big for the capsid. Other limitations of first generation vectors is the deleterious effects that viral infection typically induce in the plant as well as the high biosafety required in the process to prevent the transmission of virus to non-target crops. When the chimeric CP approach is followed, the target antigen should not exceed 25 amino acids in length to avoid interferences with proper CP folding and function.

Second generation vectors differ from the full virus vectors in the fact that only some of the viral mechanisms are preserved (e.g. genome replication and in some cases cell to cell movement), whereas other functions such as DNA delivery are provided by non-viral elements (e.g. *A. tumefaciens*). With these deconstructed virus vectors, the transformed plants do not develop severe infection symptoms that impact biomass production and heterologous protein yields. An example of a technology based on second generation vectors is the MagnICON system that relies on a Tobacco mosaic virus-based deconstructed vector, which is delivered into the plant cell by agroinfiltration. In this way, the system takes advantage of the velocity and expression level of the virus, the potent transfection mediated by Agrobacterium, the post-translational capabilities, and low production costs of plants (Gleba et al. 2004, 2005). An important feature of this system is the obtained yield (up to 80% TSP or 5 g/kg fresh weight biomass), which is higher than that attained using first generation vectors. Other advantages comprise short production time (4–15 days) and the fact that, in contrast to first generation vectors, the gene of interest can be in the 2–80 kb range as the viral genome lacks several native genes and most of the vectors do not rely on CP functions. Among the factors that should be optimized in this method are the host plant, the agrobacterium strain, and the initial agrobacteria density. Interestingly, this expression system was developed by the industry (Icon Genetics, Germany) and its use has been expanded to a number of human vaccine candidates, some of them under clinical trials. A limitation of the current methodologies based on transient expression is the fact that purification to obtain parenteral formulations is the path to be followed since bacterial residues are present in plant tissues.

In the case of veterinary vaccines, some candidates expressed by transient transformation have been reported. For instance, the Infectious Bursal Disease Virus (IBDV) is a highly contagious disease that affects young birds to which a plant-based vaccine was developed by expressing the VP2 protein transiently produced in *N. benthamiana* (Gómez et al. 2013). Other examples include the Bluetongue virus (BTV), which causes high mortality in ruminants (Thuenemann et al. 2013); the VP6 protein of rotavirus expressed in Chenopodium leaves (Zhou et al. 2010), and the SAG1 antigen from *Toxoplasma gondii* (Laguía-Becher et al. 2010).

Non-replicative vectors based on viral elements have also led to highly productive systems. For instance, a system based on a disabled version of the cowpea mosaic virus RNA-2 has been reported in which the gene of interest is flanked by the 5′ leader sequence and 3′ untranslated region (UTR) of RNA-2. At 6 days post-infiltration, the protein of interest was expressed at yields up to 20% of TSP. In addition to the reported production levels, the system provides beneficial traits for the production of antigens in plants such as the following: RNA-dependent RNA polymerases are not required, reducing the vector size; since no viral replication occurs, no restriction on the size of the insert exists; and biocontainment is facilitated since there is no risk of releasing replicative agents (Sainsbury and Lomonossoff 2008). In the case of multiple gene expression, non-replicative vectors have demonstrated their potential for heteromultimeric expression. Bluetongue virus-like particles (VLPs) have been efficiently produced in *N. benthamiana* using the Cowpea mosaic virus–based HyperTrans (CPMV-HT) vector, which requires 4 proteins to assemble the VLPs (Thuenemann et al. 2013).

5 Targeting Proteins to Organelles

Nuclear expression, whether stable or transient, allows targeting the recombinant protein towards several organelles, which can be favorable since the protein may reach the complex plant cell machinery that performs posttranslational modifications and could be protected from degradation. As an example on how organelle targeting influences accumulation levels, studies on the expression of the major subunit and immunogenic polypeptide of K99 fimbriae (FanC) from enterotoxigenic *Escherichia coli* (ETEC) revealed that chloroplast targeting in transgenic soybean (Garg et al. 2007) led to 5-fold less accumulation when compared to cytosol-targeted FanC. The same issue was observed by He et al. when targeting the domain III (DIII) of the envelope (E) protein from the West Nile Virus to the chloroplast of *N. benthamiana* plants. In fact, the obtained yield decreased 63 times when compared to endoplasmic reticulum (ER) targeting (He et al. 2014). However, in a recent study the use of a new chloroplast transit peptide (CTP) from *E. coli* was assessed in rice, leading to improved chloroplast targeting efficiency. Thus, this peptide might lead to improvements in the efficiency of chloroplast targeting (Shen et al. 2017).

Targeting to the ER is a well-known approach to avoid cytosolic toxicity of the produced protein; it also diminishes protein degradation due to low proteolytic activity and the presence of chaperones and stabilizing agents, thus enhancing the accumulation of the recombinant protein. Several proteins have been targeted into the ER by using specific signal peptides at the N-terminus and the SEKDEL signal sequence at the C-terminus. Examples of this focus include: the vaccine candidate against the Dengue virus (DENV) in which a fusion protein comprising the consensus envelope protein domain III (cEDIII) from DENV and the cholera toxin B subunit (CTB) was expressed in transgenic rice calli using the signal peptide from the luminal binding protein BiP at the N-terminus plus the SEKDEL signal sequence at the C-terminus (Kim et al. 2016), the hepatitis B surface antigen (rHBsAg) containing the soybean vegetative storage protein vspA signal peptide and the KDEL signal; which enhanced the accumulation of the protein in NT1 tobacco cells (Sojikul et al. 2003), and the SAG1 antigen from *Toxoplasma gondii* produced in tobacco leaves using the AP24 osmotin signal peptide sequence and KDEL signal (Laguía-Becher et al. 2010); among others. In the case of ER-targeted veterinary vaccines, examples include production of HA protein from avian influenza in *Arabidopsis thaliana* (Lee et al. 2015a, b) and *Nicotiana benthamiana* (Kanagarajan et al. 2012), GP antigens from porcine reproductive and respiratory syndrome (PRRS) virus in *A. thaliana* seed (Piron et al. 2014), E2 protein from bovine viral diarrhea virus in tobacco (Nelson et al. 2012), a shiga toxin subunit from pig edema disease in lettuce (Matsui et al. 2009), and FaeG from enterotoxigenic *Escherichia coli* (ETEC) in barley (Joensuu et al. 2006).

Compartmentalization to the apoplast has also been explored as a possibility to increase the protein yield since it is secreted and not accumulated in the cell endomembrane system or the cytosol. Another possible advantage is the ease for purification of proteins in this compartment with a low content of endogenous protein. For instance, it has been reported that the expression of the human epidermal growth factor (hEGF) showed a 4-fold increase when targeted to the apoplast in tobacco when compared to the accumulation under cytosolic expression (Wirth et al. 2004). Other proteins targeted to the apoplast are the following: the recombinant silk-like protein produced in Arabidopsis cells (Yang et al. 2005), the tobacco-made *Plasmodium* antigen (Ma et al. 2012), and the recombinant human papillomavirus 8 E7 protein expressed in tobacco plants (Noris et al. 2011); among others.

Targeting the protein of interest to oil bodies is another possibility associated with yield improvement since recombinant proteins are stabilized as a discrete organelle (oil-body) and are easily separated and recovered in an aqueous solution by centrifugation and density. For this purpose the target protein is fused to oleosins; which are composed of an N-terminal hydrophilic region of variable length (from 30 to 60 residues), a central hydrophobic domain of about 70 residues, and a C-terminal amphipathic region of variable length (from 60 to 100 residues). Hirudin, a potent and specific thrombin inhibitor fused to the oleosin gene from Arabidopsis, has been successfully expressed in *Brassica napus* and *B. carinata* (Parmenter et al. 1995; Chaudhary et al. 1998). The recombinant human precursor insulin (Des-B30) and the human epidermal growth factor (hEGF) have also been expressed as oleosin fusions in Arabidopsis (Nykiforuk et al. 2005; Moloney et al. 2006).

The vacuole is another organelle to consider for protein targeting since it is recognized as an organelle with low protease levels (Neuhaus and Rogers 1988) and that is able to trim glycans, obtaining recombinant proteins with depletion of terminal GlcNAc residues and exposing the terminal Man on complex glycans (Gomord and Faye 2004; Lerouge et al. 1998), which is critical for the uptake by mammalian cells of some therapeutic proteins such as glucocerebrosidase. Taking these advantages into consideration, the glucocerebrosidase (GCD) produced in carrot cell suspension was targeted to vacuole storage via the ER (Neuhaus et al. 1991; Shaaltiel et al. 2007); obtaining a highly convenient product since in contrast to the mammalian expressed enzyme the plant-made GCD does not require in vitro enzymatic treatments to display mannose residues.

6 Possible Innovations for Nuclear-Based Expression

Considering biosafety issues, the use of contained production facilities such as those based on greenhouses and bioreactors are considered the best option to alleviate concerns of transgene release. The transfer of antigen-encoding transgenes to non-target crops represents a risk that may result in serious consequences derived from the uncontrolled exposure to antigens in the diet, although such risk is considered of very low probability. Transgene containment becomes more crucial considering that plant-made vaccines are proposed to benefit developing countries in which corn or rice can be cultivated in open fields or in low technology greenhouses. In such cases, enhanced biosafety features should be built into the transformation approach. For instance, the use of inducible promoters may not only optimize protein yield while avoiding deleterious effects in the plant phenotype but also provide the possibility of avoiding antigen expression in non-target crops. Surprisingly, this expression modality has been narrowly explored in the molecular farming field and is limited to the studies by Werner et al. (2011); in which the *Aspergillus nidulans* alcohol dehydrogenase (*alcA*) promoter is used (Werner et al. 2011).

In terms of optimizing expression levels using nuclear transformation, it should be considered that random integration of the transgene is the main factor that leads to low levels, due to insertion in regions with low transcriptional activity or due to silencing derived from multigene insertion. These limitations can be overcome through the development of technologies to target the transgene to specific loci. On this subject, the CRISPR/Cas9 system has emerged as a new tool in genetic engineering that allows for the site-specific insertion of heterologous DNA. Thus, this technology promises to avoid gene silencing due to disruption of host genes, minimize the variability in expression level among different transgenic lines, and enhance the low integration efficiency in recalcitrant plants. There are reports describing the use of the CRISPR/Cas9 system, for example a 3.7-kb gene expression cassette was site-specific inserted in Chinese hamster ovary (CHO) cells and as consequence improvements in protein yields and consistent production were achieved (Lee et al. 2015a, b). The application of such technology remains a pending objective in the field of plant-made vaccines.

7 Chloroplast Transformation

Chloroplast transformation is a stable approach with unique characteristics. The chloroplast is a photosynthetic organelle belonging to the plastid family found in plant cells and eukaryotic algae in which important biosynthetic pathways take place; including those related to fatty acids, amino acids, isoprenoids, etc. (Leister 2003; Bobik and Burch-Smith 2015). The chloroplast genome is about 120–150 kb in length and a single cell contains multiple plastids. In fact, a typical photosynthetic plant cell contains approximately 100 chloroplasts and each chloroplast has about 100 genome copies, rendering up to 10,000 copies per cell (Bendich 1987; Golczyk et al. 2014; Daniell et al. 2016), in contrast to one copy of the nuclear genome per cell. This number of genomes allows a greater number of transgenes per cell, resulting in more transcription and translation of transgenes versus nuclear transformation. In combination with absent or rare silencing phenomena (Wani et al. 2010; Sidorov et al. 1999), this property generally leads to higher protein levels (Maliga 1993; Wakasugi et al. 2001; Staub et al. 2000; Guda et al. 2000). Other advantages of chloroplast transformation include the feasibility of expressing multiple proteins through polycistronic mRNAs (Daniell and Dhingra 2002), and gene containment due to the maternal inheritance of chloroplast (Daniell 2002, 2007; Daniell and Parkinson 2003; Ruf et al. 2001).

Methods to generate transplastomic plants involve the introduction of the transgene by biolistics (Sanford et al. 1993; Bock 2015) or treating the protoplasts with PEG (Maliga et al. 1993; Maliga 2004; Koop et al. 1996). Electroporation has been applied to transiently express, in the spinach chloroplast, the GUS and CAT reporter genes (To et al. 1996). Although these methodologies allow either nuclear and chloroplast transformation, they have been mainly applied to transform chloroplasts.

On the other hand for the PEG-mediated method; enzymatic removal of the cell wall is required to obtain protoplasts that are subsequently treated with PEG in the presence of the transformation vector. In contrast to the biolistics method, the PEG-mediated transformation is cheaper since no special equipment as the gen gun is required (Rivera et al. 2012). Several plant species with agronomical and not agronomical importance have been transformed using the PEG-mediated method, for instance: *Nicotiana tabacum* (Golds et al. 1993), *Nicotiana plumbaginifolia* (O'Neill et al. 1993), *Brassica oleracea* var. botrytis (cauliflower) (Nugent et al. 2005a), *Solanum lycopersicum* (tomato) (Nugent et al. 2005b), and *Lactuca sativa* (lettuce) (Lelivelt et al. 2005).

Irrespective of the employed transformation protocol, the regeneration of whole transplastomic plants from the transformed cells is a laborious goal. In the case of tobacco, a crop model easily regenerated, it takes about 5 months to obtain the primary generation (T0) while nuclear transformation requires approximately 6 weeks. This long period arises from the several rounds of selection and regeneration cycles required to achieve a homoplastomic state, meaning that all chloroplast genomes with the plant carry the transgene, a concept similar to homozygosity of nuclear genes, and which ensures the stability of transplastomic lines. Some bacterial antigens that have been expressed in tobacco chloroplasts are the following: heat-labile enterotoxin subunit B of *E. coli* responsible for diarrhea (Kang et al. 2003), tetanus toxin fragment C (Tregoning et al. 2005), and anthrax protective antigen (Watson et al. 2004; Ruhlman et al. 2007). However, the expression in tobacco has the limitation given by the content of toxic alkaloids; which makes this species questionable in terms of toxicity if used as delivery vehicle of oral vaccines, although it should be considered that cultivars having low concentration of alkaloids have been applied in the field (Li et al. 2006). Therefore edible plant species have also been explored for chloroplast expression. Some antigens that have been produced in lettuce for human use are the following: dengue-3-premembrane and envelop polyprotein, CTB-malaria antigen and CTB-proinsulin antigen, and some tuberculosis antigens such as ESAT-6, Mtb72F, and LipY (Kanagaraj et al. 2011; Davoodi-Semiromi et al. 2010; Ruhlman et al. 2007; Lakshmi et al. 2013). Plastid transformation methodologies have also been implemented for tomato (Lu et al. 2013), potato (Valkov et al. 2014), rice (Lee et al. 2006), and lettuce (Ruhlman 2014; Lelivelt et al. 2014).

As mentioned above the integration of the gene of interest is site-directed. A typical approach consists in the insertion of the transgene at the inverted repeated region (IR), which is achieved by the use of specific recombination flanks located in the transformation vector. Once inserted in one of the IR regions, a phenomenon of copy correction takes place, in which the heterologous DNA is duplicated, rendering two copies of the foreign DNA per genome; thus enhancing recombinant protein yields. The homoplastic state in the transformed lines is achieved thanks to multiple rounds of selection on antibiotic-containing media that mediate the elimination of chloroplasts carrying wild type genomes (Verma and Daniell 2007; Guda et al. 2000). Consequently, transformation plasmids targeting the IR are typically used. The commonly used insertion sites in tobacco include those between the tRNA-Ile (TrnI) and tRNA-Ala (TrnA) genes (Verma and Daniell 2007) and between the tRNA-Val (TrnV) and rps12 or rbcL and accD genes (Zoubenko et al. 1994; Jin and Daniell 2015). On the other hand, plasmids targeting genome locations outside the IR result in a single transgene copy with relatively lower protein yields; thus these are mainly used to study plastid translational and transcriptional mechanisms (Bally et al. 2009).

Innovative and adaptive vectors for plastid expression such as the "Operon-extension" vectors have been developed and rely on the extension of endogenous operons by using a DNA vector carrying untranslated regions (UTRs) from particular operons that serve as flanking regions for recombination. In this manner endogenous promoters drive the transgene expression. Since no promoter and regulatory elements are inserted along with the gene of interest, undesirable rearrangements in the plastome are diminished (Herz et al. 2005). As a result, the foreign gene inserted under the control of a highly transcribed operon is efficiently expressed leading to higher amounts of recombinant protein. Well characterized operons include those from the ATP synthase (*atpB/atpE*) and the D1 polypeptide of the photosystem II (*psbA*); these have been targeted in operon-extension approaches (Deng and Gruissem 1987).

Split transformation vectors rely in the use of at least two deconstructed DNA vectors, each of them containing one flanking region required for homologous recombination into the plastome, and a second sequence for recombination that is shared by another vector. Under this approach plants can be co-transformed and following the insertion of the individual vectors co-integrates are formed which are inherently unstable. However, an in situ rearrangement can then occur via recombination of the shared vector sequence, leading to stable integration that corresponds to a functional expression cassette, which is typically an operon like arrangement (Staub and Maliga 1994; Klaus et al. 2004). Split transformation vectors can be applied for promoter-containing or operon-extension configurations and are advantageous for the insertion of multiple genes (Arai et al. 2004; Nakashita et al. 2001).

Regarding the post-transcriptional modifications, it should be considered that glycosylation does not occur in the chloroplast; however, disulfide bonds can be generated in this organelle (Daniell et al. 2005; Bally et al. 2008). The fact that chloroplasts do not glycosylate recombinant proteins becomes an attractive trait for the expression of bacterial antigens or of therapeutic proteins in which concerns related to allergenicity, derived from the presence of plant glycans that differ from those of mammalian hosts, exist (e.g. the presence of β1,2-linked xylose, core α1,3-linked fucose, and Lewis A-type structures). However, it should be considered that glycosylation is also implicated in the stability and functionality of some target proteins, and several authors argue that plant glycans confer null or rare risks (Shaaltiel and Yl 2016). In terms of biosafety, chloroplast transformation eliminates the risk related to gene transfer through pollen since the genome of chloroplast shows maternal inheritance in most plant species (Daniell 2007; Ruf et al. 2007).

8 Approaches for Multicomponent Vaccines Production

Multicomponent vaccines can be produced in plants, and each component can accumulate in a different subcellular location in order to increase yields. For instance, co-transformation of the nucleus and plastid genome by the biolistic method has already been successfully achieved (Elghabi et al. 2011), and such an approach could be used to co-express cytokines or other proteins that serve as vaccine adjuvants along with the target vaccine antigen in a single cell. The expression organelle could be selected based on the post-translational modifications demanded by each vaccine component.

Another possibility is multigene transformation at the nuclear level while targeting individual proteins to distinct organelles according to their individual requirements in terms of post-translational modifications or their toxicity in certain subcellular localization. This can be done using multiple gene insertion, or by using viral elements to produce several proteins with different target signals from a single transcriptional unit. The 2A picornavirus sequence has been successfully applied for the first time for the expression of multiple antigens in plant cells through a single transformation event. This sequence induces a 'ribosomal skip' mechanism that results in self-cleavage events at a translational level, which allows for the production of antigens from *Taenia solium* in tobacco plants (Monreal-Escalante et al. 2015).

In the case of polycistronic expression in plastids, only few studies have focused on exploring their configuration for multicomponent vaccine production. We consider that further use of such approaches will lead to innovative plant-based vaccines, considering that the most challenging pathogens require multiepitopic vaccines able to induce broad protective immune responses targeting several components of the pathogen. The effects of expressing multiple antigens at high levels could lead to deleterious effects in plant development, thus the phenotype of the transformed plants should be evaluated case by case.

9 Conclusions

Significant advances in the field of plant transformation technologies have been achieved over the last decades to benefit the biopharmaceutical production field. Such advances can offer significant improvements in yields, biosafety, and multigene expression. Although such innovative systems have substantially improved the expression of antigenic proteins in the plant cell, the advances of the last ten years constitute a promise to further improve and diversify plant-made vaccine production. Thanks to new expression technologies, the next generation plant-made vaccines have emerged and are in the pipeline; they will lead with no doubt to new products that will positively impact the animal health field.

References

Arai Y, Shikanai T, Doi Y, Yoshida S, Yamaguchi I, Nakashita H (2004) Production of polyhydroxybutyrate by polycistronic expression of bacterial genes in tobacco plastid. Plant Cell Physiol 45:1176–1184

Bally J, Nadai M, Vitel M, Rolland A, Dumain R, Dubald M (2009) Plant physiological adaptations to the massive foreign protein synthesis occurring in recombinant chloroplasts. Plant Physiol 150:1474–1481

Bally J, Paget E, Droux M, Job C, Job D, Dubald M (2008) Both the stroma and thylakoid lumen of tobacco chloroplasts are competent for the formation of disulphide bonds in recombinant proteins. Plant Biotechnol J 6:46–61

Bendich AJ (1987) Why do chloroplasts and mitochondria contain so many copies of their genome? BioEssays 6:279–282

Bobik K, Burch-Smith TM (2015) Chloroplast signaling within, between and beyond cells. Front Plant Sci 6:781

Bock R (2015) Engineering plastid genomes: methods, tools, and applications in basic research and biotechnology. Annu Rev Plant Biol 66:211–241

Brune B, Hartzell P, Nicotera P, Orrenius S (1991) Spermine prevents endonuclease activation and apoptosis in thymocytes. Exp Cell Res 195:323–329

Chaudhary S, Parmenter DL, Moloney MM (1998) Transgenic Brassica carinata as a vehicle for the production of recombinant proteins in seeds. Plant Cell Rep 17:195–200

Dafny-Yelin M, Levy A, Tzfira T (2008) The ongoing saga of Agrobacterium–host interactions. Trends Plant Sci 13(3):102–105

Daniell H (1993) Foreign gene expression in chloroplasts of higher plants mediated by tungsten particle bombardment. Methods Enzymol 217:536–556

Daniell H (1997) Transformation and foreign gene expression in plants by microprojectile bombardment. Methods Mol Biol 62:463–489

Daniell H (2002) Molecular strategies for gene containment in transgenic crops. Nat Biotechnol 20:581–586
Daniell H (2007) Transgene containment by maternal inheritance: effective or elusive? Proc Natl Acad Sci U S A 104(17):6879–6880
Daniell H, Chebolu S, Kumar S, Singleton M, Falconer R (2005) Chloroplast-derived vaccine antigens and other therapeutic proteins. Vaccine 23:1779–1783
Daniell H, Dhingra A (2002) Multigene engineering: dawn of an exciting new era in biotechnology. Curr Opin Biotechnol 13:136–141
Daniell H, Lin CS, Yu M, Chang WJ (2016) Chloroplast genomes: diversity, evolution, and applications in genetic engineering. Genome Biol 17(1):134
Daniell H, Parkinson CL (2003) Jumping genes and containment. Nat Biotechnol 21:374–375
Davoodi-Semiromi A, Schreiber M, Nalapalli S, Verma D, Singh ND, Banks RK, Chakrabarti D, Daniell H (2010) Chloroplast-derived vaccine antigens confer dual immunity against cholera and malaria by oral or injectable delivery. Plant Biotechnol J 8:223–242
De Block M, Herrera-Estrella L, Van Montagu M, Schell J, Zambryski P (1984) Expression of foreign genes in regenerated plants and in their progeny. EMBO 3(8):1681–1689
Deng XW, Gruissem W (1987) Control of plastid gene expression during development: the limited role of transcriptional regulation. Cell 49:379–387
Earley KW, Haag JR, Pontes O, Opper K, Juehne T, Song K, Pikaard CS (2006) Gateway-compatible vectors for plant functional genomics and proteomics. Plant J 45:616–629
Elghabi Z, Ruf S, Bock R (2011) Biolistic co-transformation of the nuclear and plastid genomes. Plant J 67:941–948
Francis KE, Spiker S (2005) Identification of Arabidopsis thaliana transformants without selection reveals a high occurrence of silenced T-DNA integrations. Plant J 41(3):464–477
Garg R, Tolbert M, Oakes JL, Clemente TE, Bost KL, Piller KJ (2007) Chloroplast targeting of FanC, the major antigenic subunit of Escherichia coli K99 fimbriae, in transgenic soybean. Plant Cell Rep 26(7):1011–1023
Gelvin SB (2003) Agrobacterium-mediated plant transformation: the biology behind the "gene-jockeying" tool. Microbiol Mol Biol Rev 67(1):16–37
Gelvin SB (2010) Finding a way to the nucleus. Curr Opin Microbiol 13(1):53–58
Gleba Y, Klimyuk V, Marillonnet S (2005) Magnifection–a new platform for expressing recombinant vaccines in plants. Vaccine 23(17–18):2042–2048
Gleba Y, Klimyuk V, Marillonnet S (2007) Viral vectors for the expression of proteins in plants. Curr Opin Biotechnol 18(2):134–141
Gleba Y, Marillonnet S, Klimyuk V (2004) Engineering viral expression vectors for plants: the 'full virus' and the 'deconstructed virus' strategies. Curr Opin Plant Biol 7(2):182–188
Golczyk H, Greiner S, Wanner G, Weihe A, Bock R, Börner T, Herrmann RG (2014) Chloroplast DNA in mature and senescing leaves: a reappraisal. Plant Cell 26:847–854
Golds T, Maliga P, Koop HU (1993) Stable plastid transformation in PEG-treated protoplasts of Nicotiana tabacum. Nat Biotechnol 11:95–97
Gómez E, Lucero MS, Chimeno Zoth S, Carballeda JM, Gravisaco MJ, Berinstein A (2013) Transient expression of VP2 in Nicotiana benthamiana and its use as a plant-based vaccine against infectious bursal disease virus. Vaccine 31(23):2623–2627
Gomord V, Faye L (2004) Post-translational modification of therapeutic proteins in plants. Curr Opin Plant Biol 7:171–181
Guda C, Lee S-B, Daniell H (2000) Stable expression of a biodegradable protein-based polymer in tobacco chloroplasts. Plant Cell Rep 19:257–262
He J, Peng L, Lai H, Hurtado J, Stahnke J, Chen Q (2014) A plant-produced antigen elicits potent immune responses against West Nile virus in mice. Biomed Res Int 2014:952865
Hefferon KL (2012) Plant virus expression vectors set the stage as production platforms for biopharmaceutical proteins. Virology 433(1):1–6

Hellens RP, Edward EA, Leyland NR, Bean S, Mullineaux PM (2000) pGreen: a versatile and flexible binary Ti vector for Agrobacterium-mediated plant transformation. Plant Mol Biol 42:819–832

Herrera-Estrella L, Depicker A, Van Montagu M, Schell J (1983) Expression of chimeric genes transferred into plant cells using a Ti-plasmid-derived vector. Nature 303(5914):209–213

Herz S, Füssl M, Steiger S, Koop HU (2005) Development of novel types of plastid trans formation vectors and evaluation of factors controlling expression. Transgenic Res 14:969–982

Hoekema A, Hirsch PR, Hooykaas PJJ, Schilperoort R (1983) A binary plant vector strategy based on separation of vir- and T-region of the Agrobacterium tumefaciens Ti-plasmid. Nature 303:179–180

Jin S, Daniell H (2015) The engineered chloroplast genome just got smarter. Trends Plant Sci 20:622–640

Joensuu JJ, Kotiaho M, Teeri TH, Valmu L, Nuutila AM, Oksman-Caldentey K-M, Niklander-Teeri V (2006) Glycosylated F4 (K88) fimbrial adhesin FaeG expressed in barley endosperm induces ETEC-neutralizing antibodies in mice. Transgenic Res 15:359–373

Kanagarajan S, Tolf C, Lundgren A, Waldenström J, Brodelius PE (2012) Transient expression of hemagglutinin antigen from low pathogenic avian influenza a (H7N7) in nicotiana benthamiana. PLoS ONE 7:e33010

Kanagaraj AP, Verma D, Daniell H (2011) Expression of dengue-3 premembrane and envelope polyprotein in lettuce chloroplasts. Plant Mol Biol 76:323–333

Kang TJ, Loc NH, Jang MO, Jang YS, Kim YS, Seo JE, Yang MS (2003) Expression of the B subunit of *E. coli* heat-labile enterotoxin in the chloroplasts of plants and its characterization. Transgenic Res 12(6):683–691

Karimi M, Inze D, Depicker A (2002) GATEWAY™ vectors for Agrobacterium-mediated plant transformation. Trends Plant Sci 7:193–195

Kikkert JR, Vidal JR, Reisch BI (2005) Stable transformation of plant cells by particle bombardment/biolistics. Meth Mol Biol 286:61–78

Kim MY, Kim BY, Oh SM, Reljic R, Jang YS, Yang MS (2016) Oral immunisation of mice with transgenic rice calli expressing cholera toxin B subunit fused to consensus dengue cEDIII antigen induces antibodies to all four dengue serotypes. Plant Mol Biol 92:347–356

Klaus SM, Huang FC, Golds TJ, Koop HU (2004) Generation of marker-free plastid transformants using a transiently cointegrated selection gene. Nat Biotechnol 22:225–229

Koop HU, Steinmuller K, Wagner H, Rossler C, Eibl C, Sacher L (1996) Integration of foreign sequences into the tobacco plastome via polyethylene glycol-mediated protoplast transformation. Planta 199:193–201

Krenek P, Samajova O, Luptovciak I, Doskocilova A, Komis G, Samaj J (2015) Transient plant transformation mediated by Agrobacterium tumefaciens: Principles, methods and applications. Biotechnol Adv 1(33):1024–1042

Laguía-Becher M, Martín V, Kraemer M, Corigliano M, Yacono ML, Goldman A, Clemente M (2010) Effect of codon optimization and subcellular targeting on Toxoplasma gondii antigen SAG1 expression in tobacco leaves to use in subcutaneous and oral immunization in mice. BMC Biotechnol 10:52

Lakshmi PS, Verma D, Yang X, Lloyd B, Daniell H (2013) Low cost tuberculosis vaccine antigens in capsules: expression in chloroplasts, bio-encapsulation, stability and functional evaluation in vitro. PLoS ONE 8(1):e54708

Lee G, Na YJ, Yang B-G, Choi J-P, Seo YB, Hong C-P, Yun CH, Kim DH, Sohn EJ, Kim JH, Sung YC, Kim Y-K, Jang MH, Hwang I (2015a) Oral immunization of haemaggulutinin H5 expressed in plant endoplasmic reticulum with adjuvant saponin protects mice against highly pathogenic avian influenza A virus infection. Plant Biotechnol J 13:62–72

Lee JS, Kallehauge TB, Pedersen LE, Kildegaarda HF (2015b) Site-specific integration in CHO cells mediated by CRISPR/Cas9 and homology-directed DNA repair pathway. Sci Rep 5:8572

Lee SM, Kang K, Chung H, Yoo SH, Xu XM, Lee SB, Cheong JJ, Daniell H, Kim M (2006) Plastid transformation in the monocotyledonous cereal crop, rice (Oryza sativa) and transmission of transgenes to their progeny. Mol Cells 21(3):401–410

Leister D (2003) Chloroplast research in the genomic age. Trends Genet 19:47–56
Lelivelt CL, McCabe MS, Newell CA, Desnoo CB, van Dun KM, Birch-Machin I, Gray JC, Mills KH, Nugent JM (2005) Stable plastid transformation in lettuce (*Lactuca sativa* L.). Plant Mol Biol 58:763–774
Lelivelt CL, van Dun KM, de Snoo CB, McCabe MS, Hogg BV, Nugent JM (2014) Plastid transformation in lettuce (*Lactuca sativa* L.) by polyethylene glycol treatment of protoplasts. Methods Mol Biol 1132:317–330
Lerouge P, Cabanes-Macheteau M, Rayon C, Fischette-Laine AC, Gomord V, Faye L (1998) N-Glycoprotein biosynthesis in plants: recent developments and future trends. Plant Mol Biol 38:31–48
Li D, O'Leary J, Huang Y, Huner NP, Jevnikar AM, Ma S (2006) Expression of cholera toxin B subunit and the B chain of human insulin as a fusion protein in transgenic tobacco plants. Plant Cell Rep 25(5):417–424
Lu Y, Rijzaani H, Karcher D, Ruf S, Bock R (2013) Efficient metabolic pathway engineering in transgenic tobacco and tomato plastids with synthetic multigene operons. Proc Natl Acad Sci U S A 110(8):E623–E632
Ma C, Wang L, Webster DE, Campbell AE, Coppel RL (2012) Production, characterisation and immunogenicity of a plant-made Plasmodium antigen–the 19 kDa C-terminal fragment of Plasmodium yoelii merozoite surface protein 1. Appl Microbiol Biotechnol 94(1):151–161
Maliga P (1993) Towards plastid transformation in flowering plants. Trends Biotechnol 11:101–106
Maliga P (2004) Plastid transformation in higher plants. Annu Rev Plant Biol 55:289–313
Maliga P, Carrer H, Kanevski I, Staub J, Svab Z (1993) Plastid engineering in land plants: a conservative genome is open to change. Phil Trans R Soc Lond B 341:449–454
Matsui T, Asao H, Ki M, Sawada K, Kato K (2009) Transgenic lettuce producing a candidate protein for vaccine against edema disease. Biosci Biotechnol Biochem 73:1628–1634
Moloney MM, Van-Rooijen G, Sembiosys Genetics Inc (2006) Expression of epidermal growth factor in plant seeds. United States patent US 7091401
Monreal-Escalante E, Bañuelos-Hernández B, Hernández M, Fragoso G, Garate T, Sciutto E, Rosales-Mendoza S (2015) Expression of multiple taenia solium immunogens in plant cells through a ribosomal skip mechanism. Mol Biotechnol 57(7):635–643
Nakashita H, Arai Y, Shikanai T, Doi Y, Yamaguchi I (2001) Introduction of bacterial metabolism into higher plants by polycistronic transgene expression. Biosci Biotechnol Biochem 65:1688–1691
Nelson G, Marconi P, Periolo O, La Torre J, Alvarez MA (2012) Immunocompetent truncated E2 glycoprotein of bovine viral diarrhea virus (BVDV) expressed in Nicotiana tabacum plants: a candidate antigen for new generation of veterinary vaccines. Vaccine 30:4499–4504
Neuhaus JM, Rogers JC (1988) Sorting of proteins to vacuoles in plant cells. Plant Mol Biol 38:127–144
Neuhaus JM, Sticher L, Meins FJ, Boller T (1991) A short C-terminal sequence is necessary and sufficient for the targeting of chitinases to the plant vacuole. Proc Natl Acad Sci U S A 88 (22):10362–10366
Noris E, Poli A, Cojoca R, Rittà M, Cavallo F, Vaglio S, Matic S, Landolfo S (2011) A human papillomavirus 8 E7 protein produced in plants is able to trigger the mouse immune system and delay the development of skin lesions. Arch Virol 156(4):587–595
Nugent GD, Coyne S, Nguyen TH, Kavanagh TA, Dix PJ (2005a) Nuclear and plastid transformation of Brassica oleracea var. botrytis (cauliflower) using PEG-mediated uptake of DNA into protoplasts. Plant Sci 170:135–142
Nugent GD, ten Have M, van der Gulik A, Dix PJ, Uijtewaal BA, Mordhorst AP (2005b) Plastid transformants of tomato selected using mutations affecting ribosome structure. Plant Cell Rep 24:341–349
Nykiforuk CL, Boothe JG, Murray EW, Keon RG, Goren J, Markley NA, Moloney MM (2005) Transgenic expression and recovery of biologically active recombinant human insulin from Arabidopsis thaliana seeds. Plant Biotechnol J 4:77–85

O'Neill C, Horvath GV, Horvath E, Dix PJ, Medgyesy P (1993) Chloroplast transformation in plants: polyethylene glycol (PEG) treatment of protoplasts is an alternative to biolistic delivery systems. Plant J 3:729–738
Păcurar DI, Thordal-Christensen H, Păcurar ML, Pamfil D, Botez C, Bellini C (2011) Agrobacterium tumefaciens: from crown gall tumors to genetic transformation. Physiol Mol Plant Pathol 76:76–81
Parmenter DL, Boothe JG, Van Rooijen GJH, Yeung EC, Moloney MM (1995) Production of biologically active hirudin in plants seeds using oleosin partitioning. Plant Mol Biol 29:1167–1180
Paul M, Ma JK (2011) Plant-made pharmaceuticals: leading products and production platforms. Biotechnol Appl Biochem 58(1):58–67
Piron R, De Koker S, De Paepe A, Goossens J, Grooten J, Nauwynck H, Depicker A (2014) Boosting in planta production of antigens derived from the porcine reproductive and respiratory syndrome virus (PRRSV) and subsequent evaluation of their immunogenicity. PLoS ONE 9:e91386
Rivera AL, Gómez-Lim M, Fernández F, Loske AM (2012) Physical methods for genetic plant transformation. Phys Life Rev 9(3):308–345
Ruf S, Hermann M, Berger IJ, Carrer H, Bock R (2001) Stable genetic transformation of tomato plastids and expression of a foreign protein in fruit. Nat Biotechnol 19:870–875
Ruf S, Karcher D, Bock R (2007) Determining the transgene containment level provided by chloroplast transformation. Proc Natl Acad Sci U S A 104(17):6998–7002
Ruhlman T, Ahangari R, Devine A, Samsam M, Daniell H (2007) Expression of cholera toxin B-proinsulin fusion protein in lettuce and tobacco chloroplasts-oral administration protects against development of insulitis in non-obese diabetic mice. Plant Biotechnol J 5(4):495–510
Ruhlman TA (2014) Plastid transformation in lettuce (*Lactuca sativa* L.) by biolistic DNA delivery. Methods Mol Biol 1132:331–343
Sainsbury F, Lomonossoff GP (2008) Extremely high-level and rapid transient protein production in plants without the use of viral replication. Plant Physiol 148(3):1212–1218
Sanford JC, Smith FD, Russell JA (1993) Optimizing the biolistic process for different biological applications. Methods Enzymol 217:483–509
Schnell J, Steele M, Bean J, Neuspiel M, Girard C, Dormann N, Pearson C, Savoie A, Bourbonnière L, Macdonald P (2015) A comparative analysis of insertional effects in genetically engineered plants: considerations for pre-market assessments. Transgenic Res 24 (1):1–17
Shaaltiel Y, Bartfeld D, Hashmueli S, Baum G, Brill-Almon E, Galili G, Dym O, Boldin-Adamsky SA, Silman I, Sussman JL, Futerman AH, Aviezer D (2007) Production of glucocerebrosidase with terminal mannose glycans for enzyme replacement therapy of Gaucher's disease using a plant cell system. Plant Biotechnol J 5(5):579–590
Shaaltiel Y, Yl Tekoah (2016) Plant specific N-glycans do not have proven adverse effects in humans. Nat Biotechnol 34(7):706–708
Shen BR, Zhu CH, Yao Z, Cui LL, Zhang JJ, Yang CW, He ZH, Peng XX (2017) An optimized transit peptide for effective targeting of diverse foreign proteins into chloroplasts in rice. Sci Rep 11(7):46231
Sidorov VA, Kasten D, Pang SZ, Hajdukiewicz PTJ, Staub JM, Nehra NS (1999) Stable plastid transformation in potato: use of green fluorescent protein as a plastid marker. Plant J 19:209–216
Smith EF, Townsend CO (1907) A plant tumor of bacterial origin. Science 25:671–673
Sojikul P, Buehner N, Mason HS (2003) A plant signal peptide-hepatitis B surface antigen fusion protein with enhanced stability and immunogenicity expressed in plant cells. Proc Natl Acad Sci U S A 100(5):2209–2214
Staub JM, Garcia B, Graves J, Hajdukiewicz PT, Hunter P, Nehra N, Paradkar V, Schlittler M, Carroll JA, Spatola L, Ward D, Ye G, Russell DA (2000) High-yield production of a human therapeutic protein in tobacco chloroplasts. Nat Biotechnol 18(3):333–338

Staub JM, Maliga P (1992) Long regions of homologous DNA are incorporated into the tobacco plastid genome by transformation. Plant Cell 4:39–45

Staub JM, Maliga P (1994) Extrachromosomal elements in tobacco plastids. Proc Natl Acad Sci U S A 91:7468–7472

Svab Z, Hajdukiewicz P, Maliga P (1990) Stable transformation of plastids in higher plants. Proc Natl Acad Sci U S A 87:8526–8530

Svab Z, Maliga P (1993) High-frequency plastid transformation in tobacco by selection for a chimeric aadA gene. Proc Natl Acad Sci U S A 90:913–917

Thomas TJ, Kulkarni GD, Greenfield NJ, Shirahata A, Thomas T (1996) Structural specificity effects of trivalent polyamine analogues on the stabilization and conformational plasticity of triplex DNA. Biochem J 319:591–599

Thuenemann EC, Meyers AE, Verwey J, Rybicki EP, Lomonossoff GP (2013) A method for rapid production of heteromultimeric protein complexes in plants: assembly of protective bluetongue virus-like particles. Plant Biotechnol J 11(7):839–846

To KY, Cheng MC, Chen LF, Chen SC (1996) Introduction and expression of foreign DNA in isolated spinach chloroplasts by electroporation. Plant J 10(4):737–743

Tregoning JS, Clare S, Bowe F, Edwards L, Fairweather N, Qazi O, Nixon PJ, Maliga P, Dougan G, Hussell T (2005) Protection against tetanus toxin using a plant-based vaccine. Eur J Biotechnol 35(4):1320–1326

Tzfira T, Citovsky V (2002) Partners-in-infection: host proteins involved in the transformation of plant cells by Agrobacterium. Trends Cell Biol 12(3):121–129

Tzfira T, Citovsky V (2006) Agrobacterium-mediated genetic transformation of plants: biology and biotechnology. Curr Opin Biotechnol 17(2):147–154

Tzfira T, Li J, Lacroix B, Citovsky V (2004) Agrobacterium T-DNA integration: molecules and models. Trends Genet 20(8):375–383

Valkov VT, Gargano D, Scotti N, Cardi T (2014) Plastid transformation in potato: solanum tuberosum. Methods Mol Biol 1132:295–303

van Wordragen MF, Dons HJM (1992) Agrobacterium tumefaciens-mediated transformation of recalcitrant crops. Plant Mol Biol Rep 10:12–36

Verma D, Daniell H (2007) Chloroplast vector systems for biotechnology applications. Plant Physiol 145:1129–1143

Wakasugi T, Tsudzuki T, Sugiura M (2001) The genomics of land plant chloroplasts: gene content and alteration of genomic information by RNA editing. Photosynth Res 70:107–118

Wani SH, Haider N, Kumar H, Singh NB (2010) Plant plastid engineering. Curr Genomics 11:510–512

Watson J, Koya V, Leppla SH, Daniell H (2004) Expression of Bacillus anthracis protective antigen in transgenic chloroplasts of tobacco, a non-food/feed crop. Vaccine 22:4374–4384

Werner S, Breus O, Symonenko Y, Marillonnet S, Gleba Y (2011) High-level recombinant protein expression in transgenic plants by using a double-inducible viral vector. Proc Natl Acad Sci U S A. 108(34):14061–14066

Wirth S, Calamante G, Mentaberry A, Bussmann L, Lattanzi M, Barañao L, Bravo-Almonacid F (2004) Expression of active human epidermal growth factor (hEGF) in tobacco plants by integrative and non-integrative systems. Mol Breed 13:23

Yang J, Barr LA, Fahnestock SR, Liu ZB (2005) High yield recombinant silk-like protein production in transgenic plants through protein targeting. Transgenic Res 14(3):313–324

Zhou B, Zhang Y, Wang X, Dong J, Wang B, Han C, Yu J, Li D (2010) Oral administration of plant-based rotavirus VP6 induces antigen-specific IgAs, IgGs and passive protection in mice. Vaccine 28(37):6021–6027

Zoubenko OV, Allison LA, Svab Z, Maliga P (1994) Efficient targeting of foreign genes into the tobacco plastid genome. Nucleic Acids Res 22:3819–3824

Part I
Vaccines for Wild, Feral and Companion Animals

Rabies and Related Lyssaviruses

Charles E. Rupprecht and Rachel Chikwamba

Abstract Rabies is a significant neglected vaccine-preventable disease that is global in distribution. Multiple biologics are utilized in routine prevention and control of this zoonosis. Currently, rabies vaccines are used to interrupt a productive viral encephalitis before or after pathogen exposure in humans and animals. In addition, rabies immune globulins are used as part of prophylaxis after human exposure to a known or suspect animal. Such rabid animals are diagnosed based upon antigenic detection in the brain by selective antibody conjugates. Although experimental proof of concept has been demonstrated in a variety of systems, to date no plant-produced biologics have been licensed for such applications in rabies surveillance, prevention or control. In addition, given the breadth of the host spectrum, there are multiple domestic and wild mammalian species that lack specific vaccines and the cross reactivity of existing products is limited by considerable viral diversity. Hence, if safe, effective and inexpensive biologics may be produced in plants, especially for oral delivery, there is a considerable global niche to fill within the realms of public health, veterinary medicine and conservation biology.

Keywords Diagnosis · Dogs · Encephalitis · Lyssavirus · Neglected tropical disease · Prophylaxis · Surveillance · Rabies · Wildlife · Zoonosis

1 Introduction

Rabies is a major viral zoonosis, with a substantial veterinary and public health burden (Hampson et al. 2015). Given its significant threat, vaccines are used in humans and animals for prevention of a productive infection, both before and after

C. E. Rupprecht (✉)
LYSSA LLC, Atlanta, GA, USA
e-mail: charles_rupprecht@yahoo.com

R. Chikwamba
Biosciences Unit, Council for Scientific and Industrial Research, Pretoria, South Africa

J. MacDonald (ed.), *Prospects of Plant-Based Vaccines in Veterinary Medicine*,
https://doi.org/10.1007/978-3-319-90137-4_3

exposure. Despite being one of the oldest known infectious diseases, with sensitive diagnostic tests and highly effective biologics, rabies is neglected. This disease is enzootic in Africa, the Americas, Australia and Eurasia. Rabies has the highest case fatality of any infectious disease, but lacks true global commitment for prioritization, due to a number of inter-related biological, cultural, economic and social characteristics (Meltzer and Rupprecht 1998; Rupprecht and Burgess 2015; Rupprecht et al. 2017). The objective of this chapter is to provide an overview on rabies, regarding its etiological agents, pathobiology, diagnosis, prevention and control, with an aim to introducing opportunities where plant-based applications may be relevant. Given modern biologics, current and future uses for products generated in plants will be dependent upon developmental opportunities for improvements in viral cross reactivity, species breadth, duration of immunity, ease of application, flexibility in routes of delivery, thermostability, cost, speed of regulatory approval and overall public acceptance.

2 Taxonomy

The viral agents of rabies belong to the Genus *Lyssavirus* within the Family *Rhabdoviridae*. All lyssaviruses are highly neurotropic and cause the same disease. The most important and broadly distributed member of the genus is the type species, Rabies virus (RABV), which is the only lyssavirus found in the Americas (Rupprecht et al. 2011). New lyssavirus species have been described increasingly since the 1950s, due in part to renewed interests in pathogen detection and improvements in diagnostic techniques and characterization (Fooks et al. 2003; Kuzmin et al. 2005, 2008, 2011; Singh and Sandhu 2008; van Thiel et al. 2009; Hayman et al. 2012; Aréchiga Ceballos et al. 2013; Kgaldi et al. 2013; Liu et al. 2013; Annand and Reid 2014; Banyard et al. 2014; Horton et al. 2014; Nolden et al. 2014; Shinwari et al. 2014; Gunawardena et al. 2016; Hu et al. 2018; Nokireki et al. 2018).

No commercial veterinary or human biologics have an adequate breadth to protect against all lyssaviruses, due to considerable antigenic variation. Experimental recombinant vaccines may partially expand cross reactivity across several, but not all, lyssavirus species (Weyer et al. 2008). Lyssaviruses may be considered aligned in different phylogenetic groups (Table 1). Modern RABV vaccines protect against all Phylogroup I lyssaviruses, including multiple variants of wildlife RABV variants (Brookes et al. 2005; Malerczyk et al. 2009). Novel biologics produced from plants could fill some of these vacant niches to increase breadth of protection, particularly against Phylogroup II and other lyssaviruses.

Table 1 Known and putative members of the *Lyssavirus* genus

Virus	Occurrence	Phylogroup	Comments
Australian bat lyssavirus	Australia, and likely insular locations in the region	I	Reservoir in bats with spillover to humans and domestic animals; no reports as of yet from marsupials
Aravan virus	Eurasia	I	Bat reservoirs
Bokeloh bat lyssavirus	Europe	I	Bat reservoirs
Duvenhage virus	Africa	I	Reservoir in bats with spillover to humans
European bat lyssavirus, type 1	Europe	I	Reservoir in bats with spillover to humans
European bat lyssavirus, type 2	Europe	I	Reservoir in bats with spillover to humans
Gannoruwa bat lyssavirus	Asia	I	Bat reservoirs
Ikoma lyssavirus	Africa	III?	Expected bat reservoirs with spillover to other wildlife, e.g., civet
Irkut virus	Eurasia	I	Bat reservoirs with spillover to humans and dogs
Khujand virus	Eurasia	I	Bat reservoirs
Kotalahti bat lyssavirus	Finland	I	Bat reservoirs
Lagos bat virus	Africa	II	Bat reservoirs with spillover to other mammalian species
Lleida bat lyssavirus	Europe	III?	Bat reservoirs
Mokola virus	Africa	II	Unknown wildlife reservoir with spillover to humans and other mammalian species
Rabies virus	Global	I	Primary reservoirs among bats and carnivores, but all mammals susceptible
Shimoni bat virus	Africa	II	Bat reservoirs
Taiwan bat virus	Asia	I	Bat reservoirs
West Caucasian bat virus	Eurasia	III?	Bat reservoirs

3 Hosts

Rabies is a vaccine-preventable disease, but not a candidate for eradication, unlike rinderpest and smallpox, due in part to its broad host spectrum (Rupprecht and Kuzmin 2015). Rhabdoviruses are diverse and are found among invertebrates, plants and vertebrates, but lyssaviruses are restricted to vertebrates. All warm-blooded vertebrates are susceptible. Birds can be infected, but only mammals are important for viral perpetuation (Baby et al. 2015). Major mammalian reservoirs include carnivores (e.g., dogs, ferret badgers, foxes, mongoose, raccoons, skunks, etc.) and bats, which are responsible for disease maintenance, by bite transmission. All reservoirs, in which the virus can multiply, are vectors that can transmit the virus to others (e.g., canids), but not all vectors are reservoirs (e.g., felids). Humans, and most domestic species, are essentially victims of spill-over from infected carnivores and bats, and are dead-end infections. Rodents are often involved in human exposures and mistakenly cited as a need for rabies-specific biomedical intervention, but are unimportant epidemiologically (Fitzpatrick et al. 2014).

In general, the idea of using plants to produce rabies biologics for mammals is not so far-fetched, considering that other rhabdovirus species are adapted to replication in plant tissues. Operationally, plant-based products could broaden the taxonomic gamut from *Homo sapiens*, domestic pets and livestock, as well as exotic and wild mammalian taxa, particularly for important species used for food, fiber, transport, etc. in which other RABV vaccines may not be widely available (e.g., camelids, small ruminants, swine, wildlife, etc.).

4 Viral Transmission

Rabid animals excrete large amounts of virions in their saliva (Pépin et al. 1984; Fekadu et al. 1982; Rupprecht 2016). Viral exposure includes direct transdermal or mucosal contact with infectious material, such as saliva, salivary glands, brain or related neural tissue (ACIP 2008; WHO 2013), and almost all rabies cases are caused by a bite. Non-bite exposures include scratches, aerosols, open lesion contamination or mucosal contact with infectious materials, such as a lick on a fresh wound or contact with the nose, mouth, eyes, etc. (ACIP 2008; WHO 2013).

After infection, virions travel from the periphery to the central nervous system (CNS). The incubation period is predicated in part by viral dose, route, severity, variant and host attributes. Short incubation periods of less than 2 weeks may occur after severe exposures to the head. Unusual incubation periods in excess of a year have also been documented (Boland et al. 2014). Most periods are 4–6 weeks in length. After replication in the CNS, virions transit to other organs, including the exit portals in the salivary glands. Virus may be excreted for several days before the appearance of clinical signs (Begeman et al. 2017).

The majority of animal cases are caused by RABV. For most applications, rabies vaccines are highly efficacious before viral exposure, but are ineffective in the clinically rabid animal. A myriad of clinical signs has been described, none of which are pathognomonic and a prodromal phase of 3–4 days presents with non-specific signs of illness. Thereafter, an acute neurological phase may be manifested by predominantly paralytic or furious signs (Fig. 1). Coma and death occur in a few days. Due to the many causes of encephalitis, and because virus is excreted before obvious manifestation of abnormalities, rabies may not be suspected. In some situations, death may occur without any signs observed, whereas in

Fig. 1 The dog is the major global reservoir of rabies, which may present after a variable incubation period in the prodromal phase as fever and non-specific clinical signs such as general restlessness, progressing to an acute encephalitic phase, with cranial nerve deficits and overt aggressiveness. Vaccination is highly efficacious before clinical signs, but ineffective after the onset of illness, an unfilled niche for consideration of future plant-based therapeutics (figure adapted from the open access U.S. Public Health Image Library)

other cases signs may be dramatic (Barnard 1979; Pépin et al. 1984; Hudson et al. 1996a, b; Thiptara et al. 2011; Den et al. 2012). Clinical definition alone is inadequate, because the differential diagnosis for suspect encephalitis is broad, including other relevant infectious etiologies, toxicity and trauma.

Human exposures result directly from a rabid animal initiating contact or indirectly by persons coming to the aid of an ill animal (Gräni et al. 1978; Martin et al. 1982; CDC 1983). Affected animals may appear to be choking, prompting examination of the mouth for a suspected foreign object. Human fatality may occur after such exposures in non- or poorly vaccinated persons (Tariq et al. 1991; Brito et al. 2011; Simani et al. 2012). Besides carnivores, rabid hoofed stock, such as equids, swine, camelids, etc. may also produce severe bites (Fekadu 1982; Jiang et al. 2008; de Macedo Pessoa et al. 2011; Liu et al. 2016). With furious rabies, overt aggression itself can result in death from bites, goring, mutilation, or crushing injuries from horses, bulls and other horned, tusked or antlered stock.

Rabid animals are a frequent source of mass human exposures (Rotz et al. 1998). Often, this has involved children at petting zoos or fear of perceived exposures after consumption of unpasteurized milk or improperly cooked meat products at public gatherings, with a need for multiple vaccine doses in a short period (CDC 1999). While proper cooking or pasteurization will eliminate rabies hazards, consumption of raw animal products may be common, particularly in pastoral societies, and exposures may also occur during butchering of rabid animals. Therefore exposed, rabid or suspect animals should not enter the food chain (Gadaga et al. 2016).

Besides human deaths, expenses due to case investigations and prophylaxis, and other public health and agricultural impacts, lead to direct economic loss. Real or perceived repercussions may be smaller than with other veterinary diseases which negatively impact global trade, such as avian influenza, bovine spongiform encephalopathy (BSE) or foot and mouth disease (FMDV), but the death of even a single cow, small ruminant or horse has major uncompensated impact to the individual farmer or small community in a developing country. As such, any new plant biologics should be able to immunize an individual rapidly before exposure and protect quickly after lyssavirus infection.

5 Laboratory-Based Surveillance

Rabies should be a notifiable event. A history of suspect animal bite and compatible signs may be suitable for a presumptive clinical diagnosis. However, laboratory confirmation is essential for reporting purposes, public health applications, proper animal management and control program evaluation, using appropriate techniques, as outlined by the WHO (WHO 1996) and the OIE (http://www.oie.int/fileadmin/Home/eng/Health_standards/tahm/2.01.17_RABIES.pdf).

The brain is the primary tissue for post-mortem diagnosis. Older methods required the microscopic detection of intra-cytoplasmic inclusions in neurons based upon chemical staining. Today, the direct fluorescent antibody test (FAT) is the gold standard. While initial protocols focused upon the necessity of the hippocampus as a key region of the CNS, newer diagnostic techniques have shown the importance of the brainstem, as regards antigen detection and histopathological lesions (Bingham and van der Merwe 2002; Stein et al. 2010; Bassuino et al. 2016). Rapid point-of-care diagnostic tests, such as lateral flow assays, are desirable, but may lack adequate sensitivity, specificity and lot-to-lot consistency (Ahmad and Singh 2016; Eggerbauer et al. 2016). Where an anatomic-pathologic basis for diagnosis is desirable, but limited resources prevent use of the FAT or the application of highly sensitive molecular methods, such as real-time PCR (Gigante et al. 2018), consideration could be given to the direct rapid immunohistochemical test, in support of an enhanced or active, decentralized, laboratory-based surveillance system to confirm suspect cases and as a means to measure the success of vaccination or other management programs (Madhusudana et al. 2012; Monroe et al. 2016). Given the overt utility of antigen detection in rabies diagnosis, opportunities may be present for antibodies produced in plants for rapid and broad lyssavirus detection.

Regardless of occurrence, estimates of rabies cases are considered gross underestimates of incidence in humans and domestic animals (Memish et al. 2015). The burden in wildlife is largely unknown due to a dearth of adequate studies in free-ranging wild populations. In a developed country, the total occurrence may be comparatively small, at 1–2 human cases per year (Monroe et al. 2016). Typically, where the enzootic cycle involves rabid dogs, human cases range into the thousands per year. Outbreaks from canine rabies may involve multiple species of domestic animals in rural areas, as reported recently from China (Feng et al. 2016). Similarly, in regions where both canine and wildlife rabies perpetuate in concert with semi-nomadic cultures (e.g., Mongolia), cases reported among livestock are widespread, involving multiple species over vast areas. By contrast, in areas where canine rabies has been eliminated and rabies among wild carnivores controlled via oral vaccination, reports of rabies overall may be quite rare, except for insectivorous bats (e.g., Western Europe). As an exception, within the Americas, from Mexico to Argentina, outbreaks of bovine paralytic rabies cases may number into the tens of thousands annually, secondary to infection by hematophagous vampire bats (Pawan 1959; Lord et al. 1975; Mayen 2003; Lee et al. 2012; Johnson et al. 2014). One recent study on vampire rabies from Brazil estimated in excess of 30,000 cases of infected cattle per year (Rodrigues da Silva et al. 2000).

In addition to the basic laboratory diagnosis of rabies, molecular methods should be used to characterize the background of various lyssaviruses present and differentiate indigenous from introduced cases. In some regions, only a single lyssavirus species will predominate (e.g., RABV, Australian Bat Lyssavirus, etc.). Viral characterization may indicate the emergence or translocation of a new pathogen that could impact vaccine efficacy. Similarly, if modified-live vaccines (e.g., LEP, SAD, etc.) are used in animals, genetic sequencing would readily differentiate seed viruses used for production (and the possibility of vaccine-associated rabies) from native street viruses (Höper et al. 2015).

Serological monitoring for VNA is possible, but should not be necessary as a routine measure of primary vaccine immunogenicity. However, such serology might provide insight in the management of exposed animals. For example, after viral exposure, all vaccinated animals should be boosted immediately with a dose of vaccine, and unvaccinated animals should be euthanized or quarantined. Although records of vaccination should be readily available, if such documentation is not located, prospective serological monitoring could be considered, as suggested for dogs and cats (Moore et al. 2015; NASPHV 2016). Depending upon the product and its potency, rabies VNA may be detected or an anamnestic response could be measured within ~5–7 days of a booster vaccination, indicating the vaccination status of the animal in question. If multiple vaccines from different producers are used concomitantly, post-marketing surveillance may be useful. These can differentiate the utility of multiple products, not only on the basis of cost, but other factors, such as basic VNA response and duration of immunity under field conditions (Gilbert et al. 2015). In addition to serology, analysis of the occurrence of cases in space and time, viral characterization and subsequent ecological modeling may provide longer term information on any species-specific patterns in response to vaccination or generation of at-risk seasons and localities for different populations and at risk communities (Streicker and Allgeier 2016). By comparison, a consideration of plant-based vaccines would need to generate a minimum duration of immunity of at least 1 year in nearly all subjects.

6 Viral Antigens

Lyssaviruses consist of a single strand of non-segmented, negative-sense RNA. Virions contain 5 multi-functional, structural proteins and have a bullet shaped-morphology (Fig. 2). Internally, a helical nucleocapsid includes several

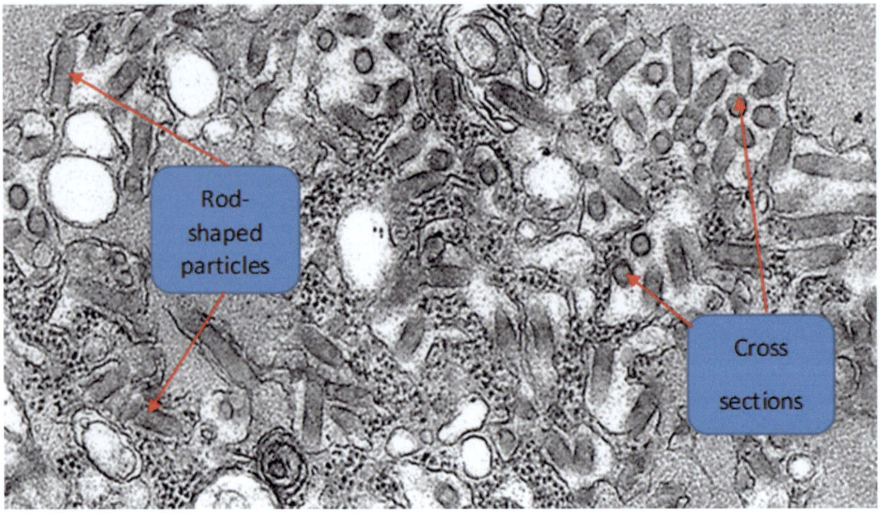

Fig. 2 Electron micrograph of lyssavirus virions. Other rhabdoviruses are adapted to plant hosts, a facet that should support future development of plant-associated biologics for rabies (figure adapted from the open access U.S. Public Health Image Library)

structural proteins and the viral RNA. The inner nucleoprotein (N) is bound tightly to the RNA, to form a stable ribonucleoprotein (RNP) complex, minimizing destruction by cellular nucleases. The RNP is the group-specific determinant and highly conserved, and has been documented to confer protective immunity (Dietzschold et al. 1987).

Associated with the RNP are the phospoprotein (P) and the RNA-dependent RNA polymerase (L), which transcribes and replicates the genome (Ogino et al. 2016). The N, P and L proteins are all involved in pathogenicity, based upon suppression of the interferon response (Tian et al. 2015). The N protein is also responsible for the formation of intracytoplasmic inclusions in neurons, which when stained appropriately can be identified by light microscopy as Negri bodies, a diagnostic hallmark of rabies (Fig. 3). Today, such inclusions are readily identified by fluorescent microscopy or immuno-histochemistry. The outer glycoprotein (G) is embedded in the viral envelope, involved in host cell reception (Fernando et al. 2016). The viral G protein is the external antigen responsible for the induction of virus-neutralizing antibodies (VNA), and is the most critical component of rabies vaccines (Piza et al. 2002; Dietzschold et al. 2008). It is the major antigen determining serotype, with considerable antigenic diversity. Finally, the matrix protein (M) forms a scaffold between the G protein and the nucleocapsid and may act as a viral immune-modulatory factor (Ben Khalifa et al. 2016).

Given the above properties and utility for prevention and control, opportunities for plant production of lyssavirus antigens would reside in expression of the G protein for vaccines or the N protein for diagnostics.

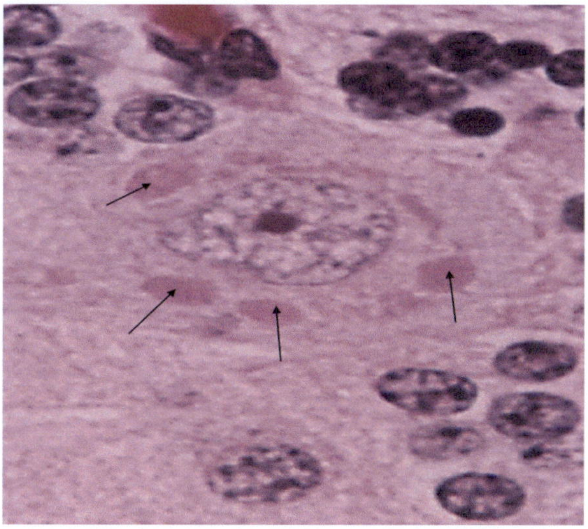

Fig. 3 Photomicrograph of intra-cytoplasmic, lyssavirus inclusions (arrows) in a neuron, stained with hematoxylin and eosin. While older, modified-live or nerve-tissue based vaccines may have led to severe adverse events (including vaccine-induced rabies), modern biologics are highly safe with much fewer untoward effects, a characteristic that any plant-produced products will need to emulate (figure adapted from the open access U.S. Public Health Image Library)

7 Immune Responses to Lyssaviruses

Natural or acquired immunity to rabies in most naïve hosts is uncommon. Evidence of prior exposure to RABV may be suggested under certain circumstances in unvaccinated animals, but such findings based upon serology are essentially irrelevant to the development of effective herd immunity and should not be used as evidence to avoid primary immunization of animals at risk of infection (Gilbert et al. 2015).

Non-specific anatomical barriers to pathogen invasion, such as thickened skin or dense fur, in part may provide a shield, as lyssaviruses cannot penetrate undamaged epidermis. However, once a transdermal or mucosal exposure has occurred, lyssaviruses avoid local removal efforts by a dual combination of stealth and suppression, as quintessential neurotropic agents (Dietzschold et al. 2008). The predominance of viral replication occurs in the CNS, as an immunologically privileged site, largely free from immune surveillance. During replication, viral products induce upregulation of host cellular processes to evade primary innate immune responses, which increases success for productive generation of viral progeny. Concomitantly, viral proteins inhibit specific immune responses, resulting in the dampening of downstream opportunities for the host to clear an active viral infection.

Limited immunity against a productive lyssavirus infection in the naïve animal is provided by a combination of both innate and adaptive responses. Innate responses are triggered by the interaction between pathogen-associated molecular patterns and host reciprocal pattern-recognition receptors, leading to secretion of pro-inflammatory cytokines (Li et al. 2011). However, this innate response alone does not promote total viral clearance, which requires adaptive responses. Adaptive immune responses develop more slowly in the unvaccinated animal. Typically, antibodies in naïve animals infected with RABV only become detectable, if at all, several days after onset of illness. In general, such antibodies, once induced, are only found in serum, but not, or only at lower titers, within the cerebrospinal fluid (CSF) of animals who succumb, as opposed to the rarity of survivors with sequellae (Bell and Moore 1979; Gnanadurai et al. 2013). This late and commonly weaker antibody response may suggest that the dose of the infecting viral inoculum is inadequate to trigger effective T and B cell activation at the site of infection. An intact blood-brain barrier minimizes the effectiveness of penetration of any antibodies that may develop in the periphery. Humoral responses are also limited by the inhibition of interferon signaling, which may prevent B cell maturation.

Long-lasting protective immunity after rabies vaccination is complex, but provided primarily via VNA. The development of rabies VNA requires assistance from $CD4^+$ cells, which are induced by inactivated rabies vaccines. Initial responses occur via T helper cell-independent, shorter-lived plasma cells, which develop outside germinal centers and produce IgM. Longer-lived plasma cells develop within germinal centers and produce switched, affinity-matured IgG antibodies.

The basic definition of a successful rabies vaccination outcome is the measurement of an international standard of VNA at a level of 0.5 IU/ml, which is not considered as 'sero-protective' per se, but rather a surrogate of an appropriate response in the individual immune-competent animal. Rather, the true measure of the ultimate benefits of rabies vaccination is efficacy against a virulent viral challenge in the surviving animal.

In general, nearly all healthy animals which receive a dose of rabies vaccine will respond with detectable VNA within 7–14 days, peak at ~day 30, begin a gradual decline by ~day 90 and start to drop below 0.5 IU/ml by ~12–15 months (Filho et al. 2012). However, rabies vaccines also induce memory B cells that may persist for life, considering they can be recalled years later, as observed in vaccinated humans (Malerczyk et al. 2007; Mansfield et al. 2016). Upon booster immunization at 12–36 months, rather than a more typical schedule of shorter intervals between doses, VNA appear more sustained. In properly vaccinated animals, the anamnestic response may result in a protective outcome, regardless of the absolute VNA level at a time of viral exposure (Rupprecht and Dietzschold 1987).

Not all animals develop VNA to the same level or duration. In any population, both low and high responders to rabies vaccination are detected (Thompson et al. 2016). Response may be associated with genetics, nutrition and overall health. Somewhat surprisingly, parasite load did not appear to significantly affect vaccine efficacy (Charlier et al. 2013). Regardless, with respect to overall response, a normal distribution of overall immunological response is expected in a population over time. In contrast to $CD4^+$cells, $CD8^+$ cells do not appear to contribute substantially to protective immunity and are not induced by conventional inactivated animal rabies vaccines. At a minimum, plant-produced immunogens could not be inferior to traditional products for maximum competiveness.

8 Brief History of Rabies Biologics

Over the past century, developments in animal vaccines have largely mirrored progress in human rabies biologics (Rupprecht et al. 2016). The gradual evolution of rabies vaccines and the history of production methods have covered the same gamut of diversity as most other veterinary biologics, from successive animal passages, to primary organ propagation, through the use of tissue culture and recombinant technology (Wu et al. 2011). For example, at the end of the 19th century, the first generation vaccines were produced locally in adult animal brains (e.g., rabbits, sheep, etc.), using methods very much akin to those originally employed by Pasteur (Rappuoli 2014). As such, Pasteur coined the term 'street' viruses, defined as wild-type RABV perpetuated in nature, in contrast to 'fixed' RABV, whose characteristics, such as relative virulence or incubation period, were altered in the laboratory by continuous passages, originally in adult animal brain. Such nerve tissue-based products contained residual, infectious RABV. The later addition of chemicals, such as ether, phenol, chloroform or formalin, were attempts to stabilize the substrate and reduce any virulence of fixed RABV, to maintain potency and minimize the risk of possible vaccine associated rabies cases. As an example, one RABV vaccine used in Latin America during the early 20th century involved passage in the brain of calves or horses, in which a phenolized emulsion of 20% brain-tissue in glycerol and water was used (Carneiro 1954). Unfortunately, such original vaccines were labor intensive and expensive to produce, and also of fairly low potency, requiring large volumes and multiple doses. Decades later, to avoid concerns related to myelin sensitization and adverse events from vaccine produced within adult animal nervous tissue, RABV was propagated successfully in suckling mouse brains with decreased myelin content, which, when supplemented with an adjuvant, appeared to provide immunity for at least one year (Fuenzalida et al. 1978).

By the mid-20th century, production using primary cells from fetal or newborn animals, avian embryos and mammalian tissue cultures (e.g., hamster, canine or porcine kidney cells) sparked a second generation of animal RABV vaccines (Abelseth 1964; Reculard 1996; Rupprecht et al. 2016). Terminology such as challenge virus standard (CVS), low egg passage (LEP), high egg passage (HEP), duck embryo vaccine (DEV), Street Alabama Dufferin and its derivatives (SAD/ ERA), Pasteur virus (PV), Kelev and many others intermingled methodology with viral strain and animal source material. These agents were isolated as street viruses, which formed the origin of common seed viruses used in attenuated virus vaccines, some still in use today. While control was often focused upon vaccination of the primary vector, such as dogs, one of the first widespread use of modified-live ERA and HEP vaccines began throughout Latin America for livestock, due to the burden there associated with vampire bat rabies. Thereafter, ERA or HEP modified-live vaccines were recommended in place of LEP, due to the possibility of LEP vaccine-induced rabies (Sikes 1970).

After the 1970s, animal rabies vaccines progressed gradually from modified-live viruses to products produced in high concentrations and inactivated with irradiation (such as ultra-violet light) or chemicals (such as beta-propiolactone). Most veterinary vaccines available today are both inactivated and adjuvanted, although several attenuated viral vaccines persist on the market. Over time, developed countries have shared seed viruses, cell cultures, protocols and production methods to assist rabies vaccine production in developing countries, such as in the case of Europe and North America to Latin America or Japan to Nepal (Devleesschauwer et al. 2016).

The transition to a third generation of animal rabies vaccines began near the end of the 20th century, with the development of recombinant technology. Several recombinant vaccines have been constructed, based upon adeno-, pox-, or RABV (Pastoret and Vanderplasschen 2003; Liu et al. 2008; Gomme et al. 2011; Fry et al. 2013). These recombinant vaccines have been tested in a variety of livestock animals for safety and several were licensed eventually for other species, such as companion animals or wildlife. Similarly, the use of nucleic acid-based vaccines have been suggested in veterinary medicine for decades, but without licensure (Biswas et al. 2001; Yang et al. 2013). Newer technology, based upon mRNA, may provide alternatives in veterinary vaccination, as shown experimentally in swine (Schnee et al. 2016). Production of RABV-like particles may also have application as future vaccines (Fontana et al. 2016). Such rabies biologics for humans, domestic animals and wildlife have evolved gradually for the better over the past century. Hopefully, the novelty and innovation of plant products will help provide a new generation of rabies vaccines, particularly for the in situ delivery of multiple antigens (Petricciani et al. 1989; Aspden et al. 2002).

Plants may enable further development of rabies vaccines by providing novel adjuvants, such as green-synthesized silver nanoparticles (Asgary et al. 2016) or saponins (Yendo et al. 2016). Currently, most inactivated animal vaccines contain adjuvants in the form of aluminum hydroxide and its derivatives. Besides the parenteral route, other applications may also be relevant. For example, oral vaccines have been successful for the control of rabies among wild carnivores (Slate et al. 2009; Müller et al. 2015), and this concept may be extended to domestic pets and livestock, including swine (Liu et al. 2008) or free-ranging hoofed stock, such as kudu (Scott et al. 2012; Yang et al. 2016). In this case, plant products may offer utility to herbivores via consumption. Other needs include development of a single dose vaccine with a longer duration of immunity, with or lacking adjuvant, minimizing the need for frequent boosters.

9 Specifications and Regulatory Concerns of Rabies Biologics

Rabies vaccines for veterinary applications are unique in a One Health context, considering that their use in domestic animals and wildlife provides an important public health benefit by creating a barrier between animal reservoirs and humans. In addition, they provide direct utility in agriculture and conservation biology.

As a result, regulatory authorities have a keen interest in ensuring that rabies vaccines meet stringent requirements for safety, purity, potency, and efficacy. Safety considerations involve not only the intended animal species, but also non-target species, such as humans. Potency is measured by evaluating relative estimates of the test vaccine compared to a reference standard, using the National Institutes of Health (NIH) test. The NIH test has been used for decades and involves large numbers of laboratory animals, although alternatives are being sought to minimize the use of animals for potency determination (Lewis et al. 2012). All reference vaccines are based on the International Standard for Rabies Vaccine provided by the WHO and should be carefully calibrated to ensure that a potency of at least 1 IU/ml is met (Hermann et al. 2012).

The true measure of rabies vaccines is the protection of the vaccinated subject from a productive lyssavirus infection. In humans, this is based upon primary safety and immunogenicity in phase I/II clinical trials and follow-up of survival of exposed individuals during phase III and post-marketing surveillance of newly licensed products. For animal vaccines, efficacy involves testing the effectiveness of the vaccine in members of the target species of the comparative age and gender representation, with a duration of immunity defined by statistically significant protection against a severe viral challenge in comparison to control animals. Viral challenge is performed in captivity, comparing vaccinates and controls after a minimum duration of immunity, usually of 1–4 years. After licensing, post-vaccination monitoring and a routine vaccine adverse event reporting system (VAERS) are parts of effective prevention and control programs and further indicators of success. All suspect rabies cases should be reported, confirmed and investigated thoroughly, including spatio-temporal data, species demographics, clinical signs, history of human and animal exposures, vaccination status, etc. (NASPHV 2016).

Besides classical inactivated products, there are several approved modified-live biotechnology-derived vaccines, which utilize a vector virus expressing the RABV G protein (Hicks et al. 2012). These are typically evaluated for potency by conducting a simple titration of the vector virus coupled with an expression assay to confirm expression of the G protein. Batch release requirements for these products are based on performance of the efficacy batch when tested in the approved test system. Besides potency and efficacy concerns, technical improvements and next generation sequencing methods will provide greater ease in the identification of viral seed strains used for production as well as genetic stability, particularly in the use of biologics intended for the vaccination of free-ranging wildlife (Höper et al. 2015).

Many authorities require that rabies vaccines be administered by or under the supervision of a veterinarian, so they are often not available for sale directly to animal owners or the general public. Especially because of the important public health implications, rabies vaccine should only be obtained from manufacturers operating under the oversight of a robust and competent regulatory authority. Desired specifications when ordering rabies vaccines are necessarily closely linked to their intended use. Most manufacturers offer single-dose and multiple-dose vials for use in domestic animals, including dogs, cats, cattle, horses, sheep and ferrets.

By contrast, products for use in wildlife are prepared as single doses in individual baits. Vaccination of wildlife is generally conducted only under the auspices of a governmental agency, and the products are not usually available to non-governmental entities or individuals.

When ordering vaccine for livestock operations or for use in kennels, catteries, animal shelters, or mass vaccination clinics, multiple-dose vials are likely to provide the most economical option. When using multiple-dose vials, once opened, any unused portion of the vial should be discarded, and not stored for future use. For vaccination of individual companion animals or "pet" livestock, such as horses, single-dose vials may be the best option to ensure that product sterility or potency are not compromised.

Revaccination interval is an important consideration. Most rabies vaccines have one or three year revaccination recommendations. Some veterinarians object to the use of three-year vaccines, for several reasons. Some believe client compliance is better when annual revaccination is the standard practice, because it's easier for clients to remember to schedule an annual visit. Others assume that the three-year vaccines have substantially more antigen and are therefore potentially more reactive. There are no data to support either of these beliefs, so the revaccination interval specification should be based on local conditions and intended use. When conducting mass vaccination campaigns in rabies endemic areas, a three-year product is generally the preferred option, as it offers the longest demonstrated protection.

Rabies vaccines for use in wildlife are available with efficacy claims for several wildlife species, such as foxes, raccoon dogs, coyotes and raccoons (Slate et al. 2009; Müller et al. 2015). When considering vaccine for use in a wildlife vaccination campaign, several factors must be considered. Ideally, the vaccine selected should have proven laboratory and field efficacy for the species being targeted. However, there are situations where it is reasonable to begin a campaign on an experimental basis, if there are data to suggest that the product will be effective in the target species.

Product shelf life and stability are other important specifications to consider when selecting rabies vaccines. Pricing is often based on quantity purchased, so buying large quantities may be an economical choice, but care should be taken to ensure that the amount purchased will be used prior to the product expiration date for the particular batch or lot. Most rabies vaccines have a shelf life of two to three years, so if large quantities are ordered, it is advisable to request a batch or lot that was recently produced, to maximize the length of time for use.

With regard to wildlife vaccines, stability of the product in the field after distribution is a critical consideration (Hermann et al. 2011). Manufacturers should have data to demonstrate field stability of products. This information should be evaluated in the context of the planned bait distribution density, target species population density, foraging behavior of the target species and ambient weather conditions.

All veterinarians and others at occupational-risk of exposure to suspect animals should be vaccinated against rabies. For the naïve person exposed to a rabid animal, life saving measures should consist of local wound care, infiltration of rabies immune globulin (RIG) and the administration of multiple doses of vaccine (Rupprecht et al. 2016). Modern human rabies vaccines are inactivated tissue-culture derived products. Detailed specifications and recommendations for pre- and post-exposure use are available at several national or international sources (ACIP 2008; WHO 2013).

Historically, the trend in developed countries has been from modified-live to inactivated rabies vaccines for use in veterinary medicine. At a minimum, such inactivated rabies vaccines have been tested for potency using the NIH test. As mentioned, this assay is problematic for a number of reasons, and several groups are investing substantially in efforts toward the development of an in vitro assay or a battery of assays to replace the NIH test (Schiffelers et al. 2014; Morgeaux et al. 2017). In the near future, ELISA-based techniques may be preferred to in vivo testing (Sigoillot-Claude et al. 2015). However, until a replacement assay is validated and adopted by the requisite regulatory authorities, potency should be conducted using the NIH test or an equivalent assay, with which plant-based products may be measured for the short term.

10 Applications of Rabies Biologics

Applications of rabies vaccination will vary, dependent in part upon species, life history stages, immunological status and exposure circumstances. Modern veterinary vaccines are highly efficacious, when used in a pre-exposure strategy and rabies is quite rare in properly vaccinated animals. Most reported cases of rabies occur in naïve animals. Vaccination of clinically suspect animals has no proven utility and is not recommended based upon a basic lack of evidence, wasted economic outlay and the public health risk of viral exposure.

Unlike canine and feline rabies vaccination, which should be mandatory, global practices for livestock vaccination vary greatly from country to country and region to region, based in part on risk perceptions and disease occurrence. Regardless of species and epidemiological conditions, the majority of domestic species are not vaccinated anywhere in the world. Historically, vaccination was considered prohibitive in cost, recognizing the primary source in dogs or wildlife, while management preferences centered upon insurance and indemnification (Korns and Zeissig 1948). Education of owners, producers and veterinarians is needed to change this perception (Miguens 2007; Swai et al. 2010; Gadaga et al. 2016; Hundal et al. 2016). At a minimum, because of the zoonotic aspect of the disease, rabies vaccination should occur for all animals in close contact with the public, such as at livestock fairs, petting zoos, etc., or those used in international competition (NASPHV 2016). Vaccination would not be needed in 'rabies-free' areas, provided that such countries truly meet the criteria for such self-declarations. Even a continent considered historically free of rabies, such as Australia, has enzootic lyssaviruses, with reservoirs in bats. Similarly, Taiwan was considered free of rabies, until surveillance was expanded to wildlife and the recognition of reservoirs among ferret badgers and bats (Chang et al. 2016; Hu et al. 2018).

Rabies vaccines are usually formulated as monovalent products. On occasion, this has most often been used in combination with FMDV vaccines in ruminants (Favre et al. 1976; Fontaine et al. 1976; Soulebot et al. 1978; Palanisamy et al. 1992). In general, immunogenicity of the monovalent or combined vaccines appeared equivalent. Another example, in horses, includes the combination of rabies vaccine and Potomac fever (NASPHV 2016).

Besides consideration for even a primary vaccination, the frequency of boosters is an economical and practical issue, especially for animals on range, not readily available for multiple parenteral immunization. Most vaccine labels call for annual boosters, but such activities may not be necessary for all products and species. For example, based upon VNA serological comparisons, a booster interval greater than 1 year may be appropriate for previously vaccinated horses, but not for naïve animals, after receiving their first dose (Harvey et al. 2016). Similarly, in cattle, priming at ~6 months gave an acceptable anamnestic response when boosted up to 3 years later (Yakobson et al. 2015). Other investigators also reported the utility of booster vaccinations in previously vaccinated animals as regards the longer term maintenance of rabies VNA in excess of 0.5 IU/ml (Monaco et al. 2006)

If vaccines are considered in the face of an outbreak, efficacy will vary if animals are already exposed and incubating. Post-exposure schedules, using vaccine only, have been used in naïve animals (Blancou et al. 1991; Basheer et al. 1997; Wilson and Clark, 2001). One schedule for buffalos suggests using vaccine within 24 h after the exposure, followed by boosters on days 3, 7, 14, 28 and 90 (http://www.buffalopedia.cirb.res.in/index.php?option=com_content&view=article&id=209&Itemid=117&lang=en). In Nepal, the annual reports of ~100 fatal cases of rabies in livestock and more than 1000 cases of rabies prophylaxis administered to livestock after exposure were felt to be gross underestimates (Devleesschauwer et al. 2016). Moreover, the efficacy of postexposure prophylaxis (PEP) in unvaccinated livestock is questionable, especially for severe exposures. Ideally, to better strive towards complete protection in such postexposure settings, a source of immune globulin should be considered in addition to vaccine, as recommended by WHO for exposed persons (Mitmoonpitak et al. 2002).

Recalling that many domestic animals, such as livestock are basically victims, vaccination is one important method of prevention but should not operate in a vacuum, irrespective of ecological and epizootiological concerns. Simply put, cases of rabies in livestock originate from either other domestic animal or wildlife sources. In developing countries, most cases of rabies in livestock are due to transmission from dogs. Clearly, the single most important element for the prevention of livestock cases in canine rabies enzootic areas is the mass vaccination of dogs (Gibson et al. 2016; Lavan et al. 2017). Where canine rabies has been controlled and other carnivores such as foxes or raccoons are the major source of infection, oral wildlife vaccination and related management strategies should be considered (Slate et al. 2009; Müller et al. 2015).

Regardless of other benefits, neither canine rabies elimination nor oral vaccination of wild carnivores will provide benefit to livestock under all scenarios. Throughout Latin America, rabies transmitted by vampire bats has a major impact upon the livestock industry, with multiple attempts aimed at control (Greenhall and Schmidt 1988). Historically, before the development of effective vaccines, rabies control was non-specific, such as by destroying suspected vampire bat roosts (and many beneficial species). In the mid-late 20th century, anticoagulants were used as a more specific method of control. Vampire bats were captured, spread with anticoagulant pastes and allowed to return to their roosts, where grooming by conspecific roost mates resulted in multiple deaths. Such methods allowed targeting to the species of interest and multiplication of effect from treating individuals, but also required staying late throughout the evening to capture vampires at affected ranches. Alternatively, livestock were injected with anti-coagulants directly, which vampire bats would ingest during a blood meal. This technique allowed veterinarians to treat multiple farms during the day, but was more expensive and required a bat bite upon a treated animal for effectiveness. Both techniques suffered the limitation of toxic environmental contamination with potential impacts to other fauna, as well as the opportunity of vampire bat population disruption, roost switching and colony dispersal, exacerbating the infectious disease issue. Neither were effective without concomitant vaccination of the herd at risk. Inconsistent use of anti-coagulants throughout the region as a whole, combined with decreasing availability of products and increasing costs has lessened the utility of population reduction as a singular long term or widespread solution to control this highly adaptive and unique vertebrate reservoir, requiring novel, more integrated solutions (Mayen 2003; Stoner-Duncan et al. 2014).

As an example of one small country in Central America that has successfully controlled the disease in dogs but has vampire bat rabies, Costa Rica reported more than 75 outbreaks in livestock over a 30 year period, with more than 780 fatal cases diagnosed in cattle (Hutter et al. 2016). Recent reports suggest that bovine paralytic rabies is spreading into areas that were previously unaffected (Bárcenas-Reyes et al. 2015). If climate change trends continue, vampire-transmitted rabies is expected to increase over widespread cattle-rearing regions, such as Mexico, Central America, Brazil and Paraguay (Lee et al. 2012). Although livestock vaccination appears highly beneficial, vampire control alone may not necessarily be efficient economically (Anderson et al. 2014). With projected growth in the livestock industry, plant-based rabies vaccines may have an especially important role, if they can compete based on safety, efficacy and cost.

11 Adverse Events Associated with Rabies Biologics

In general, modern human rabies vaccines are safe, compared to older, nerve-tissue derived products. The same is true for veterinary biologics. Local reactions may include pain, swelling, redness, alopecia, pruritus or other signs of inflammation.

Granulomas may occur, particularly with vaccines containing adjuvants. Systemic signs include fever, multi-focal vasculitis, transient lameness or rarely, anaphylaxis (Quirozr et al. 1964). In cats, vaccine injection-site sarcomas have been reported (Hartmann et al. 2015). Such non-viral components of vaccines can sensitize animals for future adverse responses.

The most serious adverse events are vaccine-associated rabies and vaccine failure. Vaccine-associated rabies cases were uncommon, even after first generation vaccines were used more widely, especially when compared to the millions of doses of modified-live vaccines used globally. The advent of viral typing by monoclonal antibodies (mAbs), and later genetic sequencing, allowed proper differentiation of laboratory strains used as seed stock from street viruses in nature (Whetstone et al. 1984; Okolo 1989; Hostnik et al. 2014; Robardet et al. 2016).

Very rare cases have been reported in association with RABV vaccines used for oral vaccination of wildlife (Fehlner-Gardiner et al. 2008; Vuta et al. 2016). Reports of cases in subjects with a history of vaccination must differentiate variables related to vaccine, the patient and the vaccinator. True failures are quite rare. Vaccinated subjects may have acquired immune compromise and fail to respond appropriately. Interference from maternal immunity may be at play. Breaks in the cold chain can lead to vaccine instability. Choice of an improper product, diluent, route, age, boosters, etc., are also variables that may be associated with rabies in an animal with a documentation of prior vaccination. Creation of national, regional or global VAERS programs related to vaccines and a searchable database for such licensed biologics would complement existing surveillance systems for the detection of acute or chronic reactions by product, species and circumstances. Any future introduction of plant products would need to compete with the strong safety record of available 21st century rabies biologics (Table 2).

12 Availability of Biologics

A global listing of all current manufacturers of rabies biologics is not feasible. However, annual reviews are useful on a national or regional basis to remain current (e.g., ACIP 2008; NASPHV 2016). Older, less potent nerve-tissue and modified-live biologics are being replaced by cell culture vaccines. Products requiring routine annual boosters are being supplemented gradually by others having a minimum duration of immunity of several years.

Rabies vaccines were produced primarily in the Americas and Eurasia, but are becoming more available on a global basis. No major shortages are forecast in the Americas or Europe. Emerging Asian markets, such as in India, may contribute to the overall diversity and number of doses needed. Elsewhere, concern has been expressed about overall quality issues (McLaughlin 2016). For example, within China, canine rabies is widespread and reports of outbreaks involving other domestic species are increasing.

Generally, rabies vaccines are available for most major domestic species, but specific rabies vaccines for some animals are not readily available, particularly in rural provinces. Under vaccine limitations and emergency settings, one 'off-label' study suggested a single injection of a 'double dose' canine vaccine was adequate for at least 1 year for application in cattle and camels, underscoring challenges throughout rural Asia (Liu et al. 2016). Parenteral off-label use of inactivated vaccines may also be used, for example, to protect valuable zoo and exotic animals in rabies enzootic regions (Miller and Fowler 2012). While such use does not prevent relevant public health measures if human exposure occurs (such as euthanasia of the biting animal), vaccination lessens the relative probability of a case in a properly immunized animal (NASPHV 2016). With the potency and safety of modern rabies vaccines, a broad margin of cross reactivity for other species is expected, based upon inference and comparative serological testing (Wallace et al. 2016). Still, use of plant-produced biologics may have utility for some specialized uses to occupy an unfilled need, where licensed products are lacking currently.

13 Plants as Sources of Rabies Biologics

Plant-based production systems are emerging as a relevant technology for production of many vaccines, antibodies and other biologics (Kristina and Luthar 2016; Topp et al. 2016). Such technology entails the integration of the desired genes encoding the antigen or antibody for the specific pathogen into the protein expression machinery of the plant. The target sequence is incorporated into an expression vector, before being transferred to a host plant expression system. These genes can be expressed transiently through the use of viral vectors, or stably, when the genes are permanently integrated into the plant genome.

Several advantages have been cited for plant-based expression systems over traditional mammalian cell-based production methods. These are especially relevant for rabies and include: rapid production via transient systems; freedom from mammalian pathogens; and, in some cases, reduced cost of goods related to the requirement for expensive fermentation infrastructure, as well as reduced handling costs for plant-based protein therapeutics. Proteins can be produced readily in large quantities, as the production systems are highly scalable (Shahaid and Daniell 2016). Plant-based rabies vaccines, antibodies and therapeutics are normally free from toxins and pathogens that are commonly produced in bacteria and yeast (Kwon et al. 2013). However, the downstream processing necessary to obtain pure therapeutic molecules is as expensive as for traditional mammalian cell culture platforms (Juarez et al. 2016).

Because this technology is still emerging, very few products have been brought to the market at large, to date. For comparison, in the veterinary field, a front runner is Newcastle Disease Virus (NDV) vaccine for poultry, approved by the United States Department of Agriculture (USDA). Another is Interberry-alpha, a biologic for the veterinary market produced in strawberries (Drake et al. 2017). A scFV mAb used in the purification process of a recombinant HBV vaccine is approved for use in Cuba. The first plant-made recombinant product approved for use in humans is Elelyso, an enzyme replacement therapy for Gaucher's disease, produced by Protalix. Thus far, this remains the only plant product licensed for use in humans (Drake et al. 2017). With the above theoretical advantages, many opportunities exist for plant-produced products in the rabies field.

14 Plant-Made Anti-RABV Antibodies

Although immune-protection is complex, VNA play a critical role in antiviral responses and in modulating lyssavirus infections (Both et al. 2012). Current PEP for bites by rabid animals involves the use of blood-derived rabies immune globulin from humans (HRIG) or horses (ERIG), accompanied by active vaccination. A weakness of current polyclonal serum-based prophylaxis is that many of the constituent virus-specific antibodies are non-neutralising (Marasco and Sui, 2007). These products also suffer a range of additional limitations, particularly in the developing world, including: limited availability; significant batch to batch variation; and challenges in obtaining immune donors in the case of HRIG (Both et al. 2012). In general, RIGs are expensive and solutions are needed to allow broader access by all segments of the population. In resource-poor settings, products also may have a higher risk of contamination with blood-borne adventitious agents, and of causing adverse reactions when administered, requiring additional screening and treatment to minimize risks associated with the use of such blood products. On the basis of these challenges, the WHO has encouraged the development and evaluation of biologics to replace RIG (WHO 2013).

While current rabies PEP is based mostly on polyclonal sera, emerging alternatives are focused increasingly on potent mAb cocktails (Tsekoa et al. 2016). Advances in antibody engineering and in the refinement of plant expression technology platforms are leading to a variety of options for new antibodies against rabies and other viral diseases. Cocktails of mAbs and bispecific constructs can be used simultaneously to target multiple viral epitopes to enhance potency and to overcome theoretical perceptions of viral variants escaping neutralization from the use of a single antibody (Both et al. 2013). Cocktails of mAbs are required to target epitopes that are not conserved in all viral species and variants, especially in infections like rabies that emerge from heterogeneous pools, circulating in various animal reservoirs. Such cocktails allow for broader coverage, and prevent viral escape mutants, issues which are important considerations in the development of passive immunotherapies.

Key to the potency of antiviral mAbs is not only that these biologics should cover a broad range of viral variants, but that they should target distinct non-overlapping epitopes and preferably should not compete for antigen binding.

In addition, the in vitro generated mAb escape mutants selected using one antibody should be neutralised by the other non-selecting mAb in the cocktail and vice versa (Marissen et al. 2005; de Kruif et al. 2007). The WHO Rabies Collaborating Centers (WHO RCC) identified 5 murine mAbs, with 4 (E559.9.14, M727-5-1, M777-16-3 and 1112-1) recognising antigenic site II of the glycoprotein, and the fifth, 62-71-3, recognising site I (Prosniak et al. 2003, Goudsmit et al. 2006; Müller et al. 2009; Both et al. 2013). For the antibodies recognising antigenic site II, E559 exhibited the broadest virus neutralisation spectrum and the greatest potency, and appears to be an ideal candidate for expression in a RIG-replacement cocktail. The Mab 62-71-3 was included for site I coverage.

As part of a strategy to develop alternative biologics to replace RIG, mAb E559 was cloned, engineered and produced in transgenic *Nicotiana tabacum* plants (van Dolleweerd et al. 2014). In this work, the murine constant domains of E559 were replaced with human IgG1κ constant domains and the resulting chimeric mouse–human genes cloned into plant expression vectors used for transformation of *N. tabacum*. The chimeric mouse-human heavy and light chain genes were introduced into separate transgenic lines. The lines were sexually crossed, resulting in progeny expressing fully assembled E559. Codon optimisation of chimeric genes enhanced expression of the chimeric antibody in transgenic plants. Sequence analysis of the heavy and light chains of the plant-made chimeric mAb E559 predicted the presence of 2 potential N-linked glycosylation sites, a conserved site in the antibody Fc region and one in the V_L domain. The in vitro effectiveness of the plant-made chimeric antibody against representative viruses from phylo groups I and II (van Dolleweerd et al. 2014) was assessed in a modified fluorescent antibody virus neutralisation (mFAVN) assay, and the plant-derived antibodies (murine and chimeric) mirrored the hybridoma-derived antibody in terms of depth of neutralisation. The efficacy of the murine-human E559 in experimental PEP was examined in hamsters administered a lethal dose of a laboratory strain of RABV (CVS-11), in the absence of vaccine. Control groups that received phosphate buffered saline (PBS only) succumbed after 14 days, while a survival rate of 11% was observed for HRIG and mouse-human groups, showing that the plant-derived chimeric E559 was at least as effective as HRIG.

In a similar but separate study, Both et al. 2013, demonstrated that a functional mouse-human chimeric mAb 62-71-3 expressed in *N. benthamiana* could be a viable candidate antibody for rabies PEP in humans. The study demonstrated potent virus neutralisation at 500 IU/mg. In addition, a critical role for antigenic site I of the G protein, as well as for two specific amino acid residues (K226 and G229) within site I, was identified with regard to mAb 62-71-3 neutralisation, thus providing a clear molecular rationale for the use of mAb 62-71-3 in a RIG-replacement cocktail (Both et al. 2013).

Thereafter, in another study, Tsekoa et al. (2016) produced a highly potent cocktail of humanized mAbs E559 and 62-71-3 in *N. benthamiana*. Recognized previously, these mAbs E559 and 62-71-3 recognise complementary sites on the

RABV G protein (Müller et al. 2009; Kuzmina et al. 2013). As in earlier studies with E559, variable domains from heavy and light chains (V_H and V_L) from murine mAbs E559 and 62-71-3 were fused to the constant domain (C_H and C_L) from human IgG_1. The codon optimised chimeric L_C and H_C for each mAb were then co-expressed to yield full, correctly folded, individual antibodies. This study employed a viral-based transient expression system to achieve high levels of expression of both E559 and 62-71-3 ($\sim$490 mg/kg of fresh leaf tissue).

Generally, plant-produced proteins are functionally similar to mammalian cell-produced versions, because plants as eukaryotic systems can add glycans as part of the protein maturation process. Protein glycosylation is an essential co- and post-translational modification of secretory and membrane proteins, critical to protein function in all eukaryotes. However, there are differences in glycosylation patterns between plants and mammals, the most notable of which involves plant-specific N-glycan modifications. In plants, α-1,3 fucose is attached to the first GlcNAc residue and the β-1,2 xylose attached to the central mannose residue. In contrast, mammalian N-glycans may be modified by α-1,6 fucoses as well as β-1,4 galactose residues, attached to terminal GlcNAc moieties (Strasser et al. 2008). With vaccines, glycosylation has been shown to enhance immunogenicity in some cases. However, in the case of antibody therapeutics, plant-type glycosylation has been a source of concern for unwanted immune responses that could potentially cause adverse events, although this has not been observed to date.

Afucosylated therapeutic mAbs (whose N-glycans lack fucose) have enhanced in vivo efficacy in different models of viral infection, due to increased antibody-dependent cell-mediated cytotoxicity (ADCC) activity (Forthal et al. 2010; Hiatt et al. 2014). For these reasons, Tsekoa and co-workers expressed two chimeric mAbs in the ΔXT/FT *N. bethamiana* host plants, which produce predominantly fucose-free GnGn glycan structures (Strasser et al. 2008). The structural integrity of the antibodies (secondary and tertiary structures) was determined using both fluorescent and circular dichroism spectroscopy.

In vitro virus neutralisation assays were conducted using a cell-based Rapid Fluorescent Focus Inhibition test (RFFIT) assay. The individual mAbs efficiently neutralised a diverse panel of RABV variants in vitro. The in vivo efficacy of the plant-based mAbs was tested in a challenge experiment with female Syrian hamsters, infected with CVS-11. The infected control group did not survive beyond 14 days post-infection, confirming lethality of inoculum. Animals were administered 2 IU of mAb or HRIG. The plant-made E559 and 62-71-3 showed 100 and 86% survival, respectively, at 14 days post-infection. At 28 days post-infection, 33% of animals given 62-71-3 and 20% of animals administered E559 survived, compared to no survival for animals given HRIG. Significantly, the cocktail of E559 and 62-71-3 exhibited higher potency against RABV compared to HRIG in a hamster model PEP challenge trial (Tsekoa et al. 2016). The methodology of expression used was selected as a highly scalable, rapid production alternative to mammalian cell (e.g., CHO Cell) culture-based production, and had a safer product profile compared to HRIG or ERIG. Collectively, results of this study laid solid ground work for the future development of a batch-consistent, safe and efficacious RIG replacement product in plants.

15 Plant-Made Vaccines

Scarcity of vaccines for prevention of fatal diseases has driven global attention towards production of safer, easier to use and more effective vaccines (Laere et al. 2016). In countries with a high incidence of rabies (which tends to coincide with low income per capita and resource limitations) efficacious, affordable vaccines are critical to preventative immunisation, preferably after a single dose (Ertl, 2009; Ortiz-Prado et al. 2014). While currently available vaccines are efficacious, modern rabies vaccines consist of inactivated RABV formulations which generally have low immunogenicity, and therefore require multiple injections. In veterinary medicine, inactivated virus vaccine often contains adjuvants to enhance potency, often with adverse events, justifying the quest for new production technologies. The vaccines are also difficult to manufacture, and are produced in such a way that there is a high degree of variation between producers (and batches from the same producers). Efficacy can be further negatively impacted by storage and transportation, especially when the cold chain cannot be maintained. In some parts of the world, vaccines are simply not readily available, and cost is a major limitation to access (Gautret et al. 2011). Other challenges include the risks of production and administration of the current whole inactivated virus vaccine and the logistic concerns of a multi-vaccination schedule for pre- and PEP vaccination. Historically, nerve tissue-based vaccines led to neurological complications, which would be prevented by plant-based applications (Swamy et al. 1984).

As described above, there are 3 basic types of rabies vaccine production methods: avian embryo-based; cell-based; and investigational new platforms. Avian-based methods require large amounts of eggs and the regulatory process associated with this platform is time consuming. As in the case on antibody therapeutics, cell-based approaches require high priced manufacturing facilities which are also limited by lengthy production processes and the possibility of contamination by human pathogens. Thus, production of vaccines using alternative technologies that address some of these challenges becomes critical. Plant expression technology is one emerging platform, relevant to develop exploratory rabies vaccine candidates. The overall benefits of plant expression technology have already been highlighted above.

Classic RABV vaccines consist of whole inactivated viruses that have the same antigenic properties as wild type viruses. However, the use of major structural proteins from disease agents in sub-unit vaccines is gaining momentum. The various RABV proteins, individually and in combination, have shown potential for use in anti-rabies subunit vaccine candidates by generating VNA, when used in vaccination of experimental animals (Conzelman et al. 1990).

16 Glycoprotein-Based Vaccines

The G protein of RABV has been identified as the major antigen that induces protective immunity (Cox et al. 1977). Vaccination with classic vaccines often results in the induction of VNA directed against the envelope G protein, activation of helper and cytotoxic T cells and protection against lethal challenge with RABV (Astray et al. 2017). When properly folded and glycosylated, the RABV G protein is fully immunogenic, bearing epitopes for humoral and cell-mediated immune responses (Benmansour et al. 1991; Johnson et al. 2010; Lafon et al. 1983; Moore et al. 2006; Macfarlan et al. 1984). The G protein monomers oligomerise in threes to form trimers that are essential for the interaction of the virus with target membrane receptors and, to a certain extent, for the induction of VNA, leading to protection against virus infection (Koraka et al. 2014; Lodmell et al. 2004). Thus, for recombinant expression systems to be effective, they must express G protein that is glycosylated (Foley et al. 2000) and folds appropriately.

Challenges of expressing this highly immunogenic protein include that the G protein is hydrophobic and unstable, and undergoes post-translational modification. To deal with problems of hydrophobicity, soluble forms of the G protein have been expressed, but they are poorly immunogenic. At least one sequon has to be glycosylated to allow the G protein to reach the cell surface (Shakin-Eshleman et al. 1992). The glycan composition is dependent on the biochemical machinery of the host cell and seems to be a determinant of immunologic properties. In a study in which the G protein was expressed in yeast, Klepfer et al. (1993) showed that the G protein was glycosylated, yet it gave protection only against intramuscular, but not intracerebral virus challenge. The lack of adequate protection against intracerebral challenge was attributed to possible hyperglycosylation from the host. Thus, suitable expression systems must be able to express high levels of functional native G protein (Dietzschold et al. 1990; Ramya et al. 2011; Benmaamar et al. 2009; Liu et al. 2014; Wang et al. 2015).

Over the past two decades, there has been a significant amount of plant-based vaccine development work focusing on the G protein (Starodubova et al. 2015), where several studies have demonstrated G protein-induced VNA and the protection of vaccinated animals against rabies. The early work of McGarvey et al. (1995) demonstrated the feasibility of expressing G proteins in the leaves and fruit of transgenic tomatoes. Following this pioneering work, the G protein has been produced in tobacco (Modelska et al. 1998; Yusibov et al. 1997; Ashraf et al. 2005), spinach (Modelska et al. 1998), carrot and maize (Loza-Rubio et al. 2012).

Ashraf et al. (2005) designed a chimeric G protein for high level expression to enhance expression in plants. They replaced G protein native signal peptide with the pathogenesis-related protein signal (PR-S) peptide of *N. tabacum* in the N-terminus, which is known to efficiently target proteins to the ER (Sijmon et al. 1990), and a SEKDEL sequence to anchor the G protein in the ER. The G protein was expressed constitutively in *N. tabacum*. Mice that were immunised intraperitoneally with the G protein isolated from the microsomal fraction elicited high levels of immune responses as compared to the inactivated commercial virus. The plant-derived G protein induced complete protective immunity against intracerebral lethal challenge with RABV (Ashraf et al. 2005). Non-glycosylated G protein prepared from bacterial cells did not confer protective immunity, as observed in studies by various groups (Lathe 1994; Malek et al. 1984). Plants were superior to bacterial expression in providing a protein with correct glycosylation and folding, leading to potent immunogenicity.

17 Nucleoprotein-Based Vaccines

Studies suggest that the N protein may also be a suitable candidate for inclusion in a subunit vaccine against rabies. This is because compared to other major RABV proteins, the N protein is more conserved across a range of viral species, and as such may provide protection against the majority of RABV variants (Dietzschold et al. 1987; da Cruz et al. 2001). Like the G protein, the N-protein can induce RABV specific helper T-cells (Ertl et al. 1989). Earlier, this protein was described by Lafon et al. (1994) as having the properties of a 'superantigen', justifying the investigation of its potential use in subunit vaccine development. These effects of the N protein include triggering RABV-specific T-cells, facilitating the production of VNA and other specific immune functions.

Perea Arango et al. (2008) transiently expressed a full length RABV N protein in transgenic tomato plants as well as transiently in *N. benthamiana* plants. Mice were immunized both intraperitoneally (i.p.) and orally with tomato fruit extracts containing the N protein. Mice immunised i.p. had an antibody titre of about 4 times that of orally immunised mice, and were partially protected from RABV challenge 60 days post immunisation, while the mice orally vaccinated with tomato extracts were not protected. In the same study, they immunised mice via i.p. or orally with *N. bethamiana* leaf extracts containing the N protein, and also demonstrated the importance of the route of immunisation when the i.p. immunised mice were partially protected from virus challenge. In both cases, mice immunised orally showed a marginal immune response, and were not protected in the RABV challenge study.

18 Oral Rabies Vaccines

Plant expression offers the possibility for oral administration of immunogenic and therapeutic proteins, negating the need for exhaustive purification (Haq et al. 1995). Such formulations are delivered as partially purified preparations that are safe for oral consumption. This approach of partially purified product is quite suitable for situations where high doses of proteins are required throughout the prophylaxis period, as in the case for oral passive immunisation. Antigens from various pathogens have been expressed in plants, with a view to administer them orally in humans or animals. Advantages of oral immunisation include removing the need to use needles for vaccine administration, thus pre-empting the safety hazards associated with needle use, particularly in resource-constrained settings. Plant-derived antigens can induce strong immune responses at the mucosal and serum levels (Chikwamba et al. 2002; Loza-Rubio et al. 2012) after administration orally or intramuscularly in mice. Additionally, they can protect against subsequent pathogen challenge. The benefits of oral vaccination with plant-derived antigens include ease of delivery and prolonged stability at room temperature, especially when expressed in seeds.

In one notable example, Loza-Rubio et al. (2012) expressed the full RABV G protein in transgenic maize, and fed a range of doses (0.5–2 mg) to sheep via the oral route. Sera of the sheep were tested for VNA against RABV CVS strain using the RFFIT. The VNA were detected at day 30 for the 120 day expression test in animals, peaking at 60–90 days, before declining in all vaccinated animals. At 120 days post-immunisation, sheep were challenged intramuscularly with RABV. Cumulative survival in groups that received 2 mg of G protein orally was comparable to the positive control (inactivated RABV commercial vaccine delivered intramuscularly), while the non-vaccine controls died after challenge. This study was a first to demonstrate the efficacy of an orally presented edible vaccine in a polygastric host model.

In another model, Rojas-Anaya et al. (2009) expressed the full-length G protein in carrots. Carrot was selected as the host expression plant to obviate the need for cooking (which denatures antigens), and for its palatability. They administered 50 μg of G protein orally in raw carrots, and 50 μg of G protein in carrot plus augmentation with 50 μg of the N protein. Orally delivered mouse chow was used as a negative control and an intramuscular dose of inactivated rabies vaccine as the positive control. Antibodies raised against RABV were detected 15 days post-immunisation, with a significant difference between the vaccinated and the non-vaccinated groups. Mice were challenged intra-cerebrally with RABV 60 days post-immunisation. All unvaccinated mice died, while 66% of mice vaccinated with carrot, and 100% of mice vaccinated with conventional vaccine, survived. The authors suggested that a higher dose of antigen in carrot (100 μg in 4 g carrot) could constitute a more viable vaccine dose. In addition, this study suggested that the carrot system could be a viable source of edible vaccine against rabies.

The cholera toxin B (CTB) has been used in several studies as a carrier for antigens, having been demonstrated to be a potent adjuvant to co-administered proteins (Dertzbaugh and Elson 1993; Song et al. 2004; Lycke et al. 1989). To enhance oral immunogenicity of the G protein, Roy et al. (2016) reported the expression of the full RABV G protein fused to a synthetic CTB subunit at its C-terminus. The chimeric protein folded into a 403 kDa pentameric structure, indicating that the fusion of the G protein into the CTB C-terminus did not interfere with the CTB pentamerisation process. The fusion of the CTB with the G protein in this study was to take advantage of the CTB as a receptor carrier to enhance T-cell response during oral immunisation. Theoretically, the amount of G protein required for oral vaccination may be reduced in the presence of a potent, receptor-specific adjuvant, like CTB, directing the antigen to immune cell receptors. The study showed that the plant-made protein had a higher affinity for GM1 receptor than the native bacterial CTB protein, possibly due to assay conditions, but also due to glycosylation of the CTB molecule in tobacco cells. The CTB-G fusion had a lower affinity for GM1 gangliosides, possibly due to stearic interference of the molecule by the G protein. The fusion protein recognised both anti-CTB and anti-G protein antibodies. Unfortunately, the immunogenicity of the fusion protein was not tested in this study, so the ultimate utility of such an approach remains unresolved.

Singh et al. (2015) expressed a fusion of the Ricin-toxin B subunit (RTB) and the RABV G protein in tomato hairy roots. The fusion protein was recognised by both anti-ricin and anti-G protein antibodies in ELISA assays. Partially purified extracts expressing the fusion were administered orally in mice. The results showed the RGP-RTB fusions were internalised through the mucosal lining, and were able to elicit lgG responses in mice. The titre of lgG1 and lgG2a antibodies in this experiment reflected the Th2 type of T-helper cell response against the RGP-RTB fusion proteins, although the titres were low. No RABV challenge studies were performed, so efficacy was not demonstrated.

19 Viral Vectors for Rabies Biologics

Unlike simple singular linear viral proteins, virus-like particles (VLPs) are formed by viral structural proteins that are stable, non-replicative and non-infective, thus making them suitable for use as potent vaccines against infectious disease (Kushnir et al. 2012). Such VLPs are a class of subunit vaccine antigens that are soluble and immunogenic but are different from viral antigens expressed by replicating microbial vectors in vivo. Unlike most conventional inactivated vaccines, these VLPs fold appropriately and stimulate innate immunity by interacting with the pathogen-associated molecular pattern and pattern receptors, eliciting adaptive immune and inflammatory responses against infection without adjuvants (Morón et al. 2002; Kurt-Jones et al. 2000). Moreover, VLPs are a potent developmental approach because of their ability to express surface proteins of interest and elicit the appropriate response in the immunised host, mimicking a natural infection with the immunogenic antigens of interest. Often, VLPs have been produced using insect viruses, but such an approach could be used in plant-based systems.

For example, Kang et al. (2015) reported the production of a chimeric VLP containing the G and M proteins of the RABV ERA strain, and the membrane-anchored granulocyte-macrophage colony-stimulating factor (GM-CSF), in an attempt to develop and improved rabies vaccine, due to the strong adjuvant activity of GM-CSF. The chimeric VLP was expressed in insect cells, and induced VNA when used to immunise mice intramuscularly. The GM-CSF had a strong effect, enhancing a larger and broader antibody subclass response compared to the use of VLPs with G and M proteins only. Similarly, in a patent application (US 2014/0178419 A1, June 26, 2014), Smith et al. (2014) described a method for the production of a RABV VLP comprising G proteins for use in the prevention and treatment of rabies. Nucleic acid-encoding G protein is expressed in insect cells using a baculovirus vector. The resulting VLP-like structures resulting from such cultures induced potent anti-RV antibodies in rabbits.

Yusibov et al. (2002) demonstrated the immunogenicity of a plant virus-based chimeric peptide containing the antigenic determinants from the RABV G protein (amino acids 253-275) and the N protein (amino acids 404-418) in a translational fusion with the alfalfa mosaic virus (AlMV) coat protein (CP). The first expression system utilised transgenic *N. tabacum* plants providing replicative functions in trans for full length RNAs of AlMV. A second system utilised *N. tabacum* and spinach (*Spinacia oleracea*), using autonomously replicating tobacco mosaic virus (TMV), lacking the native coat protein. Recombinant virus containing the chimeric RABV epitope isolated from *N. tabacum* was used to immunise mice parenterally. Animals were protected from subsequent challenge with RABV. To test basic immunogenicity, human volunteers who ingested spinach expressing the chimeric virus showed an immune response against RABV; 3 of 5 volunteers who had previously been vaccinated against rabies with a conventional vaccine specifically responded against the peptide antigen, which was unusual as subjects do not typically respond to rabies antigens per os. When 9 non-immune individuals ingested the spinach, 5 of the 9 demonstrated significant antibody responses to RABV on AlMV. When these individuals received a single dose of conventional vaccine, they showed detectable levels of RABV antibodies. The study concluded that the chimeric virus vaccine could have potential utility as a supplementary or booster vaccine. Importantly, the authors noted that the plants inoculated with the chimeric virus only showed mild signs of virus infection, and that none of the people subjected to the virus-expressing materials showed any adverse reactions.

20 Challenges and Future Opportunities in the Use of Plant-Based Products in Rabies Prevention and Control

Current human PEP involves the use of blood-derived RIG, which has several safety and efficacy limitations and may be of limited availability in the case of sudden mass exposures. The concerns arising from the use of blood-derived products could be circumvented, and consistent batches of neutralising mAbs could be produced based on a cocktail of plant-derived, potent biologics. The emerging plant expression technologies have made significant progress by addressing technical aspects of expression, yield, efficacy, scalability and reproducibility, with varying degrees of success for different products. Such attributes make plants an attractive platform for production of mAbs to replace RIG (e.g., Ko et al. 2003; van Dolleweerd et al. 2014). Similarly, potent subunit vaccines can be produced in plants. Ideally, if such biologics demonstrate utility in humans, the same constructs could be developed for domestic animals and wildlife (Rupprecht et al. 2017). For example, for the global elimination of canine rabies, a current shortfall in excess of 7 billion vaccine doses is forecast, as one significant application in which plant-produced rabies biologics could compete (Wallace et al. 2017).

Advancing plant produced anti-RABV therapeutics to mainstream use is critical, given the high human case fatality, given that there is no effective treatment once symptoms appear. However, the regulatory aspects of plant therapeutic production systems remain a work in progress. In the United States and Europe, the FDA and the USDA and European Medicines Agency (EMA) documents provide important guidance on points to consider with regards to product safety and efficacy, environmental issues and manufacturing control. The EMA published a "Guideline on the quality of biological active substances produced by stable expression in higher plants" in 2009, but this is yet to be tested (Drake et al. 2017; EMA 2016). The regulatory framework for the field cultivation of genetically modified plants is particularly demanding, especially in Europe. While these represent significant milestones in the maturation of this emerging technology, much work remains to be done before such an approach is accepted universally as a main stream source of biologic and therapeutics.

The requirement for facilities for manufacturing plant-made products to conform to current good manufacturing practice (cGMP) has made progress with several cGMP facilities established across the globe, notably Kentucky Bioprocessing LLC (Owensboro, Ky), Sigma-Aldrich Fine Chemicals (St Louis, MI), Medicago Inc (Quebec City, Canada), Protalix (Karmiel, Israel) and others (Yusibov et al. 2011). Location of such manufacturing capability in the developing world would go a long way to address the challenges of infectious disease in these regions, particularly for neglected diseases such as rabies.

The virulence and burden of rabies should be a compelling enough rationale to fast track development of replacement products, but this has not been the case. For example, the WHO has specified that plant-derived human vaccines and therapeutics have to be produced and clinically tested under Investigational New Drug (IND) applications and meet all the applicable regulatory GMP requirements (WHO 2003; van der Laan et al. 2006). Progress in bringing alternative products to market is also hampered by the complicated and as yet untried clinical development path for the replacement of current efficacious products, such as RIG with new products. The US FDA has provided an alternative animal rule pathway for development, where efficacy is established in well-controlled animal model trials and safety via the normal human trial pathway, but this is not available in developing countries. As such, the regulatory framework for plants is slow to become institutionalised and has not been fully tested thoroughly in any part of the world.

Mammalian systems are already established for rabies and have a dominant share of the market for humans, domestic animals and wildlife (Table 2). The mammalian systems are also undergoing continuous improvement to reduce the cost of goods, and wave-bag technology is a good example of such technological improvements (Singh 1999). To compete effectively, plant expression systems have

Table 2 Common commercial human and veterinary rabies biologics

Name	Producer	Type	Comments
Imovax	Sanofi	Human diploid cell vaccine (HDCV)	First human cell culture vaccine
RabAvert	Novartis	Purified chick embryo cell vaccine (PCEC)	Human vaccine produced in Germany and also in India
Verorab	Sanofi	Purified vero cell rabies vaccine (PVRV)	First human rabies vaccine produced in vero cells
Vaxirab	Zydus Cadila	Purified duck embryo vaccine (PDEV)	More potent, safe and efficacious than earlier duck embryo vaccines for humans
Speeda	Chengda Bio	Vero cell rabies vaccine	Human vaccine produced in China
Indirab	Bharat Biotech	Purified vero cell rabies vaccine	Human vaccine produced in India
HyperRAB	Grifols	Human rabies immune globulin (HRIG)	Used in human postexposure prophylaxis
KamRAB	Kamada	HRIG	Used in human postexposure prophylaxis
Imogam	Sanofi	HRIG	Used in human postexposure prophylaxis
Defensor	Zoetis	Veterinary vaccine	Used for pre-exposure immunization of domestic animals
Rabvac	Boehringer Ingelheim	Veterinary vaccine	Used for pre-exposure immunization of domestic animals
SAG-2	Virbac	Attenuated rabies virus vaccine	Used in oral wildlife rabies programs
Raboral V-RG	Merial	Recombinant vaccinia-rabies glycoprotein vaccine	Used in oral wildlife rabies programs
Onrab	Artemis	Recombinant adeno-rabies glycoprotein vaccine	Used in North American oral wildlife rabies programs

to provide additional value, as in therapeutics with additional features. The ability to manipulate the glycan profile of therapeutics offers a la carte glycosylation patterns for more effective biologics (Juarez et al. 2016)—the concept of biobetters—which could potential give an edge to plant-made products. Beyond these plant-specific issues, a number of general barriers for any new innovations in rabies prevention and control must be overcome for major impacts of novel technology to be realized over the next 5 years (van de Burgwal et al. 2017).

21 Summary

Rabies is a neglected viral zoonosis, with the highest case fatality of any infectious disease. The etiological agents are single-stranded, non-segmented, negative-sense RNA viruses. Major animal reservoirs include domestic and wild carnivores and bats, while humans are victims and dead-end hosts. Most people are infected by bite exposure via rabid domestic dogs or wild carnivores, particularly in Africa and Asia, or bats throughout the Americas, Africa, Asia, Australia and Eurasia. The viral G protein is the most important antigen for the induction of neutralizing antibodies and protective immunity. Pure, potent, safe and efficacious vaccines for pre-exposure vaccination are available for humans and most common domestic species, but limitations in supply may occur in developing countries. All humans and other mammals at risk should be considered for vaccination in enzootic areas. Priority should be given to particularly valuable animals and those in frequent contact with humans as pets or in public settings. In domestic animals, primary vaccination, followed by a booster dose a year later, leads to a long lasting immunity. As shown since the 19th century, PEP of naïve individuals is possible, if conducted in a timely and appropriate manner, but is more complicated and expensive. Therapy after the onset of clinical signs is largely futile. A detailed understanding of lyssavirus pathobiology and epidemiology is necessary if plant-produced biologics will be developed to fill remaining gaps.

Future research should entail development of: biologics with a broader spectrum of activity across the antigenic diversity of the *Lyssavirus* Genus; additional robust and safer adjuvants; methods to maximize thermo-stability under tropical conditions; relevant, humane techniques to manage rabies associated with bats, possibly by vaccination; and simpler vaccination regimens that may entail a single dose, or at most a prime-boost administration, for maximal duration of immunity, without the need for routine boosters. In a One Health context, the global strategy is to minimize the risk of rabies in humans and other animals by the mass vaccination of dogs, to eliminate the perpetuation of canine rabies as a common public good in agriculture, human health and conservation medicine. Given the above challenges, products produced in plants may have multiple diagnostic, vaccine and antibody applications using current conventional products as a model and filling unmet needs and open niches in laboratory-based surveillance, prophylaxis and therapy.

References

Abelseth MK (1964) An attenuated rabies vaccine for domestic animals produced in tissue culture. Can Vet J 5(11):279–286

Advisory Committee on Immunization Practices (ACIP) (2008) Human rabies prevention–United States, 2008: recommendations of the advisory committee on immunization practices. MMWR Recomm Rep 2008 May 23;57(RR-3):1–28

Ahmad A, Singh CK (2016) Comparison of rapid immunodiagnosis assay kit with molecular and immunopathological approaches for diagnosis of rabies in cattle. Vet World 9(1):107–112

Anderson A, Shwiff S, Gebhardt K, Ramírez AJ, Shwiff S, Kohler D, Lecuona L (2014) Economic evaluation of vampire bat (*Desmodus rotundus*) rabies prevention in Mexico. Transbound Emerg Dis 61(2):140–146

Annand EJ, Reid PA (2014) Clinical review of two fatal equine cases of infection with the insectivorous bat strain of Australian bat lyssavirus. Aust Vet J 92(9):324–332

Aréchiga Ceballos N, Vázquez Morón S, Berciano JM, Nicolás O, Aznar López C, Juste J, Rodríguez Nevado C, Aguilar Setién A, Echevarría JE (2013) Novel lyssavirus in bat. Spain Emerg Infect Dis 19(5):793–795

Asgary V, Shoari A, Baghbani-Arani F, Sadat Shandiz SA, Khosravy MS, Janani A, Bigdeli R, Bashar R, Cohan RA (2016) Green synthesis and evaluation of silver nanoparticles as adjuvant in rabies veterinary vaccine. Int J Nanomedicine 29(11):3597–3605

Ashraf S, Singh PK, Yadav DK, Shahnawaz M, Mishra S, Sawant SV, Tuli R (2005) High level expression of surface glycoprotein of rabies virus in tobacco leaves and its immunoprotective activity in mice. J Biotechnol 119(1):1–4

Aspden K, van Dijk AA, Bingham J, Cox D, Passmore JA, Williamson AL (2002) Immunogenicity of a recombinant lumpy skin disease virus (neethling vaccine strain) expressing the rabies virus glycoprotein in cattle. Vaccine 20(21–22):2693–2701

Astray RM, Jorge SA, Pereira CA (2017) Rabies vaccine development by expression of recombinant viral glycoprotein. Arch Virol 162(2):323–332

Baby J, Mani RS, Abraham SS, Thankappan AT, Pillai PM, Anand AM, Madhusudana SN, Ramachandran J, Sreekumar S (2015) Natural rabies infection in a domestic fowl (*Gallus domesticus*): a report from India. PLoS Negl Trop Dis. 9(7):e0003942

Banyard AC, Evans JS, Luo TR, Fooks AR (2014) Lyssaviruses and bats: emergence and zoonotic threat. Viruses 6(8):2974–2990

Bárcenas-Reyes I, Loza-Rubio E, Zendejas-Martínez H, Luna-Soria H, Cantó-Alarcón GJ, Milián-Suazo F (2015) Epidemiological trends in bovine paralytic rabies in central Mexico, 2001–2013. Rev Panam Salud Publica 38(5):396–402

Barnard BJ (1979) Symptoms of rabies in pets and domestic animals in South Africa and South West Africa. J S Afr Vet Assoc 50(2):109–111

Basheer AM, Ramakrishna J, Manickam R (1997) Evaluation of post-exposure vaccination against rabies in cattle. New Microbiol 20(3):289–294

Bassuino DM, Konradt G, Cruz RA, Silva GS, Gomes DC, Pavarini SP, Driemeier D (2016) Characterization of spinal cord lesions in cattle and horses with rabies: the importance of correct sampling. J Vet Diagn Invest 28(4):455–460

Begeman L, GeurtsvanKessel C, Finke S, Freuling CM, Koopmans M, Müller T, Ruigrok TJH, Kuiken T (2017) Comparative pathogenesis of rabies in bats and carnivores, and implications for spillover to humans. Lancet Infect Dis 2017 Oct 31. pii: S1473–3099(17)30574-1

Bell JF, Moore GJ (1979) Allergic encephalitis, rabies antibodies, and the blood/brain barrier. J Lab Clin Med 94(1):5–11

Ben Khalifa Y, Luco S, Besson B, Sonthonnax F, Archambaud M, Grimes JM, Larrous F, Bourhy H (2016) The matrix protein of rabies virus binds to RelAp43 to modulate NF-κB-dependent gene expression related to innate immunity. Sci Rep. 21(6):39420

Benmaamar R, Astray RM, Wagner R, Pereira CA (2009) High-level expression of rabies virus glycoprotein with the RNA-based Semliki forest virus expression vector. J Biotechnol 139(4): 283–290

Benmansour A, Leblois H, Coulon P, Tuffereau C, Gaudin Y, Flamand A, Lafay F (1991) Antigenicity of rabies virus glycoprotein. J Virol 65(8):4198–4203

Bingham J, van der Merwe M (2002) Distribution of rabies antigen in infected brain material: determining the reliability of different regions of the brain for the rabies fluorescent antibody test. J Virol Methods 101(1–2):85–94

Biswas S, Reddy GS, Srinivasan VA, Rangarajan PN (2001) Preexposure efficacy of a novel combination DNA and inactivated rabies virus vaccine. Hum Gene Ther 12(15):1917–1922

Blancou J, Baltazar RS, Molli I, Stoltz JF (1991) Effective postexposure treatment of rabies-infected sheep with rabies immune globulin and vaccine. Vaccine 9(6):432–437

Boland TA, McGuone D, Jindal J, Rocha M, Cumming M, Rupprecht CE, Barbosa TF, de Novaes Oliveira R, Chu CJ, Cole AJ, Kotait I, Kuzmina NA, Yager PA, Kuzmin IV, Hedley-Whyte ET, Brown CM, Rosenthal ES (2014) Phylogenetic and epidemiologic evidence of multiyear incubation in human rabies. Ann Neurol 75(1):155–160

Both L, Banyard AC, van Dolleweerd C, Horton DL, Ma JK, Fooks AR (2012) Passive immunity in the prevention of rabies. Lancet Infect Dis 12(5):397–407

Both L, van Dolleweerd C, Wright E, Banyard AC, Bulmer-Thomas B, Selden D, Altmann F, Fooks AR, Ma JK (2013) Production, characterization, and antigen specificity of recombinant 62-71-3, a candidate monoclonal antibody for rabies prophylaxis in humans. FASEB J 27(5): 2055–2065

Brito MG, Chamone TL, da Silva FJ, Wada MY, Miranda AB, Castilho JG, Carrieri ML, Kotait I, Lemos FL (2011) Antemortem diagnosis of human rabies in a veterinarian infected when handling a herbivore in Minas Gerais, Brazil. Rev Inst Med Trop Sao Paulo. 2011 Jan–Feb; 53(1):39–44

Brookes SM, Parsons G, Johnson N, McElhinney LM, Fooks AR (2005) Rabies human diploid cell vaccine elicits cross-neutralising and cross-protecting immune responses against European and Australian bat lyssaviruses. Vaccine. 2005 Jul 14; 23(32):4101–4109

Carneiro V (1954) Transmission of rabies by bats in Latin America. Bull World Health Organ 10(5):775–780

Centers for Disease Control (CDC) (1983) Exposure to a rabid cow–Pennsylvania. Morb Mortal Wkly Rep. 1983 Mar 11; 32(9):128–129

Centers for Disease Control and Prevention (CDC) (1999) Mass treatment of humans who drank unpasteurized milk from rabid cows–Massachusetts, 1996–1998. Morb Mortal Wkly Rep. 1999 Mar 26; 48(11):228–229

Chang SS, Tsai HJ, Chang FY, Lee TS, Huang KC, Fang KY, Wallace RM, Inoue S, Fei CY (2016) Government response to the discovery of a rabies virus reservoir species on a previously designated rabies-free island, Taiwan, 1999–2014. Zoonoses Public Health 63(5):396–402

Charlier J, Forbes A, Van Gucht S, Duchateau L, Goddeeris BM, Vercruysse J (2013) Serological evidence of *Ostertagia ostertagi* infection in dairy cows does not impact the efficacy of rabies vaccination during the housing period. Res Vet Sci 95(3):1055–1058

Chikwamba R, Cunnick J, Hathaway D, McMurray J, Mason H, Wang K (2002) A functional antigen in a practical crop: LT-B producing maize protects mice against Escherichia coli heat labile enterotoxin (LT) and cholera toxin (CT). Transgenic Res 11(5):479–493

Conzelmann KK, Cox JH, Schneider LG, Thiel HJ (1990) Molecular cloning and complete nucleotide sequence of the attenuated rabies virus SAD B19. Virology 175(2):485–499

Cox JH, Dietzschold B, Schneider LG (1977) Rabies virus glycoprotein II. Biological and serological characterization. Infect Immun. 1977 Jun; 16(3):754–759

da Cruz FW, McBride AJ, Conceição FR, Dale JW, McFadden J, Dellagostin OA (2001) Expression of the B-cell and T-cell epitopes of the rabies virus nucleoprotein in *Mycobacterium bovis* BCG and induction of an humoral response in mice. Vaccine 20(5): 731–736

de Kruif J, Bakker AB, Marissen WE, Kramer RA, Throsby M, Rupprecht CE, Goudsmit J (2007) A human monoclonal antibody cocktail as a novel component of rabies postexposure prophylaxis. Annu Rev Med 18(58):359–368

de Macedo Pessoa CR, Cristiny Rodrigues Silva ML, de Barros Gomes AA, Isabel Estévez Garcia A, Honma Ito F, Eduardo Brandão P, Riet-Correa F (2011) Paralytic rabies in swine. Braz J Microbiol 42(1):298–302

Den K, Sato T, Schneebeli M, Morita Y (2012) Acoustic characteristics of voiceless bellowing typical of bovine rabies. Am J Trop Med Hyg 86(3):528–530

Dertzbaugh MT, Elson CO (1993) Comparative effectiveness of the cholera toxin B subunit and alkaline phosphatase as carriers for oral vaccines. Infect Immun 61(1):48–55

Devleesschauwer B, Aryal A, Sharma BK, Ale A, Declercq A, Depraz S, Gaire TN, Gongal G, Karki S, Pandey BD, Pun SB, Duchateau L, Dorny P, Speybroeck N (2016) Epidemiology, impact and control of rabies in Nepal: a systematic review. PLoS Negl Trop Dis. 10(2): e0004461

Dietzschold B, Wang HH, Rupprecht CE, Celis E, Tollis M, Ertl H, Heber-Katz E, Koprowski H (1987) Induction of protective immunity against rabies by immunization with rabies virus ribonucleoprotein. Proc Natl Acad Sci U S A. 84(24):9165–9169

Dietzschold B, Gore M, Marchadier D, Niu HS, Bunschoten HM, Otvos L, Wunner WH, Ertl HC, Osterhaus AD, Koprowski H (1990) Structural and immunological characterization of a linear virus-neutralizing epitope of the rabies virus glycoprotein and its possible use in a synthetic vaccine. J Virol 64(8):3804–3809

Dietzschold B, Li J, Faber M, Schnell M (2008) Concepts in the pathogenesis of rabies. Future Virol 3(5):481–490

Drake PM, Szeto TH, Paul MJ, Teh AY, Ma JK (2017) Recombinant biologic products versus nutraceuticals from plants–a regulatory choice? Br J Clin Pharmacol 83(1):82–87

Eggerbauer E, de Benedictis P, Hoffmann B, Mettenleiter TC, Schlottau K, Ngoepe EC, Sabeta CT, Freuling CM, Müller T (2016) Evaluation of six commercially available rapid immunochromatographic tests for the diagnosis of rabies in brain material. PLoS Negl Trop Dis. 10(6):e0004776

Ertl HC (2009) Novel vaccines to human rabies. PLoS Negl Trop Dis. 3(9):e515

Ertl HC, Dietzschold B, Gore M, Otvos L Jr, Larson JK, Wunner WH, Koprowski H (1989) Induction of rabies virus-specific T-helper cells by synthetic peptides that carry dominant T-helper cell epitopes of the viral ribonucleoprotein. J Virol 63(7):2885–2892

European Medicines Agency (2016) Guideline on the quality of biological active substances produced by stable transgene expression in higher plants [online]. Available at www.ema.europa.eu/docs/en_GB/document_library/Scientific_guideline/2009/09/WC500003154.pdf. Last accessed 29 June 2016

Favre H, Valette L, Précausta P, Roulet C, Brun A, Terré J, Fontaine J, Stellmann C (1976) Anti-foot-and-mouth vaccines combined with other bacterial or viral vaccines. Dev Biol Stand 35:409–428

Fehlner-Gardiner C, Nadin-Davis S, Armstrong J, Muldoon F, Bachmann P, Wandeler A (2008) Era vaccine-derived cases of rabies in wildlife and domestic animals in Ontario, Canada, 1989–2004. J Wildl Dis 44(1):71–85

Fekadu M (1982) Rabies in Ethiopia. Am J Epidemiol 115(2):266–273

Fekadu M, Shaddock JH, Baer GM (1982) Excretion of rabies virus in the saliva of dogs. J Infect Dis 145(5):715–719

Feng Y, Shi Y, Yu M, Xu W, Gong W, Tu Z, Ding L, He B, Guo H, Tu C (2016) Livestock rabies outbreaks in Shanxi province. China Arch Virol 161(10):2851–2854

Fernando BG, Yersin CT, José CB, Paola ZS (2016) Predicted 3D model of the rabies virus glycoprotein trimer. Biomed Res Int 2016:1674580

Filho OA, Megid J, Geronutti L, Ratti J Jr, Almeida MFA, Kataoka APAG, Martorelli LFA (2012) Vaccine immune response and interference of colostral antibodies in calves vaccinated against rabies at 2, 4 and 6 months of age born from antirabies revaccinated females. Res Vet Sci 92(3):396–400

Fitzpatrick JL, Dyer JL, Blanton JD, Kuzmin IV, Rupprecht CE (2014) Rabies in rodents and lagomorphs in the United States, 1995–2010. J Am Vet Med Assoc 245(3):333–337

Foley HD, McGettigan JP, Siler CA, Dietzschold B, Schnell MJ (2000) A recombinant rabies virus expressing vesicular stomatitis virus glycoprotein fails to protect against rabies virus infection. Proc Natl Acad Sci USA. 2000 Dec19; 97(26):14680–14685

Fontaine J, Favre H, Valette L, Précausta P, Roulet C, Brun A, Terré J (1976) Combined vaccines including 1 or several anti-foot-and-mouth valences. Dev Biol Stand 35:395–407

Fontana D, Etcheverrigaray M, Kratje R, Prieto C (2016) Development of rabies virus-like particles for vaccine applications: production, characterization, and protection studies. Methods Mol Biol 1403:155–166

Fooks AR, Brookes SM, Johnson N, McElhinney LM, Hutson AM (2003) European bat lyssaviruses: an emerging zoonosis. Epidemiol Infect 131(3):1029–1039

Forthal Donald N, Gach Johannes S, Landucci Gary, Jez Jakub, Strasser Richard, Kunert Renate, Steinkellner Herta (2010) Fc-glycosylation influences Fcγ receptor binding and cell-mediated anti-HIV activity of monoclonal antibody 2G12. J Immunol 185:6876–6882

Fry TL, Vandalen KK, Duncan C, Vercauteren K (2013) The safety of ONRAB® in select non-target wildlife. Vaccine 31(37):3839–3842

Fuenzalida E, Díaz AM, Rivenson S (1978) Suckling mouse brain rabies vaccine supplemented with an adjuvant. Its use in cattle. Rev Asoc Argent Microbiol. 1978 May–Aug; 10(2):47–53

Gadaga BM, Etter EM, Mukamuri B, Makwangudze KJ, Pfukenyi DM, Matope G (2016) Living at the edge of an interface area in Zimbabwe: cattle owners, commodity chain and health workers' awareness, perceptions and practices on zoonoses. BMC Public Health 16(1):84

Gautret P, Lim PL, Shaw M, Leder K (2011) Rabies post-exposure prophylaxis in travellers returning from Bali, Indonesia, November 2008 to March 2010. Clin Microbiol Infect 17(3): 445–447

Gibson AD, Handel IG, Shervell K, Roux T, Mayer D, Muyila S, Maruwo GB, Nkhulungo EM, Foster RA, Chikungwa P, Chimera B, Bronsvoort BM, Mellanby RJ, Gamble L (2016) The vaccination of 35,000 dogs in 20 working days using combined static point and door-to-door methods in Blantyre, Malawi. PLoS Negl Trop Dis 10(7):e0004824

Gigante CM, Dettinger L, Powell JW, Seiders M, Condori REC, Griesser R, Okogi K, Carlos M, Pesko K, Breckenridge M, Simon EMM, Chu MYJV, Davis AD, Brunt SJ, Orciari L, Yager P, Carson WC, Hartloge C, Saliki JT, Sanchez S, Deldari M, Hsieh K, Wadhwa A, Wilkins K, Peredo VY, Rabideau P, Gruhn N, Cadet R, Isloor S, Nath SS, Joseph T, Gao J, Wallace R, Reynolds M, Olson VA, Li Y, Kalendar R (2018) Multi-site evaluation of the LN34 pan-lyssavirus real-time RT-PCR assay for post-mortem rabies diagnostics. PLOS ONE 13(5): e0197074

Gilbert A, Greenberg L, Moran D, Alvarez D, Alvarado M, Garcia DL, Peruski L (2015) Antibody response of cattle to vaccination with commercial modified live rabies vaccines in Guatemala. Prev Vet Med 118(1):36–44

Gnanadurai CW, Zhou M, He W, Leyson CM, Huang CT, Salyards G, Harvey SB, Chen Z, He B, Yang Y, Hooper DC, Dietzchold B, Fu ZF (2013) Presence of virus neutralizing antibodies in cerebral spinal fluid correlates with non-lethal rabies in dogs. PLoS Negl Trop Dis 7(9):e2375

Gomme EA, Wanjalla CN, Wirblich C, Schnell MJ (2011) Rabies virus as a research tool and viral vaccine vector. Adv Virus Res 79:139–164

Goudsmit J, Marissen WE, Weldon WC, Niezgoda M, Hanlon CA, Rice AB, de Kruif J, Dietzschold B, Bakker AB, Rupprecht CE (2006) Comparison of an anti-rabies human monoclonal antibody combination with human polyclonal anti-rabies immune globulin. J Infect Dis 193(6):796–801

Gräni R, Wandeler A, Steck F, Rösli R (1978) Rabies in a veterinarian. Schweiz MedWochenschr 108(16):593–597

Greenhall AM, Schmidt U (eds) (1988) The natural history of vampire bats. CRC Press, Boca Raton, Fl

Gunawardena PS, Marston DA, Ellis RJ, Wise EL, Karawita AC, Breed AC, McElhinney LM, Johnson N, Banyard AC, Fooks AR (2016) Lyssavirus in Indian Flying Foxes, Sri Lanka. Emerg Infect Dis 22(8):1456–1459

Hampson K, Coudeville L, Lembo T, Sambo M, Kieffer A, Attlan M, Barrat J, Blanton JD, Briggs DJ, Cleaveland S, Costa P, Freuling CM, Hiby E, Knopf L, Leanes F, Meslin FX, Metlin A, Miranda ME, Müller T, Nel LH, Recuenco S, Rupprecht CE, Schumacher C, Taylor L, Vigilato MA, Zinsstag J, Dushoff J (2015) Global alliance for rabies control partners for rabies prevention. Estimating the global burden of endemic canine rabies. PLoS Negl Trop Dis 9(4):e0003709

Haq TA, Mason HS, Clements JD, Arntzen CJ (1995) Oral immunization with a recombinant bacterial antigen produced in transgenic plants. Science 268:714

Hartmann K, Day MJ, Thiry E, Lloret A, Frymus T, Addie D, Boucraut-Baralon C, Egberink H, Gruffydd-Jones T, Horzinek MC, Hosie MJ, Lutz H, Marsilio F, Pennisi MG, Radford AD, Truyen U, Möstl K (2015) European advisory board on cat diseases. Feline injection-site sarcoma: ABCD guidelines on prevention and management. J Feline Med Surg 17(7):606–613

Harvey AM, Watson JL, Brault SA, Edman JM, Moore SM, Kass PH, Wilson WD (2016) Duration of serum antibody response to rabies vaccination in horses. J Am Vet Med Assoc 249(4):411–418

Hayman DT, Fooks AR, Rowcliffe JM, McCrea R, Restif O, Baker KS, Horton DL, Suu-Ire R, Cunningham AA, Wood JL (2012) Endemic Lagos bat virus infection in *Eidolon helvum*. Epidemiol Infect 140(12):2163–2171

Hermann JR, Fry AM, Siev D, Slate D, Lewis C, Gatewood DM (2011) Stability of vaccinia-vectored recombinant oral rabies vaccine under field conditions: a 3-year study. Can J Vet Res 75(4):278–284

Hermann J, Fry A, Reising M, Patterson P, Siev D, Gatewood D (2012) Rabies vaccine standards: comparison of the 5th and 6th WHO international reference standards to the USDA veterinary reference standard. Vaccine 30(48):6892–6896

Hiatt A, Bohorova N, Bohorov O, Goodman C, Kim D, Pauly MH, Velasco J et al (2014) Glycan variants of a respiratory syncytial virus antibody with enhanced effector function and in vivo efficacy. Proc Natl Acad Sci USA 111:5992–5997

Hicks DJ, Fooks AR, Johnson N (2012) Developments in rabies vaccines. Clin Exp Immunol 169(3):199–204

Höper D, Freuling CM, Müller T, Hanke D, von Messling V, Duchow K, Beer M, Mettenleiter TC (2015) High definition viral vaccine strain identity and stability testing using full-genome population data–the next generation of vaccine quality control. Vaccine 33(43):5829–5837

Horton DL, Banyard AC, Marston DA, Wise E, Selden D, Nunez A, Hicks D, Lembo T, Cleaveland S, Peel AJ, Kuzmin IV, Rupprecht CE, Fooks AR (2014) Antigenic and genetic characterization of a divergent African virus, Ikoma lyssavirus. J GenVirol 95(Pt 5):1025–1032

Hostnik P, Picard-Meyer E, Rihtarič D, Toplak I, Cliquet F (2014) Vaccine-induced rabies in a red fox (*Vulpes vulpes*): isolation of vaccine virus in brain tissue and salivary glands. J Wildl Dis 50(2):397–401

Hu SC, Hsu CL, Lee MS, Tu YC, Chang JC, Wu CH, Lee SH, Ting LJ, Tsai KR, Cheng MC, Tu WJ, Hsu WC (2018) Lyssavirus in Japanese Pipistrelle, Taiwan. Emerg Infect Dis 24(4): 782–785

Hudson LC, Weinstock D, Jordan T, Bold-Fletcher NO (1996a) Clinical features of experimentally induced rabies in cattle and sheep. Zentralbl Veterinarmed B 43(2):85–95

Hudson LC, Weinstock D, Jordan T, Bold-Fletcher NO (1996b) Clinical presentation of experimentally induced rabies in horses. Zentralbl Veterinarmed B 43(5):277–285

Hundal JS, Sodhi SS, Gupta A, Singh J, Chahal US (2016) Awareness, knowledge, and risks of zoonotic diseases among livestock farmers in Punjab. Vet World 9(2):186–191

Hutter SE, Brugger K, Sancho Vargas VH, González R, Aguilar O, León B, Tichy A, Firth CL, Rubel F (2016) Rabies in Costa Rica: documentation of the surveillance program and the endemic situation from 1985 to 2014. Vector Borne Zoonotic Dis 16(5):334–341

Jiang Y, Yu X, Wang L, Lu Z, Liu H, Xuan H, Hu Z, Tu C (2008) An outbreak of pig rabies in Hunan province, China. Epidemiol Infect 136(4):504–508

Johnson N, Cunningham AF, Fooks AR (2010) The immune response to rabies virus infection and vaccination. Vaccine 28(23):3896–3901

Johnson N, Aréchiga-Ceballos N, Aguilar-Setien A (2014) Vampire bat rabies: ecology, epidemiology and control. Viruses 6(5):1911–1928

Juarez P, Virdi V, Depicker A, Orzaez D (2016) Biomanufacturing of protective antibodies and other therapeutics in edible plant tissues for oral applications. Plant biotechnol J. 2016 Feb 1

Kang H, Qi Y, Wang H, Zheng X, Gao Y, Li N, Yang S, Xia X (2015) Chimeric rabies virus-like particles containing membrane-anchored GM-CSF enhances the immune response against rabies virus. Viruses 7:1134–1152

Kgaladi J, Wright N, Coertse J, Markotter W, Marston D, Fooks AR, Freuling CM, Müller TF, Sabeta CT, Nel LH (2013) Diversity and epidemiology of Mokola virus. PLoS Negl Trop Dis. 7(10):e2511

Klepfer SR, Debouck C, Uffelman J, Jacobs P, Bollen A, Jones EV (1993) Characterization of rabies glycoprotein expressed in yeast. Arch Virol 128(3–4):269–286

Ko K, Tekoah Y, Rudd PM, Harvey DJ, Dwek RA, Spitsin S, Hanlon CA, Rupprecht C, Dietzschold B, Golovkin M, Koprowski H (2003) Function and glycosylation of plant-derived antiviral monoclonal antibody. Proc Natl Acad Sci U S A 100(13):8013–8018

Koraka P, Bosch BJ, Cox M, Chubet R, Amerongen GV, Lövgren-Bengtsson K, Martina BE, Roose J, Rottier PJ, Osterhaus AD (2014) A recombinant rabies vaccine expressing the trimeric form of the glycoprotein confers enhanced immunogenicity and protection in outbred mice. Vaccine. 2014 Aug 6; 32(36):4644–4650

Korns RF, Zeissig A (1948) Dog, fox, and cattle rabies in New York State; evaluation of vaccination in dogs. Am J Publ Hlth Nations Hlth. 1948 Jan; 38(1 Pt1):50–65

Kristina LE, Luthar Z (2016) Production of vaccines for treatment of infectious diseases by transgenic plants. Acta agriculturae Slovenica. 2016 Apr 6; 107(1):191–217

Kurt-Jones EA, Popova L, Kwinn L, Haynes LM, Jones LP, Tripp RA, Walsh EE, Freeman MW, Golenbock DT, Anderson LJ, Finberg RW (2000) Pattern recognition receptors TLR4 and CD14 mediate response to respiratory syncytial virus. Nat Immunol 1(5):398–401

Kushnir N, Streatfield SJ, Yusibov V (2012) Virus-like particles as a highly efficient vaccine platform: diversity of targets and production systems and advances in clinical development. Vaccine 31(1):58–83

Kuzmin IV, Hughes GJ, Botvinkin AD, Orciari LA, Rupprecht CE (2005) Phylogenetic relationships of Irkut and West Caucasian bat viruses within the Lyssavirus genus and suggested quantitative criteria based on the N gene sequence for lyssavirus genotype definition. Virus Res 111(1):28–43

Kuzmin IV, Wu X, Tordo N, Rupprecht CE (2008) Complete genomes of Aravan, Khujand, Irkut and West Caucasian bat viruses, with special attention to the polymerase gene and non-coding regions. Virus Res 136(1–2):81–90

Kuzmin IV, Turmelle AS, Agwanda B, Markotter W, Niezgoda M, Breiman RF, Rupprecht CE (2011) Commerson's leaf-nosed bat (*Hipposideros commersoni*) is the likely reservoir of Shimoni bat virus. Vector Borne Zoonotic Dis 11(11):1465–1470

Kuzmina NA, Kuzmin IV, Ellison JA, Rupprecht CE (2013) Conservation of binding epitopes for monoclonal antibodies on the rabies virus glycoprotein. J Antiviral Antiretrovirals 5:37–43

Kwon KC, Verma D, Singh ND, Herzog R, Daniell H (2013) Oral delivery of human biopharmaceuticals, autoantigens and vaccine antigens bioencapsulated in plant cells. Adv Drug Deliv Rev 65(6):782–799

Laere E, Ling AP, Wong YP, Koh RY, Mohd Lila MA, Hussein S (2016) Plant-based vaccines: production and challenges. J Bot. 2016 Apr 13

Lafon M, Wiktor TJ, Macfarlan RI (1983) Antigenic sites on the CVS rabies virus glycoprotein: analysis with monoclonal antibodies. J Gen Virol 64(Pt 4):843–851

Lafon M, Scott-Algara D, Marche PN, Cazenave PA, Jouvin-Marche E (1994) Neonatal deletion and selective expansion of mouse T cells by exposure to rabies virus nucleocapsid superantigen. J Exp Med 180(4):1207–1215

Lathe CL (1994) *Neospora caninum* in British dogs. Vet Rec 134(20):532

Lavan RP, King AI, Sutton DJ, Tunceli K (2017) Rationale and support for a one health program for canine vaccination as the most cost-effective means of controlling zoonotic rabies in endemic settings. Vaccine 35(13):1668–1674

Lee DN, Papeş M, Van den Bussche RA (2012) Present and potential future distribution of common vampire bats in the Americas and the associated risk to cattle. PLoS ONE 7(8): e42466

Lewis CE, Fry AM, Hermann JR, Siev D, Dusek DM, Gatewood DM (2012) Potency testing of veterinary rabies vaccines: replacement of challenge by in vitro testing: considerations for development of alternative assays. Dev Biol (Basel) 134:29–33

Li J, Faber M, Dietzschold B, Hooper DC (2011) The role of toll-like receptors in the induction of immune responses during rabies virus infection. Adv Virus Res 79:115–126

Liu Y, Zhang S, Ma G, Zhang F, Hu R (2008) Efficacy and safety of a live canine adenovirus-vectored rabies virus vaccine in swine. Vaccine 26(42):5368–5372

Liu Y, Zhang S, Zhao J, Zhang F, Hu R (2013) Isolation of Irkut virus from a *Murina leucogaster* bat in China. PLoS Negl Trop Dis 7(3):e2097

Liu X, Yang Y, Sun Z, Chen J, Ai J, Dun C, Fu ZF, Niu X, Guo X (2014) A recombinant rabies virus encoding two copies of the glycoprotein gene confers protection in dogs against a virulent challenge. PLoS ONE 9(2):e87105

Liu Y, Zhang HP, Zhang SF, Wang JX, Zhou HN, Zhang F, Wang YM, Ma L, Li N, Hu RL (2016) Rabies outbreaks and vaccination in domestic camels and cattle in northwest China. PLoS Negl Trop Dis 10(9):e0004890

Lodmell DL, Esposito JJ, Ewalt LC (2004) Live vaccinia-rabies virus recombinants, but not an inactivated rabies virus cell culture vaccine, protect B-lymphocyte-deficient A/WySnJ mice against rabies: considerations of recombinant defective poxviruses for rabies immunization of immunocompromised individuals. Vaccine 22(25):3329–3333

Lord RD, Fuenzalida E, Delpietro H, Larghi OP, de Díaz AM, Lázaro L (1975) Observations on the epizootiology of vampire bat rabies. Bull Pan Am Health Organ 9(3):189–195

Loza-Rubio E, Rojas-Anaya E, López J, Olivera-Flores MT, Gómez-Lim M, Tapia-Pérez G (2012) Induction of a protective immune response to rabies virus in sheep after oral immunization with transgenic maize, expressing the rabies virus glycoprotein. Vaccine 30(37):5551–5556

Lycke N, Bromander AK, Holmgren J (1989) Role of local IgA antitoxin-producing cells for intestinal protection against cholera toxin challenge. Int Arch Allergy Immunol 88(3):273–279

Macfarlan RI, Dietzschold B, Wiktor TJ, Kiel M, Houghten R, Lerner RA, Sutcliffe JG, Koprowski H (1984) T cell responses to cleaved rabies virus glycoprotein and to synthetic peptides. J Immunol. 133(5):2748–2752

Madhusudana SN, Subha S, Thankappan U, Ashwin YB (2012) Evaluation of a direct rapid immunohistochemical test (dRIT) for rapid diagnosis of rabies in animals and humans. Virol Sin 27(5):299–302

Malek LT, Soostmeyer G, Garvin T, James E (1984) The rabies glycoprotein gene is expressed in *Escherichia coli* as denatured polypeptide. In: Channock RM, Lerner RA (eds) Modem approaches in vaccines: molecular and chemical basis of virus virulence and immunity. Cold Spring Harbor Laboratory Press, Cold Spring Harbor, NY. 1:203–208

Malerczyk C, Briggs DJ, Dreesen DW, Banzhoff A (2007) Duration of immunity: an anamnestic response 14 years after rabies vaccination with purified chick embryo cell rabies vaccine. J Travel Med. 2007 Jan–Feb; 14(1):63–64

Malerczyk C, Selhorst T, Tordo N, Moore S, Müller T (2009) Antibodies induced by vaccination with purified chick embryo cell culture vaccine (PCECV) cross-neutralize non-classical bat lyssavirus strains. Vaccine 27(39):5320–5325

Mansfield KL, Andrews N, Goharriz H, Goddard T, McElhinney LM, Brown KE, Fooks AR (2016) Rabies pre-exposure prophylaxis elicits long-lasting immunity in humans. Vaccine 34 (48):5959–5967

Marasco WA, Sui J (2007) The growth and potential of human antiviral monoclonal antibody therapeutics. Nat Biotechnol 25(12):1421–1434

Marissen WE, Kramer RA, Rice A, Weldon WC, Niezgoda M, Faber M, Slootstra JW, Meloen RH, Clijsters-van der Horst M, Visser TJ, Jongeneelen M (2005) Novel rabies virus-neutralizing epitope recognized by human monoclonal antibody: fine mapping and escape mutant analysis. J Virol 79(8):4672–4678

Martin RJ, Schnurrenberger PR, Walker JF (1982) Exposure to rabies—an occupational hazard for veterinarians. Can Vet J 23(11):317–322

Mayen F (2003) Haematophagous bats in Brazil, their role in rabies transmission, impact on public health, livestock industry and alternatives to an indiscriminate reduction of bat population. J Vet Med B Infect Dis Vet Public Health 50(10):469–472

McGarvey PB, Hammond J, Dienelt MM, Hooper DC, Fu ZF, Dietzschold B, Koprowski H, Michaels FH (1995) Expression of the rabies virus glycoprotein in transgenic tomatoes. Biotechnology (NY) 13(13):1484–1487

McLaughlin K (2016) Scandal clouds China's global vaccine ambitions. Science 352(6285):506

Meltzer MI, Rupprecht CE (1998) A review of the economics of the prevention and control of rabies. Part 2: Rabies in dogs, livestock and wildlife. Pharmacoeconomics 14(5):481–498

Memish ZA, Assiri AM, Gautret P (2015) Rabies in Saudi Arabia: a need for epidemiological data. Int J Infect Dis 34:99–101

Miguens L (2007) The opinion of the production sector on the role of vaccines in the control and eradication of livestock diseases in Argentina. Rev Sci Tech. 2007 Aug; 26(2):479–483, 485–488

Miller RE, Fowler M (eds) (2012) Zoo and wild animal medicine: current therapy, vol 7. Elsevier, St. Louis, Missouri

Mitmoonpitak C, Limusanno S, Khawplod P, Tepsumethanon V, Wilde H (2002) Post-exposure rabies treatment in pigs. Vaccine 20(16):2019–2021

Modelska A, Dietzschold B, Sleysh N, Fu ZF, Steplewski K, Hooper DC, Koprowski H, Yusibov V (1998) Immunization against rabies with plant-derived antigen. Proc Natl Acad Sci USA 95(5):2481–2485

Monaco F, Franchi PM, Lelli R (2006) Studies on an inactivated vaccine against rabies virus in domestic animals. Dev Biol (Basel) 125:233–239

Monroe BP, Yager P, Blanton J, Birhane MG, Wadhwa A, Orciari L, Petersen B, Wallace R (2016) Rabies surveillance in the United States during 2014. J Am Vet Med Assoc 248(7): 777–788

Moore SM, Wilkerson MJ, Davis RD, Wyatt CR, Briggs DJ (2006) Detection of cellular immunity to rabies antigens in human vaccinees. J Clin Immunol 26(6):533–545

Moore MC, Davis RD, Kang Q, Vahl CI, Wallace RM, Hanlon CA, Mosier DA (2015) Comparison of anamnestic responses to rabies vaccination in dogs and cats with current and out-of-date vaccination status. J Am Vet Med Assoc. 2015 Jan15; 246(2):205–211

Morgeaux S, Poirier B, Ragan CI, Wilkinson D, Arabin U, Guinet-Morlot F, Levis R, Meyer H, Riou P, Shaid S, Volokhov D, Tordo N, Chapsal JM (2017) Replacement of in vivo human rabies vaccine potency testing by in vitro glycoprotein quantification using ELISA—Results of an international collaborative study. Vaccine 35(6):966–971

Morón G, Rueda P, Casal I, Leclerc C (2002) CD8α2 CD11b + dendritic cells present exogenous virus-like particles to CD8 + T cells and subsequently express CD8α and CD205 molecules. J Exp Med 195(10):1233–1245

Müller T, Dietzschold B, Ertl H, Fooks AR, Freuling C, Fehlner-Gardiner C, Kliemt J, Meslin FX, Franka R, Rupprecht CE, Tordo N (2009) Correction: development of a mouse monoclonal antibody cocktail for post-exposure rabies Prophylaxis in humans. PLOS Negl Trop Dis 3 (11):10–371

Müller T, Freuling CM, Wysocki P, Roumiantzeff M, Freney J, Mettenleiter TC, Vos A (2015) Terrestrial rabies control in the European Union: historical achievements and challenges ahead. Vet J 203(1):10–17

National Association of State Public Health Veterinarians (NASPHV) (2016) Compendium of animal rabies prevention and control committee, Brown CM, Slavinski S, Ettestad P, Sidwa TJ, Sorhage FE. Compendium of animal rabies prevention and control. J Am Vet Med Assoc. 2016 Mar 1; 248(5):505–517

Nokireki T, Tammiranta N, Kokkonen UM, Kantala T, Gadd T (2018) Tentative novel lyssavirus in a bat in Finland. Transbound Emerg Dis 65(3):593–596

Nolden T, Banyard AC, Finke S, Fooks AR, Hanke D, Höper D, Horton DL, Mettenleiter TC, Müller T, Teifke JP, Freuling CM (2014) Comparative studies on the genetic, antigenic and pathogenic characteristics of Bokeloh bat lyssavirus. J Gen Virol 95(Pt 8):1647–1653

Ogino M, Ito N, Sugiyama M, Ogino T (2016) The rabies virus L protein catalyzes mRNA capping with GDP polyribonucleotidyltransferase activity. Viruses. 2016 May 21; 8(5)

Okolo MI (1989) Vaccine-induced rabies infection in rural dogs in Anambra State, Nigeria. Microbios 57(231):105–112

Ortiz-Prado E, Galarza C, Cornejo León F, Ponce J (2014) Access to drugs and the situation of the pharmaceutical market in Ecuador. Revista Panamericana de Salud Pública 36(1):57–62

Palanisamy R, Ramanna BC, Ananda Rao K, Srinivasan VA (1992) Combined vaccination of cattle against FMD and rabies. Microbiologica 15(1):45–49

Pastoret PP, Vanderplasschen A (2003) Poxviruses as vaccine vectors. Comp Immunol Microbiol Infect Dis 26(5–6):343–355

Pawan JL (1959) The transmission of paralytic rabies in Trinidad by the vampire bat (*Desmodus rotundus murinus* Wagner. Caribb Med J 21:110–136

Pépin M, Blancou J, Aubert MF (1984) Experimental bovine rabies: susceptibility, symptoms, humoral immunologic reactions, lesions and virus excretion. Ann Rech Vet. 15(3):325–333

Perea Arango I, Loza Rubio E, Rojas Anaya E, Olivera Flores T, Gonzalez de la Vara L, Gómez Lim MA (2008) Expression of the rabies virus nucleoprotein in plants at high-levels and evaluation of immune responses in mice. Plant Cell Rep 27(4):677–685

Petricciani JC, Grachev VP, Sizaret PP, Regan PJ (1989) Vaccines: obstacles and opportunities from discovery to use. Rev Infect Dis. 1989 May–Jun; 11 Suppl 3:S524–5249

Piza AT, Pieri KM, Lusa GM, Caporale GM, Terreran MT, Machado LA, Zanetti CR (2002) Effect of the contents and form of rabies glycoprotein on the potency of rabies vaccination in cattle. Mem Inst Oswaldo Cruz 97(2):265–268

Prosniak M, Faber M, Hanlon CA, Rupprecht CE, Hooper DC, Dietzschold B (2003) Development of a cocktail of recombinant-expressed human rabies virus-neutralizing monoclonal antibodies for postexposure prophylaxis of rabies. J Inf Dis 188(1):53–56

Quirozr DE, Sutmoeller P, Barroeta M (1964) Factors associated with anaphylactic reactions to chicken embryo foot-and-mouth disease vaccine and Flury Rabies vaccine in cattle of Venezuela. Am J Vet Res 25:1627–1634

Ramya R, Subramanian BM, Sivakumar V, Senthilkumar RL, Rao KS, Srinivasan VA (2011) Expression and solubilization of insect cell-based rabies virus glycoprotein and assessment of its immunogenicity and protective efficacy in mice. Clin Vaccine Immunol 18(10):1673–1679

Rappuoli R (2014) Inner Workings: 1885, the first rabies vaccination in humans. Proc Natl Acad Sci USA 111(34):12273

Reculard R (1996) Cell-culture vaccines for veterinary use. In: Meslin FX, Kaplan MM, Koprowski H (eds) Laboratory techniques in rabies, 4th edn. WHO, Geneva, pp 314–323

Robardet E, Picard-Meyer E, Dobroštana M, Jaceviciene I, Mähar K, Muižniece Z, Pridotkas G, Masiulis M, Niin E, Olševskis E, Cliquet F (2016) Rabies in the Baltic States: decoding a process of control and elimination. PLoS Negl Trop Dis 10(2):e0004432

Rodrigues da Silva AC, Caporale GM, Gonçalves CA, Targueta MC, Comin F, Zanetti CR, Kotait I (2000) Antibody response in cattle after vaccination with inactivated and attenuated rabies vaccines. Rev Inst Med Trop Sao Paulo. 2000 Mar–Apr; 42(2):95–98

Rojas-Anaya E, Loza-Rubio E, Olivera-Flores MT, Gomez-Lim M (2009) Expression of rabies virus G protein in carrots (*Daucus carota*). Transgenic Res 18:911–919

Rotz LD, Hensley JA, Rupprecht CE, Childs JE (1998) Large-scale human exposures to rabid or presumed rabid animals in the United States: 22 cases (1990–1996). J Am Vet Med Assoc 212 (8):1198–1200

Roy S, Mukherjee S, Pawar S, Chowdhary A (2016) Evaluation of in vitro antiviral activity of datura metel Linn. Against Rabies Virus. Pharmacognosy Res. 2016 Oct–Dec; 8(4):265–269

Rupprecht CE (2016) Rabies. In: Aiello SE (ed) Merck veterinary manual, 11th edn. Merck Publishing Group, Rahway, NJ, USA, pp 1302–1306

Rupprecht CE, Burgess GW (2015) Viral and vector zoonotic exploitation of a homo-sociome memetic complex. Clin Microbiol Infect 21(5):394–403

Rupprecht CE, Dietzschold B (1987) Perspectives on rabies virus pathogenesis. Lab Invest 57(6): 603–606

Rupprecht CE, Kuzmin IV (2015) Why we can prevent, control and possibly treat—but will not eradicate rabies. Future Virol 10(5):517–535

Rupprecht CE, Turmelle A, Kuzmin IV (2011) A perspective on lyssavirus emergence and perpetuation. Curr Opin Virol 1(6):662–670

Rupprecht CE, Nagarajan T, Ertl H (2016) Current status and development of vaccines and other biologics for human rabies prevention. Expert Rev Vaccines 15(6):731–749

Rupprecht C, Kuzmin I, Meslin F (2017) Lyssaviruses and rabies: current conundrums, concerns, contradictions and controversies. F1000Res. 2017 Feb 23; 6:184

Schiffelers MJ, Blaauboer B, Bakker W, Hendriksen C (2014) Replacing the NIH test for rabies vaccine potency testing: a synopsis of drivers and barriers. Biologicals 42(4):205–217

Schnee M, Vogel AB, Voss D, Petsch B, Baumhof P, Kramps T, Stitz L (2016) An mRNA vaccine encoding rabies virus glycoprotein induces protection against lethal infection in mice and correlates of protection in adult and newborn pigs. PLoS Negl Trop Dis 10(6):e0004746

Scott T, Hasse R, Nel L (2012) Rabies in kudu (*Tragelaphus strepsiceros*). Berl Munch Tierarztl Wochenschr. 2012 May–Jun; 125(5–6):236–241

Shahaid N, Daniell H (2016) Plant based oral vaccines against zoonotic and non-zoonotic diseases. Plant Biotechnol J. 2016 Jul 1

Shakin-Eshleman SH, Remaley AT, Eshleman JR, Wunner WH, Spitalnik SL (1992) N-linked glycosylation of rabies virus glycoprotein. Individual sequons differ in their glycosylation efficiencies and influence on cell surface expression. J Biol Chem 267(15):10690–10698

Shinwari MW, Annand EJ, Driver L, Warrilow D, Harrower B, Allcock RJ, Pukallus D, Harper J, Bingham J, Kung N, Diallo IS (2014) Australian bat lyssavirus infection in two horses. Vet Microbiol 173(3–4):224–231

Sigoillot-Claude C, Battaglio M, Fiorucci M, Gillet D, Vimort AS, Giraud Y, Laurent S, Vaganay A, Poulet H (2015) A versatile in vitro ELISA test for quantification and quality testing of infectious, inactivated and formulated rabies virus used in veterinary monovalent or combination vaccine. Vaccine 33(32):3843–3849

Sijmon PC, Dekker BMM, Schrammeijer B, Verwoerd TC, van den Elzen PMJ, Hoekema A (1990) Production of correctly processed human serum albumin in transgenic plants. Biotechnology 8:217–221

Sikes RK (1970) Guidelines for the control of rabies. Am J Publ Hlth, Nations Health 60(6): 1133–1138

Simani S, Fayaz A, Rahimi P, Eslami N, Howeizi N, Biglari P (2012) Six fatal cases of classical rabies virus without biting incidents, Iran 1990–2010. J Clin Virol 54(3):251–254

Singh V (1999) Disposable bioreactor for cell culture using wave-induced agitation. Cytotechnology 30:149–158

Singh CK, Sandhu BS (2008) Rabies in South Asia: epidemiological investigations and clinical perspective. Dev Biol (Basel) 131:133–136

Singh A, Srivastava S, Chouksey A, Singh Panwar B, Verma PC, Roy S, Singh PK, Saxena G, Tuli R (2015) Expression of rabies glycoprotein and ricin toxin B chain (RGP-RTB) fusion protein in tomato hairy roots: a step towards oral vaccination for rabies. Mol Biotechnol 57:359–370

Slate D, Algeo TP, Nelson KM, Chipman RB, Donovan D, Blanton JD, Niezgoda M, Rupprecht CE (2009) Oral rabies vaccination in North America: opportunities, complexities, and challenges. PLoS Negl Trop Dis. 3(12):e549

Smith CM, Fry SC, Gough KC, Patel AJ, Glenn S, Goldrick M, Roberts IS, Whitelam GC, Andrew PW (2014) Recombinant plants provide a new approach to the production of bacterial polysaccharide for vaccines. PLoS ONE 9(2):e88144

Song H, Zhou L, Fang W, Li Y, Wang X, Fang H, Li X, Wu M, Qiu B (2004) High-level expression of codon optimized foot-and-mouth disease virus complex epitopes and cholera toxin B subunit chimera in *Hansenula polymorpha*. Biochem Biophys Res Commun 315 (1):235–239

Soulebot JP, Précausta P, Brun A, Chappuis G, Terré J, Favre H, Petermann HG (1978) Prevention of animal rabies by an attenuated monovalent or mixed vaccine. Dev Biol Stand 41:389–399

Starodubova ES, Preobrazhenskaia OV, Kuzmenko YV, Latanova AA, Yarygina EI, Karpov VL (2015) Rabies vaccines: current status and prospects for development. Mol Biol 49(4):513–519

Stein LT, Rech RR, Harrison L, Brown CC (2010) Immunohistochemical study of rabies virus within the central nervous system of domestic and wildlife species. Vet Pathol 47(4):630–633

Stoner-Duncan B, Streicker DG, Tedeschi CM (2014) Vampire bats and rabies: toward an ecological solution to a public health problem. PLoS Negl Trop Dis. 2014 Jun19; 8(6):e2867

Strasser R, Stadlmann J, Schähs M, Stiegler G, Quendler H, Mach L, Glössl J, Weterings K, Pabst M, Steinkellner H (2008) Generation of glyco-engineered Nicotiana benthamiana for the production of monoclonal antibodies with a homogeneous human-like N-glycan structure. Plant Biotechnol J 6(4):392–402

Streicker DG, Allgeier JE (2016) Foraging choices of vampire bats in diverse landscapes: potential implications for land-use change and disease transmission. J Appl Ecol 53(4):1280–1288

Swai ES, Schoonman L, Daborn CJ (2010) Knowledge and attitude towards zoonoses among animal health workers and livestock keepers in Arusha and Tanga, Tanzania. Tanzan J Health Res 12(4):280–286

Swamy HS, Shankar SK, Chandra PS, Aroor SR, Krishna AS, Perumal VK (1984) Neurological complications due to beta-propiolactone (BPL)-inactivated antirabies vaccination: clinical, electrophysiological and therapeutic aspects. J Neurol Sci 63(1):111–128

Tariq WU, Shafi MS, Jamal S, Ahmad M (1991) Rabies in man handling infected calf. Lancet 337 (8751):1224

Thiptara A, Atwill ER, Kongkaew W, Chomel BB (2011) Epidemiologic trends of rabies in domestic animals in southern Thailand, 1994-2008. Am J Trop Med Hyg 85(1):138–145

Thompson MK, Fridy PC, Keegan S, Chait BT, Fenyö D, Rout MP (2016) Optimizing selection of large animals for antibody production by screening immune response to standard vaccines. J Immunol Methods 430:56–60

Tian D, Luo Z, Zhou M, Li M, Yu L, Wang C, Yuan J, Li F, Tian B, Sui B, Chen H, Fu ZF, Zhao L (2015) Critical role of K1685 and K1829 in the large protein of rabies virus in viral pathogenicity and immune evasion. J Virol 90(1):232–244

Topp E, Irwin R, McAllister T, Lessard M, Joensuu JJ, Kolotilin I, Conrad U, Stöger E, Mor T, Warzecha H, Hall JC (2016) The case for plant-made veterinary immunotherapeutics. Biotechnol Adv 34(5):597–604

Tsekoa TL, Lotter-Stark T, Buthelezi S, Chakauya E, Stoychev SH, Sabeta C, Shumba W, Phahladira B, Hume S, Morton J, Rupprecht CE (2016) Efficient in vitro and in vivo activity of glyco-engineered plant-produced rabies monoclonal antibodies E559 and 62-71-3. PLoS ONE 11(7):e0159313

van de Burgwal LH, Neevel AM, Pittens CA, Osterhaus AD, Rupprecht CE, Claassen E (2017) Barriers to innovation in human rabies prophylaxis and treatment: a causal analysis of insights from key opinion leaders and literature. Zoonoses Public Health. 2017 Mar 20. https://doi.org/10.1111/zph.12352. [Epub ahead of print]

van der Laan JW, Minor P, Mahoney R, Arntzen C, Shin J, Wood D (2006) WHO informal consultation on scientific basis for regulatory evaluation of candidate human vaccines from plants, Geneva, Switzerland, 24–25 January 2005. Vaccine 24(20):4271–4278

van Dolleweerd CJ, Teh AY, Banyard AC, Both L, Lotter-Stark HC, Tsekoa T, Phahladira B, Shumba W, Chakauya E, Sabeta CT, Gruber C (2014) Engineering, expression in transgenic plants and characterisation of e559, a rabies virus-neutralising monoclonal antibody. J Infect Dis 210(2):200–208

van Thiel PP, de Bie RM, Eftimov F, Tepaske R, Zaaijer HL, van Doornum GJ, Schutten M, Osterhaus AD, Majoie CB, Aronica E, Fehlner-Gardiner C, Wandeler AI, Kager PA (2009) Fatal human rabies due to Duvenhage virus from a bat in Kenya: failure of treatment with coma-induction, ketamine, and antiviral drugs. PLoS Negl Trop Dis. 3(7):e428

Vuta V, Picard-Meyer E, Robardet E, Barboi G, Motiu R, Barbuceanu F, Vlagioiu C, Cliquet F (2016) Vaccine-induced rabies case in a cow (*Bos taurus):* Molecular characterisation of vaccine strain in brain tissue. Vaccine. 2016 Aug 27. pii:S0264-410X(16)30670-3

Wallace RM, Niezgoda M, Waggoner EA, Blanton JD, Radcliffe RA (2016) Serologic response in eight alpacas vaccinated by extralabel use of a large animal rabies vaccine during a public health response to a rabid alpaca in South Carolina. J Am Vet Med Assoc 249(6):678–681

Wallace RM, Undurraga EA, Blanton JD, Cleaton J, Franka R (2017) Elimination of dog-mediated human rabies deaths by 2030: needs assessment and alternatives for progress based on dog vaccination. Front Vet Sci 10(4):9

Wang D, He X, Zhao Z, Feng Q, Lin R, Sun Y, Ding T, Xu F, Luo M, Zhan C (2015) Whole-brain mapping of the direct inputs and axonal projections of POMC and AgRP neurons. Front Neuroanat 27(9):40

Weyer J, Kuzmin IV, Rupprecht CE, Nel LH (2008) Cross-protective and cross-reactive immune responses to recombinant vaccinia viruses expressing full-length lyssavirus glycoprotein genes. Epidemiol Infect 136(5):670–678

Whetstone CA, Bunn TO, Emmons RW, Wiktor TJ (1984) Use of monoclonal antibodies to confirm vaccine-induced rabies in ten dogs, two cats, and one fox. J Am Vet Med Assoc 185 (3):285–288

Wilson PJ, Clark KA (2001) Postexposure rabies prophylaxis protocol for domestic animals and epidemiologic characteristics of rabies vaccination failures in Texas: 1995–1999. J Am Vet Med Assoc 218(4):522–525

World Health Organization (1996) Laboratory techniques in rabies, 4th ed. In: Meslin FX, Kaplan MM, Koprowski H (eds). WHO, Geneva

World Health Organization (2013) WHO expert consultation on Rabies. Second report. World Health Organ Tech Rep Ser 982:1–139

World Health Organization Guidelines on Good Agricultural and Collection Practices (GACP) for Medicinal Plants (2003) WHO, Geneva. Available at: http://apps.who.int/iris/bitstream/10665/42783/1/9241546271.pdf. Last accessed 29 June 2016

Wu X, Smith TG, Rupprecht CE (2011) From brain passage to cell adaptation: the road of human rabies vaccine development. Expert Rev Vaccines 10(11):1597–1608

Yakobson B, Taylor N, Dveres N, Rozenblut S, Tov BE, Markos M, Gallon N, Homer D, Maki J (2015) Cattle rabies vaccination—a longitudinal study of rabies antibody titres in an Israeli dairy herd. Prev Vet Med 121(1–2):170–175

Yang DK, Kim HH, Lee KW, Song JY (2013) The present and future of rabies vaccine in animals. Clin Exp Vaccine Res 2(1):19–25

Yang DK, Kim HH, Choi SS, Lee SH, Cho IS (2016) A recombinant rabies virus (ERAGS) for use in a bait vaccine for swine. Clin Exp Vaccine Res. 5(2):169–174

Yendo AC, de Costa F, Cibulski SP, Teixeira TF, Colling LC, Mastrogiovanni M, Soulé S, Roehe PM, Gosmann G, Ferreira FA, Fett-Neto AG (2016) A rabies vaccine adjuvanted with saponins from leaves of the soap tree (*Quillaja brasiliensis*) induces specific immune responses and protects against lethal challenge. Vaccine 34(20):2305–2311

Yusibov V, Modelska A, Steplewski K, Agadjanyan M, Weiner D, Hooper DC, Koprowski H (1997) Antigens produced in plants by infection with chimeric plant viruses immunize against rabies virus and HIV-1. Proc Natl Acad Sci USA 94(11):5784–5788

Yusibov V, Hooper DC, Spitsin SV, Fleysh N, Kean RB, Mikheeva T, Deka D, Karasev A, Cox S, Randall J, Koprowski H (2002) Expression in plants and immunogenicity of plant virus-based experimental rabies vaccine. Vaccine 20(25–26):3155–3164

Yusibov V, Streatfield SJ, Kushnir N (2011) Clinical development of plant-produced recombinant pharmaceuticals: vaccines, antibodies and beyond. Human vaccines 7(3):313–321

A Comprehensive Review of Toxoplasma Gondii Biology and Host-Cell Interaction: Challenges for a Plant-Based Vaccine

Valeria Sander, Sergio O. Angel and Marina Clemente

Abstract Toxoplasmosis is a worldwide-distributed infection caused by *Toxoplasma gondii*, which causes a wide range of clinical syndromes in humans, mammals and birds. *T. gondii* is considered a parasite of veterinary and medical importance, because it may cause abortion or congenital disease in its intermediate hosts. Despite the economic losses associated with *T. gondii* infection in farm animals and the socio-economic impact caused by this zoonotic disease in the human population, there is no effective treatment available for humans or animals able to eliminate the parasite from the host once the chronic infection has been established. The only commercial vaccine is the S48 strain of attenuated tachyzoites for use in sheep. However, this vaccine causes side effects, has a short life time and induces a short-term immunity. So far, no acellular vaccine against toxoplasmosis has been commercialized. In fact, future challenges include the development of an effective vaccine to prevent toxoplasmosis. Most parasitologists and vaccinologists agree that future efforts should be concentrated on developing multi-antigen vaccines and more efficient delivery systems able to express heterologous proteins abundantly as well as on searching for immunization schedules and adequate adjuvants to enhance the protective responses. To achieve this, platforms for the production of acellular vaccines based on the use of plants can have an important role.

Keywords Plant vaccine · Toxoplasmosis · Immune response
SAG1 · GRA4

V. Sander · S. O. Angel · M. Clemente (✉)
Instituto Tecnológico Chascomús, IIB-INTECH, CONICET-UNSAM, Chascomús, Provincia de Buenos Aires, Argentina
e-mail: mclemente@intech.gov.ar

J. MacDonald (ed.), *Prospects of Plant-Based Vaccines in Veterinary Medicine*,
https://doi.org/10.1007/978-3-319-90137-4_4

1 Toxoplasma Gondii

1.1 General Concepts

Toxoplasma gondii is an intracellular protozoan parasite member of the phylum Apicomplexa. It was found for the first time by Nicolle and Manceaux (1908) in the blood, spleen and liver of the rodent *Ctenodactylus gondii* and named with its definitive species designation in 1909. *T. gondii* is the sole etiological agent of toxoplasmosis and has been characterized as the most successful infecting parasite, since it can infect any warm-blooded animal, including livestock and humans (Dubey 2008). Toxoplasmosis causes a wide range of clinical syndromes in humans, mammals and birds. Since this infection has a worldwide distribution, virtually the whole population can be exposed to it, and it has been estimated that one third of the world's human population is seropositive (Miller et al. 2009). However, the seroprevalence and clinical severity of this disease vary according to age and is unevenly distributed across geographical areas and different socioeconomic strata of a specific population (Rosso et al. 2008); with South America exhibiting the highest burden (Bertranpetit et al. 2017).

1.2 Life Cycle and Infectious Stages

The life cycle of *T. gondii* can be divided into sexual replication, which takes place only in felids, such as the domestic cat (definitive hosts), and asexual replication, which occurs in other mammals and birds (intermediate hosts) (Dubey 1996). In nature, the parasite circulates primarily in three infectious stages: (i) the tachyzoite, a rapidly multiplying form which morphologically resembles a half-moon, (ii) the bradyzoite, a quiescent form with the ability to form tissue cysts, and (iii) the sporozoite, which is formed inside oocysts, and is shed in feces by the definitive host.

The sexual cycle occurs when the definitive host ingests environmental (water or soil) oocysts or meat contaminated with tissue cysts. After sexual reproduction, unsporulated oocysts, containing non-infective forms (sporozoites), are shed in feces, and a few days will be needed to form two infectious sporozoites and become a "sporulated oocyst" (Frenkel et al. 1970). Sporulated oocysts are highly resistant and can survive in the soil or water for several months (Dubey 1996). Like in the definitive host, in the intermediate host, the infection also occurs by ingestion of cysts or oocysts. In the intermediate host, the bradyzoites or sporozoites released in the small intestine penetrate the epithelial cells and multiply asexually, converting into tachyzoites (Fig. 1) (Blader et al. 2015). Tachyzoites may disseminate to extra-intestinal tissues through the lymph and blood by invading the same leukocytes that are recruited to respond against the parasite (Denkers et al. 2012). Tachyzoites are able to enter and multiply in any nucleated host cell, ultimately

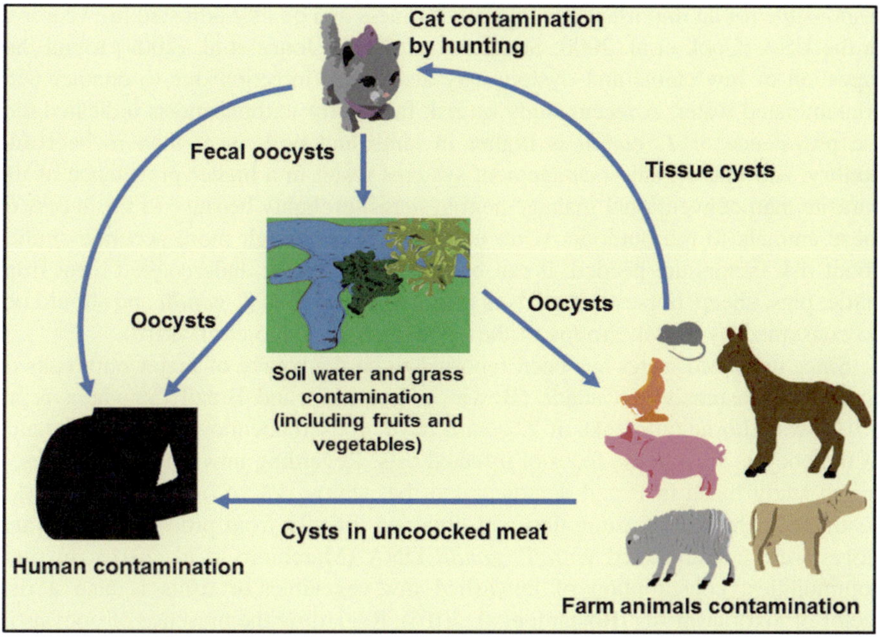

Fig. 1 Life cycle of *Toxoplasma gondii.* Felines are the definitive hosts in which the sexual replication phase of *T. gondii* occurs, resulting in the generation of thousands of highly resistant oocysts containing non-infective sporozoites that are spread by feces. Once in the environment, sporozoites become infective. Other mammals and birds are intermediate hosts. Both definitive and intermediate hosts can be infected by ingestion of oocysts present in the soil, unwashed vegetables or water. The asexual cycle of *T. gondii* occurs in the intermediate hosts, with two stages: highly replicative tachyzoites and latent bradyzoites, the latter forming tissue cysts. In addition to oocyst infection, animals and humans can be infected by ingestion of tissue cysts. Finally, vertical transmission can be observed in a variety of hosts, including humans

leading to cell death to continue dissemination to the spleen, liver, lungs, lymphoid tissue, central nervous system, retina and cardiac and skeletal muscles (Cañedo-Solares et al. 2013). This is called "the acute phase of infection", which is associated with a strong inflammatory response. Bradyzoites are the slow replicating form of the parasite, mainly in central nervous system tissues and skeletal muscles, leading to the latent or "chronic phase of infection" for the life of the host (Skariah et al. 2010).

1.3 Transmission

Transmission occurs when tissue cysts, present in raw or undercooked meat, or oocysts, present in unwashed vegetables, soil or water, are ingested. In humans, consumption of raw or undercooked meat is the main transmission route and is responsible for an important annual cost of illness and quality-adjusted life year loss in the USA (Cook et al. 2000; Scallan et al. 2011). Jones et al. (2009) found that ingestion of raw clams and oysters may also cause infection due to contact with contaminated water. A recent study on risk factors for toxoplasmosis indicated that the prevalence of *T. gondii* is higher in lamb and pork meat than in beef and poultry, and that organic management systems result in a higher prevalence of the parasite than conventional management systems, probably because of the access of these animals to the outdoors (Guo et al. 2015). Although more accurate studies about risk factors are needed, it can be noted that raw or undercooked meat from cattle, pigs, sheep, horses and goats is a potential source of *T. gondii* and should not be consumed by at-risk groups in the population (Belluco et al. 2016).

Since untreated water has been reported to be the source of major outbreaks of acute toxoplasmosis in Canada (Bowie et al. 1997) and Brazil (Vaudaux et al. 2010), waterborne outbreaks of *T. gondii* have aroused attention on the importance of the oocysts shed in the feces of infected cats. Regarding unwashed vegetables, a recent study made in Brazil demonstrated that almost 4% of fresh-leaf vegetables destined for human consumption and obtained directly from production sites and stores were contaminated with *T. gondii* DNA (Marchioro et al. 2016). In rural communities, consumption of unwashed raw vegetables or fruits is also a risk factor of toxoplasmosis (Rostami et al. 2016). Regarding the presence of oocysts in the environment, a survey made in California (USA) estimated that the annual oocyst burden is 3–434 oocysts per square foot (Dabritz et al. 2007). Considering that a single oocyst is able to cause toxoplasmosis, this oocyst burden represents a major potential public health problem.

In humans, congenital transmission involves the transmission of *T. gondii* tachyzoites from a newly infected mother (primary infection) to the fetus through the placenta or to the baby during vaginal delivery (Gras et al. 2005; Thiébaut et al. 2007). However, it has been reported that congenital transmission to the fetus also occurs from mothers chronically infected and reinfected with virulent strains of *T. gondii* during pregnancy (Elbez-Rubinstein et al. 2009), from highly immunocompromised mothers (Azevedo et al. 2010; Lindsay and Dubey 2011), and from women who acquire the infection some months before pregnancy (Hennequin et al. 1997). Congenital toxoplasmosis also occurs in many other animal species, particularly cattle, sheep, goats and rodents such as mice (Dubey 2008). In most classes of livestock, congenital toxoplasmosis has been recognized as being responsible for major economic losses through abortions, stillbirths and neonatal mortality (Raeghi et al. 2011). The clinical aspects of congenital toxoplasmosis in humans and other animals will be widely discussed in the next section.

2 Toxoplasmosis: Disease Symptoms and Occurrence of Infection

Because *Toxoplasma gondii* may cause abortion or congenital disease in its intermediate hosts, it is considered a parasite of both veterinary and medical importance (Tenter et al. 2000).

2.1 T. gondii *Infections in Humans*

The seroprevalence of *T. gondii* varies according to the age, geographical location and socio-economic strata of a specific population (Rosso et al. 2008). While one third of the world's human population is seropositive to *T. gondii*, clinical disease is largely confined to risk groups.

In the majority of immunocompetent individuals, toxoplasmosis is asymptomatic (Feustel et al. 2012). When symptoms are manifested, almost all cases are due to the tachyzoite form. Occasionally, toxoplasmosis may present with lymphadenopathy, fever, muscle aches and headaches (Dubey 1996), whereas severe manifestations such as encephalitis, myocarditis, hepatitis and sepsis syndrome very rarely occur (Tenter et al. 2000). The severity of the infection may depend on the genotype of the strain, where atypical genotypes are associated with worse clinical outcomes (Robert-Gagneaux and Dardé 2012).

In humans, congenital toxoplasmosis is one of the main clinical problems of Toxoplasma infection. Congenital toxoplasmosis can cause spontaneous abortion, as well as blindness and mental retardation in congenitally infected children (Shaapan 2016). In 2016, the World Health Organization estimated a global incidence rate of 1.5 cases of congenital toxoplasmosis per 1000 live births (Torgerson and Mastroiacovo 2013). The highest burden is found in South America (3.4 per 1000 live births) and is driven by the more pathogenic genotypes that circulate in that part of the world (Torgerson and Mastroiacovo 2013). Other regions with high incidence of congenital toxoplasmosis include parts of the Middle East (2.5 per 1000 live births) and some low income countries in Africa (2.4 per 1000 live births) (Torgerson and Mastroiacovo 2013). The risk of transplacental transmission increases with gestational age, from less than 10% in the first trimester to 30% in the second and 70–90% in the third trimester (Dunn et al. 1999). In contrast, the severity of congenital toxoplasmosis is inversely correlated with the time of gestation when maternal infection was acquired, with infection in early pregnancy displaying more severe consequences, such as stillbirths, abortions, substantial brain necrosis and hydrocephalus (Moncada and Montoya 2012; Xiao and Yolken 2015). In addition to the gestational age, the parasite genotype may also play a role in congenital disease, with atypical strains causing severe damage even when acquired during the third trimester (Delhaes et al. 2010). Near 70–90% of infants born with congenital *Toxoplasma* infection are asymptomatic at birth and therefore their infection is not recognized. However, during the first year of life or more, most of these children will develop sequelae such as ocular lesions and neurological symptoms (Wilson et al. 1980). Other factors involved in congenital toxoplasmosis transmission and clinical outcome are the genetics of the host, the size of the inocula, the infecting form of the parasite and the maternal treatment (Moncada and Montoya 2012).

Ocular toxoplasmosis, a particular form of the infection, can be postnatally acquired by immunocompetent individuals as well as by reactivation of congenital infection (Perkins 1973; Matias et al. 2014). Its incidence varies from 2–3% in the USA and Europe to nearly 18% in Southern Brazil (Glasner et al. 1992; Holland 2003). Ocular toxoplasmosis may result in inflammation in the retina, choroid, and uvea, and consequently lead to complications such as glaucoma, cataract, and posterior synechiae (Jasper et al. 2017). This type of toxoplasmosis is considered the most frequent cause of infectious posterior uveitis (Pleyer et al. 2014). Since ocular toxoplasmosis is a preventable cause of blindness, it is necessary to assess factors that have the potential to control this disease.

Different factors, including HIV infection, certain types of cancers including lymphomas, immunosuppressive therapies (Robert-Gagneaux and Dardé 2012), and chronic inflammatory diseases such as diabetes or obesity (Esch and Petersen 2013), can profoundly impair cellular immunity, leading to severe toxoplasmosis. Reactivation of a latent infection or de novo infection in immunocompromised individuals can cause the development of clinical illness with varied presentations, including encephalitis, pneumonitis, chorioretinitis, meningitis, and disseminated toxoplasmosis with multi-organ involvement (Khurana and Batra 2016). In this sense, toxoplasmosis represents one of the main opportunistic infections in HIV/AIDS, associated with high morbidity and mortality rates (Basavaraju 2016). In transplant patients, the severity of toxoplasmosis is clearly associated with the degree of induced immunosuppression (Robert-Gagneaux and Dardé 2012), as well as with the timing of appropriate anti-Toxoplasma therapy (Coster 2013). About 10–25% of transplant recipients with toxoplasmosis show central nervous system manifestations, mainly encephalitis. When toxoplasmosis is severe or disseminated, patients present myocarditis, encephalitis, pneumonitis, or multi-organ failure, and its fatality rate is >80%.

Since *T. gondii* has a strong predilection to infect the central nervous system, over the past decade, chronic *T. gondii* infection has been increasingly associated with psychiatric disorders (Torrey and Yolken 2003; Dickerson et al. 2009; Zhu 2009; Shibre et al. 2010), such as attention-deficit hyperactivity disorder, obsessive compulsive disorder, schizophrenia, major depression, and bipolar disorder (Halonen and Weiss 2013; Elsheikha et al. 2016; Del Grande et al. 2017). Three putative mechanisms for *T. gondii* infection of the central nervous system have been proposed: (i) the direct effect position, which is based on the fact that *T. gondii* bradyzoites are present mainly in neurons, affecting their function (Ferguson and Hutchison 1987); (ii) a neuroimmunological mechanism based on the fact that proinflammatory cytokines may modulate dopaminergic and glutamatergic neurotransmission (Mastropaolo et al. 1989; Emery et al. 2007); (iii) a proposal that suggests that bradyzoites are able to synthesize and release the dopamine neurotransmitter (Ayaz et al. 2016).

2.2 *Toxoplasmosis in Other Animals*

As mentioned above, almost every warm-blooded animal species can be infected with *T. gondii*. Although most of them show no clinical signs of the disease, their latent infection is still of veterinarian/zoonotic relevance, since their meat could be the source of de novo infections by carnivorous animals, including humans.

2.2.1 Toxoplasmosis in Pets

The first reported case of fatal toxoplasmosis in a domestic animal was in a dog in 1910 (Dubey 2008). In dogs, the seroprevalence of *T. gondii* varies according to the geographical location, age and gender, ranging from 9.1% in dogs from Shangai (Jiang et al. 2015) to 24% in hunting dogs from Italy (Machacova et al. 2016), 42.2% in dogs from St Kitts in the Caribbean (Dubey et al. 2016), 48.75% in dogs from Fernando de Noronha Island in Brazil (Magalhães et al. 2017) and 51.9% in dogs from Zhanjiang city in southern China (Jiang et al. 2015). As in humans, the clinical presentation of the infection varies from total lack of symptoms to affections in different organ systems, including encephalitis, and even severe disease, involving the lungs and liver, which may kill dogs within a week (Dubey et al. 2009). Transplacental transmission in dogs can cause spontaneous abortion and fetal death (Bresciani et al. 1999, 2001), and even reinfection of pregnant females can lead to clinical alterations such as fever, lymphadenopathy, miscarriage, and fetal death (Bresciani et al. 2009).

The first report of toxoplasmosis in a cat was in 1942 (Dubey 2008) and nearly 30 years later, Frenkel et al. (1970) demonstrated that cats are the definitive host of *T. gondii*. The estimated seroprevalence for *T. gondii* in domestic cats worldwide is 30–40% (Muñoz et al. 2011), but antibodies to *T. gondii* may be detected in up to 74% of adult cat populations, mainly depending on the type of feeding and whether the cat remains indoors or outdoors (Tenter et al. 2000). As in other hosts, most of the postnatally acquired infections in cats are asymptomatic and vertical transmission is infrequent (Tenter et al. 2000). However, when congenital toxoplasmosis occurs, clinical illness is common and may result in stillbirth or kitten death before weaning. The clinical presentation is often characterized by inflammation of the liver, lungs and central nervous system. In postnatally acquired infections, anorexia, lethargy and dyspnea are common (Dubey et al. 2009). The zoonotic relevance of cats, especially related to the shed of oocysts in the environment, has been previously discussed (Sect. 1.2).

2.2.2 Toxoplasmosis in Animals Destined to Human Consumption

The study and control of toxoplasmosis in meat-producing animals is of major relevance mainly for two reasons: first, because, as previously mentioned, toxoplasmosis is one of the main foodborne diseases because tissue cysts of *T. gondii* contained in meat of livestock are considered the main source of infection in humans (Cook et al. 2000), and second, because the disease is responsible for major economic losses in most classes of livestock through abortions, stillbirths, early embryonic death, resorption, mummification, and neonatal mortality or birth of live but weak offspring (Buxton et al. 2007; Raeghi et al. 2011), especially in sheep (Hiszczyńska-Sawicka 2014), goats and lambs (Dubey 2008; Dubey et al. 2009). Congenital toxoplasmosis in sheep and goats is slightly different, since, in sheep, it occurs only in primary infections of pregnant sheep, whereas, in goats, the same goat can transplacentally infect its fetuses in successive pregnancies (Dubey 1982). In the UK, congenital toxoplasmosis is responsible for 1–2% of neonatal losses in sheep and goats per annum (approximately 300,000 sheep/goats lose their fetuses) (Menzies et al. 2008), while, in the European Union, pregnancy losses in these species are between 0.7 and 1.4 million annually (Katzer et al. 2011).

In pigs, symptomatic toxoplasmosis is rare, but clinical signs include anorexia, fever, dyspnea, limb weakness, encephalitis, pneumonitis, lymph node necrosis, hepatic necrosis and even death (Dubey 2009). In addition, abortions related to congenital toxoplasmosis in pigs have also been reported (Kim et al. 2009).

In cattle and horses, there are no confirmed cases of clinical toxoplasmosis (Dubey 2007). However, a recent experimental study has shown that dams inoculated with tachyzoites of the *T. gondii* RH strain aborted on days 6 and 11 post-inoculation, which demonstrates that vertical transmission of toxoplasmosis in cattle can occur (Wiengcharoen et al. 2011). The role of toxoplasmosis in economic losses due to bovine abortion warrants further investigation.

In poultry, there is a high prevalence of *T. gondii* infection but neither chickens nor turkeys develop any clinical sign of the disease (Guo et al. 2015), and toxoplasmosis does not cause any economic losses in these species (Dhama et al. 2013). However, chickens are considered one of the most important hosts in the epidemiology of *T. gondii* infection and the importance of chicken meat as a source of human toxoplasmosis must not be ignored.

Finally, in rabbits, toxoplasmosis is one of the most important diseases in commercial herds (Dubey et al. 2011). *T. gondii* transmission in rabbits occurs through food contaminated with parasite oocysts, transplacental transmission and interaction between rabbits and domestic cats infected with *T. gondii* (Dubey et al. 1992).

3 Mechanism of Infection and Immune Response

3.1 Mechanism of Host-Cell Invasion and Egression

Toxoplasma gondii is an apicomplexan parasite, and, as any other member of the group, it is characterized by the presence of an "apical complex" consisting of a closed, truncated cone called the conoid, which is composed of unique fibers of tubulin polymers. This apical complex acts as an organizing center, and a cluster of apical secretory organelles including rhoptries, micronemes and dense granules (Hu et al. 2006; Sibley 2011; Okamoto and Keeling 2014). In addition, the apical complex is involved in the processes of host attachment and invasion, being fundamental for *T. gondii* infection (Okamoto and Keeling 2014).

Apicomplexan parasites lack cilia and flagella and their mode of locomotion, called gliding, propels them across the substrate and supports active penetration of host cells. Gliding relies on a unique form of substrate-dependent motility based on the actin-myosin motor of the parasites (Barragan and Sibley 2002; Sibley 2011).

T. gondii invades host cells in several orchestrated steps that begin with the molecular interaction between the parasite and the host cell before internalization, first mediated by transient interactions followed by sequential secretion of specialized proteins (micronemes and rhoptry proteins) to establish the "parasitophorous vacuola" (Peng et al. 2011; Blader et al. 2015). The "parasitophorous vacuola" protects *T. gondii* from being eliminated by host cells, especially by endosomal acidification and lysosomal fusion. Once inside, tachyzoites induce the host cell to arrest at the G2/M phase of the cell cycle and to provide a nutrient pool to the parasite that enables *T. gondii* proliferation and replication by endodiogeny (Fig. 2a). After several rounds of replication, the parasite activates motility and exits the host cell by rupturing the vacuola and breaching the host cell plasma membrane (Fig. 2a) (Melzer et al. 2008; Peng et al. 2011).

3.2 T. gondii *Dissemination Through the Host*

T. gondii primary infection generally occurs by the ingestion of oocysts or tissue cysts, which release sporozoites or bradyzoites to the lumen of the gut from where the parasite spreads throughout the body of the individual (Fig. 2b). The dissemination of *T. gondii* to a large variety of other organs in the body occurs via intracellular and extracellular mechanisms (Harker et al. 2015). *T. gondii* can flow freely in host fluids, migrate on cell layers (endothelium, epithelium), cross them (paracellular route) or use motile host cells (intracellular route) to achieve distant or hardly accessible organs (Długonska 2014). As a consequence of oral infection, innate immune cells such as neutrophils, monocytes and dendritic cells (DCs) are recruited to the small intestine (Cohen and Denkers 2014). During the acute phase of the infection, *T. gondii* shows a certain preference for cells of the immune

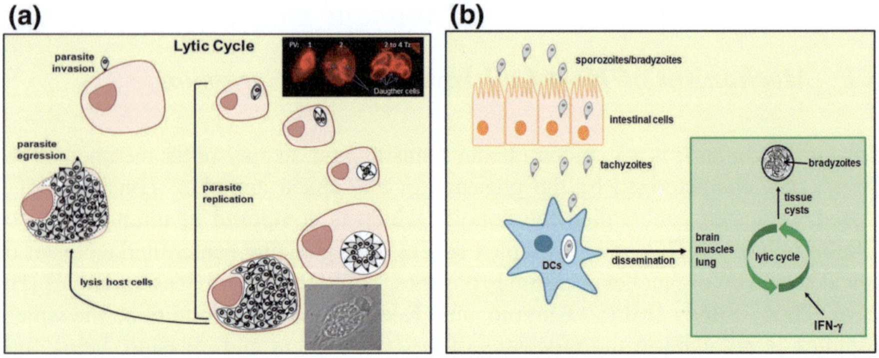

Fig. 2 Toxoplasma lytic cycle and infection. **a** *T. gondii* invades the host cells by an active process that results in the formation of an intracellular parasitophorous vacuole (PV). Once inside the cell, tachyzoites begin to replicate by a process called endodiogeny, in which the gestation of two daughter cells within the mother cell can be observed (see upper panel). After several replication rounds, it is possible to observe the presence of 2, 4, 8 and so on tachyzoites per PV until the generation of a large PV (see lower panel) near the egression step to resume the lytic cycle. Upper panel: Replicating intracellular tachyzoites were labeled with anti-tubulin antibodies and detected by epifluorescence microscopy. Lower panel: intracellular PV observed by phase contrast. **b** The parasite enters the lumen of the intestine by ingestion of cysts or oocysts releasing bradyzoites or sporozoites, respectively. The bradyzoites/sporozoites penetrate the intestinal epithelium and convert to the highly replicative tachyzoites, which disseminate within the new host. Tachyzoites invade every nucleated cell, including dendritic cells, which give them mobility, allowing the spread of *T. gondii* throughout the body of the host. *T. gondii* also has the ability to modulate the immune response in such a way as to allow its dissemination. Finally, the IFN-γ-dependent cell-mediated immune response is able to kill the intracellular parasite but also to trigger the conversion of tachyzoites to the latent form of bradyzoites, generating tissue cysts, mainly at the central nervous system and muscle. The immune response is also important to maintain the latent infection. The lack of a good control of infection may lead to serious clinical consequences, including death of the infected individual, as observed in immunocompromised patients

system, mainly DCs (Sanecka and Frickel 2012). DCs play a key role in the immune function, being the link between innate and adaptive responses, as well as in actively secreting IL-12 and a number of other cytokines that attract and activate other cells of the immune system such as pathogen-specific T cells (Sanecka and Frickel 2012; Długonska 2014). On the other hand, DCs have motile properties, resulting in a very effective vehicle to cross many biological barriers that have exaggerated immune responses (the blood brain barrier, blood-eye barrier and the placenta) to enter immune privileged organs like the brain and the eye or even to infect the fetus (Bierly et al. 2008).

3.3 Immune Control of T. gondii

The immune response against *T. gondii* strongly depends on the strain of the infecting parasite (Xiao and Yolken 2015), as well as on the genetic background and immune status of the host (Suzuki and Remington 1993). However, the salient features of this complex immune reaction have been elucidated mainly based on the study of murine experimental models of toxoplasmosis (Dupont et al. 2012).

3.3.1 Immune Response During the Acute Phase of Infection

When *T. gondii* reaches the gut of the host, monocytes, granulocytes (mainly neutrophils), DCs and lymphocytes residing in the small intestinal epithelium are rapidly recruited to the site of infection (Muñoz et al. 2011; Dupont et al. 2012). Among them, neutrophils are very important during the initial response (Bliss et al. 2001) and their recruitment depends on the chemokine receptors CXCR2 (Del Rio et al. 2001) and CCR1 (Khan et al. 2001). In *T. gondii* infection, the main pattern-recognition receptors involved are the Toll-like receptors 2, 4 (Debierre-Grockiego et al. 2007; Bereswill et al. 2014) and 11 (Yarovinsky et al. 2008). After recognition, mainly DCs but also neutrophils and activated macrophages secrete IL-12 (Muñoz et al. 2011; Dupont et al. 2012), which in turns stimulates natural killer cells (Sturge and Yarovinsky 2014). Natural killer cells are the main source of IFN-γ during the initial phase of infection and are also able to kill the parasite by apoptosis (Muñoz et al. 2011). DCs and macrophages increase the expression of major histocompatibility complex type II molecules, stimulated by IFN-γ (Denkers et al. 1996), becoming mature antigen-presenting cells. Activated macrophages and neutrophils are also able to secrete reactive oxygen species and reactive nitrogen intermediaries that help to control the parasite, limiting infection (Gazzinelli et al. 1993; Alves et al. 2013). Therefore, the innate immune response limits tachyzoite replication and drives the generation of a specific T-cell response.

As previously discussed, DCs then migrate to the lymph nodes, both helping the parasite to disseminate to other tissues and acting as antigen-presenting cells, leading to the proliferation of *T. gondii*-specific T lymphocytes (TLs). In the presence of co-stimulatory molecules and in an IL-12 context, CD4+ and CD8+ *T. gondii*-specific TLs proliferate and differentiate into populations that produce pro-inflammatory cytokines, particularly IFN-γ, TNF-α, IL-6 and IL-1 (Sibley et al. 1991; Langermans et al. 1992). In addition, CD4+ TLs secrete IL-2, which in turn helps CD8+ TLs to proliferate and become cytotoxic lymphocytes, which act as effectors lysing the parasites, promoting apoptosis by CD40-CD40L and secreting IFN-γ (Dupont et al. 2012).

The role of humoral immunity in toxoplasmosis is not so clear. Specific immunoglobulins (Igs) against *T. gondii* antigens, including IgM, IgA, IgE and IgGs, have been found in infected individuals (Robert-Gagneaux and Dardé 2012).

In this sense, in vitro assays have demonstrated that *T. gondii*-specific antibodies are able to opsonize the parasite for phagocytosis, block invasion and activate the complement (Dupont et al. 2012).

3.3.2 Limitation of the Immune Response Against *T. gondii* and Immune Evasion and Anti-apoptotic Reactions of the Parasite in the Host Cell

It is important to note that the immune response against *T. gondii* is a double-edged sword, since a potent pro-inflammatory reaction may kill the parasite, but it also can be detrimental and even fatal for the host, as in the case of *T. gondii* intestinal ileitis (Mennechet et al. 2004). The invasion of host cells by *T. gondii* triggers a potent immune response that eliminates most parasites during the acute phase of infection. However, some tachyzoites can evade this response, convert to bradyzoites (the low replicative form of the parasite), and encyst in non-replicative cells for the lifespan of the host, in the chronic phase of the infection. To achieve this delicate balance between induction and suppression of the host immune responses, *T. gondii* induces modifications in the expression and secretion of immunomodulatory cytokines and in the viability of immune cells, as well as mechanisms to abolish antimicrobial effector mechanisms (Muñoz et al. 2011). In this sense, *T. gondii* is able to significantly decrease the responsiveness of host infected cells to IFN-γ signaling by blocking STAT1 transcriptional activity (Kim et al. 2007). A recent report has also demonstrated that the secretion of ROP16 by the tachyzoite is associated with prolonged phosphorylation of the transcription factors STAT3 and STAT6, enhancing IL-4 and IL-6 production while down-regulating IL-12 production, thus avoiding a strong inflammatory response and leading to an anti-inflammatory one, which is less harmful for the parasite (Saeij et al. 2007). Moreover, *T. gondii* enhances production of anti-inflammatory cytokines, including IL-10 and TGF-β, which inhibit IFN-γ production, impair macrophage activation and inhibit the activity of natural killer cells (Muñoz et al. 2011).

Another cell signaling pathway modulated by *T. gondii* is the transcription factor NF-kB signaling pathway, critical in immune response and inflammation (Du et al. 2014). The role of *T. gondii* in regulating this transcription factor is controversial. Some studies have demonstrated that invasion of cells by *T. gondii* fail to activate NF-κB signaling in macrophages and fibroblasts (Butcher et al. 2001; Shapira et al. 2005), enabling the parasite to invade cells without triggering pro-inflammatory cytokine induction. On the other hand, other studies have demonstrated that *T. gondii* activates NF-κB in mouse embryonic fibroblasts (Molestina et al. 2003) and mouse spleen cells (Kim et al. 2006), finally leading to the up-regulation of the expression of anti-apoptotic genes resulting in the enhanced survival of the parasite. Although the mechanism of *T. gondii* in regulating the NF-κB signaling pathway needs further elucidation, parasite-specific molecules likely induce this pathway as a means of disrupting host cell immune responses (Sibley 2011).

3.3.3 Immune Response During the Chronic Phase of Infection

Tachyzoites that avoid the immune reaction during the acute phase of infection reach immune privileged sites and convert into encysted bradyzoites, initiating the chronic phase of the infection. The persistence of *T. gondii* in this phase depends on immune control, since it has been shown that, in immunocompromised individuals, reactivation of the infection may occur and result in life-threatening toxoplasmic encephalitis (Israelski and Remington 1993). However, the mechanisms by which the immune system maintains a latent *T. gondii* infection are not completely understood. Resistance during chronic toxoplasmosis involves at least secretion of IFN-γ (Suzuki et al. 1989) and TNF-α (Gazzinelli et al. 1993), and the interaction between CD40 and CD40L (Reichmann et al. 2000).

4 Comparison of Plant-Made Vaccine Candidates with Current and Alternative Treatments for Toxoplasmosis

4.1 Current Treatments for Toxoplasmosis

Despite the economic losses associated with *T. gondii* infection in farm animals and socio-economic impact caused by this zoonotic disease in the human population, treatment is not able to eliminate the parasite from the host once the chronic infection has been established (Innes 2010). An ideal drug for prophylaxis and/or treatment of toxoplasmosis should show effective penetration and concentration in the placenta, transplacental passage, parasiticidal properties against the different parasitic stages, penetration into cysts, and distribution in the main sites of infection. Unfortunately, so far, no available drug fulfills these criteria (Derouin and Santillana-Hayat 2000; Montoya and Liesenfeld 2004). Even if a drug with these characteristics were available, it could hardly be used in animals since there is a risk related to the increment of drug resistance and to drug residues entering in the food chain (Hiszczyńska-Sawicka et al. 2014). In addition, a drug-based treatment would not be as well-suited to control toxoplasmosis in farm animals, since the main goal of toxoplasmosis prevention in livestock is related to the control of abortions and to prevent the transmission of the parasite to humans through the consumption of meat. In this sense, it is the development of a prophylactic treatment based on vaccination seems to be the most effective method to avoid the spread of the disease in livestock.

The current anti-*T. gondii* chemotherapy for humans is based on the prophylactic administration of pyrimethamine and sulfadiazine (Montazeri et al. 2017). However, it is a deficient treatment since it is not well-tolerated by immunocompromised patients (Rodriguez and Szajnman 2012). In addition, the use of these drugs has several side effects, including renal calculi (McGettigan et al. 2012), neutropenia, severe drop of platelet count, thrombocytopenia, leukopenia, increase in serum creatinine and serum liver enzymes, hematological abnormalities, and hypersensitivity reactions (Montazeri et al. 2017). In congenital toxoplasmosis, spiramycine, a macrolide that has not been shown to be teratogenic, is used to treat infection. However, prenatal spiromycine treatment is still controversial (Kieffer and Wallon 2013), since multicenter studies conducted in Europe have shown conflicting results. Some studies have shown that prenatal administration of spiromycine decreases the severe neurological sequelae of congenital toxoplasmosis, whereas other studies have shown that prenatal treatment is not able to prevent fetal transmission (Gilbert and Gras 2003; Gras et al. 2005; Fricker-Hidalgo et al. 2013).

Many of the current treatment regimens are based on case series and case studies and there are no large-scale trials on drugs used to treat toxoplasmosis. Thus, there is a need to evaluate the available evidence with regard to the efficacy and safety of the drugs used to treat symptomatic toxoplasmosis in immunocompetent and immunocompromised hosts, pregnant women, and patients with ocular toxoplasmosis (Rajapakse et al. 2007).

In recent years, a highly valuable goal for global toxoplasmosis control is the development of well-tolerated and safe specific immunoprophylaxis against the disease (Lim and Othman 2014). Immunotherapeutic strategies to improve toxoplasmosis control could be a vaccine able to induce either strong protective immunity or passive immunization in cases of disease recrudescence. So far, the only commercial vaccine is the S48 strain of attenuated tachyzoites (Toxovax®, MSD Animal Health, Summit, NJ, USA) for use in sheep (Buxton et al. 1991). Toxovax vaccine is used in the UK, New Zealand, France, and Ireland (Garcia et al. 2014). Vaccination of sheep with this vaccine allows a reduction of abortions and neonatal mortality, and improves the birth weight of lambs (Buxton et al. 1991). However, this vaccine causes side effects, has a short life time, induces a short-term immunity, and has a high cost of production because it depends on the growth of the parasite in mammalian cell cultures. Moreover, there are some concerns about its safety because this type of attenuated live vaccines carries the risk of reverting to a pathogenic strain and can infect humans (Garcia et al. 2014).

4.2 *Plant-Made* T. gondii *Vaccine Candidates: Rationale, Efficacy and Yield*

Despite the potential of vaccine antigens expressed in plant systems, the possibility of using this strategy for the expression of parasite antigens has aroused greater interest in recent years (Matsumoto et al. 2009; Clemente and Corigliano 2012; Sathish et al. 2012; Jacob et al. 2013; Hernandez et al. 2014; Ganapathy et al. 2014; Monreal-Escalante et al. 2016; Kesik-Brodacka et al. 2017; Wilbers et al. 2017). Antigens of *Plasmodium* sp. (Clemente and Corigliano 2012), *Entamoeba histolytica* (Chebolu and Daniel 2007) and *T. gondii* (Clemente 2014) were the pioneer examples of parasitic antigens expressed in plants. Although the results obtained are highly promising, *T. gondii* antigen expression in plants is just beginning and only a few *T. gondii* antigens have been successfully expressed.

4.2.1 Dense Granule Protein 4 (GRA4) of *T. gondii*

The dense granule protein 4 (GRA4) of *T. gondii* belongs to the family of proteins secreted into the lumen of the parasitophorous vacuole by the tachyzoite (Labruyère et al. 1999). GRA4 is involved in the parasite-host interaction, which is associated with a intravacuolar network that participates in the transport of nutrients and proteins into the parasitophorous vacuole (Labruyère et al. 1999).

GRA4 is considered a highly feasible candidate for vaccine development against toxoplasmosis, since recombinant GRA4 protein produced in bacteria or GRA4 DNA vaccines have been shown to induce both a humoral and a cellular response during *T. gondii* oral infection (Desolme et al. 2000; Martin et al. 2004; Mévélec et al. 2005; Chen et al. 2009; Sánchez et al. 2011). In fact, results of mouse immunizations with the recombinant protein or its gene have shown that both are able to confer protection against Toxoplasma infection (Mévélec et al. 1998; Desolme et al. 2000; Martin et al. 2004, Mévélec et al. 2005). In addition, Mévélec et al. found that oral immunization of mice with truncated soluble forms of recombinant GRA4 in association with cholera toxin induces a significant Th2-like mucosal response and partial resistance to oral infection with *T. gondii* (Mévélec et al. 1998) and that the recombinant GRA4 protein is able to induce a specific Th1 humoral and cellular immune response in a murine model (Mévélec et al. 2005).

Considering that the portal of entry of *T. gondii* is the mucosa, the stimulation of an efficient local response of mucosal membranes as well as that of a systemic response constitute a priority which could be achieved by the administration of an oral/nasal vaccine. Therefore, GRA4 is an interesting protein to express in plants for the production of eukaryotic immunoprophylactic antigens and for the assessment of its effectiveness in oral vaccines against intracellular pathogens.

The region of GRA4 that contains the B-cell and T-cell epitopes ($Gra4_{163\text{-}345}$) (Mévélec et al. 1998) displays antigenic properties (Nigro et al. 2003; Altcheh et al. 2006) and confers immune protection against toxoplasmosis in mice (Martin et al. 2004). Thus, this region has been chosen for expression in tobacco plants. Ferraro et al. (2008) transiently expressed $GRA4_{163\text{-}345}$ in tobacco leaves using both a Potato Virus X (PVX)-based amplicon and an apoplast-targeting system, and found that the targeting to the secretory pathway increased the production of GRA4 (0.01% vs. 0.001% of total soluble proteins using PVX amplicon system). Although yields were not very high, $GRA4_{163\text{-}345}$ successfully accumulated into the apoplastic space. They also found that extracellular targeting allowed the recovery of recombinant products from apoplastic washing fluids (in the case of leaf infiltration systems) or from the nutrient medium (in the case of hydroponically cultivated transgenic plants), thus simplifying the extraction of fractions enriched in the recombinant protein. In addition, the authors showed that plant-derived GRA4 was immune reactive with seropositive human sera (Ferraro et al. 2008). Later, Del L Yácono et al. (2012) analyzed the expression of $GRA4_{163\text{-}345}$ by chloroplast transformation (chlGRA4) in tobacco plants, and found that this transformation allowed a significant 30-fold increase in GRA4 protein accumulation in the plant. In the work by Ferraro et al., the yields of GRA4 in tobacco infiltrated leaves were around 0.2 μg/g of fresh weight (Ferraro et al. 2008), whereas in the work by Del L Yácono et al., the chlGRA4 expression levels in transplastomic plants were up to 6 μg/g of fresh weight (or 0.2% of total proteins) (Del L Yácono et al. 2012). These results support the idea that the chloroplast would be the best compartment to express this protein. In addition, oral immunization with the chloroplast-derived GRA4 resulted in a decrease of 59% in the brain cyst load of mice and elicited both a mucosal and systemic immune response characterized by the production of specific IgA and IgG, and secretion of IFN-γ, IL-4 and IL-10. The authors suggested that the antigen expressed in the chloroplasts would have been recognized by the antigen-presenting cells and presented in the gut-associated lymphoid tissues, resulting in the activation of T helper cells that would trigger a partial protection against Toxoplasma infection (Del L Yácono et al. 2012). This protection correlates with a mucosal and systemic balanced Th1/Th2 response.

4.2.2 Main Surface Antigen 1 (SAG1) of *T. gondii*

The proteins present on the surface of the tachyzoite, called surface antigens (SAGs), are the ligands through which the tachyzoite recognizes and adheres to the host cells (Pollard et al. 2008). Since cellular invasion is essential for the survival and spread of the parasite, the proteins involved in this process have been the principal target of the study for vaccine development against *T. gondii* (Jongert et al. 2009).

SAG1 is well conserved at the immunological and amino acid sequence levels, making it an attractive antigen for immunoprophylaxis of toxoplasmosis (Wang and Yin 2014). SAG1 is able to stimulate IFN-γ production by T cells in seropositive individuals through the action of CD8 + T cells that have cytotoxic activity (Khan et al. 1988). In addition, several studies have identified B- and T-epitopes in the SAG1 coding sequence, which would be recognized by the immune system after the infection. These epitopes would be able to induce a humoral and/or cellular immune response and, specifically, stimulate cytotoxic T lymphocytes cells (Siachoque et al. 2006; Cardona et al. 2009; Wang et al. 2013). In this context, numerous studies have established the potential of SAG1 as a candidate to produce an anti-*T. gondii* vaccine (Wang and Yin 2014). Previous attempts to express SAG1 using various classical heterologous expression systems have been impaired by the low expression levels or poor antigenicity due to misfolding or an excessive glycosylation of the recombinant SAG1 protein (Burg et al. 1988; Makioka and Kobayashi 1991; Harning et al. 1996; Nigro et al. 2003). Thus, SAG1 was a good candidate to optimize an antigen production strategy using an alternative protein expression system such as plant-based protein production. SAG1 was the first antigen of *T. gondii* expressed in plants (Clemente et al. 2005).

The transient transformation of SAG1 was evaluated using a vacuum infiltration system via recombinant Agrobacterium in tobacco plants (Clemente et al. 2005). In that study, this expression system was used to test the performance of three different constructs carrying the *SAG1* gene (Clemente et al. 2005). Two of the constructs were based on a PVX amplicon, whereas in the third construct, the *SAG1* gene was fused to an apoplastic peptide signal under the CaMV 35S promoter. The expression levels of SAG1 in infiltrated leaves ranged from 0.06 to 0.1% of TSP (~6–10 µg/g of fresh weight). Contrary to the results for GRA4, the infiltrated leaves with the version of SAG1 targeting to the apoplast space showed the lowest expression levels (Clemente et al. 2005). Thus, the better replication capacity of the amplicons could explain the higher SAG1 levels compared with the construct that targeted SAG1 to the apoplast.

Later, Laguía-Becher et al. (2010) designed and synthesized a plant-codon-optimized version of SAG1, and transiently expressed it in tobacco leaves. Both plant-optimized SAG1 and native SAG1 genes were fused to the apoplast or endoplasmic reticulum peptide signals for stable protein accumulation (Laguía-Becher et al. 2010). Surprisingly, the authors observed that leaves agroinfiltrated with an unmodified SAG1 gene accumulated 5- to 10-fold more than leaves agroinfiltrated with a codon-optimized SAG1 gene, although transcript accumulation was similar. The expression level of the unmodified SAG1 was 1.3 µg/g of fresh weight, while that of the codon-optimized SAG1 was 0.4 µg/g of fresh weight. In addition, the endoplasmic reticulum localization allowed the accumulation of higher levels of unmodified SAG1 compared to localization in the apoplast (1.3 µg vs. 0.7 µg/g of fresh weight, respectively). The authors suggested that the endoplasmic reticulum provided a relatively protective environment, which resulted in an increase in protein stability and an enhanced level of protein accumulation (Laguía-Becher et al. 2010).

Finally, Albarracín et al. (2015) expressed SAG1 in transplastomic tobacco plants. In addition, to improve expression in transplastomic plants, these authors expressed the 90-kDa heat shock protein (HSP) of *Leishmania infantum* (LiHsp83) as a carrier for the SAG1 antigen. Fusion of SAG1 to LiHsp83 significantly increased the level of SAG1 accumulation in tobacco chloroplasts (by up to 500-fold). The authors showed that LiHsp83-SAG1 protein accumulation did not decrease significantly as the plant aged, yielding up to approximately 100 µg/g of fresh weight. These authors proposed that LiHsp83 is a promising candidate to function as a carrier protein for the expression of vaccine antigens in plants.

Studies on the immunogenicity of the plant-made SAG1 have shown that, when this antigen is expressed in plants, it is able to elicit an immune response by subcutaneous (s.c.) or oral vaccination in a murine model (Clemente et al. 2005; Laguía-Becher et al. 2010; Albarracín et al. 2015). These findings provide a rationale for the development of a plant-made oral vaccine against toxoplasmosis. Our group was the first to demonstrate the immunogenic properties of SAG1 (Clemente et al. 2005). We found that s.c. immunization with SAG1-infiltrated leaf extracts emulsified with Freund's incomplete adjuvant induced a significant increase in the systemic-specific antibodies and a reduction of number of cysts (55%) against *T. gondii* (Clemente et al. 2005). Then, Laguía-Becher et al. (2010) demonstrated that plant-made SAG1 is able to induce a reduction in the brain cyst burden (30%) when delivered s.c. or orally without any adjuvant. In addition, Laguía-Becher et al. (2010) showed that this protection is associated with the secretion of significant levels of IFN-γ and that the protection can be increased with a significant reduction in the parasite load (54%) when mice were intradermally boosted with rSAG1 (SAG1 + boost). Similarly, in an oral immunization assay, these authors demonstrated that the SAG1 + boost group showed a significantly lower brain cyst burden (50%) than the rest of the groups. More recently, Albarracín et al. (2015) found that oral immunization of mice using SAG1 fused to LiHsp83 elicited an effective immunity against Toxoplasma infection, suggesting that the LiHsp83-SAG1 fusion protein retains the structural integrity to elicit immunological responses in mice.

4.3 Plant-Derived Vaccines as an Alternative Treatment

Several factors, including the economic losses associated with *T. gondii* infection in farm animals, the risk of transmission of the parasite to animals and humans, the unsatisfactory chemotherapy associated with increasing drug resistance, and the drug residues entering the food chain, justify the current attempts to develop an effective prophylactic *T. gondii* vaccine for both humans and animals. For all these reasons and since a vaccine against *T. gondii* is considered the most efficient method to prevent this infection, in the last few years, much progress has been made in research on DNA vaccination, protein vaccination, live attenuated vaccination and heterologous vaccination. New vaccine candidates, including SAGs and secretory antigens (ROP, MIC, and GRA organelles), have been tested either individually or as multi-antigen vaccines and novel adjuvants. However, researchers have not been able to find a proper vaccine for prevention of toxoplasmosis in animals and/or humans (Zhang et al. 2013, 2015; Montazeri et al. 2017). One possible explanation for this is that most studies have used single or only a few antigen candidate vaccines eliciting only partial protective immunity against *T. gondii* and never allowed a complete protection against tissue cysts (Zhang et al. 2013). This could be due to the low number of T lymphocytes epitope generated after administrating vaccines with a single antigen. In this way, experimental vaccines should include many antigens (Jongert et al. 2009; Hiszczyńska-Sawicka et al. 2014).

Our laboratory is mainly focused on studying the potential of plant-produced *T. gondii* antigens as a mucosal vaccine. SAG1 and GRA4 are considered the antigens with the greatest potential to be incorporated in a multicomponent vaccine against *T. gondii* (Lim and Othman 2014). Therefore, the achievement of efficiently expressing these antigens in plants is a success per se. In this sense, GRA4 and SAG1 antigens have been correctly expressed in tobacco using various strategies

(Ferraro et al. 2008; Del L Yácono et al. 2012; Clemente et al. 2005; Laguía-Becher et al. 2010; Albarracín et al. 2015). When transiently expressed in tobacco leaved via agroinfiltration, yields of GRA4 were around 0.2 μg/g of fresh weight (Ferraro et al. 2008), while GRA4 expression levels in transplastomic plants were up to 6 μg/g of fresh weight (Del L Yácono et al. 2012). Therefore, the best yields were obtained by plastid transformation (Del L Yácono et al. 2012). Similarly, yields of SAG1 from transient expression via agroinfiltration were approximately 1.3 μg/g of fresh weight (Laguía-Becher et al. 2010), while SAG1 accumulated in transplastomic plants at around 0.2 μg/g of fresh weight (Albarracín et al. 2015). However, the best attained yields for SAG1 protein were achieved when it was fused to the carrier/adjuvant LiHsp83 chaperone and this fusion protein was expressed in chloroplasts (100 μg/g of fresh weight) (Albarracín et al. 2015). This result suggests that the fusion of the protein of interest to chaperones like Hsp90 could be implemented as a new strategy. Since plants are considered a new platform to produce drugs and vaccines, the ability of HSPs to chaperone peptides could provide stability to the recombinant protein, increasing the production yields and providing added value to plant-based platforms.

However, the protection achieved can be also improved expressing various antigens in the same plant tissue. This can be accomplished through transient expression by vacuum agroinfiltration, which has been demonstrated to be a versatile system that allows the simultaneous expression of several recombinant proteins. The main advantage of the transient expression plant platform is that antigen-encoding genes can be cloned in different constructs and then later introduced in leaves by a single event of agroinfiltration. However, the expression levels of different antigens are very dissimilar and this generates variability in the antigens ratios. An alternative is to produce each antigen in a separate plant and then combine this material to reach the required dose. In this case, different expression systems such as nuclear and/or chloroplast transformation could be used. These strategies allow several antigens of *T. gondii* to be simultaneously accumulated within the same delivery system and to obtain a multicomponent vaccine that addresses one of the main challenges for the production of a vaccine against toxoplasmosis. In the future, new strategies of immunization should be implemented to improve the degree of protection. The combination of GRA4-SAG1 mixed plant material, as well as the incorporation of new antigens and/or the use of heterologous vaccination protocols like intradermal or intranasal boost, could contribute to achieving a higher level of protection against Toxoplasma infection.

A remarkable feature of *T. gondii* is that the main route of host infection is oral. The main form of infection in herbivores is with oocysts and the route of infection in pigs and humans is through tissue cysts. So, local immunity in the gut via lymphocytes (mainly intraepithelial lymphocytes with CD8+ activity) and IgA is of fundamental importance in host resistance to the parasite (Bourguin et al. 1993). A potential strategy to resolve this problem would be the implementation of plant tissue as a vehicle for vaccine antigens (Fig. 3). The production of plant-derived vaccines has been widely assessed in the last 20 years, with several antigens from human and animal pathogens being correctly expressed and shown to produce a

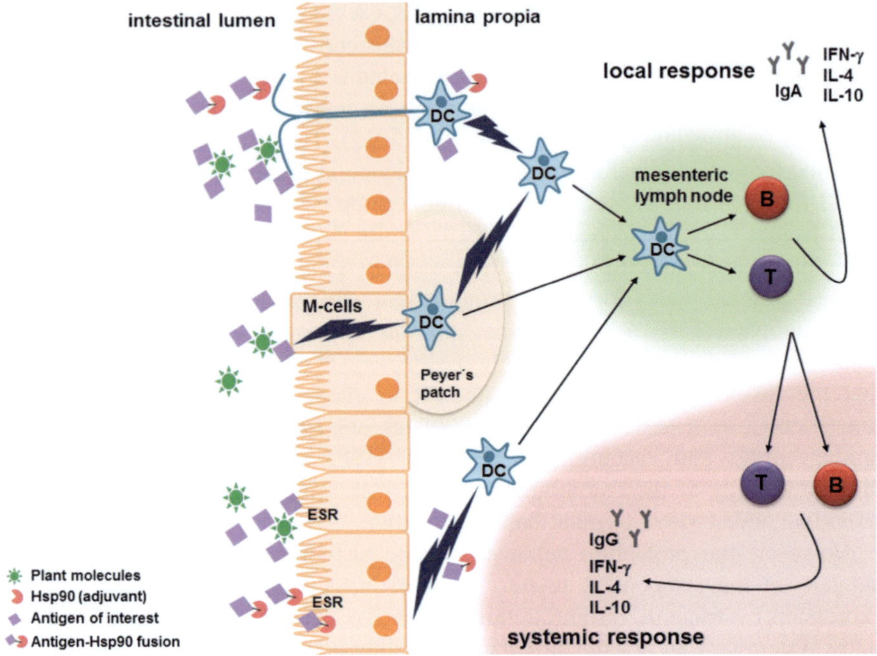

Fig. 3 Hypothetical mechanism to explain the immune response triggered by antigens of *T. gondii* expressed in plants in oral administration. Plant molecules with adjuvant properties or Hsp90 (90-kDa Heat shock protein) as a carrier enhance the antigenicity of the co-administered soluble antigen. Plant molecules and/or Hsp90 proteins as carriers/adjuvants facilitate the delivery of encapsulated antigens and can promote interactions with immune-responsive cells at mucosal surfaces through epithelial surface receptors (ESR), M cells with a subsequent processing by dendritic cells (DCs) and/or a direct uptake by DCs through dendrites that cross the epithelium. Plant-based vaccines carrying *T. gondii* antigens stimulate local immune responses, inducing the production of IgAs, IL-4, IL-10 and IFN-γ and the systemic immune responses by the stimulation of production of IgGs and IL-4, IL-10 and IFN-γ

specific humoral response, and in some cases, a protective response against infection in murine models (Kong et al. 2001; Clemente et al. 2005; Gómez et al. 2008; Santi et al. 2008; Kostrzak et al. 2009; Zhang et al. 2010; Laguía-Becher et al. 2010; Gonzalez-Rabade et al. 2011). One of the main advantages of plants is that their tissues provide a natural environment for antigen encapsulation, which protects the antigen from degradation (Limaye et al. 2006; Hayden et al. 2012). In this way, when the plant tissue is digested, a sufficient quantity of antigen can be captured from the mucous membranes and stimulate an immune response (Berinstein et al. 2005; Kapusta et al. 2010). For this reason, plants are an ideal vehicle for oral vaccine administration.

Oral immunization with chlGRA4 showed that this antigen can elicit an immune protective response, like chlLiHsp83-SAG1, when administrated without any exogenous adjuvant supplement. Oral immunization with *T. gondii* GRA4 antigen expressed in transplastomic tobacco elicits both mucosal and systemic immune responses (Del L Yácono et al. 2012). Several publications have suggested that plant tissues provide protection and prevent degradation of the antigen when it passes through the gut (Yusibov and Rabindran 2008; Paul and Ma 2010; Hayden et al. 2012). These data imply that orally delivered plant-made antigens can be processed to elicit systemic humoral and cellular responses. Evidently, additional adjuvants present in the plant material could contribute to the modulation of the immune response (Fig. 3). In fact, plant secondary metabolites including lectins, saponins, alkaloids, phenolic compounds, and flavonoids are beginning to be exploited as a source for adjuvant capacity (Granell et al. 2010; Vajdy 2011; Rosales-Mendoza and Salazar-González 2014). It has been demonstrated that some phytochemicals and proteins present in plants could synergistically affect the immunogenicity of plant-expressed antigens, acting as endogenous adjuvants (Licciardi and Underwood 2011; Buriani et al. 2011; Corigliano et al. 2011). In this context, plants are an ideal vehicle for oral vaccine administration and should continue to be explored for the development of an anti-*T. gondii* vaccine.

Another important challenge for the development of a successful vaccine against *T. gondii* infection is to find appropriate adjuvants that would facilitate the transport of the antigen from the gut lumen to gut-associated lymphoid tissues. In this context, the use of adjuvants for improving or enhancing the immune response induced by the antigen cocktail expressed in plants should be also explored. In plants, adjuvants and antigens could be expressed in the same plant as recombinant fusion proteins (Fig. 3). Thus, adjuvants could be either co-delivered with the antigen or incorporated to the delivery system with the antigen. In the *T. gondii* model, the most promising results were attained using LiHsp83-SAG1 in oral immunization since LiHSP83-SAG1-immunized mice showed a significant decrease of parasite load (60%) compared to the control group (Albarracín et al. 2015). These results suggest that oral immunization with leaf extract-expressed antigen could be improved by using efficient adjuvants (such as Hsp90) to enhance immunogenicity. Interestingly, Corigliano et al. (2013) demonstrated that the covalent linkage of plant Hsp90 (pHsp90) fused to maltose binding protein (MBP) as a reporter antigen is essential to induce anti-MBP antibodies, causing predominance of specific IgG2a isotype and IFN-γ secretion. In this context, these pHsp90 s were fused to antigenic proteins or peptides to assess their immunomodulatory properties in adjuvant-free immunizations. In addition, Buriani et al. (2012) showed that plant Hsp70 purified from plant tissue transiently expressing the influenza virus nucleoprotein is able to induce both the activation of major histocompatibility complex class I restricted polyclonal T-cell responses and antibody production in different mouse strains without the need of exogenous adjuvants addition. All these results suggest that HSPs could be used as novel carriers for vaccine antigen candidates to improve the immunogenicity property of the plants as delivery vehicles of such antigens (Fig. 3).

The choice of adjuvants for co-expression with antigens in plants depends upon the proved efficacy of the specific adjuvant with a selected antigen. In this way, the expression of adjuvant–antigen as a fusion protein would help to improve the formulations based on plant-made vaccines. The benefit of using a plant system for antigen and adjuvant production is that plants are able to express, process and assemble complex proteins, favoring the formation of recombinant immunogenic complexes inside the plant, which could be subsequently purified or partially purified to then be orally or nasally administered. In addition, adjuvants and antigens can be co-produced directly in edible plants. The main advantage of edible plant-based vaccines is that they are highly safe and cost-effective. Several researchers have demonstrated the potential of edible vegetables and fruits as vaccines against cholera, measles, hepatitis-B, Norwalk virus and rabies virus (Ahmad et al. 2012; Mason and Herbst-Kralovetz 2012). In particular, some plant-based vaccines developed in edible crops have been tested in Phase I clinical trials against diarrheal disease, hepatitis B, and against rabies (Yusibov et al. 2011). Therefore, recombinant immunogenic complexes or adjuvant-protein fusion proteins expressed in edible tissues or seeds could be directly administered by oral delivery as lyophilized tissue or as seed bioencapsulations, thus reducing delivery costs. While expression of *T. gondii* antigens in edible plants has not yet been explored, oral immunization would be the perfect way to obtain the appropriate immunity against this pathogen and an edible vaccine should therefore be pursued in the future.

5 Pathway to Commercialization

Vaccination against bacterial and viral diseases is widespread, routine, and successful, but only a few vaccines for veterinary protozoan diseases have been developed successfully (Garcia et al. 2014). Development of an effective acellular vaccine against toxoplasmosis is a great challenge for medical and veterinary science. One point in favor of using vaccination as a method of toxoplasmosis control is that after the primary infection with *T. gondii*, the host develops an effective protective immunity against the disease (Innes and Vermeulen 2006). Much progress has been made in the understanding of how to induce and regulate protective immune responses, encouraging a real optimism in developing cost-effective new vaccines that may be suitable for large-scale production.

The target antigens chosen for the development of an effective vaccine formulation should: (1) limit acute infection and protect against congenital toxoplasmosis; (2) reduce tissue cysts; and (3) reduce oocyst shedding in cats to avoid environmental contamination (Innes and Vermeulen 2006; Hiszczyńska-Sawicka et al. 2014). As a main goal of anti-*T. gondii* vaccination in animals is to reduce infection in humans, a successful vaccine is needed for both consumption animals and cats. A vaccine for farm animals should prevent parasite transmission to other animals and humans, while a vaccine for cats should prevent environmental contamination and infection risk for intermediate hosts. Generally, vaccine studies have shown that multi-antigenic formulations confer better protection than single-subunit vaccines. Several antigens have been shown to efficiently stimulate an effective response, which could be included in a multi-component vaccine. SAG1, GRA2, GRA4, GRA7, ROP2, ROP5, MIC2, MIC3, MIC4, M2AP and AMA1 are strong candidates to develop an effective acellular vaccine that prevents oocyst shedding by cats and tissue cyst formation in food animals, which combined would have great impact on environmental contamination and consequently on public health (Zhang et al. 2013). However, so far no acellular vaccine against toxoplasmosis has been commercialized, and an effective vaccine to prevent toxoplasmosis remains to be developed. Most parasitologists and vaccinologists agree that future efforts should be concentrated on developing multi-antigen vaccines, developing more efficient delivery systems able to express heterologous proteins abundantly, and determining appropriate immunization schedules and adjuvants to enhance the protective responses. To this end, the platforms for the production of acellular vaccines based on the use of plants can have an important role.

References

Ahmad P, Ashraf M, Younis M et al (2012) Role of transgenic plants in agriculture and biopharming. Biotechnol Adv 30(3):524–540. https://doi.org/10.1016/j.biotechadv.2011.09.006

Albarracín RM, Becher ML, Farran I et al (2015) The fusion of *Toxoplasma gondii* SAG1 vaccine candidate to *Leishmania infantum* heat shock protein 83-kDa improves expression levels in tobacco chloroplasts. Biotechnol J 10(5):748–759. https://doi.org/10.1002/biot.201400742

Altcheh J, Diaz NS, Pepe CM et al (2006) Kinetic analysis of the humoral immune response against 3 *Toxoplasma gondii*-recombinant proteins in infants with suspected congenital toxoplasmosis. Diagn Microbiol Infect Dis 56(2):161–165

Alves CM, Silva DA, Azzolini AE et al (2013) Galectin-3 is essential for reactive oxygen species production by peritoneal neutrophils from mice infected with a virulent strain of *Toxoplasma gondii*. Parasitology 140(2):210–219. https://doi.org/10.1017/S0031182012001473

Ayaz E, Türkoğlu ŞA, Orallar H (2016) *Toxoplasma gondii* and Epilepsy. Turkiye Parazitol Derg 40(2):90–96. https://doi.org/10.5152/tpd.2016.4708

Azevedo KM, Setúbal S, Lopes VG et al (2010) Congenital toxoplasmosis transmitted by human immunodeficiency-virus infected women. Braz J Infect Dis 14(2):186–189

Basavaraju A (2016) Toxoplasmosis in HIV infection: an overview. Trop Parasitol 6(2):129–135

Barragan A, Sibley LD (2002) Transepithelial migration of *Toxoplasma gondii* is linked to parasite motility and virulence. J Exp Med 195(12):1625–1633

Belluco S, Mancin M, Conficoni D et al (2016) Investigating the determinants of *Toxoplasma gondii* prevalence in meat: a systematic review and meta-regression. PLoS ONE 11(4): e0153856. https://doi.org/10.1371/journal.pone.0153856

Bereswill S, Kühl AA, Alutis M et al (2014) The impact of Toll-like-receptor-9 on intestinal microbiota composition and extra-intestinal sequelae in experimental *Toxoplasma gondii* induced ileitis. Gut Pathog 6:19. https://doi.org/10.1186/1757-4749-6-19

Berinstein A, Vazquez-Rovere C, Asurmendi S et al (2005) Mucosal and systemic immunization elicited by Newcastle disease virus (NDV) transgenic plants as antigens. Vaccine 23(48–49): 5583–5589

Bertranpetit E, Jombart T, Paradis E et al (2017) Phylogeography of *Toxoplasma gondii* points to a South American origin. Infect Genet Evol 48:150–155. https://doi.org/10.1016/j.meegid.2016.12.020

Bierly AL, Shufesky WJ, Sukhumavasi W et al (2008) Dendritic cells expressing plasmacytoid marker PDCA-1 are Trojan horses during *Toxoplasma gondii* infection. J Immunol 181 (12):8485–8491

Blader IJ, Coleman BI, Chen CT et al (2015) Lytic cycle of *Toxoplasma gondii*: 15 years later. Annu Rev Microbiol 69:463–485. https://doi.org/10.1146/annurev-micro-091014-104100

Bliss SK, Gavrilescu LC, Alcaraz A et al (2001) Neutrophil depletion during *Toxoplasma gondii* infection leads to impaired immunity and lethal systemic pathology. Infect Immun 69(8):4898–4905

Bourguin I, Chardès T, Bout D (1993) Oral immunization with *Toxoplasma gondii* antigens in association with cholera toxin induces enhanced protective and cell-mediated immunity in C57BL/6 mice. Infect Immun 61(5):2082–2088

Bowie WR, King AS, Werker DH et al (1997) Outbreak of toxoplasmosis associated with municipal drinking water. The BC Toxoplasma Investigation Team. Lancet 350(9072): 173–177

Bresciani KDS, Costa AJ, Toniollo GH et al (1999) Experimental toxoplasmosis in pregnant bitches. Vet Parasitol 86:143–145

Bresciani KDS, Toniollo GH, Costa AJ et al (2001) Clinical, parasitological and obstetric observations in pregnant bitches with experimental toxoplasmosis. Ciênc Rur 31:1039–1043

Bresciani KD, Costa AJ, Toniollo GH et al (2009) Transplacental transmission of *Toxoplasma gondii* in reinfected pregnant female canines. Parasitol Res 104(5):1213–1217. https://doi.org/10.1007/s00436-008-1317-5

Burg JL, Perelman D, Kasper LH et al (1988) Molecular analysis of the gene encoding the major surface antigen of *Toxoplasma gondii*. J Immunol 141(10):3584–3591

Buriani G, Mancini C, Benvenuto E et al (2011) Plant heat shock protein 70 as carrier for immunization against a plant-expressed reporter antigen. Transgenic Res 20(2):331–344. https://doi.org/10.1007/s11248-010-9418-1

Buriani G, Mancini C, Benvenuto E et al (2012) Heat-shock protein 70 from plant biofactories of recombinant antigens activate multiepitope-targeted immune responses. Plant Biotechnol J 10 (3):363–371. https://doi.org/10.1111/j.1467-7652.2011.00673.x

Butcher BA, Kim L, Johnson PF et al (2001) *Toxoplasma gondii* tachyzoites inhibit proinflammatory cytokine induction in infected macrophages by preventing nuclear translocation of the transcription factor NF-kappa B. J Immunol 167(4):2193–2201

Buxton D, Thomson K, Maley S (1991) Vaccination of sheep with a live incomplete strain (S48) of *Toxoplasma gondii* and their immunity to challenge when pregnant. Vet Rec 129(5):89–93

Buxton D, Maley SW, Wright SE et al (2007) *Toxoplasma gondii* and ovine toxoplasmosis: new aspects of an old story. Vet Parasitol 149(1–2):25–28

Cañedo-Solares I, Calzada-Ruiz M, Ortiz-Alegría LB et al (2013) Endothelial cell invasion by *Toxoplasma gondii*: differences between cell types and parasite strains. Parasitol Res 112(8): 3029–3033. https://doi.org/10.1007/s00436-013-3476-2

Cardona N, de-la-Torre A, Siachoque H (2009) *Toxoplasma gondii*: P30 peptides recognition pattern in human toxoplasmosis. Exp Parasitol 123(2):199–202. https://doi.org/10.1016/j.exppara.2009.06.017

Chebolu S, Daniell H (2007) Stable expression of Gal/GalNAc lectin of Entamoeba histolytica in transgenic chloroplasts and immunogenicity in mice towards vaccine development for amoebiasis. Plant Biotechnol J 5(2):230–239

Chen R, Lu SH, Tong QB, Lou D, Shi DY, Jia BB, Huang GP, Wang JF (2009) Protective effect of DNA-mediated immunization with liposome-encapsulated GRA4 against infection of Toxoplasma gondii. J Zhejiang Univ Sci B 10(7):512–521

Clemente M, Curilovic R, Sassone A et al (2005) Production of the main surface antigen of *Toxoplasma gondii* in tobacco leaves and analysis of its antigenicity and immunogenicity. Mol Biotechnol 30:41–50

Clemente M (2014) Overview of plant-made vaccine antigens against toxoplasmosis. In: Rosales-Mendoza S (ed) Genetically engineered plants as a source of vaccines against wide spread diseases. Springer, New York, pp 215–241. https://doi.org/10.1007/978-1-4939-0850-9_11

Clemente M, Corigliano MG (2012) Overview of plant-made vaccine antigens against malaria. J Biomed Biotechnol 2012:206918. https://doi.org/10.1155/2012/206918

Cohen SB, Denkers EY (2014) Border maneuvers: deployment of mucosal immune defenses against *Toxoplasma gondii*. Mucosal Immunol 7(4):744–752. https://doi.org/10.1038/mi.2014.25

Cook AJC, Gilbert RE, Buffolano W et al (2000) Sources of toxoplasma infection in pregnant women: European multicentre case-control study. BMJ 321:142–147

Corigliano MG, Maglioco A, Becher ML, Goldman A, Martín V, Angel SO, Clemente M, Kanellopoulos J (2011) Plant Hsp90 proteins interact with B-Cells and stimulate their proliferation. PLoS ONE 6(6):e21231

Corigliano MG, Fenoy I, Sander V et al (2013) Plant heat shock protein 90 as carrier-adjuvant for immunization against a reporter antigen. Vaccine 31(49):5872–5878. https://doi.org/10.1016/j.vaccine.2013.09.047

Coster LO (2013) Parasitic infections in solid organ transplant recipients. Infect Dis Clin North Am 27(2):395–427. https://doi.org/10.1016/j.idc.2013.02.008

Dabritz HA, Miller MA, Atwill ER et al (2007) Detection of *Toxoplasma gondii*-like oocysts in cat feces and estimates of the environmental oocyst burden. J Am Vet Med Assoc 231(11):1676–1684

Debierre-Grockiego F, Campos MA, Azzouz N et al (2007) Activation of TLR2 and TLR4 by glycosylphosphatidylinositols derived from *Toxoplasma gondii*. J Immunol 179(2):1129–1137

Del Grande C, Galli L, Schiavi E (2017) Is *Toxoplasma gondii* a trigger of bipolar disorder? Pathogens 10;6(1): pii E3. https://doi.org/10.3390/pathogens6010003

Del Rio L, Bennouna S, Salinas J et al (2001) CXCR2 deficiency confers impaired neutrophil recruitment and increased susceptibility during *Toxoplasma gondii* infection. J Immunol 167 (11):6503–6509

Del Yácono LM, Farran I, Becher ML et al (2012) Chloroplast-derived *Toxoplasma gondii* GRA4 antigen used as an oral vaccine protects against toxoplasmosis in mice. Plant Biotechnol J 10(9):1136–1144. https://doi.org/10.1111/pbi.12001

Delhaes L, Ajzenberg D, Sicot B et al (2010) Severe congenital toxoplasmosis due to a *Toxoplasma gondii* strain with an atypical genotype: case report and review. Prenat Diagn 30(9):902–905. https://doi.org/10.1002/pd.2563

Denkers EY, Scharton-Kersten T, Barbieri S et al (1996) A role for CD4+ NK1.1+ T lymphocytes as major histocompatibility complex class II independent helper cells in the generation of CD8 + effector function against intracellular infection. J Exp Med 184(1):131–139

Denkers EY, Schneider AG, Cohen SB, Butcher BA (2012) Phagocyte responses to protozoan infection and how *Toxoplasma gondii* meets the challenge. Publ Libr Sci Pathog 8:e1002794

Derouin F, Santillana-Hayat M (2000) Anti-toxoplasma activities of antiretroviral drugs and interactions with pyrimethamine and sulfadiazine *in vitro*. Antimicrob Agents Chemother 44(9):2575–2577

Desolme B, Mévélec MN, Buzoni-Gatel D et al (2000) Induction of protective immunity against toxoplasmosis in mice by DNA immunization with a plasmid encoding *Toxoplasma gondii* GRA4 gene. Vaccine 18(23):2512–2521

Dhama K, Rajagunalan S, Chakraborty S et al (2013) Food-borne pathogens of animal origin-diagnosis, prevention, control and their zoonotic significance: a review. Pak J Biol Sci 16(20):1076–1085

Dickerson FB, Stallings CR, Boronow JJ et al (2009) A double-blind trial of adjunctive azithromycin in individuals with schizophrenia who are seropositive for *Toxoplasma gondii*. Schizophr Res 112(1–3):198–199. https://doi.org/10.1016/j.schres.2009.05.005

Długonska H (2014) *Toxoplasma gondii* and the host cells. Ann Parasitol 60(2):83–88

Du J, An R, Chen L et al (2014) *Toxoplasma gondii* virulence factor ROP18 inhibits the host NF-κB pathway by promoting p65 degradation. J Biol Chem 289(18):12578–12592. https://doi.org/10.1074/jbc.M113.544718

Dubey JP (1982) Repeat transplacental transfer of *Toxoplasma gondii* in dairy goats. J Am Vet Med Assoc 180(10):1220–1221
Dubey JP, Brown CA, Carpenter JL et al (1992) Fatal toxoplasmosis in domestic rabbits in the USA. Vet Parasitol 44(3–4):305–309
Dubey JP (1996) Infectivity and pathogenicity of *Toxoplasma gondii* oocysts for cats. J Parasitol 82(6):957–961
Dubey JP (2007) The history and life cycle of *Toxoplasma gondii*. In: Weiss LM, Kim K (ed) *Toxoplasma gondii*. The model apicomplexan: perspectives and methods. Academic Press, New York, pp 1–1
Dubey JP (2008) The history of *Toxoplasma gondii*-the first 100 years. J Eukaryot Microbiol 55 (6):467–475. https://doi.org/10.1111/j.1550-7408.2008.00345.x
Dubey JP (2009) Toxoplasmosis in pigs–the last 20 years. Vet Parasitol 164(2–4):89–103. https://doi.org/10.1016/j.vetpar.2009.05.018
Dubey JP, Lindsay DS, Lappin MR (2009) Toxoplasmosis and other intestinal coccidial infections in cats and dogs. Vet Clin North Am Small Anim Pract 39(6):1009–1034. https://doi.org/10.1016/j.cvsm.2009.08.001
Dubey JP, Passos LM, Rajendran C et al (2011) Isolation of viable *Toxoplasma gondii* from feral guinea fowl (Numida meleagris) and domestic rabbits (Oryctolagus cuniculus) from Brazil. J Parasitol 97(5):842–845. https://doi.org/10.1645/GE-2728.1
Dubey JP, Verma SK, Villena I et al (2016) Toxoplasmosis in the Caribbean islands: literature review, seroprevalence in pregnant women in ten countries, isolation of viable *Toxoplasma gondii* from dogs from St. Kitts, West Indies with report of new *T. gondii* genetic types. Parasitol Res 115(4):1627–1634. https://doi.org/10.1007/s00436-015-4900-6
Dunn D, Wallon M, Peyron F, Petersen E, Peckham C, Gilbert R (1999) Mother-to-child transmission of toxoplasmosis: risk estimates for clinical counselling. Lancet 353(9167):1829–1833
Dupont CD, Christian DA, Hunter CA (2012) Immune response and immunopathology during toxoplasmosis. Semin Immunopathol 34(6):793–813. https://doi.org/10.1007/s00281-012-0339-3
Elsheikha HM, Büsselberg D, Zhu XQ (2016) The known and missing links between *Toxoplasma gondii* and schizophrenia. Metab Brain Dis 31(4):749–759. https://doi.org/10.1007/s11011-016-9822-1
Esch KJ1, Petersen CA (2013) Transmission and epidemiology of zoonotic protozoal diseases of companion animals. Clin Microbiol Rev 26(1):58–85. https://doi.org/10.1128/cmr.00067-12
Elbez-Rubinstein A, Ajzenberg D, Dardé ML et al (2009) Congenital toxoplasmosis and reinfection during pregnancy: case report, strain characterization, experimental model of reinfection, and review. J Infect Dis 199(2):280–285
Emery S, Kelleher AD, Workman C et al (2007) Influence of IFNgamma co-expression on the safety and antiviral efficacy of recombinant fowlpox virus HIV therapeutic vaccines following interruption of antiretroviral therapy. Hum Vaccin 3(6):260–267
Ferguson DJ, Hutchison WM (1987) An ultrastructural study of the early development and tissue cyst formation of *Toxoplasma gondii* in the brains of mice. Parasitol Res 73(6):483–491
Ferraro G, Laguía Becher M, Angel SO et al (2008) Efficient expression of a *Toxoplasma gondii* dense granule Gra4 antigen in tobacco leaves. Exp Parasitol 120:118–122. https://doi.org/10.1016/j.exppara.2008.06.002
Feustel SM, Meissner M, Liesenfeld O (2012) *Toxoplasma gondii* and the blood-brain barrier. Virulence 3(2):182–192. https://doi.org/10.4161/viru.19004
Frenkel JK, Dubey JP, Miller NL (1970) *Toxoplasma gondii* in cats: fecal stages identified as coccidian oocysts. Science 167:893–896
Fricker-Hidalgo H, Cimon B, Chemla C et al (2013) Toxoplasma seroconversion with negative or transient immunoglobulin M in pregnant women: myth or reality? A French multicenter retrospective study. J Clin Microbiol 51(7):2103–2111. https://doi.org/10.1128/JCM.00169-13

Ganapathy M1, Perumal A, Mohan C et al (2014) Immunogenicity of *Brugia malayi* Abundant Larval Transcript-2, a potential filarial vaccine candidate expressed in tobacco. Plant Cell Rep 33(1):179–188. https://doi.org/10.1007/s00299-013-1521-3

Garcia JL, Innes EA, Katzer F (2014) Current progress toward vaccines against *Toxoplasma gondii*. Vaccine: Develop Ther 4:23–37

Gazzinelli RT, Eltoum I, Wynn TA et al (1993) Acute cerebral toxoplasmosis is induced by in vivo neutralization of TNF-alpha and correlates with the down-regulated expression of inducible nitric oxide synthase and other markers of macrophage activation. J Immunol 151 (7):3672–3681

Gilbert R, Gras L (2003) European multicentre study on congenital toxoplasmosis. Effect of timing and type of treatment on the risk of mother to child transmission of *Toxoplasma gondii*. BJOG 110(2):112–120

Glasner PD, Silveira C, Kruszon-Moran D et al (1992) An unusually high prevalence of ocular toxoplasmosis in southern Brazil. Am J Ophthalmol 114(2):136–144

Gómez E, Chimeno Zoth S, Carrillo E et al (2008) Mucosal immunity induced by orally administered transgenic plants. Immunobiology 213(8):671–675. https://doi.org/10.1016/j.imbio.2008.02.002

Gonzalez-Rabade N, McGowan EG, Zhou F et al (2011) Immunogenicity of chloroplast-derived HIV-1 p24 and a p24-Nef fusion protein following subcutaneous and oral administration in mice. Plant Biotechnol J 9(6):629–638. https://doi.org/10.1111/j.1467-7652.2011.00609.x

Granell A, Fernández del-Carmen A, Orzáez D (2010) In planta production of plant-derived and non-plant-derived adjuvants. Expert Rev Vaccines 9(8):843–858. https://doi.org/10.1586/erv.10.80

Gras L, Wallon M, Pollak A et al (2005) Association between prenatal treatment and clinical manifestations of congenital toxoplasmosis in infancy: a cohort study in 13 European centres. Acta Paediatr 94(12):1721–1731

Guo M, Dubey JP, Hill D et al (2015) Prevalence and risk factors for *Toxoplasma gondii* infection in meat animals and meat products destined for human consumption. J Food Prot 78(2):457–476. https://doi.org/10.4315/0362-028X.JFP-14-328

Halonen SK1, Weiss LM (2013) Toxoplasmosis. Handb Clin Neurol 114:125–145. https://doi.org/10.1016/b978-0-444-53490-3.00008-x

Harning D, Spenter J, Metsis A et al (1996) Recombinant *Toxoplasma gondii* surface antigen 1 (P30) expressed in Escherichia coli is recognized by human *Toxoplasma*-specific immunoglobulin M (IgM) and IgG antibodies. Clin Diagn Lab Immunol 3(3):355–357

Harker KS, Ueno N, Lodoen MB (2015) *Toxoplasma gondii* dissemination: a parasite's journey through the infected host. Parasite Immunol 37(3):141–149. https://doi.org/10.1111/pim.12163

Hayden CA, Egelkrout EM, Moscoso AM et al (2012) Production of highly concentrated, heat-stable hepatitis B surface antigen in maize. Plant Biotechnol J 10(8):979–984. https://doi.org/10.1111/j.1467-7652.2012.00727.x

Hennequin C, Dureau P, N'Guyen L et al (1997) Congenital toxoplasmosis acquired from an immune woman. Pediatr Infect Dis J 16(1):75–77

Hernandez M, Rosas G, Cervantes J et al (2014) Transgenic plants: a 5-year update on oral antipathogen vaccine development. Expert Rev Vaccines 13(12):1523–1536. https://doi.org/10.1586/14760584.2014.953064

Hiszczynska-Sawicka E, Gatkowska JM, Grzybowski MM et al (2014) Veterinary vaccines against toxoplasmosis. Parasitology 141(11):1365–1378. https://doi.org/10.1017/S0031182014000481

Holland GN (2003) Ocular toxoplasmosis: a global reassessment. Part I: epidemiology and course of disease. Am J Ophthalmol 136(6):973–988

Hu K, Johnson J, Florens L et al (2006) Cytoskeletal components of an invasion machine-the apical complex of *Toxoplasma gondii*. PLoS Pathog 2(2):e13. https://doi.org/10.1371/journal.ppat.0020013

Innes EA, Vermeulen AN (2006) Vaccination as a control strategy against the coccidial parasites *Eimeria Toxoplasma* and *Neospora*. Parasitology 133(Suppl):S145–S168

Innes EA (2010) Vaccination against *Toxoplasma gondii*: an increasing priority for collaborative research? Expert Rev Vaccines 9(10):1117–1119. https://doi.org/10.1586/erv.10.113

Israelski DM, Remington JS (1993) Toxoplasmosis in the non-AIDS immunocompromised host. Curr Clin Top Infect Dis 13:322–356

Jacob SS, Cherian S, Sumithra TG et al (2013) Edible vaccines against veterinary parasitic diseases—current status and future prospects. Vaccine 31(15):1879–1885. https://doi.org/10.1016/j.vaccine.2013.02.022

Jasper S, Vedula SS, John SS et al (2017) Corticosteroids as adjuvant therapy for ocular toxoplasmosis. Cochrane Database Syst Rev 1:CD007417. https://doi.org/10.1002/14651858.cd007417

Jiang HH, Li MW, Xu MJ et al (2015) Prevalence of *Toxoplasma gondii* in dogs in Zhanjiang, Southern China. Korean J Parasitol 53(4):493–496. https://doi.org/10.3347/kjp.2015.53.4.493

Jones JL, Dargelas V, Roberts J et al (2009) Risk factors for *Toxoplasma gondii* infection in the United States. Clin Infect Dis 49(6):878–884. https://doi.org/10.1086/605433

Jongert E, Roberts CW, Gargano N (2009) Vaccines against *Toxoplasma gondii*: challenges and opportunities. Mem Inst Oswaldo Cruz 104(2):252–266. Erratum in: Mem Inst Oswaldo Cruz. 2010 Feb; 105

Kapusta J, Pniewski T, Wojciechowicz J et al (2010) Nanogram doses of alum-adjuvanted HBs antigen induce humoral immune response in mice when orally administered. Arch Immunol Ther Exp (Warsz) 58(2):143–151. https://doi.org/10.1007/s00005-010-0065-2

Katzer F, Brülisauer F, Collantes-Fernández E et al (2011) Increased *Toxoplasma gondii* positivity relative to age in 125 Scottish sheep flocks; evidence of frequent acquired infection. Vet Res 42:121. https://doi.org/10.1186/1297-9716-42-121

Kesik-Brodacka M, Lipiec A, Kozak Ljunggren M et al (2017) Immune response of rats vaccinated orally with various plant-expressed recombinant cysteine proteinase constructs when challenged with Fasciola hepatica metacercariae. PLoS Negl Trop Dis 11(3):e0005451. https://doi.org/10.1371/journal.pntd.0005451

Khan IA, Smith KA, Kasper LH (1988) Induction of antigen-specific parasiticidal cytotoxic T cell splenocytes by a major membrane protein (P30) of *Toxoplasma gondii*. J Immunol 141(10):3600–3605

Khan IA, Murphy PM, Casciotti L et al (2001) Mice lacking the chemokine receptor CCR1 show increased susceptibility to *Toxoplasma gondii* infection. J Immunol 166(3):1930–1937

Khurana S, Batra N (2016) Toxoplasmosis in organ transplant recipients: evaluation, implication, and prevention. Trop Parasitol 6(2):123–128

Kieffer F, Wallon M (2013) Congenital toxoplasmosis. Handb Clin Neurol 112:1099–1101. https://doi.org/10.1016/B978-0-444-52910-7.00028-3

Kim JY, Ahn MH, Jun HS et al (2006) *Toxoplasma gondii* inhibits apoptosis in infected cells by caspase inactivation and NF-kappaB activation. Yonsei Med J 47(6):862–869

Kim SK, Fouts AE, Boothroyd JC (2007) *Toxoplasma gondii* dysregulates IFN-gamma-inducible gene expression in human fibroblasts: insights from a genome-wide transcriptional profiling. J Immunol 178(8):5154–5165

Kim JH, Kang KI, Kang WC et al (2009) Porcine abortion outbreak associated with *Toxoplasma gondii* in Jeju Island. Korea J Vet Sci 10:147–151

Kong Q, Richter L, Yang YF (2001) Oral immunization with hepatitis B surface antigen expressed in transgenic plants. Proc Natl Acad Sci U S A 98(20):11539115–11539144

Kostrzak A, Cervantes Gonzalez M, Guetard D et al (2009) Oral administration of low doses of plant-based HBsAg induced antigen-specific IgAs and IgGs in mice, without increasing levels of regulatory T cells. Vaccine 27(35):4798–47807. https://doi.org/10.1016/j.vaccine.2009.05.092

Labruyere E, Lingnau M, Mercier C et al (1999) Differential membrane targeting of the secretory proteins GRA4 and GRA6 within the parasitophorous vacuole formed by *Toxoplasma gondii*. Mol Biochem Parasitol 20;102(2):311–324

Laguía-Becher M, Martín V, Kraemer M et al (2010) Effect of codon optimization and subcellular targeting on *Toxoplasma gondii* antigen SAG1 expression in tobacco leaves to use in

subcutaneous and oral immunization in mice. BMC Biotechnol 10:52. https://doi.org/10.1186/1472-6750-10-52
Langermans JA, van der Hulst ME, Nibbering PH et al (1992) Endogenous tumor necrosis factor alpha is required for enhanced antimicrobial activity against *Toxoplasma gondii* and *Listeria* monocytogenes in recombinant gamma interferon-treated mice. Infect Immun 60(12):5107–5112
Licciardi PV, Underwood JR (2011) Plant-derived medicines: a novel class of immunological adjuvants. Int Immunopharmacol 11(3):390–398. https://doi.org/10.1016/j.intimp.2010.10.014
Limaye A, Koya V, Samsam M et al (2006) Receptor-mediated oral delivery of a bioencapsulated green fluorescent protein expressed in transgenic chloroplasts into the mouse circulatory system. FASEB J 20(7):959–961
Lim SS, Othman RY (2014) Recent advances in *Toxoplasma gondii* immunotherapeutics. Korean J Parasitol 52(6):581–593. https://doi.org/10.3347/kjp.2014.52.6.581
Lindsay DS, Dubey JP (2011) *Toxoplasma gondii*: the changing paradigm of congenital toxoplasmosis. Parasitology 138(14):1829–1831. https://doi.org/10.1017/S0031182011001478
Machacova T, Bartova E, Sedlak K (2016) Seroprevalence and risk factors of infections with *Neospora caninum* and *Toxoplasma gondii* in hunting dogs from Campania region, southern Italy. Folia Parasitol (Praha) 6;63. pii:2016.012. https://doi.org/10.14411/fp.2016.012
Magalhães FJ, Ribeiro-Andrade M, Souza FM et al (2017) Seroprevalence and spatial distribution of *Toxoplasma gondii* infection in cats, dogs, pigs and equines of the Fernando de Noronha Island. Brazil Parasitol Int 66(2):43–46. https://doi.org/10.1016/j.parint.2016.11.014
Makioka A, Kobayashi A (1991) Toxoplasmacidal activity of macrophages activated by recombinant major surface antigen (P30) of *Toxoplasma gondii*. Infect Immun 59(8): 2851–2852
Marchioro AA, Tiyo BT, Colli CM et al (2016) First Detection of *Toxoplasma gondii* DNA in the fresh leafs of vegetables in South America. Vector Borne Zoonotic Dis 16(9):624–626. https://doi.org/10.1089/vbz.2015.1937
Martin V, Supanitsky A, Echeverria P et al (2004) Recombinant GRA4 or ROP2 protein combined with alum or the gra4 gene provides partial protection in chronic murine models of toxoplasmosis. Clin Lab Immunol 11:704–710
Mason HS, Herbst-Kralovetz MM (2012) Plant-derived antigens as mucosal vaccines. Curr Top Microbiol Immunol 354:101–120. https://doi.org/10.1007/82_2011_158
Mastropaolo JP, Moskowitz KH, Dacanay RJ et al (1989) Conditioned taste aversions as a behavioral baseline for drug discrimination learning: an assessment with phencyclidine. Pharmacol Biochem Behav 32(1):1–8
Matias M, Gomes A, Marques T, Fonseca AC (2014) Ocular toxoplasmosis: a very rare presentation in an immunocompetent patient. BMJ Case Rep. https://doi.org/10.1136/bcr-2014-205846
Matsumoto Y, Suzuki S, Nozoye T et al (2009) Oral immunogenicity and protective efficacy in mice of transgenic rice plants producing a vaccine candidate antigen (As16) of Ascaris suum fused with cholera toxin B subunit. Transgenic Res 18(2):185–192. https://doi.org/10.1007/s11248-008-9205-4
McGettigan BD, Hew M, Phillips E et al (2012) Sulphadiazine-induced renal stones in a 63-year-old HIV-infected man treated for toxoplasmosis. BMJ Case Rep pii:bcr2012006638. https://doi.org/10.1136/bcr-2012-006638
Melzer T, Duffy A, Weiss LM et al (2008) The gamma interferon (IFN-gamma)-inducible GTP-binding protein IGTP is necessary for toxoplasma vacuolar disruption and induces parasite egression in IFN-gamma-stimulated astrocytes. Infect Immun 76(11):4883–4894. https://doi.org/10.1128/iai.01288-07
Mennechet FJ, Kasper LH, Rachinel N et al (2004) Intestinal intraepithelial lymphocytes prevent pathogen-driven inflammation and regulate the Smad/T-bet pathway of lamina propria CD4+ T cells. Eur J Immunol 34(4):1059–1067
Menzies FM, Henriquez FL, Roberts CW (2008) Immunological control of congenital toxoplasmosis in the murine model. Immunol Lett 115(2):83–89

Mévélec MN, Mercereau-Puijalon O, Buzoni-Gatel D et al (1998) Mapping of B epitopes in GRA4, a dense granule antigen of *Toxoplasma gondii* and protection studies using recombinant proteins administered by the oral route. Parasite Immunol 20(4):183–195

Mévélec MN, Bout D, Desolme B et al (2005) Evaluation of protective effect of DNA vaccination with genes encoding antigens GRA4 and SAG1 associated with GM-CSF plasmid, against acute, chronical and congenital toxoplasmosis in mice. Vaccine 23:4489–4499

Miller CM, Boulter NR, Ikin RJ, Smith NC (2009) The immunobiology of the innate response to *Toxoplasma gondii*. Int J Parasitol 39(1):23–39. https://doi.org/10.1016/j.ijpara.2008.08.002

Molestina RE, Payne TM, Coppens I et al (2003) Activation of NF-kappaB by *Toxoplasma gondii* correlates with increased expression of antiapoptotic genes and localization of phosphorylated IkappaB to the parasitophorous vacuole membrane. J Cell Sci 116(Pt 21):4359–4371

Moncada PA, Montoya JG (2012) Toxoplasmosis in the fetus and newborn: an update on prevalence, diagnosis and treatment. Expert Rev Anti Infect Ther 10(7):815–828. https://doi.org/10.1586/eri.12.58

Monreal-Escalante E, Govea-Alonso DO, Hernández M et al (2016) Towards the development of an oral vaccine against porcine cysticercosis: expression of the protective HP6/TSOL18 antigen in transgenic carrots cells. Planta 243(3):675–685. https://doi.org/10.1007/s00425-015-2431-0

Montazeri M, Sharif M, Sarvi S et al (2017) A systematic review of in vitro and in vivo activities of anti-Toxoplasma drugs and compounds (2006-2016). Front Microbio 8:25. https://doi.org/10.3389/fmicb.2017.00025

Montoya JG, Liesenfeld O (2004) Toxoplasmosis. Lancet 363(9425):1965–1976

Muñoz M, Liesenfeld O, Heimesaat MM (2011) Immunology of *Toxoplasma gondii*. Immunol Rev 240(1):269–285. https://doi.org/10.1111/j.1600-065X.2010.00992.x

Nicolle C, Manceaux L (1908) Sur une infection à corps de Leishman (ou organismes voisins) du gondi. C R Hebd Seances Acad Sci 147:763–766

Nigro M, Gutierrez A, Hoffer AM et al (2003) Evaluation of *Toxoplasma gondii* recombinant proteins for the diagnosis of recently acquired toxoplasmosis by an immunoglobulin G analysis. Diagn Microbiol Infect Dis 47:609–613

Okamoto N, Keeling PJ (2014) The 3D structure of the apical complex and association with the flagellar apparatus revealed by serial TEM tomography in *Psammosa pacifica*, a distant relative of the Apicomplexa. PLoS ONE 9(1):e84653. https://doi.org/10.1371/journal.pone.0084653

Paul M, Ma JK (2010) Plant-made immunogens and effective delivery strategies. Expert Rev Vaccines 9(8):821–833. https://doi.org/10.1586/erv.10.88

Peng HJ, Chen XG, Lindsay DS (2011) A review: competence, compromise, and concomitance-reaction of the host cell to *Toxoplasma gondii* infection and development. J Parasitol 97(4):620–628. https://doi.org/10.1645/GE-2712.1

Perkins ES (1973) Ocular toxoplasmosis. Br J Ophthalmol 57(1):1–17

Pleyer U, Schlüter D, Mänz M (2014) Ocular toxoplasmosis: recent aspects of pathophysiology and clinical implications. Ophthalmic Res 52(3):116–123. https://doi.org/10.1159/000363141

Pollard AM, Onatolu KN, Hiller L et al (2008) Highly polymorphic family of glycosylphosphatidylinositol-anchored surface antigens with evidence of developmental regulation in *Toxoplasma gondii*. Infect Immun 76(1):103–110

Raeghi S, Akaberi A, Sedeghi S (2011) Seroprevalence of *Toxoplasma gondii* in sheep, cattle and horses in Urmia North-West of Iran. Iran J Parasitol 6(4):90–94

Rajapakse R, Uring-Lambert B, Andarawewa KL (2007) 1,25(OH)2D3 inhibits in vitro and in vivo intracellular growth of apicomplexan parasite *Toxoplasma gondii*. J Steroid Biochem Mol Biol 103(3–5):811–814

Reichmann G, Walker W, Villegas EN (2000) The CD40/CD40 ligand interaction is required for resistance to toxoplasmic encephalitis. Infect Immun 68(3):1312–1318

Rodriguez JB, Szajnman SH (2012) New antibacterials for the treatment of toxoplasmosis; a patent review. Expert Opin Ther Pat 22(3):311–333. https://doi.org/10.1517/13543776.2012.668886

Robert-Gangneux F, Dardé ML (2012) Epidemiology of and diagnostic strategies for toxoplasmosis. Clin Microbiol Rev 25(2):264–96. doi:0.1128/CMR.05013-11. Erratum in: Clin Microbiol Rev 2012 25(3):583

Rosales-Mendoza S, Salazar-González JA (2014) Immunological aspects of using plant cells as delivery vehicles for oral vaccines. Expert Rev Vaccines 13(6):737–749. https://doi.org/10.1586/14760584.2014.913483

Rosso F, Les JT, Agudelo A et al (2008) Prevalence of infection with *Toxoplasma gondii* among pregnant women in Cali, Colombia, South America. Am J Trop Med Hyg 78(3):504–508

Rostami A, Seyyedtabaei SJ, Aghamolaie S et al (2016) Seroprevalence and risk factors associated with *Toxoplasma gondii* infection among rural communities in northern Iran. Rev Inst Med Trop Sao Paulo 22(58):70. https://doi.org/10.1590/S1678-9946201658070

Sathish K, Sriraman R, Subramanian BM et al (2012) Plant expressed coccidial antigens as potential vaccine candidates in protecting chicken against coccidiosis. Vaccine 30(30):4460–4464. https://doi.org/10.1016/j.vaccine.2012.04.076

Saeij JP, Coller S, Boyle JP et al (2007) *Toxoplasma* co-opts host gene expression by injection of a polymorphic kinase homologue. Nature 445(7125):324–327

Sánchez VR, Pitkowski MN, Fernández Cuppari AV, Rodríguez FM, Fenoy IM, Frank FM, Goldman A, Corral RS, Martin V (2011) Combination of CpG-oligodeoxynucleotides with recombinant ROP2 or GRA4 proteins induces protective immunity against Toxoplasma gondii infection. Exp Parasitol 128(4):448–453. https://doi.org/10.1016/j.exppara.2011.04.004

Sanecka A, Frickel EM (2012) Use and abuse of dendritic cells by *Toxoplasma gondii*. Virulence 3 (7):678–689. https://doi.org/10.4161/viru.22833

Santi L, Batchelor L, Huang Z et al (2008) An efficient plant viral expression system generating orally immunogenic Norwalk virus-like particles. Vaccine 26(15):1846–1854. https://doi.org/10.1016/j.vaccine.2008.01.053

Scallan E, Hoekstra RM, Angulo FJ et al (2011) Foodborne illness acquired in the United States-major pathogens. Emerg Infect Dis 17:7–15

Shaapan RM (2016) The common zoonotic protozoal diseases causing abortion. J Parasit Dis 40 (4):1116–1129

Shapira S, Harb OS, Margarit J et al (2005) Initiation and termination of NF-kappaB signaling by the intracellular protozoan parasite *Toxoplasma gondii*. J Cell Sci 118(Pt 15):3501–3508

Shibre T, Alem A, Abdulahi A et al (2010) Trimethoprim as adjuvant treatment in schizophrenia: a double-blind, randomized, placebo-controlled clinical trial. Schizophr Bull 36(4):846–851. https://doi.org/10.1093/schbul/sbn191

Siachoque H, Guzman F, Burgos J et al (2006) *Toxoplasma gondii*: immunogenicity and protection by P30 peptides in a murine model. Exp Parasitol 114(1):62–65

Sibley LD, Adams LB, Fukutomi Y et al (1991) Tumor necrosis factor-alpha triggers antitoxoplasmal activity of IFN-gamma primed macrophages. J Immunol 147(7):2340–2345

Sibley LD (2011) Invasion and intracellular survival by protozoan parasites. Immunol Rev 240 (1):72–91. https://doi.org/10.1111/j.1600-065X.2010.00990.x

Skariah S, McIntyre MK, Mordue DG (2010) *Toxoplasma gondii*: determinants of tachyzoite to bradyzoite conversion. Parasitol Res 107(2):253–260. https://doi.org/10.1007/s00436-010-1899-6

Sturge CR, Yarovinsky F (2014) Complex immune cell interplay in the gamma interferon response during *Toxoplasma gondii* infection. Infect Immun 82(8):3090–3097. https://doi.org/10.1128/IAI.01722-14

Suzuki Y, Conley FK, Remington JS (1989) Importance of endogenous IFN-gamma for prevention of toxoplasmic encephalitis in mice. J Immunol 143(6):2045–2050

Suzuki Y, Remington JS (1993) Toxoplasmic encephalitis in AIDS patients and experimental models for study of the disease and its treatment. Res Immunol 144(1):66–67

Tenter AM, Heckeroth AR, Weiss LM (2000) *Toxoplasma gondii*: from animals to humans. Int J Parasitol 30(12–13):1217–1258. Erratum in: Int J Parasitol 2001 31(2):217–220

Thiébaut R, Leproust S, Chêne G et al (2007) Effectiveness of prenatal treatment for congenital toxoplasmosis: a meta-analysis of individual patients' data. Lancet 369(9556):115–122

Torgerson PR, Mastroiacovo P (2013) The global burden of congenital toxoplasmosis: a systematic review. Bull World Health Organ 91(7):501–508. https://doi.org/10.2471/BLT.12.111732

Torrey EF, Yolken RH (2003) *Toxoplasma gondii* and schizophrenia. Emerg Infect Dis 9 (11):1375–1380

Vajdy M (2011) Immunomodulatory properties of vitamins, flavonoids and plant oils and their potential as vaccine adjuvants and delivery systems. Expert Opin Biol Ther 11(11):1501–1513. https://doi.org/10.1517/14712598.2011.623695

Vaudaux JD, Muccioli C, James ER et al (2010) Identification of an atypical strain of *Toxoplasma gondii* as the cause of a waterborne outbreak of toxoplasmosis in Santa Isabel do Ivai, Brazil. J Infect Dis 202(8):1226–1233

Wang Y, Wang G, Zhang D et al (2013) Screening and identification of novel B cell epitopes of *Toxoplasma gondii* SAG1. Parasit Vectors 6:125. https://doi.org/10.1186/1756-3305-6-125

Wang Y, Yin H (2014) Research progress on surface antigen 1 (SAG1) of *Toxoplasma gondii*. Parasit Vectors 7:180. https://doi.org/10.1186/1756-3305-7-180

Wiengcharoen J, Thompson RC, Nakthong C et al (2011) Transplacental transmission in cattle: is *Toxoplasma gondii* less potent than *Neospora caninum*? Parasitol Res 108(5):1235–1241. https://doi.org/10.1007/s00436-010-2172-8

Wilbers RH, Westerhof LB, van Noort K et al (2017) Production and glyco-engineering of immunomodulatory helminth glycoproteins in plants. Sci Rep 7:45910. https://doi.org/10.1038/srep45910

Wilson CB, Remington JS, Stagno S et al (1980) Development of adverse sequelae in children born with subclinical congenital Toxoplasma infection. Pediatrics 66(5):767–774

Xiao J, Yolken RH (2015) Strain hypothesis of *Toxoplasma gondii* infection on the outcome of human diseases. Acta Physiol (Oxf) 213(4):828–845. https://doi.org/10.1111/apha.12458

Yarovinsky F, Hieny S, Sher A (2008) Recognition of *Toxoplasma gondii* by TLR11 prevents parasite-induced immunopathology. J Immunol 181(12):8478–8484

Yusibov V, Rabindran S (2008) Recent progress in the development of plant derived vaccines. Expert Rev Vaccines 7(8):1173–1183. https://doi.org/10.1586/14760584.7.8.1173

Yusibov V, Streatfield SJ, Kushnir N (2011) Clinical development of plant-produced recombinant pharmaceuticals: vaccines, antibodies and beyond. Hum Vaccin 7(3):313–321

Zhang H, Liu M, Li Y et al (2010) Oral immunogenicity and protective efficacy in mice of a carrot-derived vaccine candidate expressing UreB subunit against *Helicobacter pylori*. Protein Expr Purif 69(2):127–131. https://doi.org/10.1016/j.pep.2009.07.016

Zhang NZ, Chen J, Wang M (2013) Vaccines against *Toxoplasma gondii*: new developments and perspectives. Expert Rev Vaccines 12(11):1287–1299. https://doi.org/10.1586/14760584.2013.844652

Zhang NZ, Wang M, Xu Y (2015) Recent advances in developing vaccines against *Toxoplasma gondii*: an update. Expert Rev Vaccines 14(12):1609–1621. https://doi.org/10.1586/14760584.2015.1098539

Zhu S (2009) Psychosis may be associated with toxoplasmosis. Med Hypotheses 3(5):799–801. https://doi.org/10.1016/j.mehy.2009.04.013

Vaccines Against West Nile Virus

Haiyan Sun and Qiang Chen

Abstract Despite the availability of two veterinary vaccines against West Nile virus (WNV), there remains a desperate need for a more efficient, safer, cheaper WNV vaccine that can be delivered conveniently to animals. The global threat of WNV epidemics with increasingly severe neuroinvasive infections makes this need even more urgent. To date, vaccine candidates based on inactivated, live-attenuated, or chimeric virus, and viral DNA and protein subunits have been developed. However, commercialization of a WNV veterinary vaccine may largely depend on the economics of vaccine production, as only novel low-cost production platforms would produce vaccines that outcompete the cost of clinical treatment for animals. In this chapter, we review the progress of using plants to develop effective WNV vaccines and address the economic challenges of WNV vaccine production. The status of current WNV vaccine development is summarized. The advantages of plant-based platforms for WNV vaccine production in cost, speed and scalability are briefly discussed. The progress in developing WNV subunit vaccines in plants is reviewed within the context of their expression, characterization, downstream processing, and potency in animal models. The development of WNV vaccines based on virus-like particles is also highlighted. We are confident that plants are one of the platforms that offer potent, safe and affordable veterinary vaccines against WNV.

Keywords West nile virus (WNV) · Flavivirus · Zika virus (ZIKV)
Plant-made vaccine · Virus-like particle (VLP) · Plant-made biologics (PMB)
Plant-made pharmaceutical (PMP) · Downstream processing · Oral vaccine

---

H. Sun · Q. Chen (✉)
Center for Immunotherapy, Vaccines & Virotherapy, Biodesign Institute, Tempe, AZ, USA
e-mail: qiang.chen.4@asu.edu

Q. Chen
School of Life Sciences, Arizona State University, Tempe, AZ 85287-5401, USA

J. MacDonald (ed.), *Prospects of Plant-Based Vaccines in Veterinary Medicine*,
https://doi.org/10.1007/978-3-319-90137-4_5

1 Introduction

West Nile virus (WNV) is a single-stranded RNA virus that belong to the genus *Flavivirus* of the family *Flaviviridae*. This genus includes many arthropod-borne pathogens such as yellow fever virus (YFV), Japanese encephalitis virus (JEV), Dengue virus (DENV1-4) tick-borne encephalitis virus (TBEV), and Zika virus (ZIKV). In nature, WNV is mostly maintained in sylvatic cycles between birds and ornithophilic arthropods, i.e. mosquitos and occasionally ticks. Although not all bird species develop illness from WNV infection, high mortality rate has been observed in some species (Komar et al. 2003). WNV can easily spread to humans and domestic mammals, especially horses. Epidemics of WNV fever were first reported in the 1950s in Israel (Zeller and Schuffenecker 2004). Since then, WNV outbreaks both in humans and horses have occurred in Africa, Europe and eventually North America in 1999 (Zeller and Schuffenecker 2004). WNV has become an important health concern for both humans and horses, including pet horses, because of the increased frequency of outbreaks and increased neurological disease cases (Castillo-Olivares and Wood 2004).

Due to its global health impact, WNV infection has been under intensified surveillance in humans, horses, birds and mosquitos since the 1990s. This has led to the identification of many WNV strains. WNV strains can be classified in up to seven different genetic lineages based on phylogenetic analysis (Rizzoli et al. 2015). Among all the known WNV strains, Lineage 1 is the most popularly spread lineage which was responsible for the outbreak in New York in 1999 (Rizzoli et al. 2015). Lineage 1 can be further divided into three sublineages: 1a, 1b and 1c. WNV-1a has been found in Africa, Europe, North America and Asia. WNV-1b contains Kunjin virus strains identified in Australia and WNV-1c was only found in India (Rizzoli et al. 2015). Lineage 2 is considered the oldest WNV strain and the second most widely spread (Rizzoli et al. 2015). Lineage 3, also known as Rabensburg virus, was isolated from mosquitos found in Czech Republic in 1997 (Rizzoli et al. 2015). Lineage 4 includes three sublineages with 4a first isolated from ticks then in mosquitos and frogs in Russia while 4b and 4c were detected in mosquitos from Spain and Australia, respectively. Kunjin virus KUN MP502-66 isolated from Malaysia, Koutango virus and a putative new lineage of WNV isolated from Senegal in Africa were classified as Lineage 5, 6 and 7 (Rizzoli et al. 2015). Most circulating WNV strains belong to either lineage 1 or 2 and are associated with outbreaks among humans and horses, while other lineages have been identified, so far, mainly in mosquitos and birds (Rizzoli et al. 2015).

In humans, most of the WNV infections are asymptomatic, but 15–20% of the cases reported mild symptoms including fever, nausea, headache and vomiting and, in less than 1% cases, it leads to neuroinvasive diseases, even death (Zeller and Schuffenecker 2004). In horses, various clinical signs have been reported for WNV encephalomyelitis that includes fever, ataxia, partial paralysis, recumbency, and behavioral changes. Compared to humans, the incidence of WNV infection in horses is much higher and the disease is more severe with almost one third of infections resulting in fatality and 40% of the surviving horses have neurological sequelae (Aharonson-Raz et al. 2014). Unlike in humans, the severity of neurological diseases in horses is not associated with aging (Castillo-Olivares and Wood 2004).

WNV has an 11 kb single-stranded RNA genome that encodes three structural proteins: Capsid protein (C), Pre-membrane/membrane protein (prM/M) and envelope protein (E), as well as seven non-structural proteins (NS1, NS2A, NS2B, NS3, NS4A, NS4B and NS5). The functions of the non-structural proteins are not yet fully determined, but they are known to be involved in WNV RNA replication, translation, and the assembly of viral replication complexes in infected cells (Brinton 2013). In addition to its role for viral RNA replication and packaging, the WNV Capsid protein was found to interact with multiple host cell proteins (Yang et al. 2002; van Marle et al. 2007; Medigeshi et al. 2009; Brinton 2013) that determines virulence and plays an important role in pathogenesis. The prM and E protein are the two main immunogens and they form heterodimers in immature viral particles (Xu et al. 2011) while the prM acts as chaperone for the maturation of E protein. Cleavage of the pr peptide from PrM by the cellular protease furin and the resulting conformational change of E protein yields mature virions. Co-expression of the prM and E proteins in vitro has been shown to generate noninfectious virus like particles (VLPs) that have been tested as WNV vaccine candidates (Ohtaki et al. 2010). The E protein is the major surface component of the mature WNV viral particle and it is responsible for receptor binding and viral entry into host cells through membrane fusion. Similar to DENV and TBEV, the WNV E protein consists of three distinct domains DI, DII and DIII (Kanai et al. 2006). The DI contains a β barrel core structure and connects the DII and DIII. DII is built mostly of β strands and contains the well-conserved fusion loop among flaviviruses. DIII has an Ig-like structure and contains the putative receptor binding sites and epitopes for WNV-specific neutralizing antibodies (Kanai et al. 2006). There is no definitive conclusion on which WNV E domain(s) displays the most diversity across WNV strains, while it is clear that DIII is the most diverse domain among the E protein of different flaviviruses.

2 Current Development of WNV Vaccines

Currently, there is no effective therapeutic for WNV related diseases; all treatments for both humans and horses are supportive, which include providing fluid, nutritional care and reducing inflammation in neurological diseases (Castillo-Olivares and Wood 2004). Several vaccine candidates are under clinical investigations, but no WNV vaccine has been yet approved for human use (Table 1). For veterinary applications, vaccines currently available are poorly immunogenic, unsafe, or not economical to produce (Table 1). Therefore, development of effective, safe, and cost-effective vaccines is urgently needed for preventing WNV infection and controlling its outbreaks in human or horse populations.

Table 1 Current WNV veterinary vaccines and vaccines under clinical trials

Vaccine name and sponsor	Vaccine type	Antigen	Development stage	References
West Nile innovator (Pfizer)	Inactivated whole virus vaccine	Formalin inactivated WNV-NY99	Licensed for veterinary use	Ng et al. (2003)
Recombitek (Merial)	Chimeric/ recombinant vaccine	Canarypox expressing PrM/E	Licensed for veterinary use	Minke et al. (2004)
Prevenile (Intervet)	Chimeric/ recombinant vaccine	YF17D backbone expressing PrM/E	Licensed for veterinary use in 2005 then recalled in 2010	Arroyo et al. (2004)
West Nile innovator DNA (Pfizer)	DNA vaccine	DNA plasmid encodes PrM/E	Licensed for veterinary use in 2005, discontinued	Martin et al. (2007)
ChimeriVax-WN02	Chimeric/ recombinant vaccine	YF17D backbone expressing PrM/E	Phase II trial	Monath et al. (2006), Dayan et al. (2013)
rWN/DEN4Δ30	Chimeric/ recombinant vaccine	DV4 expressing PrM/E	Phase I trial	Durbin et al. (2013)
VRC-WNVDNA017-00-VP	DNA vaccine	DNA plasmid encodes PrM/E	Phase I trial	Martin et al. (2007), Ledgerwood et al. (2011)
WN-80E (Hawaii Biotech)	Recombinant subunit vaccine	Truncated WNV E protein	Phase I trial	Coller et al. (2012)

2.1 Inactivated Whole WNV Vaccines

Using an inactivated whole virus as a vaccine is well established and has been continuously used for a variety of viral diseases. In fact, the first WNV vaccine was based on a formalin-inactivated WNV-NY99 that was developed by Fort Dodge Animal Health (commercialized by Pfizer) shortly after the 1999 outbreak in New York (Ng et al. 2003). This inactivated whole virus vaccine induces protective immunity against WNV infection in horses and other animal models (Ng et al. 2003). However, the requirement of handling highly pathogenic WNV strains in a Biosafety Level 3 (BSL-3) environment in large-scale during the manufacturing process is costly and labor intensive. Moreover, the risk of incomplete inactivation of the live virus presents another public safety concern. Recently, a new WNV vaccine candidate has been developed by inactivation of a naturally attenuated

WNV Kunjin strain using hydrogen peroxide (Amanna et al. 2012). Compared to formalin, inactivation with hydrogen peroxide minimizes the damage to WNV antigenic structures. As a result, this vaccine candidate may have improved immunogenicity over the first WNV vaccine, as well as a better safety profile due to the attenuated nature of the Kunjun strain. Animal studies have shown that two doses of this WNV vaccine candidate can protect mice and non-human primates from WNV infection by inducing broadly neutralizing WNV specific antibody responses (Pinto et al. 2013; Poore et al. 2017).

2.2 *Chimeric/Recombinant Vaccines*

Chimeric/recombinant vaccines are another important approach for WNV vaccine development. These vaccines use a non-WNV virus as the backbone and replace the PrM/E coding sequences of the carrier virus with the corresponding WNV genes to create chimeric viruses. Two such vaccines have been licensed for veterinary use using the canarypox and YFV 17D as the backbone, respectively (Arroyo et al. 2004; Minke et al. 2004). Further mutations in the E protein were introduced to reduce the potential side effect of neurovirulence in immunized subjects (Arroyo et al. 2004; Monath et al. 2006). The YFV-based vaccine (ChimeriVax-WN02) has been evaluated in numerous pre-clinical studies and three clinical trials (Dayan et al. 2013). The potential safety risk is a major disadvantage of these chimeric vaccines. Since they are attenuated live viruses, there is always a concern for their potential reversion to virulent strains. In addition, new viruses could be generated as the result of recombination between the chimeric virus and another flavivirus, raising further safety concerns. A similar strategy was used to develop another live and attenuated chimeric/recombinant vaccine for WNV using DENV4 backbone with 30 neucleotides deleted in the 3′ non-coding region (rWN/DEN4Δ30) (Durbin et al. 2013). This vaccine candidate is well tolerated and immunogenic. However, it can replicate in a mosquito vector that carries wild-type WNV and DENV, which raises safety concerns for its further development.

2.3 *DNA Based WNV Vaccines*

DNA based vaccine approaches can provide efficient and cost-effective vaccine development using modern genetic technology. The first WNV DNA vaccine was developed in 2001 with one single plasmid encoding the WNV prM/E region (Davis et al. 2001). This DNA vaccine was shown to protect against WNV by inducing neutralizing antibodies in both mice and horses. It was licensed by USDA for preventing WNV infection in horses in 2005 (Martin et al. 2007), but later discontinued by Pfizer (Brandler and Tangy 2013). Subsequently, a similar DNA plasmid vaccine and its updated version with a modified promoter were assessed in phase I clinical trials (Martin et al. 2007; Ledgerwood et al. 2011). These studies showed that these DNA-based vaccines were safe and somewhat immunogenic in both young (18–50) and elder (50–65) human groups. However, this vaccine candidate required three 4-mg doses via intramuscular injection to achieve detectable immunogenicity, rendering it a suboptimal vaccine candidate due to its poor immunogenicity. Another DNA vaccine for WNV used a plasmid vector to direct in vivo transcription of the full length Kunjin viral RNA in mice (Hall et al. 2003). Attenuated Kunjin virus was detected in the sera of immunized mice after 3–4 days. The vaccinated mice showed full protection against a lethal challenge of virulent WNV strain NY99.

For most DNA vaccines, the biggest challenge is their low immunogenicity. Over the years several approaches have been applied to improve the stimulation of immune responses. One approach was to co-express the capsid protein from a separate promoter along with a capsid-deleted DNA vaccine to allow the formation of secreted single-round infectious particles (SRIPs) (Chang et al. 2008), thereby mimicking the live virus infection to induce protective immune response but without the usage of infectious virus. Other approaches include co-immunization of an E protein DIII-based DNA vaccine with an optimized IL-15 plasmid to stimulate antibody secreting B cells, which enhanced the overall immune response by four to five folds (Ramanathan et al. 2009). While promising, these strategies use more than one plasmid in their formulation, complicating the manufacturing process and increasing the production cost. More recently, nanoparticles were used as carriers for DNA vaccine delivery (De Filette et al. 2014); however, it failed to induce the desired humoral immune response when applied alone. Furthermore, efforts have been made to develop infectious yet safe WNV DNA vaccines by designing chimeric WNV W1806 DNA with mutations in the E protein to further attenuate its virulence (Yamshchikov et al. 2017).

2.4 Recombinant Subunit Protein Vaccines

Recombinant subunit WNV proteins have been researched as vaccine candidates against WNV infection since 2001. The WNV E protein was expressed in *E. coli* and purified and used to immunize mice, which protected mice against a lethal WNV infection (Wang et al. 2001). However, Flavivirus E or its subdomains produced in *E. coli* are often recovered in insoluble inclusion bodies, requiring a cumbersome and unscalable refolding process to obtain native antigenic structures (Yang et al. 2017a). Later, a truncated form of the E protein was expressed in Drosophila S2 cells to produce an antigen that resembled its native conformation (Ledizet et al. 2005). Vaccination of mice with this truncated E protein, using aluminum hydroxide as an adjuvant, protected animals from a WNV lethal challenge. Similarly, another truncated WNV E protein without the C-terminal membrane anchor (WN-80E) was also expressed in Drosophila S2 cells. When co-delivered with adjuvants, this subunit vaccine candidate elicited both humoral and cellular immune responses in mice and a nonhuman primate model (Lieberman et al. 2007, 2009). WN-80E was eventually tested in a Phase I clinical trial for safety and immunogenicity. Human subjects were vaccinated with three injections of 5, 15 or 50 μg of WN-80E with adjuvant or 50 μg of WN-80E without adjuvant (Coller et al. 2012). The vaccine was well tolerated and all subjects developed neutralizing antibodies after the third injection with the PRNT50 (plaque reduction neutralization test) ranging from 1:10 to 1:100. These results indicate that insect cells provide an improved production platform for producing soluble antigens. Like all cell culture-based systems, however, it has limitations in high production cost and scalability (Chen 2011a, b).

VLPs are another group of vaccine candidates in the recombinant subunit protein vaccine category. VLPs containing the WNV prM and E protein were first produced in the baculovirus expression system and elicited neutralizing antibody responses in immunized BALB/c mice (Qiao et al. 2004). Subsequent studies in a Chinese hamster ovary (CHO) cell expression system with serum free culture media showed that two types of VLPs differing in size and maturation stage were produced. The larger VLP was shown to contain the mature M protein and have better protective efficacy in immunized mice (Ohtaki et al. 2010). In another study, WNV prM-E VLPs were expressed with a herpes simplex virus 1 recombinant vector. Furthermore, they were efficiently released from host cells and induced a protective immune response against WNV upon vaccination in mice (Taylor et al. 2016). While efficacious, these mammalian cell-produced vaccine candidates are costly. Further development and production of these candidates in a more cost-effective system will facilitate the realization of their full potential in preventing WNV global epidemics (Chen 2011a, b).

Since DIII of the E protein has been shown to contain the receptor-binding motif and epitopes for WNV-specific neutralizing antibodies, VLPs with DIII displayed on the surface were also explored (Spohn et al. 2010; Chua et al. 2013; Taylor et al. 2016). A conjugated vaccine with recombinant DIII chemically cross-linked to bacteriophage AP205-derived VLPs induced higher titers of DIII-specific neutralizing antibodies in mice compared to those induced by recombinant DIII protein alone (Spohn et al. 2010). Similarly, DIII-carrying mosaic AP205 VLPs were also developed by genetic fusion of the DIII sequence to the C-terminal of AP205 coat protein under codon suppression condition, which potently induced the production of WNV-neutralizing antibodies (Cielens et al. 2014). Another variation of DIII-displaying VLPs were designed by fusing DIII with the gamma chain of IgE receptor. When this fusion construct was expressed in insect cells, VLPs were actively secreted by host cells into the media and DIII protein was found on the surface of these extracellular VLPs (Chua et al. 2013). However, these DIII displayed VLPs only generated modest neutralizing immune responses in mice.

3 Plants as an Optimal Platform to Produce WNV Veterinary Vaccines

As WNV has become a serious health concern for humans and animals, with effective therapeutics still not available, there is an urgent need to develop vaccines not only for humans but also for other susceptible animals. WNV has been found in about 300 bird species and many non-avian vertebrates including horses, who are also susceptible to and commonly infected by WNV (Iyer and Kousoulas 2013). Low cost and effective vaccines that can be delivered via flexible routes would protect animals against WNV infection and significantly reduce its transmission. While two WNV vaccines are currently commercially available for use in horses, these vaccines require at least two injections and an annual boost to ensure protection. Adverse effects have been widely-reported for these vaccines in horses, indicating the current veterinary vaccines are far from ideal for WNV prevention.

Besides the difficulty in balancing the immunogenicity and safety of the vaccines, development of new WNV candidate vaccines is often haunted by the high cost of large-scale recombinant protein expression, purification and storage. However, plants represent an alternative system that may address these limitations. Tobacco and sunflower plants have been used to express recombinant proteins for more than 30 years (Barta et al. 1986). Plant based production systems have drawn more attention as Ebola virus-infected patients showed dramatic improvement after receiving ZMapp, a plant-made antibody cocktail against EBOV (Lyon et al. 2014). Plant-based expression systems may yield large amounts of recombinant proteins in a relatively short production period without the risk of contamination by animal pathogens, compared to using mammalian cell culture (Chen 2011a, b). Unlike bacteria or other prokaryotic systems, plants share similar endomembrane and secretory pathways with mammalian cells, thereby allowing proper assembly of recombinant proteins and the necessary post translational modifications. Moreover, plant-made recombinant proteins are more cost effective than mammalian cell expressed recombinant proteins (Tuse et al. 2014; Chen and Davis 2016) and can be easily scaled for manufacturing (Chen 2018; Chen and Lai 2015; Chen et al. 2016). For orally delivered drugs, the cost of production can be further reduced when edible plants are used for production as no intensive purification steps are needed (Chen and Davis 2016). Therefore, plant based expression systems may serve as a useful alternative for developing low cost WNV recombinant veterinary vaccines (Chen et al. 2018).

3.1 Plant Expression System

Traditionally, foreign genes were incorporated into the plant nuclear genome to generate transgenic plants for recombinant protein production (Chen 2008, 2011a, b). This typically includes using *Agrobacterium tumefaciens* to deliver the gene of interest randomly into the plant genome and selection of positive clones. While establishing a permanent genetic source of an antigen is attractive, this procedure is often time consuming and can be complicated by unexpected gene silencing (Takeyama et al. 2015). More recently, transplastomic plants have been developed for recombinant protein production. This approach uses particle bombardment to deliver the gene of interest to the chloroplast of the plants. While it can produce high protein yield and prevent transgene outcrossing through pollen, this has been limited to the production of subunit vaccines containing one polypeptide and those that do not need post-translational modification (Takeyama et al. 2015).

Recent advances in plant viral vectors made it possible for transient vaccine expression in plants with high protein yield in a short time period. Many plant viruses have been used for this purpose, including tobacco mosaic virus (TMV), Potato virus X (PVX), Cucumber mosaic virus (CMV), Cowpea mosaic virus (CPMV), geminivirus and Cauliflower mosaic virus (CaMV) (Lico et al. 2008; Hefferon 2014). Historically, full virus vectors carrying the gene of interest were used; these vectors retain infectivity and produce viral particles from host cells that may spread to other plants. The second generation of plant viral vectors use a "deconstructed virus" strategy in which minimal sections of the viral genome important for replication and non-viral sections were integrated together to form a complete replicon (Peyret and Lomonossoff 2015). One such popularly used viral vector system is the magnICON system, which contains a 5′ module that has the TMV-based elements necessary for replication, a 3′ module that has the gene of interest, and a recombinase module containing a streptomyces phage 31 integrase (Peyret and Lomonossoff 2015). The magnICON system also relies on the co-infiltration of the three Agrobacterium strains with each containing one of the three modules. Once delivered to plant cells, the integrase fuses the 5′ and 3′ modules into a complete replicon and targets protein production (Peyret and Lomonossoff 2015).

Another example of the destructed virus vectors is the geminivirus-based expression systems. Unlike the single stranded RNA viruses, geminivirus-based vectors have a circular DNA genome that is replicated through a rolling-circle mechanism using double stranded DNA as intermediates (Chen et al. 2011). The geminivirus genome contains the long and short intergenic regions (LIR and SIR). The DNA can be circularized by joining two LIRS by Rep/RepA proteins which is particularly useful for handling large gene sequence or expressing multiple genes at the same time. Using a bean yellow dwarf virus (BeDV) based replicon system, DIII of WNV E protein was fused to Hepatitis B core antigen (HBcAg) gene and successfully produced in *Nicotiana benthamiana* plants (Chen et al. 2011).

The pEAQ vectors are another series of small binary vectors that use the CPMV hyper-translational expression system to facilitate recombinant protein expression in the plants (Peyret and Lomonossoff 2013). The vector backbone has the RNA-2 which encodes the viral coat protein and movement protein deleted from the original CPMV genome and it was replaced with foreign gene of interest. P19, a suppressor of gene silencing from Tomato bushy stunt virus was inserted into the pEAQ vector to enhance the recombinant protein expression. After the genes of interest are inserted into the pEAQ vectors, the plasmid DNA is transformed into *A.*

tumefaciens and then inoculated into *N. benthamiana* through infiltration. The pEAQ vectors have been successfully used for producing VLPs as candidate vaccines.

3.2 Plant-Produced WNV Subunit Vaccines

Since the first report of using a plant system to express the DIII of DENV E protein (Saejung et al. 2007), plant expression systems have been used to successfully express several flavivirus subunit vaccine candidates against DENV (Saejung et al. 2007; Martinez et al. 2010; Coconi-Linares et al. 2013; Kim et al. 2016), JEV (Appaiahgari et al. 2009; Wang et al. 2009) and WNV (He et al. 2012; He et al. 2014). Our laboratory has been interested in developing plant made flavivirus proteins as vaccine candidates and diagnostic reagents. The WNV E protein DIII sequence was cloned into a pICH11599 vector of the MagnICON system and then transformed into *A. tumefaciens* prior to the agroinfiltration of *N. benthamiana* using the syringe method (Leuzinger et al. 2013). The leaves were harvested 3–8 days following infiltration (dpi). Similar to the *E. coli*-expressed WNV DIII protein (Chu et al. 2005), a 13.5 kD DIII protein band from samples of plant extracts infiltrated with the DIII construct was detected by western blot analysis using an anti-WNV DIII polyclonal antibody (He et al. 2012, 2014). The expression of the plant-derived WNV DIII protein was quantified by ELISA using two WNV DIII specific antibodies. The expression level reaches its peak four days after infiltration to an average about 0.1 mg/g of leaf fresh weight (LFW), which was the highest expression level of a flavivirus antigen ever reported. Subsequent studies demonstrated that the WNV DIII expression level is subcellular compartment dependent (He et al. 2014). Using three different 5′ modules of the MagnICON system to specifically direct the DIII into ER, chloroplast or the cytosol, Western blot detected the 13.5 kD DIII protein band only from leaves infiltrated with ER targeted DIII construct, but not from leaves infiltrated with chloroplast or cytosol targeted DIII constructs (Fig. 1a). ELISA then showed that the ER targeted DIII reached an average level of 73 μg/g LFW, while the maximum level of DIII targeted to chloroplast or cytosol is only about 1.16 μg/g LFW (Fig. 1b). However, the level of the ER targeted WNV DIII is lower than the levels of other viral antigens or antibodies we produced in *N. benthamiana* using MagnICON-based plant expression vectors ranging from 100 to 700 μg/g LFW (Santi et al. 2008; Lai et al. 2010; Phoolcharoen et al. 2011; Dent et al. 2016). This could be the result of leaf necrosis observed after 4 dpi which significantly shortened the accumulation time for DIII expression. Nevertheless, the plant-produced WNV DIII was rapidly expressed as a soluble protein in the ER and can be directly extracted and purified to more than 95% purity using a simple procedure based on metal chelation chromatography (He et al. 2014). Recombinant WNV E protein DIII has been expressed in *E. coli* (Chu et al. 2005) and in insect cells (Alonso-Padilla et al. 2011). In *E. coli*, DIII is expressed as an insoluble protein in the inclusion bodies

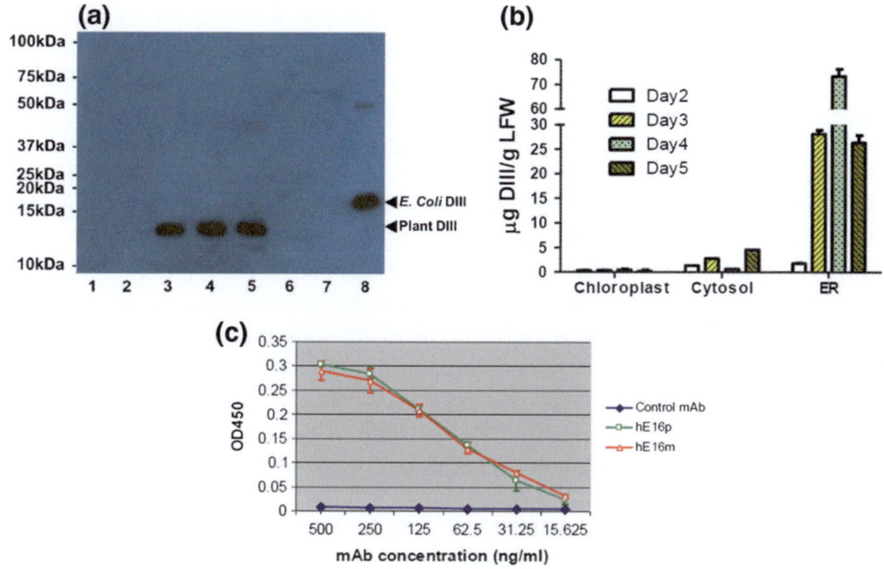

Fig. 1 Expression and characterization of plant-produced WNV E DIII. **a** WNV E DIII was extracted from *N. benthamiana* leaves and separated on 15% SDS-PAGE gels and transferred onto PVDF membranes. A WNV E DIII-specific antibody was used to detect DIII. Lane 1, Protein sample extracted from un-infiltrated leaves as a negative control; Lanes 2–5, Sample collected 2, 3, 4, and 5 days post Agro-infiltration (DPI) from leaves infiltrated with ER-targeted DIII construct; Lane 6: Sample collected 5 DPI from leaves infiltrated with chloroplast-targeted DIII construct; Lane 7: Sample collected 5 DPI from cytosol-targeted DIII leaves; Lane 8: *E. coli*-produced DIII as a positive control. **b** Total protein from plant leaves infiltrated with chloroplast, cytosol or ER-targeted DIII construct was extracted 2–5 DPI and analyzed by an ELISA with an anti-WNV E DIII antibody. Mean ± SD of samples from several independent experiments are presented. **c** Serial dilutions of hE16 purified from mammalian or plant cells were incubated in sample wells coated with plant-produced WNV E DIII and detected with an HRP-conjugated anti-human gamma antibody. A commercial generic human IgG was used as a negative control. Mean ± SD of samples from three independent experiments is presented

and it requires a time-consuming solubilization and refolding process to produce the recombinant DIII in its native conformation. The insect expression system needs several weeks to generate recombinant baculovirus of high titer prior to infection of insect cells for protein production. Our plant-produced DIII is folded properly as it can be recognized by WNV DIII specific antibodies (Oliphant et al. 2005; He et al. 2014). Moreover, the production and purification procedure for plant-produced DIII is fast, cost-effective and highly scalable.

ELISA was used to measure the binding affinity of plant-produced DIII to the mammalian monoclonal neutralizing antibody E16 (hE16m) against WNV E protein (Nybakken et al. 2005; Oliphant et al. 2005) and the plant-made E16 (hE16p) which has been shown to protect mice from lethal WNV infection (Lai et al. 2010). The results indicate plant-produced-DIII resembles the native DIII conformation displayed on WNV viral particles (Fig. 1c). BALB/c mice were treated with four doses of plant-produced DIII subcutaneously over 8 weeks to evaluate the immunogenicity of plant-produced DIII protein (He et al. 2014). Two groups of mice (n = 6) were treated with either 5 or 25 μg of plant-produced DIII, respectively, and were compared with mice treated with the same doses of DIII produced in *E. coli* or negative control (PBS + alum) (Fig. 2). DIII specific antibody responses were detected in the two groups of mice treated with 25 μg of DIII two weeks after the first injection. All groups immunized with DIII showed DIII specific antibody response after the third injection. The geometric mean titers (GMT) calculated from ELISA data of mouse sera indicate plant-produced DIII has equivalent potency as the DIII produced from *E. coli* in inducing DIII specific antibody responses. To evaluate which type of immune response was induced by DIII, IgG1 and IgG2a concentrations in mouse sera collected from groups immunized with 25 μg plant-produced or *E. coli*-produced DIII proteins were measured by ELISA. Apparently, IgG1 was more than 99% of the total DIII specific IgGs in both groups indicating a predominant Th2-type response. It is not surprising that this result was different from the early studies with DIII produced in *E. coli* which shows a Th1-type response (Chu et al. 2007). Previous studies with different adjuvants have shown that alum induces primarily Th2-type response (Cribbs et al. 2003) while CpG induces Th1 based responses when using the same antigen (Demento et al. 2010). Flow cytometry analysis using yeast cells that displayed WNV DIII on the surface demonstrated the anti-DIII sera showed positive binding to DIII, like that of hE16, but not the sera collected from the negative control group. This further confirms plant-produced DIII induces WNV DIII specific immune responses in mice. Competitive ELISA showed that pre-incubation with

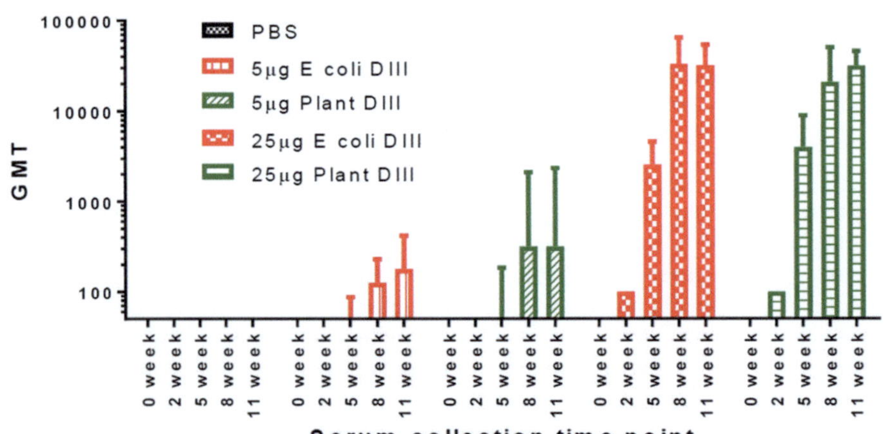

Fig. 2 Time course of WNV E DIII-specific antibody responses in mice upon subcutaneous delivery of plant-derived WNV E DIII. BALB/c mice were injected on weeks 0, 3, 6 and 9 with the indicated dosage of antigen. Blood samples were collected on the indicated weeks and serum IgG was measured by ELISA. The y axis shows the geometric means titers (GMT) and the error bars show the 95% level of confidence of the mean

anti-DIII sera inhibited the subsequent hE16 binding to the immobilized plant-produced DIII. This result indicates at least some of the antibodies in the anti-DIII sera bind to DIII in a hE16-like fashion suggesting they may have neutralizing and protective effects against WNV infection (Lai et al. 2018). These results demonstrated that *N. benthamiana* can be used to efficiently produce immunogenic WNV vaccine candidates with low cost and scalability.

3.3 Plant-Produced VLP-Based Vaccines

In recent years, VLPs have attracted great interest for vaccine development as they can resemble the native virus particles yet remain non-infectious due to deletion of the viral genetic components. Since the 1980s several VLP-based vaccines have been developed and approved for human use (Zhao et al. 2013). VLP-based vaccines have several advantages over inactivated whole virus vaccines or simple subunit vaccines. First, VLP vaccines are non-infectious which means fewer safety concerns for manufacturing or application to humans and animals. Second, they are very immunogenic and elicit both humoral and cellular immune responses. As VLPs mimic the native virus with epitopes displayed on the surface and their particulate nature, they can be easily recognized and processed by antigen presenting cells to trigger T cell activation and proliferation. In addition, VLPs can directly activate B cells, triggering epitope-specific immunoglobulin secretion. Furthermore, VLPs are more stable than subunit vaccines, allowing them to induce long-lasting antibody responses, thereby providing long term protection against viral infection. The stability of VLPs may also extend shelf life of VLP-based vaccines. These features of VLPs make them particularly useful for the development of veterinary vaccines, as VLP vaccines can be produced with lower cost and fewer doses would be needed for animal immunization, minimizing the complications from vaccine side effects.

Currently there are several expression systems being used for production of VLP-based vaccines including bacteria, yeast, insect cell, mammalian cell and plant expression systems. Bacterial expression is simple and easy to use with low cost. Several VLP based vaccines have been produced in this system including a commercialized VLP vaccine against Hepatitis E (Zhao et al. 2013). However, bacteria are prokaryotes that do not have the necessary machinery for post-translational modifications required for production of VLPs that maximally mimic the native viruses. Compared to bacteria, yeast is a better expression system for producing VLP based vaccines as yeast can perform post-translational modification. Several VLP based vaccines have been successfully produced in yeast and commercialized, such as Recombivax against Hepatitis B and Gardasil against human papillomavirus (HPV), both from Merck (Zhao et al. 2013). But expression of VLPs in yeast has its own unique challenges. For example, the HPV VLPs are not secreted by yeast cells but instead are expressed intracellularly, which increases costs for purification. Also, optimization of fermentation conditions and purification processes were required for production of highly purified VLPs for vaccine use (Bryan et al. 2016). Furthermore, neither bacteria nor yeast can be used to produce enveloped VLPs, which means they may not be suitable for WNV VLPs as WNV is an enveloped virus. Insect cells and mammalian cells are more suitable for expressing enveloped VLPs as they can perform more complex post-translational modifications. Despite the advantages of using insect or mammalian expression systems, production of VLPs from these systems tend to be time consuming and expensive as they require large amounts of culture media and multiple steps of downstream processing.

Plant expression systems have emerged as a useful alternative for production of VLP-based vaccines, not only because of its low cost and high efficiency but also the capability to perform necessary post-translational modifications, especially the flexibility to tolerate manipulation of glycosylation patterns (Chen 2016). Our laboratory has long been interested in producing VLP-based vaccine candidates in plants which includes non-enveloped VLPs such as VLPs derived from Norwalk virus capsid protein (NVCP) or HBcAg (Santi et al. 2008; Huang et al. 2009; Lai et al. 2012), and enveloped VLPs displaying WNV prM and E protein (Chen and Lai 2013) or chimeric VLPs displaying WNV DIII of the E protein (Chen et al. 2011; Chen and Lai 2013). To explore the feasibility of producing WNV VLPs in plants, we fused the plant codon optimized WNV prM and E protein DNA sequence into the deconstructed TMV vectors and transiently expressed the construct in *N. benthamiana* (Chen and Lai 2013). Leaf proteins were extracted 7 dpi and purified for western blot analysis using an antibody against WNV E protein. The results showed that the WNV prM and E proteins were expressed at expected sizes in plants and both unprocessed prM and processed mature M proteins were detected (Fig. 3). The relative band intensity of the plant-produced prM and M protein was comparable to that of gradient-purified WN virion proteins (Chen and Lai 2013). This indicates WNV M protein was processed by the plant cell machinery similarly to the native WNV virion M protein. Assembly of prM and E protein-containing VLPs was confirmed by sucrose gradient centrifugation (Chen and Lai 2013).

HBcAg has been used as a carrier protein to display foreign antigens since the 1980s because of its excellent immunogenicity and the capability to accommodate different antigens in both prokaryotic and eukaryotic expression systems (Roose et al. 2013). To explore the feasibility of producing chimeric HBcAg VLPs displaying WNV epitopes in plants, we fused the WNV DIII coding sequence to the 3′ end of HBcAg using a BeYDV-based expression vector (Chen et al. 2011). The HBcAg-DIII fusion protein was expressed in the leaves of *N. benthamiana* and reached the highest level of accumulation ($\sim$350 μg/g LFW) at six days post infiltration (Fig. 4a). Transmission electron microscopy confirmed the assembly of HBcAg-DIII chimeric VLPs with consistent size after extraction and purification (Fig. 4b). Western blot analysis showed that the HBcAg-DIII fusion protein was detected by both anti-HBcAg and anti-WNV DIII antibodies at expected size, about 27kD (Fig. 4c, d). Display of WNV DIII on the chimeric VLP surface was confirmed by competitive ELISA that the HBcAg-DIII fusion protein can effectively compete with soluble DIII to bind the anti-DIII E16. The HBcAg-DIII chimeric VLPs also

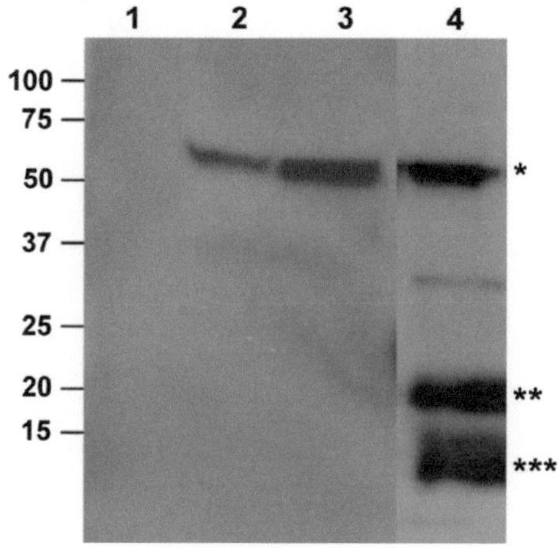

Fig. 3 Production of West Nile virus enveloped VLP based on the prM/M and the E protein in *N. benthamiana* plants. Leaf tissue was infiltrated with *Agrobacterium* harboring the WNV prM-E construct. Leaf proteins were extracted 7 DPI. PrM/M-E VLP was isolated by PEG precipitation and analyzed on 4–12% SDS PAGE gels and transferred onto PVDF membranes. The membranes were incubated with an anti-WNV E antibody (Lanes 1–3) or an anti-WNV M-E antibody (Lane 4). Lane 1, Protein sample from buffer-infiltrated leaves; Lane 2, Purified WNV E protein as positive control; Lanes 3–4, Samples from prM-E construct-infiltrated plants. *: E protein; **: Unprocessed prM protein; ***: Processed M protein

induced stronger DIII-specific immune responses in mice injected with a single dose (25 μg) of the chimeric VLPs than that of non-fused DIII protein (Chen 2015). Using the MagnICON vectors the WNV chimeric VLPs were expressed with an even higher level of accumulation (>1000 μg/g LFW) in *N. benthamiana* leaves. Analysis of these chimeric VLPs demonstrated that they are similar to those expressed using Geminiviral vectors, both structurally and immunologically (Chen 2015).

Our laboratory has also demonstrated the advantage of plant-made VLPs in improving the safety of vaccines against flaviviruses, including WNV and the recently emerged ZIKV. Several ZIKV vaccine candidates are being developed using inactivated whole virus and DNA or RNA that express the E protein of ZIKV. These vaccine candidates are successful in eliciting the production of ZIKV-targeted antibodies and in protecting animals against ZIKV challenges (Abbink et al. 2016; Larocca et al. 2016; Pardi and Weissman 2017). However, their use may potentially predispose vaccinated subjects to infection by related flaviviruses including DENV and WNV due to phenomenon called antibody-dependent enhancement (ADE) (Sun et al. 2017). ADE is caused by cross-reactive, but non- or sub-neutralizing antibodies, elicited by closely-related viruses or vaccines against related viruses. Instead of neutralizing the infecting virus, these cross-reactive antibodies form complexes with the infecting virus that bind to Fc gamma receptor (FcγR)-bearing cells, resulting in increased viral uptake and infection (Morens 1994). For example, previous infection or vaccination

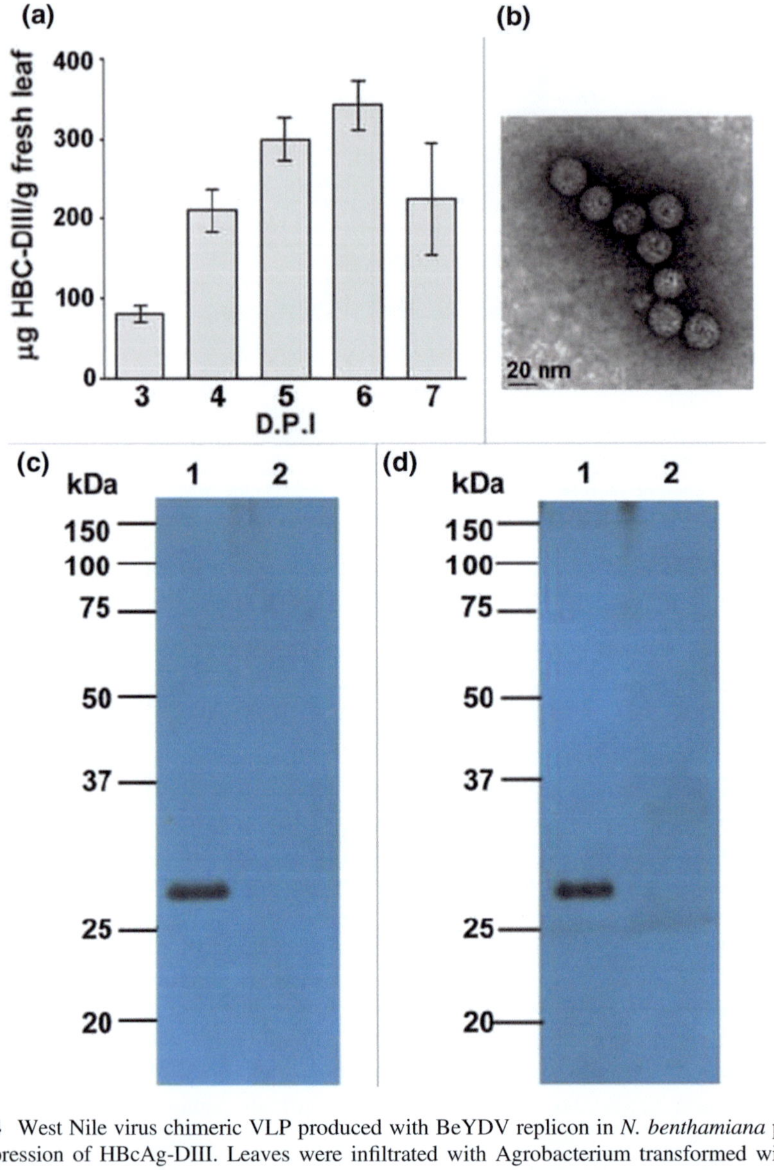

Fig. 4 West Nile virus chimeric VLP produced with BeYDV replicon in *N. benthamiana* plants. **a** Expression of HBcAg-DIII. Leaves were infiltrated with Agrobacterium transformed with the HBcAg-DIII construct. Proteins were extracted on days 3–7 DPI and were analyzed with an ELISA that detects HBcAg. Mean ± standard error (SEM) of samples from three independent infiltration experiments are presented. **b** Electron microscopy of purified HBcAg-DIII VLPs. HBcAg-DIII VLP was purified and subject to negative staining with 0.2% aqueous uranyl acetate, and transmission electron microscopy with a Philips CM-12 microscope. **c–d** Western blot analysis of HBcAg-DIII. Leaf proteins were separated on 12% SDS-PAGE gels under reducing condition and blotted onto PVDF membranes. The membranes were incubated with an anti-HBc antibody (**c**) or an anti-DIII antibody (**d**). Lane 1, Protein sample extracted from leaves infiltrated with the HBc-DIII construct; lane 2, Extract from un-infiltrated leaves

against one serotype of DENV may predispose these individuals to develop the more severe dengue hemorrhagic fever/dengue shock syndrome (DHF/DSS) through ADE if they are exposed to another serotype of DENV subsequently (Halstead 2014). The recent outbreaks of ZIKV further complicate vaccine development for flaviviruses as antibodies against DENV and ZIKV have been shown to cross-react and enhance the replication of each other in mice (Barba-Spaeth et al. 2016; Dejnirattisai et al. 2016; Stettler et al. 2016; Bardina et al. 2017). Another study also demonstrated that previous WNV infection can also enhance subsequent secondary ZIKV infection (Bardina et al. 2017), suggesting that ADE may also occur between WNV and ZIKV. This raises safety concerns for current WNV vaccine candidates in promoting heterologous flavivirus infection via ADE. In response, our research group has developed several protein subunit vaccines (Yang et al. 2017a, b, c). We demonstrated that an HBcAg VLP that displays the ZIKV E DIII (HBcAg-zDIII VLP) can be robustly produced and easily purified in large quantities from plants. When tested in mice, plant-produced HBcAg-zDIII VLPs evoked potent humoral and cellular responses against ZIKV. Notably, the neutralization potency exceeds the threshold correlated with protective immunity against multiple strains of ZIKV (Yang et al. 2017b). Remarkably, antibodies induced by HBcAg-zDIII VLPs neither cross-react with DENV, nor do they enhance the infection of DENV in FcγR-expressing cells (Fig. 5). We also demonstrated that a plant-produced WNV DIII protected mice from a lethal challenge of WNV infection but without enhancing ZIKV or DENV infectivity (Lai et al. 2018). These results highlight the potential of plant-made vaccines in offsetting the concern of WNV vaccines in sensitizing people to subsequent DENV or ZIKV infection. Overall, these results indicate plants can rapidly produce high levels of immunogenic WNV-specific VLPs.

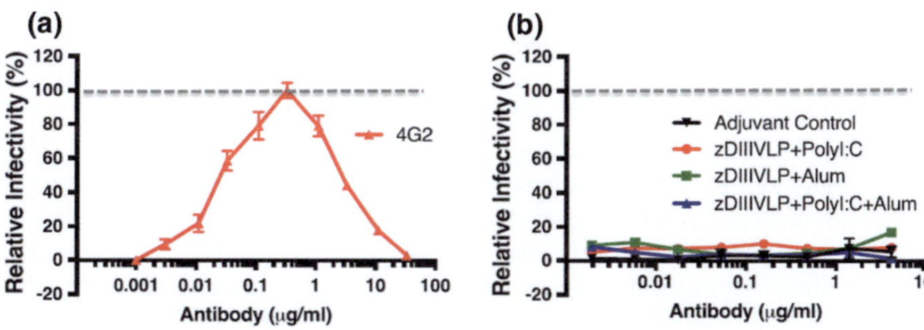

Fig. 5 Lack of enhancement of DENV infection by antibodies in serum from mice immunized with HBcAg-zEDIII VLPs. IgGs were isolated from week 5 pooled sera of mice receiving PBS + Adjuvant (adjuvant control) or HBcAg-zDIII VLP + indicated adjuvant. Serial dilutions of IgGs were mixed with DENV-2 and incubated with FcγR-expressing K562 cells. Forty-eight hour later, cells were fixed, permeabilized and analyzed by flow cytometry for DENV infection. Anti-DENV-2 E mAb 4G2 was used as an ADE positive control with its maximum infectivity defined as 100%. Enhancement by IgGs from anti-HBcAg-zEDIII sera is expressed as a % relative to that of the 4G2

3.4 Downstream Processing of Plant-Produced WNV Vaccines

To produce highly purified WNV vaccines, cost-effective and scalable downstream processing is required after recombinant expression in plants. Currently, downstream processing is a major barrier, not only for plant-made biologics (PMBs), but for those produced by conventional platforms as well (Sabalza et al. 2014). Downstream processing typically includes two phases: primary recovery from the plant host and purification of the recombinant protein (Wilken and Nikolov 2012). The primary recovery process usually includes homogenization of the plant tissues and extraction of recombinant protein from plant homogenate. Though the recovery phase of downstream processing varies depending on the plant host used for expression, the purification step is like that of conventional production platforms. Our laboratory has developed a scalable scheme for extraction and purification of plant-produced NVCP VLPs with the combination of a low pH precipitation step, ultrafiltration/diafiltration with a polyethersulfone tangential flow (PES TFF) membrane, and anion exchange chromatography (Chen 2008; Lai and Chen 2012). The feasibility of the NVCP VLP production and purification process was demonstrated under current Good Manufacturing Practices (cGMP) regulations to enrich the VLPs to more than 95% purity (Lai and Chen 2012). Such a downstream processing scheme was applicable to our WNV chimeric VLPs (Chen et al. 2011; Chen and Lai 2013); it yields highly purified HBcAg-DIII VLPs that are assembled with consistent size (Fig. 4b). The WNV enveloped prM and E VLPs were produced in a similar process including leaf homogenization, centrifugation and multiple chromatographic steps for purification. These results demonstrated that the downstream processing of plant-produced WNV VLP vaccines is scalable and efficient.

4 Challenges and Future Development

The development of WNV vaccines in plants has transformed the landscape of vaccine-production economics and contributed to the optimism of licensing an efficacious, safe and low-cost veterinary WNV vaccine in the future. To achieve this goal, we envision novel approaches combining the advancements in discovering more potent adjuvants with a deeper understanding of the biology of drug targeting to immune cells and stimulating systemic immune responses via oral delivery.

Currently, WNV vaccines are injected into animals through at least two doses and an annual boost to ensure protection. It would be desirable to develop oral WNV vaccines for animals. For example, vaccines produced in plant tissue can be simply fed to horses for immunization. Such oral vaccines will further reduce the production cost by eliminating the extraction and purification process and obviating costs associated with cold-storage, transportation and sterile injection.

Orally-deliverable vaccines have been elusive due to the challenges of denaturation and degradation in the digestive system, and their inability to cross the gut epithelium and be delivered to target cells. Not surprisingly, plant cells may be the best vehicle to overcome these challenges. It has been shown that plant cells can protect the vaccines they express from acids and enzymes in the stomach by bioencapsulation, due to the inability of animal digestive enzymes in hydrolyzing the glycosidic bonds in the plant cell wall. This allows the delivery of plant-made vaccines to the gut lumen where they are enzymatically released by commensal bacteria (Kwon and Daniell 2015). Recent studies also demonstrated that when tagged with a specific receptor-binding peptide, plant cell-encapsulated proteins can either be targeted to the gut immune system or cross the gut epithelium to reach circulation (Kwon and Daniell 2015; Su et al. 2015a, b). Furthermore, encapsulated protein drugs in plant cells have been found to maintain their pharmacological efficacy several years after they have been stored at room temperature (Lakshmi et al. 2013). These findings suggest that plant cell-encapsulated vaccines may present an ambient, temperature-stable product that can be delivered to animals by simply feeding, thereby, circumventing logistical costs and allowing practical implementation of vaccination programs to wild susceptible animals.

Indeed, oral immunization of a plant expressed recombinant antigen was first demonstrated in 1995 (Haq et al. 1995). Oral immunization of mice with plant-made *E. coli* heat-labile enterotoxin binding subunit (LT-B) induced production of neutralizing antibodies in serum and gut mucosal against enterotoxin. Later studies using the same foreign antigen expressed in distinct plant tissues demonstrated that the leaf-based vaccines transiently expressed in *N. benthamiana* have higher efficacy than root-based vaccines through oral immunization in mice (Pelosi et al. 2011, 2012). When LT-B was fused with a heat stable (ST) toxin and produced in transgenic tobacco plants, this fusion protein induced similar immune response to that of LT-B produced in bacteria at a much lower dose when orally administered to mice (Rosales-Mendoza et al. 2011). Furthermore, oral delivery of plant-made vaccines has shown promising results in preventing various animal diseases including plague, cholera, swine and bird flu, and porcine reproductive and respiratory syndrome (Shahid and Daniell 2016) with oral vaccines for veterinary use produced in potato, rice, maize, tobacco and other edible plants against various pathogens (Takeyama et al. 2015). These studies indicate plant-produced oral vaccines can induce protective mucosal and systemic immune responses against pathogens. Relevant to WNV vaccines, Kim et al. recently showed that a fusion protein containing the consensus DENV E DIII and M cell-targeting peptide ligand can be produced in transgenic rice calli and the fusion protein effectively binds to the target mucosal cells (Kim et al. 2013). Very recently, an oral vaccine candidate based on hepatitis C virus (HCV) E protein has been developed in lettuce plant. Feeding mice with E protein-containing lettuce powder induced both systemic and mucosal humoral responses against HCV. Since HCV is in the same *Flaviviridae* family with WNV, this study has demonstrated the feasibility of using oral vaccines to prevent WNV in horses and other animals (Liu Clarke et al. 2017).

The remaining challenge is to effectively induce strong systemic responses by oral vaccines alone, as many oral vaccines still rely on injectable priming to achieve protective potency (Sequeira and Harrison 1982). New approaches in vaccine development are required to overcome this challenge. This, in turn, needs a more thorough understanding of the biology of immune cell stimulation by orally-delivered antigens. We expect that the focus of the development of novel WNV vaccines over the next 5 years will be on not just improvements of vaccine potency and production optimization in edible plants, but also on identifying the best combination of antigen design and oral adjuvant to induce optimal protective immunity.

Acknowledgements The authors thank the current and past members of the Chen laboratory, especially Dr. J. He, H. Lai, J. Hurtado, M. Yang, and L. Peng for the data presented in this chapter. The contribution of numerous undergraduate students to the WNV project is also greatly appreciated. We also thank C. Jugler for the critical reading of the chapter. The research relevant to this chapter in the authors' laboratory was supported in part by NIAID grants number U01 AI075549 and R21/R33 AI101329 to Q. Chen.

References

Abbink P, Larocca RA, De La Barrera RA, Bricault CA, Moseley ET, Boyd M, Kirilova M, Li Z, Ng'ang'a D, Nanayakkara O, Nityanandam R, Mercado NB, Borducchi EN, Agarwal A, Brinkman AL, Cabral C, Chandrashekar A, Giglio PB, Jetton D, Jimenez J, Lee BC, Mojta S, Molloy K, Shetty M, Neubauer GH, Stephenson KE, Peron JPS, Zanotto PMDA, Misamore J, Finneyfrock B, Lewis MG, Alter G, Modjarrad K, Jarman RG, Eckels KH, Michael NL, Thomas SJ, Barouch DH (2016) Protective efficacy of multiple vaccine platforms against Zika virus challenge in rhesus monkeys. Science 353(6304):1129

Aharonson-Raz K, Lichter-Peled A, Tal S, Gelman B, Cohen D, Klement E, Steinman A (2014) Spatial and temporal distribution of West Nile virus in horses in Israel (1997–2013)–from endemic to epidemics. PLoS ONE 9(11):e113149

Alonso-Padilla J, de Oya NJ, Blazquez AB, Escribano-Romero E, Escribano JM, Saiz JC (2011) Recombinant West Nile virus envelope protein E and domain III expressed in insect larvae protects mice against West Nile disease. Vaccine 29(9):1830–1835

Amanna IJ, Raue HP, Slifka MK (2012) Development of a new hydrogen peroxide-based vaccine platform. Nat Med 18(6):974–979

Appaiahgari MB, Abdin MZ, Bansal KC, Vrati S (2009) Expression of Japanese encephalitis virus envelope protein in transgenic tobacco plants. J Virol Methods 162(1–2):22–29

Arroyo J, Miller C, Catalan J, Myers GA, Ratterree MS, Trent DW, Monath TP (2004) ChimeriVax-West Nile virus live-attenuated vaccine: preclinical evaluation of safety, immunogenicity, and efficacy. J Virol 78(22):12497–12507

Barba-Spaeth G, Dejnirattisai W, Rouvinski A, Vaney M-C, Medits I, Sharma A, Simon-Lorière E, Sakuntabhai A, Cao-Lormeau V-M, Haouz A, England P, Stiasny K, Mongkolsapaya J, Heinz FX, Screaton GR, Rey FA (2016) Structural basis of potent Zika–dengue virus antibody cross-neutralization. Nature 536(7614):48–53

Bardina SV, Bunduc P, Tripathi S, Duehr J, Frere JJ, Brown JA, Nachbagauer R, Foster GA, Krysztof D, Tortorella D, Stramer SL, Garcia-Sastre A, Krammer F, Lim JK (2017) Enhancement of Zika virus pathogenesis by preexisting antiflavivirus immunity. Science 356(6334):175–180

Barta A, Sommergruber K, Thompson D, Hartmuth K, Matzke MA, Matzke AJ (1986) The expression of a nopaline synthase—human growth hormone chimaeric gene in transformed tobacco and sunflower callus tissue. Plant Mol Biol 6(5):347–357

Brandler S, Tangy F (2013) Vaccines in development against West Nile virus. Viruses 5(10):2384–2409

Brinton MA (2013) Replication cycle and molecular biology of the West Nile virus. Viruses 6(1):13–53

Bryan JT, Buckland B, Hammond J, Jansen KU (2016) Prevention of cervical cancer: journey to develop the first human papillomavirus virus-like particle vaccine and the next generation vaccine. Curr Opin Chem Biol 32:34–47

Castillo-Olivares J, Wood J (2004) West Nile virus infection of horses. Vet Res 35(4):467–483

Chang DC, Liu WJ, Anraku I, Clark DC, Pollitt CC, Suhrbier A, Hall RA, Khromykh AA (2008) Single-round infectious particles enhance immunogenicity of a DNA vaccine against West Nile virus. Nat Biotechnol 26(5):571–577

Chen Q (2008) Expression and purification of pharmaceutical proteins in plants. Biol Eng 1(4):291–321

Chen Q (2011a) Expression and manufacture of pharmaceutical proteins in genetically engineered horticultural plants. Transgenic Horticultural Crops: Challenges and Opportunities - Essays by Experts, Mou B, Scorza R. Taylor & Francis, Boca Raton, pp 83–124

Chen Q (2011b) Turning a new leaf. Eur Biopharm Rev 2(56):64–68

Chen Q (2015) Plant-made vaccines against West Nile virus are potent, safe, and economically feasible. Biotechnol J 10(5):671 680

Chen Q (2016) Glycoengineering of plants yields glycoproteins with polysialylation and other defined N-glycoforms. Proc Natl Acad Sci U S A 113(34):9404–9406

Chen Q (2018) Recombinant therapeutic molecules produced in plants. Adv Bota Res 86:207–244. https://doi.org/10.1016/bs.abr.2017.11.006

Chen Q, Davis K (2016) The potential of plants as a system for the development and production of human biologics. F1000Research 5(912): https://doi.org/10.12688/f11000research.18010.12681

Chen Q, Lai H (2013) Plant-derived virus-like particles as vaccines. Hum Vaccin Immunother 9(1):26–49

Chen Q, Lai H (2015) Gene delivery into plant cells for recombinant protein production. Biomed Res Int 2015:932161

Chen Q, He J, Phoolcharoen W, Mason HS (2011) Geminiviral vectors based on bean yellow dwarf virus for production of vaccine antigens and monoclonal antibodies in plants. Hum Vaccin 7(3):331–338

Chen Q, Dent M, Hurtado J, Stahnke J, McNulty A, Leuzinger K, Lai H (2016) Transient protein expression by agroinfiltration in lettuce. Methods Mol Biol 1385:55–67

Chen Q, Dent M, Mason H (2018) Plant-made vaccines. In: Kermode A, Jiang L (eds) Molecular pharming: Applications, challenges, and emerging areas. John Wiley & Sons, pp 231–273. https://doi.org/10.1002/9781118801512.ch10

Chu JJ, Rajamanonmani R, Li J, Bhuvanakantham R, Lescar J, Ng ML (2005) Inhibition of West Nile virus entry by using a recombinant domain III from the envelope glycoprotein. J Gen Virol 86(Pt 2):405–412

Chu JH, Chiang CC, Ng ML (2007) Immunization of flavivirus West Nile recombinant envelope domain III protein induced specific immune response and protection against West Nile virus infection. J Immunol 178(5):2699–2705

Chua AJ, Vituret C, Tan ML, Gonzalez G, Boulanger P, Ng ML, Hong SS (2013) A novel platform for virus-like particle-display of flaviviral envelope domain III: induction of Dengue and West Nile virus neutralizing antibodies. Virol J 10:129

Cielens I, Jackevica L, Strods A, Kazaks A, Ose V, Bogans J, Pumpens P, Renhofa R (2014) Mosaic RNA phage VLPs carrying domain III of the West Nile virus E protein. Mol Biotechnol 56(5):459–469

Coconi-Linares N, Ortega-Davila E, Lopez-Gonzalez M, Garcia-Machorro J, Garcia-Cordero J, Steinman RM, Cedillo-Barron L, Gomez-Lim MA (2013) Targeting of envelope domain III protein of DENV type 2 to DEC-205 receptor elicits neutralizing antibodies in mice. Vaccine 31(19):2366–2371

Coller BA, Pai V, Weeks-Levy C, Ogata S (2012) United States patent application No. US20120141520 A1: recombinant subunit West Nile virus vaccine for protection of human subjects

Cribbs DH, Ghochikyan A, Vasilevko V, Tran M, Petrushina I, Sadzikava N, Babikyan D, Kesslak P, Kieber-Emmons T, Cotman CW, Agadjanyan MG (2003) Adjuvant-dependent modulation of Th1 and Th2 responses to immunization with beta-amyloid. Int Immunol 15 (4):505–514

Davis BS, Chang GJ, Cropp B, Roehrig JT, Martin DA, Mitchell CJ, Bowen R, Bunning ML (2001) West Nile virus recombinant DNA vaccine protects mouse and horse from virus challenge and expresses in vitro a noninfectious recombinant antigen that can be used in enzyme-linked immunosorbent assays. J Virol 75(9):4040–4047

Dayan GH, Pugachev K, Bevilacqua J, Lang J, Monath TP (2013) Preclinical and clinical development of a YFV 17 D-based chimeric vaccine against West Nile virus. Viruses 5 (12):3048–3070

De Filette M, Soehle S, Ulbert S, Richner J, Diamond MS, Sinigaglia A, Barzon L, Roels S, Lisziewicz J, Lorincz O, Sanders NN (2014) Vaccination of mice using the West Nile virus E-protein in a DNA prime-protein boost strategy stimulates cell-mediated immunity and protects mice against a lethal challenge. PLoS ONE 9(2):e87837

Dejnirattisai W, Supasa P, Wongwiwat W, Rouvinski A, Barba-Spaeth G, Duangchinda T, Sakuntabhai A, Cao-Lormeau V-M, Malasit P, Rey FA, Mongkolsapaya J, Screaton GR (2016) Dengue virus sero-cross-reactivity drives antibody-dependent enhancement of infection with zika virus. Nat Immunol advance online publication

Demento SL, Bonafe N, Cui W, Kaech SM, Caplan MJ, Fikrig E, Ledizet M, Fahmy TM (2010) TLR9-targeted biodegradable nanoparticles as immunization vectors protect against West Nile encephalitis. J Immunol 185(5):2989–2997

Dent M, Hurtado J, Paul AM, Sun H, Lai H, Yang M, Esqueda A, Bai F, Steinkellner H, Chen Q (2016) Plant-produced anti-dengue virus monoclonal antibodies exhibit reduced antibody-dependent enhancement of infection activity. J Gen Virol 97(12):3280–3290

Durbin AP, Wright PF, Cox A, Kagucia W, Elwood D, Henderson S, Wanionek K, Speicher J, Whitehead SS, Pletnev AG (2013) The live attenuated chimeric vaccine rWN/DEN4Delta30 is well-tolerated and immunogenic in healthy flavivirus-naive adult volunteers. Vaccine 31 (48):5772–5777

Hall RA, Nisbet DJ, Pham KB, Pyke AT, Smith GA, Khromykh AA (2003) DNA vaccine coding for the full-length infectious Kunjin virus RNA protects mice against the New York strain of West Nile virus. Proc Natl Acad Sci U S A 100(18):10460–10464

Halstead SB (2014) Dengue antibody-dependent enhancement: knowns and unknowns. Microbiol Spectr 2(6)

Haq TA, Mason HS, Clements JD, Arntzen CJ (1995) Oral immunization with a recombinant bacterial antigen produced in transgenic plants. Science 268(5211):714–716

He J, Lai H, Brock C, Chen Q (2012) A novel system for rapid and cost-effective production of detection and diagnostic reagents of West Nile virus in plants. J Biomed Biotechnol 2012:106783

He J, Peng L, Lai H, Hurtado J, Stahnke J, Chen Q (2014) A plant-produced antigen elicits potent immune responses against West Nile virus in mice. Biomed Res Int 2014:952865

Hefferon K (2014) Plant virus expression vector development: new perspectives. Biomed Res Int 2014:785382

Huang Z, Chen Q, Hjelm B, Arntzen C, Mason H (2009) A DNA replicon system for rapid high-level production of virus-like particles in plants. Biotechnol Bioeng 103(4):706–714

Iyer AV, Kousoulas KG (2013) A review of vaccine approaches for West Nile virus. Int J Environ Res Public Health 10(9):4200–4223

Kanai R, Kar K, Anthony K, Gould LH, Ledizet M, Fikrig E, Marasco WA, Koski RA, Modis Y (2006) Crystal structure of West Nile virus envelope glycoprotein reveals viral surface epitopes. J Virol 80(22):11000–11008

Kim TG, Kim MY, Huy NX, Kim SH, Yang MS (2013) M cell-targeting ligand and consensus dengue virus envelope protein domain III fusion protein production in transgenic rice calli. Mol Biotechnol 54(3):880–887

Kim MY, Li JY, Tien NQ, Yang M. S. (2016). Expression and assembly of cholera toxin B subunit and domain III of dengue virus 2 envelope fusion protein in transgenic potatoes. Protein Expr Purif

Komar N, Langevin S, Hinten S, Nemeth N, Edwards E, Hettler D, Davis B, Bowen R, Bunning M (2003) Experimental infection of North American birds with the New York 1999 strain of West Nile virus. Emerg Infect Dis 9(3):311–322

Kwon K-C, Daniell H (2015) Low-cost oral delivery of protein drugs bioencapsulated in plant cells. Plant Biotechnol J 13(8):1017–1022

Lai H, Chen Q (2012) Bioprocessing of plant-derived virus-like particles of Norwalk virus capsid protein under current good manufacture practice regulations. Plant Cell Rep 31(3):573–584

Lai H, Engle M, Fuchs A, Keller T, Johnson S, Gorlatov S, Diamond MS, Chen Q (2010) Monoclonal antibody produced in plants efficiently treats West Nile virus infection in mice. Proc Natl Acad Sci U S A 107(6):2419–2424

Lai H, He J, Engle M, Diamond MS, Chen Q (2012) Robust production of virus-like particles and monoclonal antibodies with geminiviral replicon vectors in lettuce. Plant Biotechnol J 10(1).95–104

Lai H, Paul AM, Sun H, He J, Yang M, Bai F, Chen Q (2018) A plant-produced vaccine protects mice against lethal west nile virus infection without enhancing zika or dengue virus infectivity. Vaccine 36(14):1846–1852

Lakshmi PS, Verma D, Yang X, Lloyd B, Daniell H (2013) Low cost tuberculosis vaccine antigens in capsules: expression in chloroplasts, bio-encapsulation, stability and functional evaluation in vitro. PLoS ONE 8(1):e54708

Larocca RA, Abbink P, Peron JPS, de A. Zanotto PM, Iampietro MJ, Badamchi-Zadeh A, Boyd M, Ng'ang'a D, Kirilova M, Nityanandam R, Mercado NB, Li Z, Moseley ET, Bricault CA, Borducchi EN, Giglio PB, Jetton D, Neubauer G, Nkolola JP, Maxfield LF, De La Barrera RA, Jarman RG, Eckels KH, Michael NL, Thomas SJ, Barouch DH (2016) Vaccine protection against Zika virus from Brazil. Nature 536(7617): 474–478

Ledgerwood JE, Pierson TC, Hubka SA, Desai N, Rucker S, Gordon IJ, Enama ME, Nelson S, Nason M, Gu W, Bundrant N, Koup RA, Bailer RT, Mascola JR, Nabel GJ, Graham BS, Team VRCS (2011) A West Nile virus DNA vaccine utilizing a modified promoter induces neutralizing antibody in younger and older healthy adults in a phase I clinical trial. J Infect Dis 203(10):1396–1404

Ledizet M, Kar K, Foellmer HG, Wang T, Bushmich SL, Anderson JF, Fikrig E, Koski RA (2005) A recombinant envelope protein vaccine against West Nile virus. Vaccine 23(30):3915–3924

Leuzinger K, Dent M, Hurtado J, Stahnke J, Lai H, Zhou X, Chen Q (2013) Efficient agroinfiltration of plants for high-level transient expression of recombinant proteins. J Vis Exp (77)

Lico C, Chen Q, Santi L (2008) Viral vectors for production of recombinant proteins in plants. J Cell Physiol 216(2):366–377

Lieberman MM, Clements DE, Ogata S, Wang G, Corpuz G, Wong T, Martyak T, Gilson L, Coller BA, Leung J, Watts DM, Tesh RB, Siirin M, Travassos da Rosa A, Humphreys T, Weeks-Levy C (2007) Preparation and immunogenic properties of a recombinant West Nile subunit vaccine. Vaccine 25(3):414–423

Lieberman MM, Nerurkar VR, Luo H, Cropp B, Carrion R Jr, de la Garza M, Coller BA, Clements D, Ogata S, Wong T, Martyak T, Weeks-Levy C (2009) Immunogenicity and protective efficacy of a recombinant subunit West Nile virus vaccine in rhesus monkeys. Clin Vaccine Immunol 16(9):1332–1337

Liu Clarke J, Paruch L, Dobrica MO, Caras I, Tucureanu C, Onu A, Ciulean S, Stavaru C, Eerde A, Wang Y, Steen H, Haugslien S, Petrareanu C, Lazar C, Popescu I, Bock R, Dubuisson J, Branza-Nichita N (2017) Lettuce-produced hepatitis C virus E1E2 heterodimer triggers immune responses in mice and antibody production after oral vaccination. Plant Biotechnol J

Lyon GM, Mehta AK, Varkey JB, Brantly K, Plyler L, McElroy AK, Kraft CS, Towner JS, Spiropoulou C, Stroher U, Uyeki TM, Ribner BS, U. Emory Serious Communicable Diseases (2014) Clinical care of two patients with Ebola virus disease in the United States. N Engl J Med 371(25): 2402–2409

Martin JE, Pierson TC, Hubka S, Rucker S, Gordon IJ, Enama ME, Andrews CA, Xu Q, Davis BS, Nason M, Fay M, Koup RA, Roederer M, Bailer RT, Gomez PL, Mascola JR, Chang GJ, Nabel GJ, Graham BS (2007) A West Nile virus DNA vaccine induces neutralizing antibody in healthy adults during a phase 1 clinical trial. J Infect Dis 196(12):1732–1740

Martinez CA, Topal E, Giulietti AM, Talou JR, Mason H (2010) Exploring different strategies to express Dengue virus envelope protein in a plant system. Biotechnol Lett 32(6):867–875

Medigeshi GR, Hirsch AJ, Brien JD, Uhrlaub JL, Mason PW, Wiley C, Nikolich-Zugich J, Nelson JA (2009) West Nile virus capsid degradation of claudin proteins disrupts epithelial barrier function. J Virol 83(12):6125–6134

Minke JM, Siger L, Karaca K, Austgen L, Gordy P, Bowen R, Renshaw RW, Loosmore S, Audonnet JC, Nordgren B (2004) Recombinant canarypoxvirus vaccine carrying the prM/E genes of West Nile virus protects horses against a West Nile virus-mosquito challenge. Arch Virol Suppl 18:221–230

Monath TP, Liu J, Kanesa-Thasan N, Myers GA, Nichols R, Deary A, McCarthy K, Johnson C, Ermak T, Shin S, Arroyo J, Guirakhoo F, Kennedy JS, Ennis FA, Green S, Bedford P (2006) A live, attenuated recombinant West Nile virus vaccine. Proc Natl Acad Sci U S A 103 (17):6694–6699

Morens DM (1994) Antibody-dependent of enhancement of infection and the pathogenesis of viral disease. Clin Inf Dis 19:500–512

Ng T, Hathaway D, Jennings N, Champ D, Chiang YW, Chu HJ (2003) Equine vaccine for West Nile virus. Dev Biol (Basel) 114:221–227

Nybakken GE, Oliphant T, Johnson S, Burke S, Diamond MS, Fremont DH (2005) Structural basis of West Nile virus neutralization by a therapeutic antibody. Nature 437(7059):764–769

Ohtaki N, Takahashi H, Kaneko K, Gomi Y, Ishikawa T, Higashi Y, Kurata T, Sata T, Kojima A (2010) Immunogenicity and efficacy of two types of West Nile virus-like particles different in size and maturation as a second-generation vaccine candidate. Vaccine 28(40):6588–6596

Oliphant T, Engle M, Nybakken GE, Doane C, Johnson S, Huang L, Gorlatov S, Mehlhop E, Marri A, Chung KM, Ebel GD, Kramer LD, Fremont DH, Diamond MS (2005) Development of a humanized monoclonal antibody with therapeutic potential against West Nile virus. Nat Med 11(5):522–530

Pardi N, Weissman D (2017) Nucleoside modified mRNA vaccines for infectious diseases. In: Kramps T, Elbers K (eds) RNA vaccines: methods and protocols. Springer, New York, NY, pp 109–121

Pelosi A, Shepherd R, Guzman GD, Hamill JD, Meeusen E, Sanson G, Walmsley AM (2011) The release and induced immune responses of a plant-made and delivered antigen in the mouse gut. Curr Drug Deliv 8(6):612–621

Pelosi A, Piedrafita D, De Guzman G, Shepherd R, Hamill JD, Meeusen E, Walmsley AM (2012) The effect of plant tissue and vaccine formulation on the oral immunogenicity of a model plant-made antigen in sheep. PLoS ONE 7(12):e52907

Peyret H, Lomonossoff GP (2013) The pEAQ vector series: the easy and quick way to produce recombinant proteins in plants. Plant Mol Biol 83(1–2):51–58

Peyret H, Lomonossoff GP (2015) When plant virology met Agrobacterium: the rise of the deconstructed clones. Plant Biotechnol J 13(8):1121–1135

Phoolcharoen W, Bhoo SH, Lai H, Ma J, Arntzen CJ, Chen Q, Mason HS (2011) Expression of an immunogenic Ebola immune complex in Nicotiana benthamiana. Plant Biotechnol J 9(7):807–816

Pinto AK, Richner JM, Poore EA, Patil PP, Amanna IJ, Slifka MK, Diamond MS (2013) A hydrogen peroxide-inactivated virus vaccine elicits humoral and cellular immunity and protects against lethal West Nile virus infection in aged mice. J Virol 87(4):1926–1936

Poore EA, Slifka DK, Raue HP, Thomas A, Hammarlund E, Quintel BK, Torrey LL, Slifka AM, Richner JM, Dubois ME, Johnson LP, Diamond MS, Slifka MK, Amanna IJ (2017) Pre-clinical development of a hydrogen peroxide-inactivated West Nile virus vaccine. Vaccine 35(2):283–292

Qiao M, Ashok M, Bernard KA, Palacios G, Zhou ZH, Lipkin WI, Liang TJ (2004) Induction of sterilizing immunity against West Nile Virus (WNV), by immunization with WNV-like particles produced in insect cells. J Infect Dis 190(12):2104–2108

Ramanathan MP, Kutzler MA, Kuo YC, Yan J, Liu H, Shah V, Bawa A, Selling B, Sardesai NY, Kim JJ, Weiner DB (2009) Coimmunization with an optimized IL15 plasmid adjuvant enhances humoral immunity via stimulating B cells induced by genetically engineered DNA vaccines expressing consensus JEV and WNV E DIII. Vaccine 27(32):4370–4380

Rizzoli A, Jimenez-Clavero MA, Barzon L, Cordioli P, Figuerola J, Koraka P, Martina B, Moreno A, Nowotny N, Pardigon N, Sanders N, Ulbert S, Tenorio A (2015) The challenge of West Nile virus in Europe: knowledge gaps and research priorities. Euro Surveill 20(20)

Roose K, De Baets S, Schepens B, Saelens X (2013) Hepatitis B core-based virus-like particles to present heterologous epitopes. Expert Rev Vaccines 12(2):183–198

Rosales-Mendoza S, Soria-Guerra RE, Moreno-Fierros L, Govea-Alonso DO, Herrera-Diaz A, Korban SS, Alpuche-Solis AG (2011) Immunogenicity of nuclear-encoded LTB:ST fusion protein from Escherichia coli expressed in tobacco plants. Plant Cell Rep 30(6):1145–1152

Sabalza M, Christou P, Capell T (2014) Recombinant plant-derived pharmaceutical proteins: current technical and economic bottlenecks. Biotechnol Lett 36(12):2367–2379

Saejung W, Fujiyama K, Takasaki T, Ito M, Hori K, Malasit P, Watanabe Y, Kurane I, Seki T (2007) Production of dengue 2 envelope domain III in plant using TMV-based vector system. Vaccine 25(36):6646–6654

Santi L, Batchelor L, Huang Z, Hjelm B, Kilbourne J, Arntzen CJ, Chen Q, Mason HS (2008) An efficient plant viral expression system generating orally immunogenic Norwalk virus-like particles. Vaccine 26(15):1846–1854

Sequeira JC, Harrison BD (1982) Serological studies on the cassave latent virus. Ann Appl Biol 101:33–42

Shahid N, Daniell H (2016) Plant-based oral vaccines against zoonotic and non-zoonotic diseases. Plant Biotechnol J 14(11):2079–2099

Spohn G, Jennings GT, Martina BE, Keller I, Beck M, Pumpens P, Osterhaus AD, Bachmann MF (2010) A VLP-based vaccine targeting domain III of the West Nile virus E protein protects from lethal infection in mice. Virol J 7:146

Stettler K, Beltramello M, Espinosa DA, Graham V, Cassotta A, Bianchi S, Vanzetta F, Minola A, Jaconi S, Mele F, Foglierini M, Pedotti M, Simonelli L, Dowall S, Atkinson B, Percivalle E, Simmons CP, Varani L, Blum J, Baldanti F, Cameroni E, Hewson R, Harris E, Lanzavecchia A, Sallusto F, Corti D (2016) Specificity, cross-reactivity, and function of antibodies elicited by Zika virus infection. Science 353(6301):823

Su J, Sherman A, Doerfler PA, Byrne BJ, Herzog RW, Daniell H (2015a) Oral delivery of acid alpha Glucosidase epitopes expressed in plant chloroplasts suppresses antibody formation in treatment of Pompe mice. Plant Biotechnol J 13(8):1023–1032

Su J, Zhu L, Sherman A, Wang X, Lin S, Kamesh A, Norikane JH, Streatfield SJ, Herzog RW, Daniell H (2015b) Low cost industrial production of coagulation factor IX bioencapsulated in lettuce cells for oral tolerance induction in hemophilia B. Biomaterials 70:84–93

Sun H, Chen Q, Lai H (2017) Development of antibody therapeutics against faviviruses. Int J Mol Sci 19(1):54–84. https://doi.org/10.3390/ijms19010054

Takeyama N, Kiyono H, Yuki Y (2015) Plant-based vaccines for animals and humans: recent advances in technology and clinical trials. Ther Adv Vaccines 3(5–6):139–154

Taylor TJ, Diaz F, Colgrove RC, Bernard KA, DeLuca NA, Whelan SP, Knipe DM (2016) Production of immunogenic West Nile virus-like particles using a herpes simplex virus 1 recombinant vector. Virology 496:186–193

Tuse D, Tu T, McDonald K (2014) Manufacturing economics of plant-made biologics: case studies in therapeutic and industrial enzymes. BioMed Research International 2014, 10. https://doi.org/10.1155/2014/256135

van Marle G, Antony J, Ostermann H, Dunham C, Hunt T, Halliday W, Maingat F, Urbanowski MD, Hobman T, Peeling J, Power C (2007) West Nile virus-induced neuroinflammation: glial infection and capsid protein-mediated neurovirulence. J Virol 81(20):10933–10949

Wang T, Anderson JF, Magnarelli LA, Wong SJ, Koski RA, Fikrig E (2001) Immunization of mice against West Nile virus with recombinant envelope protein. J Immunol 167(9): 5273–5277

Wang Y, Deng H, Zhang X, Xiao H, Jiang Y, Song Y, Fang L, Xiao S, Zhen Y, Chen H (2009) Generation and immunogenicity of Japanese encephalitis virus envelope protein expressed in transgenic rice. Biochem Biophys Res Commun 380(2):292–297

Wilken LR, Nikolov ZL (2012) Recovery and purification of plant-made recombinant proteins. Biotechnol Adv 30(2):419–433

Xu Z, Anderson R, Hobman TC (2011) The capsid-binding nucleolar helicase DDX56 is important for infectivity of West Nile virus. J Virol 85(11):5571–5580

Yamshchikov V, Manuvakhova M, Rodriguez E, Hebert C (2017) Development of a human live attenuated West Nile infectious DNA vaccine: Identification of a minimal mutation set conferring the attenuation level acceptable for a human vaccine. Virology 500:122–129

Yang JS, Ramanathan MP, Muthumani K, Choo AY, Jin SH, Yu QC, Hwang DS, Choo DK, Lee MD, Dang K, Tang W, Kim JJ, Weiner DB (2002) Induction of inflammation by West Nile virus capsid through the caspase-9 apoptotic pathway. Emerg Infect Dis 8(12):1379–1384

Yang M, Dent M, Lai H, Sun H, Chen Q (2017a) Immunization of Zika virus envelope protein domain III induces specific and neutralizing immune responses against Zika virus. Vacccine 35(33):4287–4294

Yang M, Lai H, Sun H, Chen Q (2017b) Virus-like particles that display Zika virus envelope protein domain III induce potent neutralizing immune responses in mice. Scientific Reports 7(1):7679

Yang M, Sun H, Lai H, Hurtado J, Chen Q (2017c) Plant-produced Zika virus envelope protein elicits neutralizing immune responses that correlate with protective immunity against Zika virus in mice. Plant Biotechnol J 16:572–580. https://doi.org/10.1111/pbi.12796

Zeller HG, Schuffenecker I (2004) West Nile virus: an overview of its spread in Europe and the Mediterranean basin in contrast to its spread in the Americas. Eur J Clin Microbiol Infect Dis 23(3):147–156

Zhao Q, Li S, Yu H, Xia N, Modis Y (2013) Virus-like particle-based human vaccines: quality assessment based on structural and functional properties. Trends Biotechnol 31(11):654–663

Part II
Vaccines for Poultry

Plant-Made Veterinary Vaccines for Newcastle Disease Virus

David R. Thomas and Amanda M. Walmsley

Abstract Newcastle disease is a word-wide, highly problematic, infectious, acute respiratory disease of domesticated and wild avian species. Newcastle disease virus proved an ideal target for a proof-of-concept study investigating the ability of a plant-made vaccines to successfully navigate the US veterinary regulatory system because it has: a single dominant viral surface antigen that is protective; a well-defined disease challenge model; readily available positive controls; and standardized diagnostic assays (Mihaliak et al. in Development of plant cell produced vaccines for animal health applicaions. United States Animal Health Association, Greensboro, 2004). However despite the Proof-of Concept study proving successful and an additional study proving an orally delivered, plant-made NDV vaccine could protect against challenge (Guerrero-Andrade et al. in Transgenic Res, 15(4): 455–463, 2006) a plant-made NDV vaccine has not been marketed. This is most likely due to the competition faced from numerous NDV vaccines already on the market that have proven successful for many years. Now that plant-made vaccines (for animal and human use) have successfully made it through regulatory systems, the target of future, commercial, plant-made vaccine investigations should either target a niche disease or a disease that has little or weaker competition already on the market.

Keywords Newcastle disease · Plant-made vaccines · Plant cell culture
Commercially licensed

D. R. Thomas
Department of Biochemistry and Molecular Biology, Monash University,
Clayton, VIC, Australia

A. M. Walmsley (✉)
Institute of Vector-Borne Disease, Monash University, Clayton, VIC, Australia
e-mail: amanda.walmsley@eliminatedengue.com

J. MacDonald (ed.), *Prospects of Plant-Based Vaccines in Veterinary Medicine*,
https://doi.org/10.1007/978-3-319-90137-4_6

1 History

Newcastle disease (ND) derives its name from one of the first outbreaks identified in New Castle upon Tyne, United Kingdom, in 1927 (Doyle 1927). Since then ND has rapidly spread worldwide (Lancaster 1962, 1977). Newcastle Disease Virus (NDV) is known to infect over 241 species of birds, and it is suspected that all avian species are susceptible to varying degrees (Kaleta and Baldauf 1988). NDV has also been found to infect animals other than birds, including reptiles and humans (Lancaster 1966). Its presence in wild bird populations makes it difficult to manage, and may also facilitate its spread nationally and internationally (Ramey et al. 2013). Outbreaks have severe economic impacts through not only flock loss, but also through trade and quarantine restrictions at outbreak regions (Leslie 2000). The costs of ND are especially pronounced in developing countries where ND is endemic in most village flocks (Martin 1992, Awan et al. 1994a, b). NDV is classified as a list A pathogen by the Office International des Epizooties (OIE), and is considered one of the most important avian diseases alongside avian influenza (Aldous and Alexander 2001).

2 Classification

NDV is classified in the family Paramyxoviridae, genus Avulavirus. Avulaviruses possess a single stranded negative sense monopartite RNA genome (Rima et al. 1995). Avulaviruses are classified into 12 species designated Avian paramyxovirus 1 through to 12 by their activity in haemagglutination and neuraminidase assays (Gogoi et al. 2017), with NDV known as Avian paramyxovirus 1. While all strains of NDV are contained in a single serotype of Avulavirus, strains can be divided into two classes (I and II) (Seal et al. 1995). Most class I strains are isolated from wild bird populations and largely possess low virulence, while class II strains range from low to high virulence, and are generally isolated from chickens. However, the geographic distribution and host range can vary between genotypes in class II strains (Dimitrov et al. 2016b). The ancestor of class II NDV colonized chickens multiple times, producing 16 genotypes to date, while class I consists of a single genotype (Taylor et al. 1996, 2017; Hao et al. 2016).

NDV infects birds through ingestion or inhalation, allowing rapid transmission. It is exceptionally contagious, and when the virus is introduced into a flock, the majority will be infected within two to six days. The virus remains infectious as airborne particles and large amounts are excreted via droppings, so ingestion of feces provides a major method of infection (Awan et al. 1994a, b). While its stability depends on environmental factors such as heat and humidity, it has been reported to remain viable for up to 255 days in a henhouse at temperatures ranging from −11 to 36 °C, and 10 to 14 days at 23–29 °C (Yune and Abdela 2017). This allows it to spread rapidly through contaminated people, equipment, food, and water, as well as the movement of infected birds (Burridge et al. 1975; Alexander 1995; Davis-Fields et al. 2014). NDV spread can also occur through live vaccines which are routinely found to spillover into local wildlife populations (Ayala et al. 2016).

3 Viral Characteristics

The pleomorphic NDV virions are enveloped and generally 200–300 nm in diameter (Fig. 1) (Nagai et al. 1989). NDV isolates contain one of three genome sizes; 15,192, 15,186, or 15,198 nucleotides long (Czegledi et al. 2006). The genome also follows the 'rule of six', whereby viral replication is most effective when the sequence is a multiple of six nucleotides long (Peeters et al. 2000). The RNA genome facilitates cytoplasmic replication and contains six genes encoding the following eight proteins.

4 Nucleocapsid Protein (N)

N is a 55 kDa protein that binds the genomic, negative sense RNA to protect against ribonucleases (Hugh-Jones et al. 1973). Polymers of N form a hollow helical nucleocapsid that contains the viral RNA (Kho et al. 2003). N is the most abundant protein in the virion, and is involved in forming the ribonucleoprotein complex that facilitates RNA synthesis (Kho et al. 2003). The amino region binds

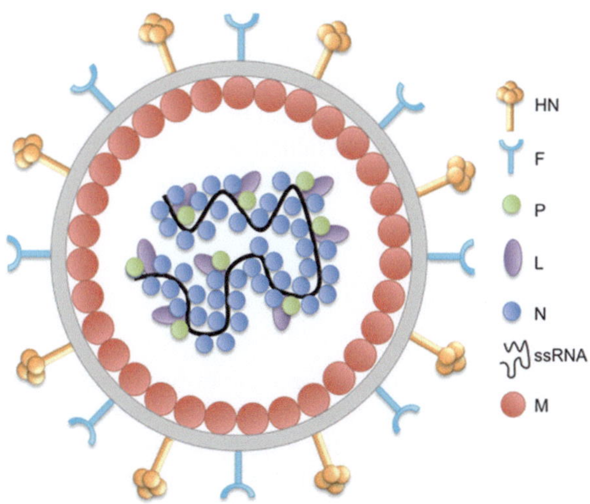

Fig. 1 Diagram of NDV. The ssRNA genome is encapsulated by nucleocapsid (N), and also associates with the phosphoprotein (P) and large polymerase protein (L) to form the ribonucleoprotein complex for RNA synthesis. The matrix protein (M) separates the host derived envelope (grey ring) from the nucleocapsid, while the haemagglutinin-neuraminidase (HN) and fusion (F) proteins are present at the viral surface

the Phosphoprotein (P), while the carboxy-terminus is thought to be responsible for polymerisation, enabling nucleocapsid structure formation (Kho et al. 2004).

5 Phosphoprotein (P)

P exists in multiple forms between 50 and 55 kDa, depending on its phosphorylation state (Smith and Hightower 1981). Involved in viral replication and transcription, P assists in stabilizing the Large protein (L) which function as an RNA-dependent RNA polymerase (Hamaguchi et al. 1983a, b). This complex produces full length anti-genomes from the N-RNA complex from which negative stranded viral genomes can be produced (Lamb and Kolakofsky 2001). P is also involved with preventing the encapsulation of non-viral RNA, and binding to unassembled N protein to regulate a shift from transcription to replication (Errington and Emmerson 1997).

6 V and W PROTEINS

The V and W proteins are truncations of the P protein produced through RNA editing. The proportions of mRNA encoding P,V, and W proteins in infected cells has been measured at ~ 68, 29, and 2%, respectively (Mebatsion et al. 2001). Interferons (IFN) are produced by the host cell in response to viral infection, which activate pathways leading to the expression of antiviral proteins. V protein inhibits the IFN activated pathway, and when V is absent viral replication is efficiently blocked by the host cell (Park et al. 2003). This is achieved by the V protein targeting STAT1 (a transcription factor activated by IFNs) for degradation, however the W protein was found to have little impact on pathogenicity (Huang et al. 2003). The V protein has also been found to inhibit the activity of MDA5, a protein that detects viral dsRNA to induce IFN-β production (Childs et al. 2007). Currently, the role of W in NDV is unknown.

7 Matrix Protein (M)

The M protein is highly conserved amongst paramyxoviruses, and mutations in the sequence are used to classify and track strains (Seal et al. 2000). The 40 kDa M protein is involved in assembling the virions on the host cell membrane and facilitating budding of mature viral particles (Pantua et al. 2006; Harrison et al. 2010). Despite functioning largely in the cytoplasm, the M protein contains a nuclear localization signal (NLS) that results in it accumulating in the nucleolus (Peeples et al. 1992). The M protein is thought to be involved in inhibiting host protein synthesis, and the pathogenicity of NDV is greatly reduced when the M protein is unable to enter the nucleus (Duan et al. 2014).

8 Fusion Protein (F)

The F protein is a surface glycoprotein that mediates NDV fusion with the host cell membrane for infection, as well as the fusion of the plasma membranes of two neighboring cells (McGinnes and Morrison 1986). It is synthesized in an inactive form that requires cleavage into two disulfide linked active subunits (F_1 and F_2) (Scheid and Choppin 1974). The amino acid sequence at the cleavage site varies between strains, and alters its susceptibility to cellular proteases, with an insensitivity to cellular proteases associated with greatly reduced virulence (Fujii et al. 1999; Panda et al. 2004). The cleavage site sequence is a major factor in pathogenicity, and is able to rapidly revert from non-pathogenic to pathogenic sequences after intracerebral passaging in chickens (Peeters et al. 1999; de Leeuw et al. 2003).

9 Hemagglutinin-Neuraminidase Protein (HN)

Another surface protein, the 74 kDa HN protein, functions as a homotetramer in NDV (Corey et al. 2003). HN performs several functions including receptor recognition on the host cell, receptor removal, and interacting with the F protein to facilitate cell entry (Yuan et al. 2011). While HN proteins of 571 aa or 616 aa have been found only in virulent or avirulent strains respectively, the length of the protein was not found to modulate pathogenicity (Romer-Oberdorfer et al. 2003). However, sequence specific differences in HN are able to modify virulence through altered tissue tropism, neuraminidase activity, and receptor binding (Huang et al. 2004).

10 Large Polymerase Protein (L)

L is the largest protein in the NDV genome at 250 kDa. As its name suggests, L synthesises viral mRNA and functions in genomic replication (Hamaguchi et al. 1983a, b). It also performs post-transcriptional modifications on newly produced mRNA, including 5′ capping, methylation, and polyadenylation (Dortmans et al. 2010). The L protein is also a factor in pathogenicity, with an increase in pathogenicity likely resulting from an increase in viral replication (Rout and Samal 2008).

11 Pathogenicity

ND is classified into three groups based on virulence; velogenic (high), mesogenic (medium), and lentogenic (low, or no clinical signs) (Alexander and Allan 1974). Velogenic strains are highly virulent, resulting in up to 100% mortality and are divided into viscerotropic and neurotropic strains based on their pathology (Alexander and Allan 1974; Chulan et al. 1982). Infection produces several broad symptoms, including depression, anorexia, ruffled feathers and hyperthermia, while viscerotropic strains produce haemorrhagic gut lesions, and neurotropic strains produce nervous signs such as ataxia and paralysis (Cattoli et al. 2011; Kapczynski et al. 2013). Both velogenic and mesogenic strains are also able to induce respiratory difficulties, while lentogenic strains generally show little or no clinical signs of disease (Brown et al. 1999; Cattoli et al. 2011).

Infection with NDV is recognized by Toll-like receptors (TLR) and nucleotide-binding oligomerization domain proteins (NOD), which lead to the release of interferons and cytokines (Kapczynski et al. 2013). Interestingly, velogenic strains of NDV produce a greater increase in expression of INF-α, IFN-γ, IL-1b, and IL-6 genes than lentogenic strains, although both viruses contain the V protein which suppresses IFN signaling (Kapczynski et al. 2013). It is possible that this strong immune response to the velogenic strains is actually deleterious to the host and contributes to the pathological effects (Rue et al. 2011).

Antibodies are detected against NDV six days after infection, peaking at 21–28 days post infection, and these function by preventing viral release from infected cells, and inhibiting infection from free virions (Al-Garib et al. 2003a, b). While vaccination is an important part of NDV control, it is not without shortcomings. Virulent ND strains are still able to replicate in vaccinated birds, although the clinical signs are attenuated (Kapczynski and King 2005). The efficacy of the vaccine is also dependent on how closely related the vaccine strain is to the challenge strain, potentially resulting in sub-optimal outcomes in field conditions (Miller et al. 2013). Further, the use of live vaccines maintains a reservoir of NDV which may develop into virulent strains (de Leeuw et al. 2003; Chong et al. 2010; Ayala et al. 2016). As such, there is still a strong need for more sustainable vaccination methods.

12 New Castle Disease Virus Vaccines and Vaccine Candidates

12.1 *Traditional NDV Vaccines*

Stringent vaccination of commercial flocks, use of rapid diagnostic assays, and culling of infected flocks are the current NDV control methods used by the poultry industry. If the disease is identified in a previously disease free region, many countries practice a stamping out policy involving strict isolation or quarantine of outbreaks, humane destruction of infected and exposed birds, and depopulation followed by 21 days without poultry.

From the early 1950's to the late 1990's the only NDV vaccines available were based on live or inactivated ND strains (Gallili and Ben-Nathan 1998).

Currently, the most commonly used NDV vaccines (LaSota, B1, and VG/GA vaccines) (Orsi et al. 2009) are live vaccine viruses formulated from strains isolated in the 1940's and 1960's. These live vaccines provide both mucosal and humoral immunity and can be administered using mass application techniques (Table 1). However considering they are approaching 60–80 years old, it comes as no surprise that there is some 18.3–26.6% nucleotide distance between the vaccines strains and the currently circulating, virulent NDV strains (Dimitrov et al. 2016b). While still affording protection, this genetic diversity prevents the effective reduction of shedding of virulent virus from vaccinated birds (Miller et al. 2007, 2009). Hens vaccinated with live NDV strains can transfer antibodies via the egg to offspring that can partially neutralize the live ND vaccines (Dimitrov et al. 2016a) and since long term immunity is not gained through immunization with live NDV strains, continued vaccinations are necessary through the lifetime of layers and breeders.

Although inactivated NDV vaccines produce high humoral antibody levels, unlike vaccination with live NDV strains, mass immunization is not possible thus resulting in increased cost and labor (Table 1). Immunization with inactivated NDV vaccines also does not result in a strong cell mediated response (Schijns et al. 2013) mainly invoking a circulating antibody response (Grimes 2002), and allowing the shedding of larger amounts of virulent virus compared to birds vaccinated with live ND vaccines (Miller et al. 2009, 2013). Since chickens are food animals, long withdrawal periods are also required before vaccinated birds can be processed for human consumption.

Although the live and inactivated vaccines protect against clinical disease in Specific Pathogen Free (SPF) chickens, there are continuous reports of vaccine failures under field conditions (Perozo et al. 2012; Rehmani et al. 2015). Dimitrov et al. (2016) summarize the main properties of live, inactivated and vectored vaccines, the most widely used ND vaccines.

12.2 Vectored Vaccines

12.2.1 Fowl Pox Virus Vectored Vaccines

Recombinant, Turkey Herpes Virus (HVT) vectored NDV vaccines are the most commonly used recombinant NDV vaccines. Currently, two bivalent commercial recombinant HVT (rHVT) vaccines have been registered and used in the poultry industry. HVT-vectored vaccines protect chickens when challenged with virulent NDV (Sonoda et al. 2000) and are hindered by maternal antibodies (Le Gros et al. 2009) though only mildly. Although not able to be mass immunized, the rHVT-NDV vaccines can be administered *in ovo* or subcutaneously after hatch. *In ovo* immunization has the advantage of using high tech machines. These machines, enable the volume and concentration of the administered vaccine to be standardized. They also reduce human error and labor cost when compared to vaccination of chickens later in life. HVT-vectored NDV vaccines also produce long-term

Table 1 Summary of virus-based vaccines

Vaccine	Ease of administration	Induced immune response	Challenge results	Frequency of immunization	Comments	References
Live NDV strain vaccines	+++	Mucosal and humoral	Protection	Multiple times across life span	Interfered with by pre-existing NDV antibodies (maternal or previous immunisation); shedding occurs	Miller et al. (2007, 2009), Dimitrov et al. (2016a, b)
Inactive NDV strain vaccines	+	Mainly humoral	Protection	Multiple times across life span	High shedding; long withdrawal period	Grimes (2002), Miller et al. (2009, 2013), Schijns et al. (2013)
Fowl Pox vectored	+	Strong cell-mediated immunity (CMI) in addition to humoral immune response	Protection	Long lasting response	Interference by vector specific antibodies; not heat stable; delayed generation of protective immune response; specialised use	Grimes (2002), Miller et al. (2009, 2013), Hghihghi et al. (2010), Schijns et al. (2013), Van den Berg (2013), Sanchez-Sampedro et al. (2015)
Turkey Herpes vectored	++	Humoral immune response	Protection	Long term immunisation	Not heat stable; often used in combination with other vaccines	Sonoda et al. (2000), Le Gros et al. (2009), Palya et al. (2012), Esaki et al. (2013)
Antigen-antibody complex	++	Humoral immune response	Protection		Quick induction of immune response; reduced shedding	Terres and Wolins (1959, 1961), Kapczynski et al. (2012)
Nanoparticle delivery	++	Strong humoral, cellular and mucosal	Protection	Released sustainably	No clinical signs of infection	Dai et al. (2015), Chahal et al. (2016), Zhao et al. (2016b)
NDV VLPs	+	NDV-specific antibodies and cell mediated response	Protection in mouse		No infection risk or risk of conversion back to virulence	McGinnes et al. (2010), Park et al. (2014), Dimitrov et al. (2016b)

immunity (Esaki et al. 2013). Unfortunately these vaccines are not temperature stable, requiring not only storage in liquid nitrogen, but administration within an hour of being thawed. Four weeks is also required before full immunity is reached (Palya et al. 2012). Since heat sensitivity and the time delay of effectiveness decrease their ease of use, recombinant HVT vaccines have only been widely used in countries where minimum viral challenges exist. In endemic countries, these vaccines need to be used in combination with other NDV vaccines to confer acceptable protection.

There are two commercial NDV vaccines using Fowl Pox Virus (FPV) as a vector. While these vaccines protect chickens from a challenge with virulent NDV (Boursnell et al. 1990; Taylor et al. 1996; Karaca et al. 1998) their effectiveness can be decreased by interference from vector-specific antibodies either transferred maternally (Faulkner et al. 2013), or induced through previous vaccination (Bublot et al. 2006). Since they also cannot be applied through mass methods they are also only usually used in specialized instances.

12.2.2 Antigen-Antibody Complex Vaccine

Antigen-antibody complexes or Immune complexes (ICs) were at first thought as indicators of an undesirable immune response due to their frequent presence in sites of autoimmunity and inflammation (Wen et al. 2016). However studies as early as the 1950s demonstrated that ICs enabled faster sensitization of a host to an antigen than delivery of the antigen alone or in combination with a control antiserum (Terres and Wolins 1959, 1961). NDV IC vaccines can be delivered *in ovo* at 18 or 19 days of embryonation when the eggs are moved into hatching trays (Kapczynski et al. 2012). This delivery allows mucosal immunization of the respiratory and gastrointestinal tracts and enables the newly hatched chickens to develop an early immune response. Unlike *in ovo* delivery of live NDV, *in ovo* NDV IC delivery using live NDV does not decrease hatchability or weaken chicks (Kapczynski et al. 2012).

12.3 Nanoparticle Vaccines

Nanoparticle systems are often used as vaccine carriers as they protect vaccine antigens from maternal antibodies and nucleases (Dai et al. 2015; Chahal et al. 2016; Zhao et al. 2016b) and disruption, and induce higher antigen uptake, controlled release, and increased duration of responses (Dai et al. 2015; Zhao et al. 2016b). They also have flexible delivery methods, being able to be administered mucosally (including oral), and parenterally (Dai et al. 2015; Zhao et al. 2016b). Two chitosan derivatives, O-2′-hydroxypropyltrimethyl ammonium chloride chitosan and N-2-hydroxypropyl trimethyl ammonium chloride chitosan, have been used to mucosally deliver live attenuated NDV vaccines (Dai et al. 2015; Zhao et al. 2016b). The chitosan-derived nanoparticle carriers effectively released NDV in a sustainable manner and induced strong cellular, humoral, and mucosal immune responses that resulted in protection after challenge with virulent NDV. No clinical signs or microscopic lesions were observed in the chitosan delivered NDV vaccinated birds, while 20% mortality and some hyperplastic changes were observed in the chickens vaccinated with a traditional commercial NDV vaccine.

Silver based nanoparticles have also been successfully used to vaccinate against NDV with @SiO_2 and double hydroxide @SiO_2 nanoparticles intranasally delivering DNA NDV vaccine candidates (Zhao et al. 2015, 2016a). The SiO_2 nanoparticles showed low toxicity in SPF chickens, were released sustainably after an initial burst, and induced stronger cellular, humoral, and mucosal immune responses than intramuscular delivery of the same vaccine and naked DNA. The @SiO_2 and double hydroxide @SiO_2 nanoparticle delivered NDV DNA vaccine candidates demonstrated 100% protection in chickens after challenge with a virulent NDV strain.

12.4 *Virus-like Particle Vaccines*

Virus-Like Particles (VLPs) are potent immunogens since they authentically present repeating units of protective antigen on their surface and within the core of particles that are similar in size to the viral pathogen. They also reduce the risk involved in vaccination since live virus is not involved in their manufacture. Without infectious material they can't infect the vaccinated host, even if the host is immunocompromised, nor can they revert to virulence or recombine with endemic viruses. Pre-clinical studies into NDV VLPs performed in a murine model compared immune responses induced with NDV VLPs with those stimulated by UV-inactivated, vaccine strain of Newcastle disease virus (McGinnes et al. 2010). Soluble NDV-specific antibodies induced after immunization with NDV VLPs were as high or higher than those resulting from immunization with the inactivated vaccine virus (McGinnes et al. 2010). In addition, NDV VLPs stimulated significantly higher T-cell responses than those stimulated by the vaccine virus (McGinnes et al. 2010).

NDV VLPs have also proven effective in clinical studies. Park and colleagues (Park et al. 2014), produced recombinant NDV VLPs expressing the NDV fusion (F) protein along with influenza virus matrix 1 (M1) protein using an insect cell expression system. SPF chickens were immunized with oil emulsion NDV VLPs of increasing dosages (0.4, 2, 10, or 50 μg of VLPs/0.5-ml dose VLPs). Three weeks after immunization, the chickens were challenged with a highly virulent NDV strain. The NDV VLPs elicited NDV-specific antibodies and provided protection against lethal challenge in a dose-dependent manner. A single immunization with a 10 or 50 μg of NDV VLPs was found capable of fully protecting chickens from lethal challenge and greatly reduced virus shedding. Like their infectious viral particle counterparts, NDV VLPs have been used as antigen carriers for viral antigens including those from respiratory syncytial virus (Murawski et al. 2010) and avian influenza (Shen et al. 2013; Noh et al. 2016).

13 Plant-Made Vaccine Candidates

Oral delivery has long been the holy grail of many vaccination regimens. However, the original idea of recombinant (transgenic), edible plant-made vaccines (that a reliable and robust immune response would result after ingesting transgenic plant material that produced a vaccine antigen within its tissues) has always been questioned and has been difficult to prove. Deservedly rigid guidelines for human health have hindered progress of mucosal delivery of Plant-made Vaccines (PMVs) in humans. Purified, injectable PMVs and Plant-made Therapeutics (PMTs) are therefore the mainstream accepted delivery route.

13.1 Plant Cell Culture-Made HN

Following this trend, the first PMV to be commercially licensed (Katsnelson et al. 2006), a veterinary vaccine to protect poultry from NDV was partially purified and injectable. The study was undertaken to demonstrate the ability of a recombinant plant system to produce a protective antigen that could then be approved through the existing US regulatory system for animal health. Newcastle disease virus was chosen as the target for the proof-of-concept study because: there is a single dominant viral surface antigen that is protective, the HN protein; there is a well-defined disease challenge model; there are positive controls that are readily available in the form of commercialized viral vaccine strains; and there are standardized haemagglutination inhibition assays that measure antibody titre (Mihaliak et al. 2004).

The developed NDV vaccine was produced using a contained, *Nicotiana benthamiana* (NT1) cell culture system (Table 2). This production system circumvented proposed regulatory problems of whole plant production systems (namely pollen). The HN antigen was expressed in the NT1 cells after stable transformation using *Agrobacterium*. The NT1 cell-produced HN was found to retain the size (western analysis) and immunoreactivity (ELISA) of the native antigen. There was no degradation of the HN in the plant cells and it was bioactive in red blood cell haemagglutination assays.

As part of the regulatory requirements, a master seed established from the cell line was used to produce a vaccine batch that was used in a disease challenge study in chickens (Mihaliak et al. 2004). Non-transgenic, negative controls plus four different formulations of the recombinant vaccine were prepared from plant cells to test in a chicken NDV challenge trial. On days 0 and 14, SPF chicks were subcutaneously vaccinated into the loose skin of the neck with the partially purified, plant cell material or control material. Individual blood samples were collected from each bird on day 24, and on day 28 the chicks (except unchallenged controls) were challenged with NDV Texas GB strain. Antigen specific responses were determined by haemagglutination-inhibition assays and daily clinical observations were made

Table 2 Plant-made NDV antigen immunogenicity trials in animals

NDV antigen	Production system	Delivery	Immune response	Challenge trials	References
HN protein	Stably transformed NT1 cells	Subcutaneous delivery to chickens	NDV-HN specific antibodies	Protected against NDV challenge	Mihaliak et al. (2004)
F protein and HN	Stably transformed potato	Oral deliver to mice	NDV-specific IgG and IgA antibodies	Not performed	Berinstein et al. (2005), Gomez et al. (2008)
F protein	Stably transformed Maize	Oral delivery to chickens	Antigen-specific immune responses	Protected against NDV challenge	Guerrero-Andrade et al. (2006)
F Protein	Stably transformed rice	Intraperitoneal delivery to mice	Antigen-specific immune responses	Not performed	Yang et al. (2007)
HN protein	Stably transformed tobacco	Oral delivery to chickens	Low HN-specific IgG immune responses	Not performed	Hahn et al. (2007)

for 14 days post-challenge. Birds vaccinated subcutaneously with the plant-made, subunit HN antigen from NDV proved to be protected against lethal challenge to NDV. The plant-made vaccine when provided between 3 and 33 μg/dose resulted in an overall average protection of 95% (Mihaliak et al. 2004). Thus a plant-cell produced vaccine serologically converted birds and provided protective immunity against NDV without the risk of shedding or spreading the disease. Since the vaccine only contained the HN antigen there was also the possibility of differentiation of diseased and vaccinated birds. A formulation was successfully advanced through the USDA Center for Veterinary Biologics' regulatory approval process therefore demonstrating that a plant-made vaccine could be developed and approved within the existing regulatory framework.

13.2 Whole Plant-Made NDV Antigens

The only other plant-made vaccine to be tested in NDV challenge trials was produced in stably transformed (particle bombardment) maize plants (Guerrero-Andrade et al. 2006) (Table 2). The NDV F protein was found to accumulate in the maize kernels which were then fed to chickens. The immune responses of 45 day old, SPF chickens was compared between treatments of feeding with transformed or non-transformed ground maize kernel; feeding LaSota commercial vaccine mixed in with non-transformed ground kernel; and commercial vaccine administered intranasally. F protein antibodies were detected in all chickens after ingestion of the transformed maize and treatment with LaSota (oral and nasal). The F protein specific antibodies reached a maximum level by day 45 with all immunized groups having approximately the same antibodies titre by day 60.

All of the SPF chickens immunized with transgenic maize or LaSota virus survived (100% protection) while none of the chickens fed non-transgenic maize survived.

While additional studies have expressed either the HN and F proteins together in potato (Berinstein et al. 2005; Gomez et al. 2008) or separately in rice (F protein) (Yang et al. 2007) or tobacco (HN)(Hahn et al. 2007) (Table 2), no challenge trials were performed within these studies.

14 The Future of Plant-Made NDV Vaccines

The current preferred method of vaccination against NDV is largely through live, attenuated LaSota, B1, and VG/GA vaccines. Commercial NDV vaccines other than live vaccines are available but they are usually only used in niche areas. The live NDV vaccines have a strong hold on the NDV vaccine market being inexpensive, easy to apply and a time proven vaccine. The plant cell culture-made NDV vaccine was Dow AgroScience's chosen model vaccine. They achieved their proof-of-concept goal by demonstrating that a plant-made injectable vaccine could be produced at a cost comparable to its already commercialized competition and could be commercially licensed through existing US regulations. Their plant-made NDV vaccine however was never put into production and out to market due to the competition faced from the many NDV vaccines already in existence that had proven successful for many years. Even the possibility of oral delivery by Guerrero-Andrade and colleagues (Guerrero-Andrade et al. 2006) was not enough to provide a clear advantage over the existing market.

Proof-of-concept has been provided for plant-made vaccines. It must be remembered however that with a comparatively new technology such as plant-made vaccines, actual commercialization efforts should focus on areas where a clear advantage is held. Focus should either be on niche disease targets (such as those tackled by commercialized plant-made vaccines for human use) or diseases that have little or weaker competition already on the market. These targets should ideally have a known protective antigen(s); a well-defined disease challenge model; available positive controls and other diagnostics including a standardized assays to measure induced immune response.

References

Al-Garib SO, Gielkens AL, Gruys DE, Hartog L, Koch G (2003a) Immunoglobulin class distribution of systemic and mucosal antibody responses to Newcastle disease in chickens. Avian Dis 47(1):32–40

Al-Garib SO, Gielkens ALJ, Gruys E, Kochi G (2003b) Review of Newcastle disease virus with particular references to immunity and vaccination. World's Poult Sci J 59(02):185–200

Aldous EW, Alexander DJ (2001) Detection and differentiation of Newcastle disease virus (avian paramyxovirus type 1). Avian Pathol 30(2):117–128

Alexander DJ (1995) The epidemiology and control of avian influenza and Newcastle disease. J Comp Pathol 112(2):105–126

Alexander DJ, Allan WH (1974) Newcastle disease virus pathotypes. Avian Pathol 3(4):269–278

Awan MA, Otte MJ, James AD (1994a) The epidemiology of Newcastle disease in rural poultry: a review. Avian Pathol 23(3):405–423

Awan MA, Otte MJ, James AD (1994b) The epidemiology of Newcastle-disease in rural poultry—a review. Avian Pathol 23(3):405–423

Ayala AJ, Dimitrov KM, Becker CR, Goraichuk IV, Arns CW, Bolotin VI, Ferreira HL, Gerilovych AP, Goujgoulova GV, Martini MC, Muzyka DV, Orsi MA, Scagion GP, Silva RK, Solodiankin OS, Stegniy BT, Miller PJ, Afonso CL (2016) Presence of vaccine-derived Newcastle disease viruses in wild birds. PLoS ONE 11(9):e0162484

Berinstein A, Vazquez-Rovere C, Asurmendi S, Gomez E, Zanetti F, Zabal O, Tozzini A, Conte Grand D, Taboga O, Calamante G, Barrios H, Hopp E, Carrillo E (2005) Mucosal and systemic immunization elicited by Newcastle disease virus (NDV) transgenic plants as antigens. Vaccine 23(48–49):5583–5589

Boursnell ME, Green PF, Samson AC, Campbell JI, Deuter A, Peters RW, Millar NS, Emmerson PT, Binns MM (1990) A recombinant fowlpox virus expressing the hemagglutinin-neuraminidase gene of Newcastle disease virus (NDV) protects chickens against challenge by NDV. Virology 178(1):297–300

Brown C, King DJ, Seal BS (1999) Pathogenesis of Newcastle disease in chickens experimentally infected with viruses of different virulence. Vet Pathol 36(2):125–132

Bublot M, Pritchard N, Swayne DE, Selleck P, Karaca K, Suarez DL, Audonnet JC, Mickle TR (2006) Development and use of fowlpox vectored vaccines for avian influenza. Ann N Y Acad Sci 1081:193–201

Burridge MJ, Riemann HP, Utterback WW (1975) Methods of spread of velogenic viscerotropic Newcastle disease virus in the Southern Californian epidemic of 1971–1973. Avian Dis 19(4): 666–678

Cattoli G, Susta L, Terregino C, Brown C (2011) Newcastle disease: a review of field recognition and current methods of laboratory detection. J Vet Diagn Invest 23(4):637–656

Chahal JS, Khan OF, Cooper CL, McPartlan JS, Tsosie JK, Tilley LD, Sidik SM, Lourido S, Langer R, Bavari S, Ploegh HL, Anderson DG (2016) Dendrimer-RNA nanoparticles generate protective immunity against lethal Ebola, H1N1 influenza, and Toxoplasma gondii challenges with a single dose. Proc Natl Acad Sci USA 113(29):E4133–E4142

Childs K, Stock N, Ross C, Andrejeva J, Hilton L, Skinner M, Randall R, Goodbourn S (2007) mda-5, but not RIG-I, is a common target for paramyxovirus V proteins. Virology 359(1): 190–200

Chong YL, Padhi A, Hudson PJ, Poss M (2010) The effect of vaccination on the evolution and population dynamics of avian paramyxovirus-1. PLoS Pathog 6(4):e1000872

Chulan U, Ibrahim AL, Mustaffa Babjee A, SheikhOmar AR (1982) vaccination against Newcastle disease. Trop Anim Health Prod 14(3):177–184

Corey EA, Mirza AM, Levandowsky E, Iorio RM (2003) Fusion deficiency induced by mutations at the dimer interface in the Newcastle disease virus hemagglutinin-neuraminidase is due to a temperature-dependent defect in receptor binding. J Virol 77(12):6913–6922

Czegledi A, Ujvari D, Somogyi E, Wehmann E, Werner O, Lomniczi B (2006) Third genome size category of avian paramyxovirus serotype 1 (Newcastle disease virus) and evolutionary implications. Virus Res 120(1–2):36–48

Dai C, Kang H, Yang W, Sun J, Liu C, Cheng G, Rong G, Wang X, Wang X, Jin Z, Zhao K (2015) O-2'-hydroxypropyltrimethyl ammonium chloride chitosan nanoparticles for the delivery of live Newcastle disease vaccine. Carbohydr Polym 130:280–289

Davis-Fields MK, Allison AB, Brown JR, Poulson RL, Stallknecht DE (2014) Effects of temperature and pH on the persistence of avian paramyxovirus-1 in water. J Wildl Dis 50(4): 998–1000 de Leeuw OS, Hartog L, Koch G, Peeters BP (2003) Effect of fusion protein cleavage site mutations on virulence of Newcastle disease virus: non-virulent cleavage site mutants revert to virulence after one passage in chicken brain. J Gen Virol 84(Pt 2):475–484

Dimitrov KM, Afonso CL, Yu Q, Miller PJ (2016) Newcastle disease vaccines-A solved problem or a continuous challenge? Vet Microbiol

Dimitrov KM, Ramey AM, Qiu X, Bahl J, Afonso CL (2016b) Temporal, geographic, and host distribution of avian paramyxovirus 1 (Newcastle disease virus). Infect Genet Evol 39:22–34

Dortmans JC, Rottier PJ, Koch G, Peeters BP (2010) The viral replication complex is associated with the virulence of Newcastle disease virus. J Virol 84(19):10113–10120

Doyle TM (1927) A hitherto unrecorded disease of fowls due to a filter passing virus. J Comp Pathol Therap 40:144–169

Duan Z, Li J, Zhu J, Chen J, Xu H, Wang Y, Liu H, Hu S, Liu X (2014) A single amino acid mutation, R42A, in the Newcastle disease virus matrix protein abrogates its nuclear localization and attenuates viral replication and pathogenicity. J Gen Virol 95(Pt 5):1067–1073

Errington W, Emmerson PT (1997) Assembly of recombinant Newcastle disease virus nucleocapsid protein into nucleocapsid-like structures is inhibited by the phosphoprotein. J Gen Virol 78(9):2335–2339

Esaki M, Godoy A, Rosenberger JK, Rosenberger SC, Gardin Y, Yasuda A, Dorsey KM (2013) Protection and antibody response caused by Turkey herpesvirus vector Newcastle disease vaccine. Avian Dis 57(4):750–755

Faulkner OB, Estevez C, Yu Q, Suarez DL (2013) Passive antibody transfer in chickens to model maternal antibody after avian influenza vaccination. Vet Immunol Immunopathol 152(3–4): 341–347

Fujii Y, Sakaguchi T, Kiyotani K, Yoshida T (1999) Comparison of substrate specificities against the fusion glycoprotein of virulent newcastle disease virus between a chick embryo fibroblast processing protease and mammalian subtilisin-like proteases. Microbiol Immunol 43(2): 133–140

Gallili GE, Ben-Nathan D (1998) Newcastle disease vaccines. Biotechnol Adv 16(2):343–366

Gogoi P, Ganar K, Kumar S (2017) Avain paramyxovirus: a brief review. Transbound Emerg Dis 64(1):53–67

Gomez E, Chimeno Zoth S, Carrillo E, Estela Roux M, Berinstein A (2008) Mucosal immunity induced by orally administered transgenic plants. Immunobiology 213(8):671–675

Grimes SE (2002) A basic laboratory manual for the small-scale production and testing of I-2

Guerrero-Andrade O, Loza-Rubio E, Olivera-Flores T, Fehervari-Bone T, Gomez-Lim MA (2006) Expression of the Newcastle disease virus fusion protein in transgenic maize and immunological studies. Transgenic Res 15(4):455–463

Hahn BS, Jeon IS, Jung YJ, Kim JB, Park JS, Ha SH, Kim KH, Kim HM, Yang JS, Kim YH (2007) Expression of hemagglutinin-neuraminidase protein of Newcastle disease virus in transgenic tobacco. Plant Biotechnol Rep 1(2):85–92

Hamaguchi M, Yoshida T, Nishikawa K, Naruse H, Nagai Y (1983a) Transcriptive complex of Newcastle disease virus. Virology 128(1):105–117

Hamaguchi M, Yoshida T, Nishikawa K, Naruse H, Nagai Y (1983b) Transcriptive complex of Newcastle disease virus. I. Both L and P proteins are required to constitute an active complex. Virology 128(1):105–117

Hao HF, Chen SL, Liu P, Ren SH, Gao XL, Wang YP, Wang XL, Zhang SX, Yang ZQ (2016) Genetic variation in V gene of class II Newcastle disease virus. Infect Genet Evol 37:14–20

Harrison MS, Sakaguchi T, Schmitt AP (2010) Paramyxovirus assembly and budding: building particles that transmit infections. Int J Biochem Cell Biol 42(9):1416–1429

Hghihghi HR, Read LR, Mohammadi H, Pei Y, Ursprung C, Nagy E, Behboudi S, Haeryfar SM, Sharif S (2010) Characterization of host responses against a recombinant fowlpox virus-vectored vaccine expressing the hemagglutinin antigen of an avian influenza virus. Clin Vaccine Immunol 17(3):454–463

Huang Z, Krishnamurthy S, Panda A, Samal SK (2003) Newcastle disease virus V protein is associated with viral pathogenesis and functions as an alpha interferon antagonist. J Virol 77(16):8676–8685

Huang Z, Panda A, Elankumaran S, Govindarajan D, Rockemann DD, Samal SK (2004) The hemagglutinin-neuraminidase protein of Newcastle disease virus determines tropism and virulence. J Virol 78(8):4176–4184

Hugh-Jones M, Allan WH, Dark FA, Harper GJ (1973) The evidence for the airborne spread of Newcastle disease. J Hyg (Lond) 71(2):325–339

Kaleta EF, Baldauf C (1988) Newcastle disease in free-living and pet birds. Newcastle Disease. D. J. Alexander. Springer, Boston, pp 197–246

Kapczynski DR, Afonso CL, Miller PJ (2013) Immune responses of poultry to Newcastle disease virus. Dev Comp Immunol 41(3):447–453

Kapczynski DR, King DJ (2005) Protection of chickens against overt clinical disease and determination of viral shedding following vaccination with commercially available Newcastle disease virus vaccines upon challenge with highly virulent virus from the California 2002 exotic Newcastle disease outbreak. Vaccine 23(26):3424–3433

Kapczynski DR, Martin A, Haddad EE, King DJ (2012) Protection from clinical disease against three highly virulent strains of Newcastle disease virus after in ovo application of an antibody-antigen complex vaccine in maternal antibody-positive chickens. Avian Dis 56(3): 555–560

Karaca K, Sharma JM, Winslow BJ, Junker DE, Reddy S, Cochran M, McMillen J (1998) Recombinant fowlpox viruses coexpressing chicken type I IFN and Newcastle disease virus HN and F genes: influence of IFN on protective efficacy and humoral responses of chickens following in ovo or post-hatch administration of recombinant viruses. Vaccine 16(16):1496–1503

Katsnelson A, Ransom J, Vermij P, Waltz E (2006) USDA approves the first plant-based vaccine. News In Brief. Retrieved 13 Feb 2017, from http://www.nature.com/nbt/journal/v24/n3/full/nbt0306-233.html

Kho CL, Tan WS, Tey BT, Yusoff K (2003) Newcastle disease virus nucleocapsid protein: self-assembly and length-determination domains. J Gen Virol 84(Pt 8):2163–2168

Kho CL, Tan WS, Tey BT, Yusoff K (2004) Regions on nucleocapsid protein of Newcastle disease virus that interact with its phosphoprotein. Arch Virol 149(5):997–1005

Lamb RA, Kolakofsky D (2001) Paramyxoviridae: the viruses and their replication. Fields Virol 4:1305–1340 (B. N. Fields, D. M. Knipe and P. M. Howley. Philadelphia, PA, Lippincott Williams & Wilkins)

Lancaster JE (1962) World distribution of Newcastle disease—1961. Can J Comp Med Vet Sci 26(10):244–245

Lancaster JE (1966) Newcastle disease. A review of some of the literature published between 1926 and 1964. Ottawa, Health of Animals Branch, Canada Department of Agriculture: x + 188

Lancaster JE (1977) Newcastle-disease—review of geographical incidence and epizootiology. Worlds Poult Sci J 33(3):155–165

Le Gros FX, Dancer A, Giacomini C, Pizzoni L, Bublot M, Graziani M, Prandini F (2009) Field efficacy trial of a novel HVT-IBD vector vaccine for 1-day-old broilers. Vaccine 27(4): 592–596

Leslie J (2000) Newcastle disease: outbreak losses and control policy costs. Vet Rec 146(21): 603–606

Martin PAJ (1992) The epidemiology of Newcastle-disease in village chickens. Newcastle Dis Village Chickens 39:40–45

McGinnes LW, Morrison TG (1986) Nucleotide sequence of the gene encoding the Newcastle disease virus fusion protein and comparisons of paramyxovirus fusion protein sequences. Virus Res 5(4):343–356

McGinnes LW, Pantua H, Laliberte JP, Gravel KA, Jain S, Morrison TG (2010) Assembly and biological and immunological properties of Newcastle disease virus-like particles. J Virol 84(9):4513–4523

Mebatsion T, Verstegen S, De Vaan LT, Romer-Oberdorfer A, Schrier CC (2001) A recombinant newcastle disease virus with low-level V protein expression is immunogenic and lacks pathogenicity for chicken embryos. J Virol 75(1):420–428
Mihaliak CA, Webb S, Miller T, Fanton M, Kirk DD, Cardineau GA, Mason HS, Walmsley A, Arntzen C, Van Eck JH (2004) Development of plant cell produced vaccines for animal health applicaions. In: Proceedings of the 108th Annual meeting of the United States Animal Health Association, Greensboro, North Carolina, United States Animal Health Association
Miller PJ, Afonso CL, El Attrache J, Dorsey KM, Courtney SC, Guo Z, Kapczynski DR (2013) Effects of Newcastle disease virus vaccine antibodies on the shedding and transmission of challenge viruses. Dev Comp Immunol 41(4):505–513
Miller PJ, Estevez C, Yu Q, Suarez DL, King DJ (2009) Comparison of viral shedding following vaccination with inactivated and live Newcastle disease vaccines formulated with wild-type and recombinant viruses. Avian Dis 53(1):39–49
Miller PJ, King DJ, Afonso CL, Suarez DL (2007) Antigenic differences among Newcastle disease virus strains of different genotypes used in vaccine formulation affect viral shedding after a virulent challenge. Vaccine 25(41):7238–7246
Murawski MR, McGinnes LW, Finberg RW, Kurt-Jones EA, Massare MJ, Smith G, Heaton PM, Fraire AE, Morrison TG (2010) Newcastle disease virus-like particles containing respiratory syncytial virus G protein induced protection in BALB/c Mice, with no evidence of immunopathology. J Virol 84(2):1110–1123
Nagai Y, Hamaguchi M, Toyoda T (1989) Molecular biology of Newcastle disease virus. Prog Vet Microbiol Immunol 5:16–64
Noh JY, Park JK, Lee DH, Yuk SS, Kwon JH, Lee SW, Lee JB, Park SY, Choi IS, Song CS (2016) Chimeric bivalent virus-like particle vaccine for H5N1 HPAI and ND confers protection against a lethal challenge in chickens and allows a strategy of Differentiating Infected from Vaccinated Animals (DIVA). Plos One 11(9)
Orsi MA, Doretto L, Reischak D, da Silva LHA, Spilki FR, Buzinaro MG, Arns CW (2009) Newcastle disease virus vaccine strains: immunogenicity is not influenced by ICPI. Braz J Poult Sci 11(2):129–133
Palya V, Kiss I, Tatar-Kis T, Mato T, Felfoldi B, Gardin Y (2012) Advancement in vaccination against Newcastle disease: recombinant HVT NDV provides high clinical protection and reduces challenge virus shedding with the absence of vaccine reactions. Avian Dis 56(2): 282–287
Panda A, Huang Z, Elankumaran S, Rockemann DD, Samal SK (2004) Role of fusion protein cleavage site in the virulence of Newcastle disease virus. Microbial Pathog 36(1):1–10
Pantua HD, McGinnes LW, Peeples ME, Morrison TG (2006) Requirements for the assembly and release of Newcastle disease virus-like particles. J Virol 80(22):11062–11073
Park JK, Lee DH, Yuk SS, Tseren-Ochir EO, Kwon JH, Noh JY, Kim BY, Choi SW, Kang SM, Lee JB, Park SY, Choi IS, Song CS (2014) Virus-like particle vaccine confers protection against a lethal newcastle disease virus challenge in chickens and allows a strategy of differentiating infected from vaccinated animals. Clin Vaccine Immunol 21(3):360–365
Park MS, Shaw ML, Munoz-Jordan J, Cros JF, Nakaya T, Bouvier N, Palese P, Garcia-Sastre A, Basler CF (2003) Newcastle Disease Virus (NDV)-based assay demonstrates interferon-antagonist activity for the NDV V protein and the Nipah virus V, W, and C proteins. J Virol 77(2):1501–1511
Peeples ME, Wang C, Gupta KC, Coleman N (1992) Nuclear entry and nucleolar localization of the Newcastle disease virus (NDV) matrix protein occur early in infection and do not require other NDV proteins. J Virol 66(5):3263–3269
Peeters BP, Gruijthuijsen YK, de Leeuw OS, Gielkens AL (2000) Genome replication of Newcastle disease virus: involvement of the rule-of-six. Arch Virol 145(9):1829–1845
Peeters BPH, de Leeuw OS, Koch G, Gielkens ALJ (1999) Rescue of Newcastle disease virus from cloned cDNA: evidence that cleavability of the fusion protein is a major determinant for virulence. J Virol 73(6):5001–5009

Perozo F, Marcano R, Afonso CL (2012) Biological and phylogenetic characterization of a genotype VII Newcastle disease virus from Venezuela: efficacy of field vaccination. J Clin Microbiol 50(4):1204–1208

Ramey AM, Reeves AB, Ogawa H, Ip HS, Imai K, Bui VN, Yamaguchi E, Silko NY, Afonso CL (2013) Genetic diversity and mutation of avian paramyxovirus serotype 1 (Newcastle disease virus) in wild birds and evidence for intercontinental spread. Arch Virol 158(12):2495–2503

Rehmani SF, Wajid A, Bibi T, Nazir B, Mukhtar N, Hussain A, Lone NA, Yaqub T, Afonso CL (2015) Presence of virulent Newcastle disease virus in vaccinated chickens in farms in Pakistan. J Clin Microbiol 53(5):1715–1718

Rima B, Alexander D, Billeter M (1995) Family paramyxoviridae. Virus taxonomy: classification and nomenclature of viruses. M. FA, F. CM and B. DHL. Springer, New York, pp 268–274

Romer-Oberdorfer A, Werner O, Veits J, Mebatsion T, Mettenleiter TC (2003) Contribution of the length of the HN protein and the sequence of the F protein cleavage site to Newcastle disease virus pathogenicity. J Gen Virol 84(Pt 11):3121–3129

Rout SN, Samal SK (2008) The large polymerase protein is associated with the virulence of Newcastle disease virus. J Virol 82(16):7828–7836

Rue CA, Susta L, Cornax I, Brown CC, Kapczynski DR, Suarez DL, King DJ, Miller PJ, Afonso CL (2011) Virulent Newcastle disease virus elicits a strong innate immune response in chickens. J Gen Virol 92(Pt 4):931–939

Sanchez-Sampedro L, Perdiguero B, Mejias-Perez E, Garcia-Arriaza J, Di Pilato M, Esteban M (2015) The evolution of poxvirus vaccines. Viruses 7(4):1726–1803

Scheid A, Choppin PW (1974) Identification of biological activities of paramyxovirus glycoproteins. Activation of cell fusion, hemolysis, and infectivity of proteolytic cleavage of an inactive precursor protein of Sendai virus. Virology 57(2):475–490

Schijns VEJC, van de Zande S, Lupiani B, Reddy SM (2013) Practical aspects of poultry vaccination. Avian Immunol 345–362 (K. A. Schat, Kaspers, B., Kaiser, P., Elsevier Science)

Seal BS, King DJ, Bennett JD (1995) Characterization of Newcastle-disease virus isolates by reverse transcription Pcr coupled to direct nucleotide sequencing and development of sequence database for pathotype prediction and molecular epidemiologic analysis. J Clin Microbiol 33(10):2624–2630

Seal BS, King DJ, Meinersmann RJ (2000) Molecular evolution of the Newcastle disease virus matrix protein gene and phylogenetic relationships among the paramyxoviridae. Virus Res 66(1):1–11

Shen H, Xue C, Lv L, Wang W, Liu Q, Liu K, Chen X, Zheng J, Li X, Cao Y (2013) Assembly and immunological properties of a bivalent virus-like particle (VLP) for avian influenza and Newcastle disease. Virus Res 178(2):430–436

Smith GW, Hightower LE (1981) Identification of the P proteins and other disulfide-linked and phosphorylated proteins of Newcastle disease virus. J Virol 37(1):256–267

Sonoda K, Sakaguchi M, Okamura H, Yokogawa K, Tokunaga E, Tokiyoshi S, Kawaguchi Y, Hirai K (2000) Development of an effective polyvalent vaccine against both Marek's and Newcastle diseases based on recombinant Marek's disease virus type 1 in commercial chickens with maternal antibodies. J Virol 74(7):3217–3226

Taylor J, Christensen L, Gettig R, Goebel J, Bouquet JF, Mickle TR, Paoletti E (1996) Efficacy of a recombinant fowl pox-based Newcastle disease virus vaccine candidate against velogenic and respiratory challenge. Avian Dis 40(1):173–180

Taylor TL, Dimitrov KM, Afonso CL (2017) Genome-wide analysis reveals class and gene specific codon usage adaptation in avian paramyxoviruses 1. Infect Genet Evol 50:28–37

Terres G, Wolins W (1959) Enhanced sensitization in mice by simultaneous injection of antigen and specific rabbit antiserum. Proc Soc Exp Biol Med 102:632–635

Terres G, Wolins W (1961) Enhanced immunological sensitization of mice by the simultaneous injection of antigen and specific antiserum. I. Effect of varying the amount of antigen used relative to the antiserum. J Immunol 86:361–368

Van den Berg T (2013) Vector vaccines and immunity. New Technology Vaccine Symposium, Miami

Wen YM, Mu L, Shi Y (2016) Immunoregulatory functions of immune complexes in vaccine and therapy. EMBO Mol Med 8(10):1120–1133

Yang ZQ, Liu QQ, Pan ZM, Yu HX, Jiao XA (2007) Expression of the fusion glycoprotein of Newcastle disease virus in transgenic rice and its immunogenicity in mice. Vaccine 25(4): 591–598

Yuan P, Swanson KA, Leser GP, Paterson RG, Lamb RA, Jardetzky TS (2011) Structure of the Newcastle disease virus hemagglutinin-neuraminidase (HN) ectodomain reveals a four-helix bundle stalk. Proc Natl Acad Sci U S A 108(36):14920–14925

Yune N, Abdela N (2017) Update on epidemiology, diagnosis and control technique of Newcastle disease. J Vet Sci Technol 8(429):2

Zhao K, Rong G, Hao Y, Yu L, Kang H, Wang X, Wang X, Jin Z, Ren Z, Li Z (2016a) IgA response and protection following nasal vaccination of chickens with Newcastle disease virus DNA vaccine nanoencapsulated with Ag@SiO_2 hollow nanoparticles. Sci Rep 6:25720

Zhao K, Rong GY, Guo C, Luo XM, Kang H, Sun YW, Dai CX, Wang XH, Wang X, Jin Z, Cui SJ, Sun QS (2015) Synthesis, characterization, and immune efficacy of layered double hydroxide@SiO_2 nanoparticles with shell-core structure as a delivery carrier for Newcastle disease virus DNA vaccine. Int J Nanomed 10:2895–2911

Zhao K, Sun YW, Chen G, Rong GY, Kang H, Jin Z, Wang XH (2016b) Biological evaluation of N-2-hydroxypropyl trimethyl ammonium chloride chitosan as a carrier for the delivery of live Newcastle disease vaccine. Carbohydr Polym 149:28–39

Infectious Bursal Disease Virus

Evangelina Gómez, María Soledad Lucero, Matías Richetta, Silvina Chimeno Zoth and Analía Berinstein

Abstract Infectious bursal disease is an acute, highly contagious, immunosuppressive disease that affects young birds causing important economic losses in the poultry industry. Its etiological agent is the *Infectious bursal disease virus* (IBDV), a non-enveloped bi-segmented double stranded RNA virus which belongs to the Genus *Avibirnavirus* from the Family *Birnaviridae*. Currently, control of IBDV is normally achieved by vaccination programs with inactivated and live attenuated viruses. However, conventional vaccines have a number of disadvantages due to their viral nature and, in many cases, fail to provide sufficient protection against very virulent and variant strains of IBDV. Several new vaccines have been developed as alternatives to solve these problems. Among these rationally designed vaccines live viral-vectored, immune complex and subunit vaccines are found. In this chapter, the contribution of these new technologies to the field will be addressed, with special focus on plant-made vaccines candidates against IBDV. The rationale, efficacy, and yield of these plant-based developments, as well as the comparison to established vaccines or alternatives will be discussed.

Keywords Molecular farming · IBDV · Chicken · VP2 · Recombinant vaccine Transient expression

1 Infectious Bursal Disease Virus and the Disease It Causes

Infectious bursal disease virus (IBDV) is the causative agent of Infectious bursal disease (IBD), also known as Gumboro disease. It is a non-enveloped bi-segmented double stranded RNA (dsRNA) virus which belongs to the Genus *Avibirnavirus*

E. Gómez · M. S. Lucero · M. Richetta · S. Chimeno Zoth · A. Berinstein (✉)
Instituto de Biotecnología, CICVyA, INTA, Buenos Aires, Argentina
e-mail: berinstein.analia@inta.gob.ar

E. Gómez · S. Chimeno Zoth · A. Berinstein
CONICET, Buenos Aires, Argentina

J. MacDonald (ed.), *Prospects of Plant-Based Vaccines in Veterinary Medicine*,
https://doi.org/10.1007/978-3-319-90137-4_7

from the Family *Birnaviridae* (Dobos et al. 1979; Müller et al. 1979). From the two segments that compose the viral genome, segment B is the shortest and encodes for the viral protein VP1, a RNA-dependent-RNA-polymerase (Morgan et al. 1988; von Einem et al. 2004). Segment A, on the other hand, is slightly larger and consists of two partially overlapping open reading frames (ORFs). One of them encodes for a polyprotein (PP) that undergoes an autoproteolytic cleavage in early stages of an infection giving place to a precursor of VP2 (pVP2), VP3 and VP4. pVP2 is further cleaved into mature VP2 (VP2), the main capsid protein, and several C-terminal peptides. VP4 is the viral protease that cleaves the PP (Da Costa et al. 2000). VP3 is a scaffolding protein that interacts with VP1, pVP2, VP2 and with itself during morphogenesis. VP3 also has RNA-binding activity and is responsible of capsid stability (Mertens et al. 2015). The other ORF of segment A encodes the smallest of the viral proteins, VP5 (Mundt et al. 1995), which has been assigned with a role in viral progeny release (Wu et al. 2009; Méndez et al. 2017).

There are two serotypes of IBDV and, while both can infect chickens, only serotype I is pathogenic in this species (Jackwood et al. 1985; Ismail et al. 1988). The strains belonging to serotype I are traditionally classified as classical (cIBDV), variant (varIBDV) and very virulent (vvIBDV), although there also exists the "vaccine strain" category, which comprises classical strains with different degrees of attenuation for their use as vaccines against IBDV. It is well characterized that less attenuated vaccine strains are able to overcome higher levels of anti-IBDV maternally derived antibodies (MDA), but also to cause immunosuppression in vaccinated chicks (Müller et al. 2012). These vaccine strains have also been implicated in the generation of reassortant IBDV strains (Chen et al. 2012a; Raja et al. 2016).

IBD is a highly contagious disease which is regarded as endemic throughout the world (Fig. 1), causing considerable economic losses both directly, through clinical signs and mortality, and indirectly, due to failure in vaccination programs and incremented susceptibility to other pathogens (Kegne and Chanie 2014; Alkie and Rautenschlein 2016). IBD affects mainly chicks between 3 and 6 weeks of age, although the virus can also infect younger chicks. Because IBDV targets IgM-bearing B-lymphocytes, the infection will cause different degrees of immunosuppression (Sharma et al. 2000). The age and breed sensitivity of the birds, the virulence of the viral strain and the level of maternal antibodies constitute the main factors that will determine the outcome of an IBDV infection (Ahmed and Akhter 2003; Aricibasi et al. 2010; Tippenhauer et al. 2013; Zhao et al. 2016).

Three disease forms are most common in the field: classical, immunosuppressive and acute (Van den Berg et al. 2000). The classical form is associated with the presence of cIBDV and usually comes after a decline in maternal antibodies titers in vaccinated flocks. It has worldwide distribution, being endemic in most of the regions. It is often subclinical and courses with low specific mortality (0–5%). The signs, when a clinical manifestation occurs, include vent picking, trembling, ruffled feathers, watery diarrhoea, anorexia, depression, severe prostration and death. As there are no characteristic signs of IBDV infection, necropsy is where most of the information can be obtained (Van den Berg et al. 2000; Eterradossi and Saif 2008; Kegne and Chanie 2014). The immunossuppresive form is related to the emergence

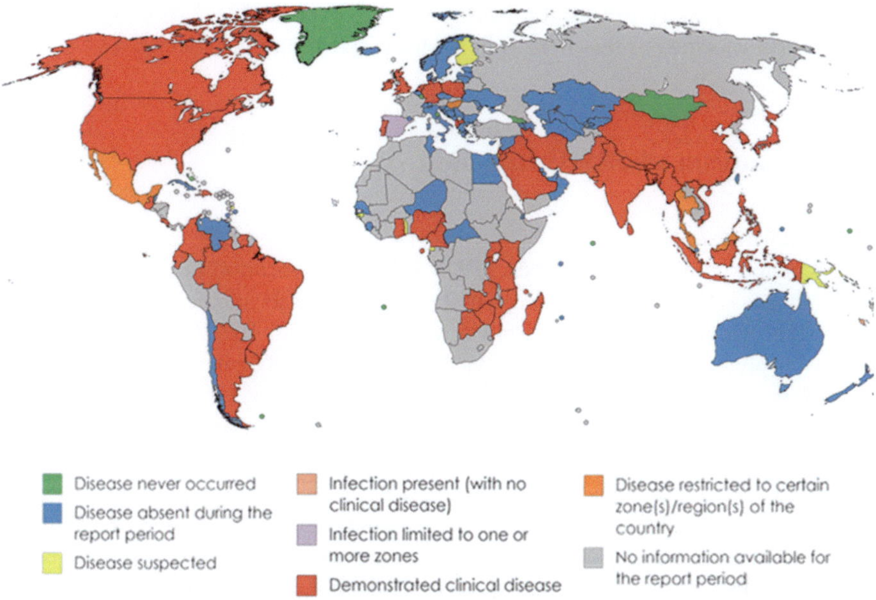

Fig. 1 Worldwide distribution of IBD and status of the disease according to the latest available reports (July–December, 2016) in the World Animal Health Information System (WAHIS) from the World Organisation for Animal Health (OIE). http://www.oie.int/wahis_2/public/wahid.php/Diseaseinformation/statuslist. Image created with https://mapchart.net

of varIBDV strains, which are able to evade the circulating neutralizing antibodies (Snyder et al. 1992), mainly found in USA, Canada and Australia (Snyder et al. 1992; Sapats and Ignjatovic 2000; Kurukulsuriya et al. 2016) because of the predominance of varIBDV strains in those regions. Although the immunocompetence of the chickens is severely diminished, this form is asymptomatic in the majority of the cases (Van den Berg et al. 2000; Kegne and Chanie 2014). However, the economic losses associated to varIBDV-related immunodeficient flocks are considerable (Zachar et al. 2016), mainly due to decreased effectiveness of vaccination programs and increased susceptibility to opportunistic pathogens (Ingrao et al. 2013). Finally, the acute form is caused by vvIBDV strains, which are able to infect in the presence of maternal antibodies and cause higher-than-expected mortality rates (Van den Berg 2000). The first reports of this form of disease took place in Europe, but vvIBDV is currently present in many regions mainly from Africa, Asia and South America (Eterradossi and Saif 2008). The clinical signs are those from the classical form, but they are described to have a more intense manifestation and to be more generalized within the affected flock. The mortality rates can range between 50 and 100% (Van den Berg 2000; Eterradossi and Saif 2008).

As there is no specific treatment for IBD, the attention should be focused on preventive measures. Hygienic measures are important but often insufficient. This makes the vaccination of the flocks the most important action to prevent IBDV entry into any poultry production facility (Müller et al. 2012). The contribution of plant-based vaccines to this field will be discussed in the next sections.

2 Mechanism of IBDV Infection

Horizontal transmission of IBDV occurs through the ingestion of food and water contaminated with infectious feces. IBDV initiates replication in lymphocytes and macrophages of the gut-associated lymphoid tissues. This stage of viral replication marks the primary viraemia. Within 5 h post-infection, IBDV reaches the liver, where it is phagocytized by resident macrophages. Virus then enters the bloodstream where it is distributed to other tissues including the bursa of Fabricius. The bursa of Fabricius is an oval sac located dorsally to the cloaca, exclusive to avian species, and it is the site where B-cell lymphopoiesis, lymphocyte maturation and differentiation and development of the antibody repertoire take place. Bursal follicles represent the structural, functional and pathological bursal unit (Oláh et al. 2013). Virus ability to spread from the bursa to other lymphoid organs depends on the virulence of the infecting IBDV strain (Alkie and Rautenschlein 2016). By 13 h post-inoculation (hpi), most bursal follicles are positive for the virus and by 16 hpi a second, more pronounced, viraemia occurs, with secondary replication in other B-lymphocyte-containing tissues leading to disease. Clinical signs and death may result from the acute phase (7–10 days) of IBD. As previously mentioned, factors such as pathogenicity and virulence of a strain, as well as the chicken's age, breed, and immune status can influence the outcome and severity of the infection (Van den Berg et al. 2000; Harris 2010). The virus infects and destroys actively dividing immunoglobulin M (IgM)–bearing B cells in the bursa of Fabricius resulting in a prolonged suppression of the primary antibody response (Rodenberg et al. 1994; Sharma et al. 2000). In chickens that survive the acute disease, virus replication subsides and almost all bursal follicles become repopulated with IgM + B cells. The primary antibody response is gradually restored to near normal levels. Although the destruction of B lymphocytes may be one of the main inhibitors of humoral immunity, the involvement of other mechanisms such as altered antigen-presenting and helper T cell functions has been also proposed (Sharma et al. 2000). Together with B lymphocyte depletion in the bursa, an infiltration of activated CD4 + and CD8 + T lymphocytes occurs. Although T-cells are not susceptible to IBDV infection, the cellular immune response is also compromised (Sharma et al. 2000). Evidences suggest that T cells may modulate IBDV immunopathogenesis by restricting IBDV replication in the bursa in the early stage of the disease. Through their release of cytokines and cytotoxic effects, T-cells may enhance bursal tissue destruction, suppress immunity and delay recovery of bursa follicles. At the same time, T-cells may promote clearance of IBDV (Sharma et al. 2000).

Cells of the monocyte-macrophage lineage can be infected in a persistent and productive manner and play a crucial role on dissemination of the virus (Burkhardt and Müller 1987; Inoue et al. 1994) and on the onset of the disease

(Kim and Sharma 2000). Increased macrophage infiltration into the bursa may cause higher expression of proinflammatory cytokines [interleukin (IL)-6, IL-1β and IL-18] and inducible nitric oxide synthase, playing a specific role in the pathology of the disease (Khatri et al. 2005; Lee et al. 2015).

Mechanisms and strategies involved in IBDV life cycle are not clear enough. Nevertheless, it is well known that the virus early life cycle comprises cell surface attachment, internalization and penetration, leading to virus replication in the cytoplasm. Different host cell receptors or structures such as N-glycosilated polypeptides (Luo et al. 2010), heat shock proteins like cHSP90 (Lin et al. 2007), α4β1 integrin (Delgui et al. 2009) or lipid raft endocytic pathways (Yip et al. 2012) have been proposed as putative receptors for IBDV. Also, it has been proposed that endocytosis is required for IBDV entry and internalization, followed by the release of Pep46, a capsid-associated peptide which induces pores in the endosomal membrane allowing the release of viral ribonucleoproteins (RNPs) into the cytosol (Galloux et al. 2007). Then, RNPs would associate with the endosomal membrane, through the VP3 membrane-targeting ability, where viral genome replication occurs. Afterwards, the RNPs associated with the endocytic vesicles could traffic along microtubules to reach the perinuclear region, establishing physical contact with the Golgi complex where viral assembly takes place (Delgui et al. 2013). Finally, two independent releasing mechanisms were proposed. The first one, dependent on VP5 expression, allows the non-lytic release of infectious particles from live and metabolically active cells. The second one is associated to cell lysis and facilitates the release of the remaining progeny together with the intracellular content (Méndez et al. 2017).

3 Plant-Made Vaccine Candidates Against IBDV

When IBDV infects a chicken, a humoral response against structural proteins VP2 and VP3 is mostly found. pVP2/VP2 and VP3 are the major proteins present in 780 and ~450 copies per capsid respectively, while VP1 is present in approximately 12 copies (Luque et al. 2009). Although an antibody response against VP3 exists and neutralizing epitopes were identified (Whetzel and Jackwood 1995), VP3 fails in promoting a protective response (Pitcovski et al. 1999). Conversely, antibodies raised against VP2 have neutralizing capability and elicit protective immunity. Therefore, along the last years, several attempts have been made to generate VP2 subunit vaccines (Ghafari et al. 2010). VP2 neutralizing epitopes are located in the hypervariable region of the protein, between amino acids 206 and 350 (Bayliss et al. 1990). The hypervariable region is highly conformational and comprises four loops named P_{BC} (aa 219–224), P_{HI} (aa 316–324), P_{DE} (aa 249–254) and P_{FG} (aa 279–284) (Coulibaly et al. 2010). P_{DE} and P_{FG} loops are responsible for virus-cell receptor binding and virulence, whereas P_{HI} and P_{BC} loops contain the neutralizing epitopes and have been proved to be suitable sites for foreign peptides display (Brandt et al. 2001; van Loon et al. 2002; Qi et al. 2009).

After infection with IBDV or recombinant expression of VP2, icosahedral T = 1 subviral particles (SVP) of ~23–26 nm in diameter formed by 20 trimers of VP2 are found (Coulibaly et al. 2005; Garriga et al. 2006; Taghavian et al. 2013). These particles were produced in different expression systems like *Pichia pastoris*, *Escherichia coli* and insect cells, for many purposes such as: subunit vaccines against IBDV (Rogel et al. 2003; Ho et al. 2010; Taghavian et al. 2013; Jackwood 2013), carrier of epitopes of non-related viruses (Remond et al. 2009; Caballero et al. 2012; Pascual et al. 2015), serological diagnosis (Dey et al. 2009) and life viral cycle studies (Lin et al. 2007; Delgui et al. 2009).

Due to the importance of the disease worldwide and the beneficial features of plant expression of valuable molecules, some groups have reported the production of VP2 in model plants and cereal crop evaluating in each case their performance as vaccine against IBD.

The first study appeared in 2004 when molecular farming, in particular the conception of edible vaccines, was in full swing. A*rabidopsis thaliana* expressing VP2 in the foliar area was the plant species of choice. Authors reported a percentage of total soluble protein (TSP) for VP2 ranging between 0.5 and 4.8%. These % TSP values were probably underestimated as authors considered in the calculations that VP2 represented 20% of IBDV TSP (Wu et al. 2004a), when nowadays it is known to be about 60% according to crystallographic and stoichiometry analyses (Luque et al. 2009). Nevertheless, these percentages are among the highest obtained in stably transformed plants for subunit vaccines.

Crude extract from leaves of the best transgenic line was evaluated by oral and subcutaneous routes in a prime/boost scheme at 1 and 3 weeks of age. Those animals immunized orally with VP2 received 5 doses at 3-days intervals. Chickens receiving subcutaneous immunization had moderate levels of antibodies compared to the live intermediate commercial vaccine (Bursine-2) group, and protection after challenge with a variant strain was 60%, measured as a bursa-to-body weight ratio. In spite of generating similar antibody levels, the oral route seemed to be more efficient than the subcutaneous route with 80% of protection. Chickens primed with the commercial vaccine at 1 wk followed by an oral booster with VP2 expressed in plants at 3 wk of age showed 90% protection while animals receiving two doses of Bursine-2 at the same time interval had 78% protection (Wu et al. 2004b).

Overall, these first approaches to a plant derived vaccine indicated that plants were capable of synthesizing IBDV VP2 and that both routes of vaccination were effective in generating protective response. Moreover, VP2 expressed in plants could be effectively used to prime or boost a previous response.

Later, another group drove the expression of VP2 to rice endosperm with the aim of producing a mucosal vaccine for IBDV (Wu et al. 2007). The strategy was to clone the coding sequence of VP2 under the promoter of Glutelin A, a very strong and specific promoter leading the expression of the most abundant protein in rice seeds.

The average of VP2 protein in the highest expressing transgenic line was 4.521% of seed TSP, which accumulated up to 56.12 μg of VP2 per grain, while the lowest presented 0.678%

Two-week old chickens fed with 1, 3 or 5 g of seeds of a line expressing 40.21 μg of VP2 per grain on days 0, 7, 14 and 21 developed a specific immune response. Unfortunately, the vaccination doses (μg VP2/g seeds) have not been informed although it was estimated elsewhere that 5 g contained 10 mg of VP2 (Mason and Herbst-Kralovetz 2012). The protection was recorded as the number of chickens with a bursal score of zero. The bursal score is a measure of how affected the bursa of Fabricius becomes after infection with IBDV and it is based on the percentage of follicles with lymphoid depletion in addition to the observation of specific lesions. The scale ranges between 0 and 4/5 (depending on the literature source) with lower score meaning less bursal damage. Results demonstrated a dose dependent response as animals fed with 5 g showed the highest rate of protection (83.3% vs. 33.33% showed by animals intranasally inoculated with a commercial attenuated vaccine strain B87 at days 0 and 21) after challenge with a very virulent strain. The neutralizing antibody levels were similar to that of the group vaccinated with the commercial vaccine and was also influenced by the vaccination dose.

Altogether, results indicated that VP2 was resistant to gut degradation and that the use of adjuvants was unnecessary. Moreover it showed an effective, safe and inexpensive vaccine with no requirements of needle/syringe or a cold chain to its commercialization. This work showed for the first time the efficacy of a rice-based vaccine in the natural host. Nevertheless, more detailed studies regarding stability of VP2 over time, storage conditions, mucosal response and the possibility of inclusion in balanced diet, would have been interesting towards the obtainment of an edible vaccine against IBDV.

Rice provides little energy per cost unit which makes it a very expensive cereal to use in birds feeding. Conversely, maize is the cereal that provides the highest amounts of metabolizable energy/kg. Also, it is a source of zeaxanthin and lutein, two carotenoids that provide color to the egg yolk and the chicken' skin, very desirable characteristics in the poultry industry. For these reasons maize would have been a better choice as an edible vaccine for chickens.

Gradually, plant transient expression of vaccine candidates has taken a prevailing place over stable expression mainly because the developing time and yields were improved with the arrival of new technologies. In this sense, Chen et al. described in 2012 the generation and immunogenicity of a chimeric Bamboo mosaic virus harboring the coding sequence of loop P_{BC} (18 aa), of a vvIBDV VP2, fused to the N terminal of the viral coat protein (CP) (Chen et al. 2012b). Bamboo mosaic virus is a filamentous potexvirus consisting of 1300 identical CP subunits so it was expected that the chimeric virus also contained 1300 IBDV epitopes. Authors reported a production of recombinant CP of 2.6–2.8 μg/mg of total soluble protein that represent 0.26 and 0.28% of TSP, respectively. Chimeric viruses were produced in *Chenopodium quinoa* and then purified for animal experiment. Three SPF chickens received an intramuscular injection of isolated chimeric viruses (600 μg) formulated with Freund's incomplete adjuvant and twenty eight days later they were challenged with a very virulent strain of IBDV. Two out of three animals of the control group died while chickens of the chimeric or commercial vaccine survived after challenge. Also, specific antibodies of the chimeric virus vaccinated animals reached similar levels to the commercial vaccine group. This work demonstrated that a single region of VP2, the loop containing the neutralizing epitopes, was able to induce a specific response even in a single dose. A large scale experiment in field conditions without adjuvant would have been desirable to prove efficacy of the vaccine.

Our group has focused on the transiently production of VP2 in *Nicotiana benthamiana* plants and its subsequent application in different vaccination approaches in susceptible chickens (Gómez et al. 2013; Lucero et al. 2016; Richetta et al. 2017) (Fig. 2). This plant expression system has allowed us the obtainment of high levels of VP2 in a short time period. In addition, this plant expression system is suitable for a rapid response in case of a field outbreak where other sequences of VP2 might be required.

We have chosen the expression of mature VP2 (VP2 of 441 aa) because it has already proved to be immunogenic and to form subviral particles in other

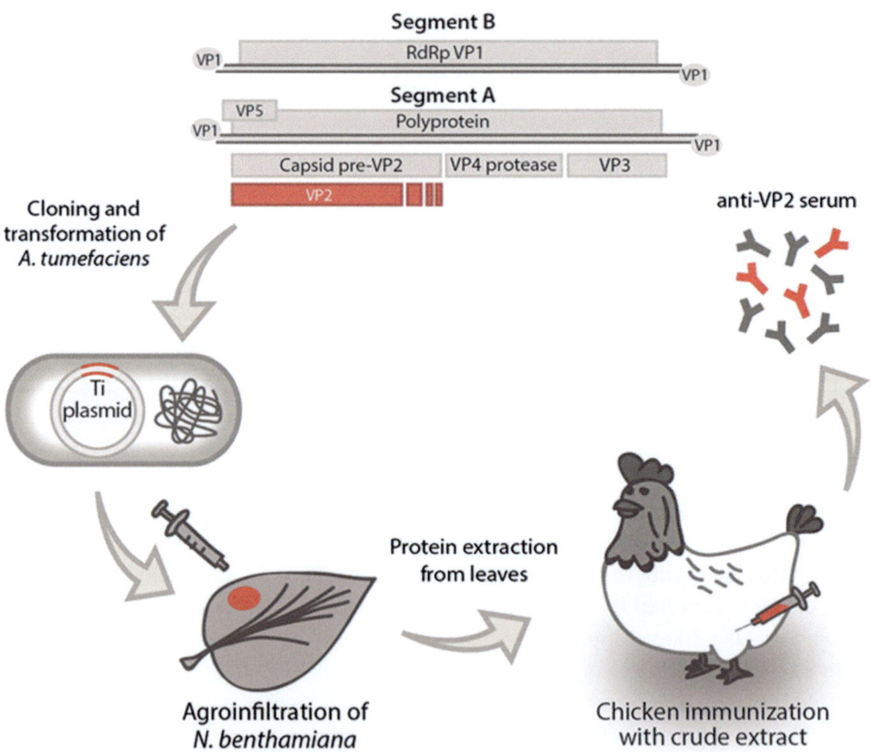

Fig. 2 The coding sequence of mature VP2 is cloned in a binary vector under a strong promoter for plant expression and introduced into *Agrobacterium tumefaciens* by electroporation. Transient expression is performed by infiltrating *Nicotiana benthamiana* leaves with a suspension of the recombinant bacteria. Four or five days later, agroinfiltrated leaves are harvested and total proteins are extracted in chilled buffer. This crude extract is used to vaccinate chickens in a prime/boost scheme which induces a humoral response against VP2

expression systems. Firstly, the strategy was to clone the coding sequence under the control of the rubisco small subunit promoter, said to be 8 times stronger than the 35S promoter. Then leaves of *N. benthamiana* were infiltrated with a suspension of recombinant agrobacteria harboring the sequence of interest followed by the collection of the leaves 4 days later. The yields were in average 1% TSP. Five animals were vaccinated intramuscularly with 200 μl of crude extract containing 12 μg, formulated with Freund's adjuvant, on days 0, 22 and 35 post inoculation. Animals were weekly bled and 18 days after the last vaccination they were challenged with a high dose of an intermediate IBDV strain. Results demonstrated that the extract was able to elicit a humoral response as early as 15 days, with neutralizing activity reaching high titers by the end of the experiment. Also, chickens vaccinated with VP2 and challenged showed a decrease in the frequency of T-cell infiltration into the bursa of Fabricius, from 2.7 to 22.6 times lower than the control group, indicating that the humoral response prompted by the experimental vaccine was efficacious in preventing the entrance of the virus in that organ (Gómez et al. 2013). Later, we showed that a more welfare-friendly immunization scheme with fewer injections and without adjuvant was also able to elicit a protective response. In four out of six animals primed and boosted with 7.5 μg of VP2, IBDV was not detected in the bursa of Fabricius while the other two animals presented a reduced viral titer of approximately 10^5 times regarding the control group. In addition, animals vaccinated with VP2 presented a bursa with normal morphology and nine times fewer infiltrating T cells than the control group (Lucero et al. 2016). We believe that the success of our antigen by parenteral route is not only related to the physical and chemical properties of the protein but also to the adjuvant capacity of the plant extract. It is possible that the plant extract contains PAMPS of toll like receptors (like LPS from *Agrobacterium*) and vegetal compounds that might help to prompt an innate response which in turn contributes to the establishment of the adaptive response (Licciardi and Underwood 2011). It is worth mentioning that when we assayed the mucosal vaccination with the same dosage, we found less encouraging results. Neither intranasal nor oral vaccinations were able to produce an effective immune response in chickens. Specific antibodies were not detected and chickens were not protected from IBDV challenge (Lucero et al. 2016). One possibility is that the immunization scheme applied was inappropriate as mucosal stimulation might require more and/or frequent boosts. Another option is that an adequate mucosal adjuvant or higher doses of immunogen could also be needed.

We also performed vaccinations with the extract in prime/boost schemes along with vectored vaccines based on recombinant Modified Ankara Virus harboring the coding region of VP2 (Richetta et al. 2017). Results showed that the extract can be used alone, as demonstrated earlier, and to prime or boost a vaccination with other types of recombinant immunogens. Finally, using the pEAQ vectors (Sainsbury et al. 2009) for VP2 plant expression we recovered SVP from plant material indicating that these nanoparticles can also be produced in plant-based expression systems (unpublished results).

4 Plant-Made Vaccines Against IBDV Versus Established Vaccines and Other Developments

IBDV was identified for the first time more than 50 years ago; still, this virus remains a significant threat to commercial poultry worldwide. Many advances have been made in the development of new recombinant vaccines, however, live-attenuated and inactivated vaccines, along with strict hygiene management of poultry farm, continue to be the most common practices to control IBDV.

Live viral vaccines can replicate and are effective in inducing both cellular and humoral immunity without the use of adjuvant (Müller et al. 2012). Besides, they are suitable for mass administration to chickens since they can be given with the drinking water. However, they present a number of disadvantages due to their viral nature. They can revert to virulence (Yamaguchi et al. 2000; He et al. 2009; Jackwood 2012), they usually produce a period of immunosuppression in young chickens and might interfere with response to other vaccines (Mazariegos et al. 1990; El-Yuguda et al. 2007); they exhibit poor efficacy in the presence of certain levels of maternally derived antibodies (MDA) (Kumar et al. 2000; Rautenschlein et al. 2005); and most importantly, they may not fully protect chickens against infection by the very virulent and variant IBDV strains (Rautenschlein et al. 2005; Alkhalaf 2009). Furthermore, although drinking water vaccination would seem to be the least labor intensive, there are major concerns regarding inconsistencies of vaccine dosage depending on water consumption within the flock and viral inactivation by traces of disinfectants or chlorine in the drinking water.

Live viral-vectored and immune complex vaccines, seem to be attractive candidates to replace the traditional live attenuated one and are already being commercialized by different animal healthcare companies. The two viral vectored vaccines available, VAXXITEK® HVT + IBD (Merial) and Vectormune® HVT IBD (Ceva) use the turkey herpesvirus (HVT) carrying IBDV antigens to stimulate immunity against Marek's disease and IBD simultaneously. On the other hand, immune complex Gumboro vaccine, Cevac® Transmune IBD (Ceva) consists of a well-defined mixture of IBDV-specific antibodies and infectious IBD vaccine virus. Both types of vaccines have proven to be effective in the presence of maternally derived antibodies (MDA) causing protection against different pathotypes of IBDV when inoculated in a single dose *in ovo* or subcutaneously to 1 day old chickens (Haddad et al. 1997; Kelemen et al. 2000; Perozo et al. 2009; Prandini et al. 2016; Gelb et al. 2016). Although *in ovo* or subcutaneous vaccination allow a more automated and systematic administration of these new vaccines, they require egg-injection machines, which are not available in all farms, or trained personnel. This, together with the fact that both live viral-vectored and immune complex vaccines are more expensive than the live attenuated ones, might be some of the reasons why the last ones have not been replaced yet.

IBD inactivated vaccines consist of virus that has been rendered incapable of replicating, so they cannot cause disease, but maintain the ability to induce a protective immune response. They are costly due to the treatment processes involved in inactivating the virus and require strict quality control to ensure that killed organisms are fully inactivated and harmless before used for vaccination (Delrue et al. 2012). Moreover, they lack efficient immunogenicity unless they are combined with adjuvants and administered in repeated injections, or follow a prime with a replicating antigen (Müller et al. 2012). Since inactivated IBD vaccines are mostly formulated as water-in-oil emulsions and inoculated through the intramuscular route, they do not stimulate mucosal immunity. Normally, their use is constrained to breeder birds just before laying in order to provide passive immunity to the offspring by means of MDA (Maas et al. 2001).

As an alternative, many proteic subunit vaccines based on recombinant VP2 expression have been assessed in the laboratory against IBDV infection with diverse results. Generally, as they are non-replicating antigens, they have similar disadvantages to inactivated vaccines regarding efficiency. To date, there is only one commercially available recombinant subunit vaccine against IBDV, Gumbin®VP2 (Phibro Animal Health Corporation), which consists of inactivated NDV and recombinant VP2 produced in yeast (Pitcovski et al. 2003). It is intended for vaccination of all chicken breed after priming with live Newcastle disease and Infectious bursal disease vaccines. Nonetheless, these subunit vaccines exhibit great promise since the lack of inactivation procedures might render them cheaper (depending on the expression system), than inactivated whole virus formulations. In addition, some of these recombinant vaccines, depending on how they are formulated, could be administrated through the oral route to stimulate mucosal immunity.

VP2 has been expressed in a number of heterologous systems such as *E. coli*, yeast, baculovirus/insect cells and plants (reviewed in Lucero et al. 2012). All of them have different characteristics that are summarized in Table 1. Although mammalian/avian cultures have not been used to express recombinant VP2, they are used to propagate live attenuated or live viral-vectored vaccines; hence, they are also included in the table for comparison.

When compared to other expression systems, the disadvantages of plant expression systems have been related to protein yield and time of development. It is difficult to compare VP2 protein yield obtained in plants vs. other expression systems since not all reports include this information or is expressed in different units. Still, some of the higher yields were obtained in *E. coli* (1.178 g/l of culture or 0.19 g/g of bacteria) (Rong et al. 2007) and yeast (0.5 g/l of culture) (Pitcovski et al. 2003). Achieving this kind of VP2 yields in plants is one of the challenges of this platform, however, their almost unlimited scale up capacity has the potential to provide plenty biomass in order to accumulate sufficient quantity of the antigenic protein even when expression levels are not very high. On the other hand, although the development and selection of suitable transgenic lines expressing adequate amounts of VP2, like the case of *Arabidopsis* and rice (Wu et al. 2004b, 2007) its laborious and can take many months, VP2 transient expression approaches using *Agrobacterium* (Gómez et al. 2013; Lucero et al. 2016) and/or plant viral vectors (Chen et al. 2012b) are able to reduce developing times and can be as fast as producing a recombinant bacteria or yeast.

Table 1 Comparison of different production systems for expression of recombinant proteins

Expression system	Bacteria	Yeast	Plants	Plant viral vectors	Baculovirus/insect cell cultures	Mammalian/avian cell cultures
Time effort	Low	Medium	High (stable transformation) Low (transient expression)	Low	High	High
Production cost	Medium	Medium	Low	Low	High	High
Scale up capacity	High	High	Very high	Very high	Medium	Low
Production scale	Limited	Limited	Worldwide	Worldwide	Limited	Limited
Cost of maintenance	Inexpensive	Inexpensive	Inexpensive	Inexpensive	Expensive	Expensive
Protein yield	High	High	Medium	Very high	High	Medium-high
Gene protein size	Unknown	Unknown	Not limited	Limited	Limited	Limited
Contamination risk	Endotoxins	Low	Low	Low	Low	Viruses

As described in detail in the previous section, VP2 has been produced in different plant platforms. The efficacy as a subunit vaccine has been demonstrated both by intramuscular (Chen et al. 2012b; Gómez et al. 2013; Lucero et al. 2016) and oral routes (Wu et al. 2004b, 2007) although several doses of the experimental plant-based vaccines were needed in order to achieve a protective immune response against IBDV. These are some of the few studies in which VP2 protein delivered orally has been successful in achieving protection against IBDV challenge. While oral administration of *Kluyveromyces lactis* expressing VP2 (1–3 mg of recombinant protein in total) in a 2/2/2 scheme (two weeks feeding, two weeks break, two weeks feeding) only achieved a 10% rate protection (Arnold et al. 2012), 4 doses of orally-administered *Pichia pastoris* producing VP2, containing 400 μg or 4 mg of viral protein, induced a protective immune response against IBDV in chickens which increased survival rates to 60–100% compared to 40% of the control groups (Taghavian et al. 2013).

Subunit vaccines are safer than traditional IBDV vaccines; however, they are less immunogenic than live attenuated, viral-vectored or immune complex vaccines which can induce a strong immune response with only one dose in young birds. However, it may be beneficial to use a plant-derived VP2 as a booster vaccine in chickens that have been primed with live vaccines. Hence, taking into account the problems with inactivated vaccines and the benefits of plants as expression systems, we believe that a plant-based subunit vaccine against IBDV represents a viable alternative to the inactivated vaccine given to breeder hens before the laying period.

5 How Far Are We from an Anti-IBDV Commercial Vaccine?

Developments of plant-based vaccines against IBDV are nowadays in early stages, however they seem to be very promising strategies. An edible vaccine seems feasible since VP2 expressed in Arabidopsis and rice invoked an immune response when given orally, showing that this protein is resistant to gut degradation (Wu et al. 2004b, 2007). Additionally, as mentioned before, the rice-based vaccine does not need any type of protein extraction or purification prior to delivery and cold chain is not required. However, both experimental vaccines implicate the use of transgenic plants, so they would have to gain regulatory approval from the corresponding organism as all genetically modified (GM) crops in order to reach the market. Moreover, the production of an IBDV vaccine in food crops such as rice might have to overcome concerns regarding the safety of the food chain from cross-contamination with the GM organism (Naderi and Fakheri 2015). On the other hand, transient expressions of VP2 by means of *Agrobacterium* (Gómez et al. 2013; Lucero et al. 2016) or plant viruses (Chen et al. 2012b) have become attractive manufacturing systems since they might be able to overcome some of the regulatory issues and public concerns for genetically modified organisms (Chen and

Lai 2015). Besides, these systems are very rapid, generating recombinant proteins within days and avoiding the long times required to generate a transgenic plant. It has been speculated that it may be difficult to maintain a high efficacy of live vaccines against IBDV due to the complication of adapting them to cover the emergence of highly virulent or variant strains of the virus. In this sense, transient approaches, which allow a fast replacement of the VP2 gene, would be the most appropriate vaccines to deal with this situation (Saif 2004). For many years transient expression systems remained restricted by laborious scale-up limitations. However, with development of new platforms optimized to facilitate a scale-up production in a short time period, transient expression of large quantities of recombinant proteins in plants may become feasible (Gleba et al. 2005; Peyret and Lomonossoff 2013; Jin et al. 2015).

There is still a way to go from the evaluation of the products in the laboratory to the achievement of a finished plant based vaccine against IBDV. To date none of these vaccines have been scaled up and evaluated in large field trials with broiler chickens. Both safety and efficacy tests need to be carried out before an IBDV plant-based vaccine reaches the market. Nonetheless, we believe that a vaccine with the characteristics enumerated before would be of easy adoption in the veterinary field.

References

Ahmed Z, Akhter S (2003) Role of maternal antibodies in protection against infectious bursal disease in commercial broilers. Int J Poult Sci 2:251–255. https://doi.org/10.3923/ijps.2003.251.255

Alkhalaf AN (2009) Detection of variant strains of infectious bursal disease virus in broiler flocks in Saudi Arabia using antigen capture enzyme-linked immunosorbent assay. Pak Vet J 29:161–164

Alkie TN, Rautenschlein S (2016) Infectious bursal disease virus in poultry: current status and future prospects. Vet Med Res Reports 7:9–18

Aricibasi M, Jung A, Heller DE, Rautenschlein S (2010) Differences in genetic background influence the induction of innate and acquired immune responses in chickens depending on the virulence of the infecting infectious bursal disease virus (IBDV) strain. Vet Immunol Immunopathol 135:79–92. https://doi.org/10.1016/j.vetimm.2009.11.005

Arnold M, Durairaj V, Mundt E et al (2012) Protective vaccination against infectious bursal disease virus with whole recombinant Kluyveromyces lactis yeast expressing the viral VP2 subunit. PLoS ONE 7:e42870. https://doi.org/10.1371/journal.pone.0042870

Bayliss CD, Spies U, Shaw K et al (1990) A comparison of the sequences of segment A of four infectious bursal disease virus strains and identification of a variable region in VP2. J Gen Virol 71:1303–1312. https://doi.org/10.1099/0022-1317-71-6-1303

Brandt M, Yao K, Liu M et al (2001) Molecular determinants of virulence, cell tropism, and pathogenic phenotype of infectious bursal disease virus. J Virol 75:11974–11982

Burkhardt E, Müller H (1987) Susceptibility of chicken blood lymphoblasts and monocytes to infectious bursal disease virus (IBDV). Arch Virol 94:297–303

Caballero JM, Garzón A, González-Cintado L et al (2012) Chimeric infectious bursal disease virus-like particles as potent vaccines for eradication of established HPV-16 E7–dependent tumors. PLoS ONE 7:e52976

Chen Q, Lai H (2015) Gene Delivery into Plant Cells for Recombinant Protein Production. Biomed Res Int 2015:1–10. https://doi.org/10.1155/2015/932161

Chen F, Liu J, Yan Z et al (2012a) Complete Genome Sequence Analysis of a Natural Reassortant Infectious Bursal Disease Virus in China. J Virol 86:11942–11943. https://doi.org/10.1128/JVI.02043-12

Chen T-H, Chen T-H, Hu C-C et al (2012b) Induction of protective immunity in chickens immunized with plant-made chimeric bamboo mosaic virus particles expressing very virulent infectious bursal disease virus antigen. Virus Res 166:109–115

Coulibaly F, Chevalier C, Gutsche I et al (2005) The birnavirus crystal structure reveals structural relationships among icosahedral viruses. Cell 120:761–772. https://doi.org/10.1016/j.cell.2005.01.009

Coulibaly F, Chevalier C, Delmas B, Rey FA (2010) Crystal structure of an Aquabirnavirus particle: insights into antigenic diversity and virulence determinism. J Virol 84:1792–1799

Da Costa B, Delmas B, Lejal N, Huet J-C (2000) Role of Ser-652 and Lys-692 in the protease activity of infectious bursal disease virus VP4 and identification of its substrate cleavage sites. J Gen Virol 81:983–992. https://doi.org/10.1099/0022-1317-81-4-983

Delgui LR, Oña A, Gutiérrez S et al (2009) The capsid protein of infectious bursal disease virus contains a functional $\alpha 4\beta 1$ integrin ligand motif. Virology 386:360–372

Delgui LR, Rodriguez JF, Colombo MI (2013) The endosomal pathway and the golgi complex are involved in the infectious bursal disease virus life cycle. J Virol 87:8993–9007. https://doi.org/10.1128/JVI.03152-12

Delrue I, Verzele D, Madder A, Nauwynck HJ (2012) Inactivated virus vaccines from chemistry to prophylaxis: merits, risks and challenges. Expert Rev Vaccines 11:695–719. https://doi.org/10.1586/erv.12.38

Dey S, Upadhyay C, Mohan CM et al (2009) Formation of subviral particles of the capsid protein VP2 of infectious bursal disease virus and its application in serological diagnosis. J Virol Methods 157:84–89

Dobos P, Hill BJ, Hallett R et al (1979) Biophysical and biochemical characterization of five animal viruses with bi-segemented double stranded RNA genomes. J Virol 32:593–605

El-Yuguda AD, Wachida N, Baba SS (2007) Interference of infectious bursal diseases (IBD) virus and vaccine with the immune responses of Guinea fowls to newcastle disease Lasota vaccination

Eterradossi N, Saif YM (2008) Infectious Bursal Disease. Diseases of poultry, 12th edn. Blackwell Publishing, Iowa, USA, pp 185–209

Galloux M, Libersou S, Morellet N et al (2007) Infectious bursal disease virus, a non-enveloped virus, possesses a capsid-associated peptide that deforms and perforates biological membranes. J Biol Chem 282:20774–20784

Garriga D, Querol-Audí J, Abaitua F et al (2006) The 2.6-Angstrom structure of infectious bursal disease virus-derived T = 1 particles reveals new stabilizing elements of the virus capsid. J Virol 80:6895–6905

Gelb J, Jackwood DJ, Brannick EM, Ladman BS (2016) Efficacy of recombinant HVT-IBD vaccines administered to broiler chicks from a single breeder flock at 30 and 60 weeks of age. Avian Dis 60:603–612. https://doi.org/10.1637/11344-120815-Reg.1

Ghafari S, Seyfiabad Shapouri MR, Moatamedi H et al (2010) Cloning and secretory expression of VP2 gene of infectious bursal disease virus in eukaryotic cells. Iran J Vet Res 11:72–77

Gleba Y, Klimyuk V, Marillonnet S (2005) Magnifection—a new platform for expressing recombinant vaccines in plants. Vaccine 23:2042–2048. https://doi.org/10.1016/j.vaccine.2005.01.006

Gómez E, Lucero MS, Chimeno Zoth S et al (2013) Transient expression of VP2 in Nicotiana benthamiana and its use as a plant-based vaccine against Infectious Bursal Disease Virus. Vaccine 31:2623–2627. https://doi.org/10.1016/j.vaccine.2013.03.064

Haddad EE, Whitfill CE, Avakian AP et al (1997) Efficacy of a novel infectious bursal disease virus immune complex vaccine in broiler chickens. Avian Dis 41:882–889

Harris J (2010) Characterization of infectious bursal disease viruses isolated from commercial chickens. University of Delaware

He C-Q, Ma L-Y, Wang D et al (2009) Homologous recombination is apparent in infectious bursal disease virus. Virology 384:51–58. https://doi.org/10.1016/j.virol.2008.11.009

Ho J-Y, Lee L-H, Lin Y-C et al (2010) Vaccine development through terminal deletions of an infectious bursal disease virus protein 2 precursor variant. Process Biochem 45:786–793

Ingrao F, Rauw F, Lambrecht B, Van den Berg TP (2013) Infectious bursal disease: a complex host-pathogen interaction. Dev Comp Immunol 41:429–438. https://doi.org/10.1016/j.dci.2013.03.017

Inoue M, Fukuda M, Miyano K (1994) Thymic lesions in chicken infected with infectious bursal disease virus. Avian Dis 38(4):839–846

Ismail NM, Saif YM, Moorhead PD (1988) Lack of pathogenicity of five serotype 2 infectious bursal disease viruses in chickens. Avian Dis 32:757–759. https://doi.org/10.2307/1590995

Jackwood DJ (2012) Molecular epidemiologic evidence of homologous recombination in infectious bursal disease viruses. Avian Dis 56:574–577. https://doi.org/10.1637/10053-010912-ResNote.1

Jackwood DJ (2013) Multivalent virus-like–particle vaccine protects against classic and variant infectious bursal disease viruses. Avian Dis 57:41–50. https://doi.org/10.1637/10312-080212-Reg.1

Jackwood DJ, Saif YM, Moorhead PD (1985) Immunogenicity and antigenicity of infectious bursal disease virus serotypes I and II in chickens. Avian Dis 29:1184–1194. https://doi.org/10.2307/1590472

Jin T, Wang J, Zhu X et al (2015) A new transient expression system for large-scale production of recombinant proteins in plants based on air-brushing an agrobacterium suspension. Biotechnol Reports 6:36–40. https://doi.org/10.1016/j.btre.2015.01.004

Kegne T, Chanie M (2014) Review on the incidence and pathology of infectious bursal disease. Br J Poult Sci 3:68–77. https://doi.org/10.5829/idosi.bjps.2014.3.3.8556

Kelemen M, Forgách K, Iván J et al (2000) Pathological and immunological study of an in ovo complex vaccine against infectious bursal disease. Acta Vet Hung 48:443–454

Khatri M, Palmquist JM, Cha RM, Sharma JM (2005) Infection and activation of bursal macrophages by virulent infectious bursal disease virus. Virus Res 113:44–50

Kim I-J, Sharma JM (2000) IBDV-induced bursal T lymphocytes inhibit mitogenic response of normal splenocytes. Vet Immunol Immunopathol 74:47–57

Kumar K, Singh KC, Prasad CB (2000) Immune responses to intermediate strain IBD vaccine at different levels of maternal antibody in broiler chickens. Trop Anim Health Prod 32:357–360

Kurukulsuriya S, Ahmed KA, Ojkic D et al (2016) Circulating strains of variant infectious bursal disease virus may pose a challenge for antibiotic-free chicken farming in Canada. Res Vet Sci 108:54–59. https://doi.org/10.1016/j.rvsc.2016.08.002

Lee C-C, Kim B-S, Wu CC, Lin TL (2015) Bursal transcriptome of chickens protected by DNA vaccination versus those challenged with infectious bursal disease virus. Arch Virol 160:69–80. https://doi.org/10.1007/s00705-014-2232-y

Licciardi PV, Underwood JR (2011) Plant-derived medicines: a novel class of immunological adjuvants. Int Immunopharmacol 11:390–398. https://doi.org/10.1016/j.intimp.2010.10.014

Lin T-W, Lo C-W, Lai S-Y et al (2007) Chicken heat shock protein 90 is a component of the putative cellular receptor complex of infectious bursal disease virus. J Virol 81:8730–8741

Lucero MS, Gómez E, Carballeda JM et al (2012) Recombinant vaccines and infectious bursal disease virus. CAB Rev Perspect Agric Vet Sci Nutr Nat Resour 7:215. https://doi.org/10.1079/PAVSNNR20127070

Lucero MS, Gómez E, Richetta M et al (2016) Effect of different routes of inoculation on plant-derived VP2 immunogenicity and ability to confer protection against infectious bursal disease

Luo J, Zhang H, Teng M et al (2010) Surface IgM on DT40 cells may be a component of the putative receptor complex responsible for the binding of infectious bursal disease virus. Avian Pathol 39:359–365

Luque D, Rivas G, Alfonso C et al (2009) Infectious bursal disease virus is an icosahedral polyploid dsRNA virus. Proc Natl Acad Sci 106:2148–2152

Maas RA, Venema S, Oei HL et al (2001) Efficacy of inactivated infectious bursal disease (IBD) vaccines: comparison of serology with protection of progeny chickens against IBD virus strains of varying virulence. Avian Pathol 30:345–354. https://doi.org/10.1080/03079450120066359

Mason HS, Herbst-Kralovetz MM (2012) Plant-derived antigens as mucosal vaccines. In: Kozlowski PA (ed) Mucosal vaccines: modern concepts, strategies, and challenges. Springer, Berlin, Heidelberg, pp 101–120

Mazariegos LA, Lukert PD, Brown J (1990) Pathogenicity and immunosuppressive properties of infectious bursal disease "intermediate" strains. Avian Dis 34:203–208

Méndez F, Romero N, Cubas LL et al (2017) Non-lytic egression of infectious bursal disease virus (IBDV) particles from infected cells. PLoS ONE 12:e0170080. https://doi.org/10.1371/journal.pone.0170080

Mertens J, Casado S, Mata CP et al (2015) A protein with simultaneous capsid scaffolding and dsRNA-binding activities enhances the birnavirus capsid mechanical stability. Sci Rep 5:13486. https://doi.org/10.1038/srep13486

Morgan MM, Macreadie IG, Harley VR et al (1988) Sequence of the small double-stranded RNA genomic segment of infectious bursal disease virus and its deduced 90-kDa product. Virology 163:240–242. https://doi.org/10.1016/0042-6822(88)90258-9

Müller H, Scholtissek C, Becht H (1979) The genome of infectious bursal disease virus consists of two segments of double-stranded RNA. J Virol 31:584–589

Müller H, Mundt E, Eterradossi N, Islam MR (2012) Current status of vaccines against infectious bursal disease. Avian Pathol 41:133–139. https://doi.org/10.1080/03079457.2012.661403

Mundt E, Beyer J, Müller H (1995) Identification of a novel viral protein in infectious bursal disease virus-infected cells. J Gen Virol 76:437–443

Naderi S, Fakheri B (2015) Overview of plant-based vaccines. Res J Fish Hydrobiol 10:275–289

Oláh I, Nagy N, Vervelde L (2013) Structure of the avian lymphoid system. In: Schat K, Kaspers B, Kaiser P (eds) Avian immunology, 2nd edn. Elsevier, San Diego, CA, pp 11–44

Pascual E, Mata CP, Gómez-Blanco J et al (2015) Structural basis for the development of avian virus capsids that display influenza virus proteins and induce protective immunity. J Virol 89:2563–2574

Perozo F, Villegas P, Fernandez R et al (2009) Efficacy of single dose recombinant herpesvirus of Turkey infectious bursal disease virus (IBDV) vaccination against a variant IBDV strain. Avian Dis 53:624–628. https://doi.org/10.1637/8687-31009RESNOTE.1

Peyret H, Lomonossoff GP (2013) The pEAQ vector series: the easy and quick way to produce recombinant proteins in plants. Plant Mol Biol 83:51–58. https://doi.org/10.1007/s11103-013-0036-1

Pitcovski J, Levi BZ, Maray T et al (1999) Failure of viral protein 3 of infectious bursal disease virus produced in prokaryotic and eukaryotic expression systems to protect chickens against the disease. Avian Dis 43(1):8–15

Pitcovski J, Gutter B, Gallili G et al (2003) Development and large-scale use of recombinant VP2 vaccine for the prevention of infectious bursal disease of chickens. Vaccine 21:4736–4743

Prandini F, Simon B, Jung A et al (2016) Comparison of infectious bursal disease live vaccines and a HVT-IBD vector vaccine and their effects on the immune system of commercial layer pullets. Avian Pathol 45:114–125. https://doi.org/10.1080/03079457.2015.1127891

Qi X, Gao H, Gao Y et al (2009) Naturally occurring mutations at residues 253 and 284 in VP2 contribute to the cell tropism and virulence of very virulent infectious bursal disease virus. Antiviral Res 84:225–233

Raja P, Senthilkumar TMA, Parthiban M et al (2016) Complete genome sequence analysis of a naturally reassorted infectious bursal disease virus from India. Genome Announc 4:e00709–16. https://doi.org/10.1128/genomeA.00709-16

Rautenschlein S, Ch Kraemer H, Vanmarcke J, Montiel E (2005) Protective efficacy of intermediate and intermediate plus infectious bursal disease virus (IBDV) vaccines against very

virulent IBDV in commercial broilers. Avian Dis 49:231–237. https://doi.org/10.1637/7310-112204R
Remond M, Da Costa B, Riffault S et al (2009) Infectious bursal disease subviral particles displaying the foot-and-mouth disease virus major antigenic site. Vaccine 27:93–98
Richetta M, Gómez E, Lucero MS et al (2017) Comparison of homologous and heterologous prime-boost immunizations combining MVA-vectored and plant-derived VP2 as a strategy against IBDV. Vaccine 35:142–148
Rodenberg J, Sharma JM, Belzer SW et al (1994) Flow cytometric analysis of B cell and T cell subpopulations in specific-pathogen-free chickens infected with infectious bursal disease virus. Avian Dis 38(1):16–21
Rogel A, Benvenisti L, Sela I et al (2003) Vaccination with *E. coli* recombinant empty viral particles of infectious bursal disease virus (IBDV) confer protection. Virus Genes 27:169–175
Rong J, Jiang T, Cheng T et al (2007) Large-scale manufacture and use of recombinant VP2 vaccine against infectious bursal disease in chickens. Vaccine 25:7900–7908. https://doi.org/10.1016/j.vaccine.2007.09.006
Saif YM (2004) Control of infectious bursal disease virus by vaccination. Dev Biol (Basel) 119:143–146
Sainsbury F, Thuenemann EC, Lomonossoff GP (2009) pEAQ: versatile expression vectors for easy and quick transient expression of heterologous proteins in plants. Plant Biotechnol J 7:682–693. https://doi.org/10.1111/j.1467-7652.2009.00434.x
Sapats SI, Ignjatovic J (2000) Antigenic and sequence heterogeneity of infectious bursal disease virus strains isolated in Australia. Arch Virol 145:773–785. https://doi.org/10.1007/s007050050670
Sharma JM, Kim I-J, Rautenschlein S, Yeh HY (2000) Infectious bursal disease virus of chickens: pathogenesis and immunosuppression. Dev Comp Immunol 24:223–235. https://doi.org/10.1016/S0145-305X(99)00074-9
Snyder DB, Vakharia VN, Savage PK (1992) Naturally occurring-neutralizing monoclonal antibody escape variants define the epidemiology of infectious bursal disease viruses in the United States. Arch Virol 127:89–101. https://doi.org/10.1007/BF01309577
Taghavian O, Spiegel H, Hauck R et al (2013) Protective oral vaccination against infectious bursal disease virus using the major viral antigenic protein VP2 produced in Pichia pastoris. PLoS ONE. https://doi.org/10.1371/journal.pone.0083210
Tippenhauer M, Heller DE, Weigend S, Rautenschlein S (2013) The host genotype influences infectious bursal disease virus pathogenesis in chickens by modulation of T cells responses and cytokine gene expression. Dev Comp Immunol 40:1–10. https://doi.org/10.1016/j.dci.2012.10.013
Van den Berg TP (2000) Acute infectious bursal disease in poultry: a review. Avian Pathol 29:175–194. https://doi.org/10.1080/03079450050045431
van den Berg TP1, Eterradossi N, Toquin D, Meulemans G (2000) Infectious bursal disease (Gumboro disease). Rev Sci Tech 19(2):509–543
van Loon AAMW, De Haas N, Mundt E (2002) Alteration of amino acids in VP2 of very virulent infectious bursal disease virus results in tissue culture adaptation and attenuation in chickens. J Gen Virol 83:121–129. https://doi.org/10.1099/0022-1317-83-1-121
von Einem UI, Gorbalenya AE, Schirrmeier H et al (2004) VP1 of infectious bursal disease virus is an RNA-dependent RNA polymerase. J Gen Virol 85:2221–2229. https://doi.org/10.1099/vir.0.19772-0
Whetzel PL, Jackwood DJ (1995) Comparison of neutralizing epitopes among infectious bursal disease viruses using radioimmunoprecipitation. Avian Dis 39(3):499–506
Wu H, Singh NK, Locy RD et al (2004a) Expression of immunogenic VP2 protein of infectious bursal disease virus in Arabidopsis thaliana. Biotechnol Lett 26:787–792
Wu H, Singh NK, Locy RD et al (2004b) Immunization of chickens with VP2 protein of infectious bursal disease virus expressed in Arabidopsis Thaliana. Avian Dis 48:663–668. https://doi.org/10.1637/7074

Wu J, Yu L, Li L et al (2007) Oral immunization with transgenic rice seeds expressing VP2 protein of infectious bursal disease virus induces protective immune responses in chickens. Plant Biotechnol J 5:570–578. https://doi.org/10.1111/j.1467-7652.2007.00270.x

Wu Y, Hong L, Ye J et al (2009) The VP5 protein of infectious bursal disease virus promotes virion release from infected cells and is not involved in cell death. Arch Virol 154:1873–1882. https://doi.org/10.1007/s00705-009-0524-4

Yamaguchi T, Setiyono A, Kobayashi M et al (2000) Infectious bursal disease live vaccines: changes in the virus population during serial passage in chickens and chicken embryo fibroblast cells. Avian Dis 44:284. https://doi.org/10.2307/1592541

Yip CW, Hon CC, Zeng F, Leung FCC (2012) Cell culture-adapted IBDV uses endocytosis for entry in DF-1 chicken embryonic fibroblasts. Virus Res 165:9–16

Zachar T, Popowich S, Goodhope B et al (2016) A 5-year study of the incidence and economic impact of variant infectious bursal disease viruses on broiler production in Saskatchewan, Canada. Can J Vet Res 80:255–261

Zhao S, Jia Y, Han D et al (2016) Influence of the structural development of bursa on the susceptibility of chickens to infectious bursal disease virus. Poult Sci 95:2786–2794. https://doi.org/10.3382/ps/pew192

Plant-Produced Avian Influenza Antigens

Yanaysi Ceballo, Alina Lopez, Kenia Tiel and Abel Hernandez

Abstract Avian influenza (AI) is a highly contagious respiratory disease that can also affect the enteric and nervous systems, causing a high degree of morbidity and mortality in animals and even in humans. Although current vaccines are effective against virus infection, new strategies need to be developed to satisfy the global demand for an AI vaccine. Plant-based expression systems can function as inexpensive platforms for the large scale production of recombinant pharmaceuticals or subunit vaccines. During the last decade, successful cases of influenza antigens production have been reported in plants, using both transient and stable expression systems. Full-length hemagglutinin (HA), as well as subunits thereof, has been produced in different compartments of the cell fused or not to other polypeptides. Immunizations of animals (mice, ferrets, rabbit, chickens and human) were performed with some of these plant-derived HA variants. These results demonstrate that plant-produced HA protein is antigenic and can induce immune response that correlate with protection against lethal AI virus. This paper reviews studies developed by several groups of researchers to improve the production of plant-based AI vaccines.

Keywords Avian Influenza · Plant molecular farming · Hemagglutinin
Veterinary vaccines

1 Avian Influenza

AI is a highly contagious respiratory disease that can also affect the enteric and nervous systems, causing a high degree of morbidity and mortality in animals and even in humans (Ferguson et al. 2006; Medina and García-Sastre 2011). Migratory birds act as reservoirs carrying the virus in the intestinal tract, and are generally asymptomatic, but domesticated poultry and other birds can also be infected.

Y. Ceballo (✉) · A. Lopez · K. Tiel · A. Hernandez
Plant Biotechnology Department, Center for Genetic Engineering and Biotechnology (CIGB), PO Box 6162, 10600 Havana, Cuba
e-mail: yanaysi.ceballo@cigb.edu.cu

J. MacDonald (ed.), *Prospects of Plant-Based Vaccines in Veterinary Medicine*,
https://doi.org/10.1007/978-3-319-90137-4_8

Infections in poultry may be more severe, and turkeys are more commonly infected than chickens. The symptoms in birds include lethargy, respiratory distress, facial swelling, decreased egg production and sudden death without clinical signs. Between January 2014 and November 2016, AI has been identified in 77 countries and 13 strains have been detected (http://www.oie.int/en/animal-health-in-the-world/update-on-avian-influenza). While most AI strains cause only mild disease in poultry, and are called low pathogenic AI (LPAI), highly pathogenic AI (HPAI) viruses can kill up to 90–100% of the flock and cause large epidemics, which can spread rapidly, generating significant impacts on the poultry industry and trade (Iowa State University Center for Food Security and Public Health 2016). (http://lib.dr.iastate.edu/cfsph_factsheets/10).

2 Avian Influenza Virus

AI virus (AIV) belongs to the family Orthomyxoviridae, genus Influenza A virus (type A), and contains a negative-sense, segmented RNA genome. Structurally, the virus is of spherical shape, wrapped with two kinds of glycoprotein spikes on its surface arranged at regular intervals, called hemagglutinin (HA) and neuraminidase (NA). HA and NA are set in a double lipid membrane (envelope) located on a coat protein (M protein or matrix) that gives form and stability to the housing (Fig. 1). Type A influenza strains are classified according to the primary viral surface proteins, the HA and NA. The HA has 16 subtypes (H1–H16) and the NA has 9

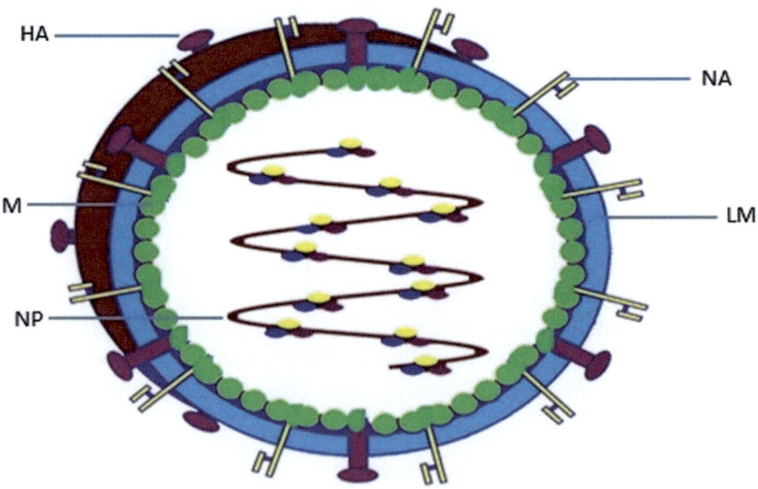

Fig. 1 Schematic views of the structure of Influenza A virus. The virus proteins are denoted as HA hemagglutinin; NA neuraminidase; M matrix protein; NP nucleoprotein; LM lipid bilayer membrane

subtypes (N1–N9). The H and N subtypes seem to be able to assort into any combination, and many of the 144 possible arrangements have been found in natural reservoir species, although some combinations are more common than others (Spackman 2008).

Hemagglutinin is the major viral surface protein that induces protective immunity to influenza infection and is synthesized as a full-length protein containing a signal peptide, an extracellular domain (HA1 and HA2 regions), and a transmembrane domain and cytoplasmic domain at the C-terminus of the HA2 region. This protein is cleaved by proteases to create two subunits, HA1 and HA2, linked by a disulfide bond (Pietrzak et al. 2016). HA proteins form trimeric structures with non-covalent bonds, and play an important role in binding to the cellular receptor sialic acid, preferably with an (alpha)-2,3 galactose linkage. This binding induces fusion of the viral envelope with the endosome membrane. (Floyd et al. 2008; Mao et al. 2017; Stephenson et al. 2003).

The NA is a homotetrameric protein, less abundant on the viral surface. However, its enzymatic activity is important for cleaving sialic acid from the viral envelope, facilitating virus release from the surface of the infected cell (Matrosovich et al. 2004). Both HA and NA glycoproteins are prone to mutation due to antigenic shift and antigenic drift (Medina and García-Sastre 2011; Du et al. 2012). These mechanisms have allowed the virus to continuously develop new strains that may or may not result in cross-species transmission.

3 Transmission

Several factors can contribute to the spread of AI viruses, including: movement of people, equipment and goods, marketing practices (live bird markets), farming practices and the presence of the viruses in migratory wild birds that come into contact with poultry, as well as other animals that act as vectors of transmission (Vittecoq et al. 2017). It has been estimated that AI viruses have greatly affected avian health and poultry production worldwide. More than 500 million poultry deaths have been associated with avian influenza infections (Ben Embarek et al. 2009; Harfoot and Webby 2017).

Wild birds play important roles in the circulation of AI viruses and are reservoirs for low pathogenicity strains. In general, AI viruses in wild birds can be transmitted to and from poultry, and potentially to and from other domestic animals and humans. The transmission is through feces and respiratory secretions of birds, although this depends on the specific virus, host and other factors (Harfoot and Webby 2017; Acha and Szyfres 2003). Waterfowl feces contain large amounts of the virus (Olsen et al. 2006).

Once an AIV has entered a herd of poultry flock, it can spread on the farm via the fecal-oral route and through aerosols, due to the proximity of the birds. The potential role of rodents in AIV transmission from wild birds to poultry and between poultry houses suggests that rodents are likely to act as mechanical vectors

(Velkers et al. 2017). Similarly, blow flies are likely candidates for mechanical transmission of HPAI, and H5N1 viral DNA has been detected in house flies (Sievert et al. 2006; Wanaratana et al. 2011) and mosquitoes (Barbazan et al. 2008). Blow flies prefer to lick carcasses and droppings of not only chickens but also pigs. This finding from Indonesia strongly suggest that it is important to pay attention to pigpens as well as poultry farms within 2–3 km, where viable H5N1 viruses are transmitted by blow flies. The H5N1 virus surveillance conducted in Indonesia suggested that pigs are at risk of infection during outbreaks and pigs can serve as intermediate hosts in which this avian virus can adapt to mammals (Nidom et al. 2010).

Currently, due to ongoing circulation of various strains (H5N1, H5N2, H5N6, H5N8 and H7N9), outbreaks of avian influenza continue to be a global public health concern. The OIE's World Animal Health Information System provides the framework for Veterinary Services to implement effective surveillance, reporting and controls for AI. Table 1 shows an overview of current avian disease events reported to the OIE.

The majority of human cases of influenza A (H5N1) and A (H7N9) virus infection have been associated with direct or indirect contact with infected live or dead poultry, so the control of the disease in the animal source is critical to decrease risk to humans. Since 2003, the H5N1 virus has spread from Asia to Europe and Africa, and has become endemic in poultry populations in some countries. Outbreaks have resulted in millions of poultry infections, having serious impacts on livelihoods, economy, and international trade in affected countries. Approximately, 65 countries on five continents have reported outbreaks in birds and globally, the number of cases in humans detected from 2003 to 2017 amounts to 860, of which 454 have died. (http://www.who.int/influenza/human_animal_interface/HAI_Risk_Assessment/en/).

In March 2013, the Chinese Centre for Disease Control and Prevention confirmed the first reported case of human infection with an AI H7N9 virus. Since then, the virus has spread in the poultry population across the country and resulted in over 1565 reported human cases and many human deaths (about 40%). (http://www.who.int/mediacentre/factsheets/avian_influenza/en/).

Table 1 Current global situation of avian influenza strain (ongoing outbreak as of 18 of September 2017)

Strain	Count of countries	Region	Count of poultry destroyed
H5N1	8	Africa, Asia and the Pacific	131,168
H5N2	1	Asia and the Pacific	1,617,816
H5N5	2	Europe	0
H5N6	3	Asia and the Pacific	25,096,648
H5N8	16	Africa, Asia and the Pacific, Europe	4,355,879
H7N9	1	Asia and the Pacific	831,087
Total			32,032,598

4 Vaccines for Influenza Virus

The threat of a major human influenza pandemic such as the avian H5N1 or the 2009 H1N1 has emphasized the need for effective prevention strategies to combat these diseases. Hundreds of millions of chickens and ducks died or were killed in order to prevent the spread of H5N1 (Sims et al. 2005).

The influenza prevention strategies are two methods; culling of potential threat populations, and vaccination against currently circulating virus strains. Vaccination can be of two types: emergency or prophylactic. In the case of emergency vaccination, it prevents the spread of the disease and the mass sacrifice of millions of birds. Prophylactic vaccination for H5 and H7 highly pathogenic AI viruses is a recommended measure for countries or regions exposed to a high risk of contracting the disease. In either case, vaccination must always be accompanied by biosecurity measures (Marangon and Capua 2005; FAO 2016). Existing commercial vaccines are predominantly of the inactivated, whole virus type, delivered with a water-in-oil-emulsion which requires parenteral administration. Generally, inactivated vaccines render protection against AIV through antibody-mediated immune responses. Unlike inactivated vaccines that only induce humoral immune responses, live-attenuated vaccines (LAVs) can stimulate humoral, cellular and mucosal immune responses, which provide greater cross-protection and longer-lasting immunity (Suguitan et al. 2006; Joseph et al. 2008).

In general, influenza vaccines are made from AI seeds that grow well on embryonated chicken eggs. Seed viruses are generated by World Health Organization Collaborating Centers and distributed to vaccine manufacturers (Hickling and D'Hondt 2006). For decades these vaccines proved to be safe and effective. However, their production system has limitations, including a need for specific pathogen-free embryonated eggs, time needed for adaptation to the new circulating virus subtypes, long production time, and limited capacity (Cummings et al. 2014). These limitations became evident when H1N1 influenza pandemic emerged unexpectedly, while the main attention was focused on the control of the H5N1 virus.

To address such limitations, extensive research has been carried out on the production of recombinant antigens expressed from mammalian or insect cells (Stachyra et al. 2017; Hu et al. 2017; Pushko et al. 2017; Pose et al. 2011; Treanor et al. 2017). The majority of target antigens are HA and NA which are the major antigenic influenza viral proteins, and the secondary targets are matrix (M) or nucleoprotein (NP), which confers potentially broad protective effect through cellular immunity (Kolpe et al. 2017; Yang et al. 2017; Zheng et al. 2014).

Two recombinant vaccines against the AIV of serotypes H5 and H7 have been described (Veits et al. 2006), the first developed using a modified Newcastle disease virus, as a vector, in which has been inserted the H5N2 hemagglutinin gene H5N2. The tests carried out showed that the vaccinated animals are able to resist the highly pathogenic H5 subtype and produce antibodies against both H5 and H7 viruses, so it could be used as a bivalent vaccine against both avian diseases. The second vaccine uses as a vector the modified virus of infectious larigotracheitis, on which the H7 gene, serotype H7N1, has been inserted. Vaccinated animals developed antibodies against the virus of and were protected against the inoculation of a lethal dose of the H7N1 virus (Veits et al. 2006). However, production of subunit vaccines on a large scale using these systems is still expensive.

5 Plants as Bioreactors

Plants are a very versatile system for producing recombinant proteins for pharmaceutical or industrial uses. Among the many advantages offered by plants is the low cost of biomass production compared to the industrial bioreactor infrastructure used for microorganisms and mammalian cells, facilities to scale up in greenhouse-based manufacturing, the capacity of efficiently producing complex proteins at high yields, and low risk of contamination with pathogens of animals and humans (Topp et al. 2016; Marsian and Lomonossoff 2016; Fahad et al. 2015; Sack et al. 2015; Chan and Daniell 2015).

Recombinant proteins in plants can be produced by either stable genomic integration or transient expression (Fig. 2). While the accumulation level of the target protein is usually low, many strategies have been proposed for the enhancement of recombinant protein expression including; chloroplast transformation (Clarke et al. 2011), seed-specific promoters (De Jaeger et al. 2002; Hernández-Velázquez et al. 2015), untranslated leader sequences (Sainsbury et al. 2009), and deconstructed viral vectors (Azhakanandam et al. 2007; Peyret and Lomonossoff 2015). The long time required for the generation of transgenic plants is a limitation for the production of recombinant proteins. To address this concern, transient expression is more flexible method to express proteins in short time without chromosomal positional effects. The technique is based in the use of *Agrobacterium tumefaciens,* plant viruses, or hybrid vectors with components of both, and exploits the abilities of these plant pathogens to infect plant tissues in a short time. In all cases, *A. tumefaciens* is either infiltrated using a syringe into the underside of leaves, or vacuum infiltrated using the whole plant or detached leaf (Fujiuchi et al. 2016). This procedure has been carried out in a variety of plants with different experimental purposes, with *Nicotiana benthamiana* as the most common host because it can be grown in high density to produce large amounts of biomass in a matter of weeks (Goodin et al. 2008).

Based on these advantages, the production of AI vaccines in plants has been the focus of many researchers worldwide, mainly by the production of influenza HA-based subunit vaccines, because of this antigen's ability to generate neutralizing antibodies by the host immune system. Other targets include M and NP antigens, which confer a potentially broad protective effect through cellular immunity. During the last years, clinical trials have been carried out to evaluate the immunogenicity and safety of plant-made influenza vaccines with positive outcomes (Cummings et al. 2014; Landry et al. 2014; Chichester et al. 2012).

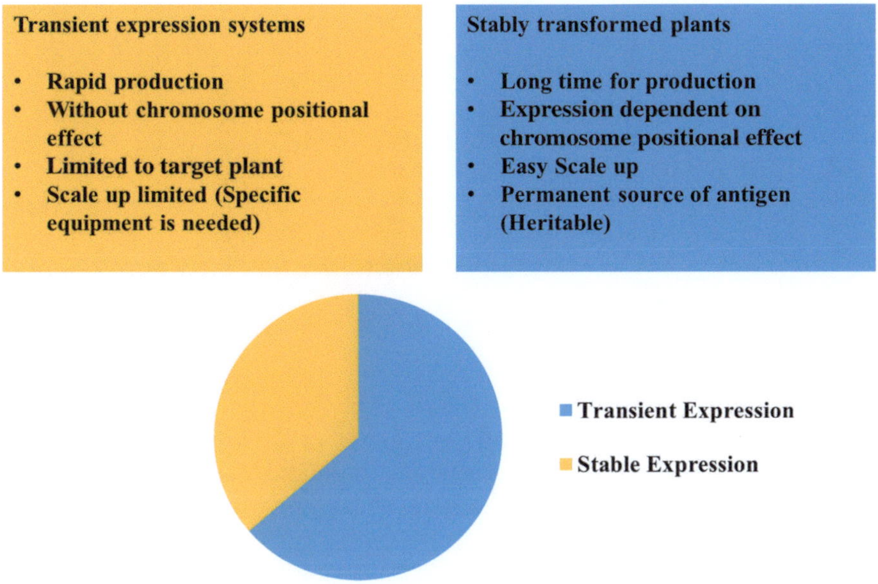

Fig. 2 Avian influenza antigens produced in plants based on Table 2. Advantages and disadvantages of plant expression strategies

6 Influenza Virus Antigen Production by Plant Transient Expression

New strains of influenza rapidly emerge every year, which can cause pandemics, which, in turn, demands a faster way of vaccine production than is currently possible using embryonated eggs (Nemchinov and Natilla 2007; Pua et al. 2017). The transient expression platform allow the rapid production of large amounts of protein without the need to develop transgenic lines, with minimal costs and workload (Buyel 2015). Over the past decade, many vaccine antigens for various subtypes and strains of influenza A virus have been expressed transiently in plants, including the humans H1N1 (Cummings et al. 2014; Shoji et al. 2011, 2013; D'Aoust et al. 2008; Neuhaus et al. 2014; Jul-Larsen et al. 2012; Iyer et al. 2012; Pillet et al. 2016; Ward et al. 2014), and H3N2 (Pillet et al. 2016; Ward et al. 2014; Mett et al. 2008; Shoji et al. 2008).

In the same way, antigens for protection to AI have been produced with different vectors, keeping in mind the possession of intellectual property on the expression vectors and the current ongoing outbreaks of AI among other factors (Table 2).

One of these systems is mediated by deconstructed viral vectors, that combine the elements of plant RNA viral vectors (tobacco mosaic virus, potato virus X and several others) and *Agrobacterium* binary plasmids (Peyret and Lomonossoff 2015). The plant transient expression system MagnICON® (ICON Genetics GmbH, Halle, Germany) is based on *in planta* assembly of functional viral vectors from two

Table 2 Avian Influenza antigens produced in plant

Antigen	Strain	Expression system	Expression levels	Immune response	Animal	References
H5HA	AIV A/Indonesia/5/05 (H5N1)	*N. benthamiana*/transient	50 mg/kg of FW	IR+ P +	Mice, Ferrets and Adults 18–60 years	D'Aoust et al. (2008), Ward et al. (2014), Landry et al. (2010), Le Mauff et al. (2015)
H5HA	AIV A/Indonesia/05/05 (H5N1)	*N. benthamiana*/transient	50–60 mg/kg of FW	IR+ P +	Mice, New Zealand White rabbits, Ferret and Adults 18–49 years	Chichester et al. (2012), Shoji et al. (2011, 2009), Major et al. (2015)
H5HA	AIV A/Indonesia/5/05 (H5N1)	*N. benthamiana*/transient	300 mg/kg of FW	IR + P +	White Leghorn chickens	Kalthoff et al. (2010)
H5HA	AIV A/Indonesia/05/05 (H5N1)	*N. benthamiana*/transient	Up to 23 mg/kg of FW	RI + P +	Mice	Ling et al. (2012)
H5HA	AIV A/Indonesia/05/05 (H5N1)	*N. tabacum* LAFC 53 *N. glauca*, *N. tabacum* LAFC-HA X *N. glauca* A622-RNAi/transgenic	5.4–9.6 mg/kg of FW	RI + P +	Mice	Ling et al. (2012)
H5HA1	AIV A/Indonesia/05/05 (H5N1)	*N. tabacum*/transient	8,8 of TSP	No data	No data	Farsad et al. (2017)
H5HA1 fusion to Fc	AIV A/VietNam/1203/2004 (H5N1)	*N. benthamiana*/transient	1 mg/kg of FW	IR+ P −	Mice	Spitsin et al. (2009)
H5HA1 fusion to Fc	AIV A/VietNam/1203/2004 (H5N1)	*N. tabacum*/transgenic	4 mg/kg of FW	IR+ P −	Mice	Spitsin et al. (2009)

(continued)

Table 2 (continued)

Antigen	Strain	Expression system	Expression levels	Immune response	Animal	References
H5HA	AIV A/Viet Nam/1194/2004 (H5N1)	*N. benthamiana*/transient	62–675 mg/kg of FW	IR+ P ± not conclusive	Mice and Rhode Island Red chickens	Mortimer et al. (2012)
H5HA	AIV A/VietNam/1194/2004 (H5N1)	*N. tabacum*/transgenic	160–1440 mg/kg of FW	IR+ P ±	Mice Rhode Island Red chickens	Mortimer et al. (2012)
H5HA	AIV A/VietNam/1203/2004 (H5N1)	*N. tabacum*/transgenic seed	3000 mg/kg	IR+ P +	Leghorn chickens	Ceballo et al. (2017)
H5HA	AIV A/Bar-headed Goose/Qinghai/1A/05 (H5N1) and A/Anhui/1/05 (H5N1)	*N. benthamiana*/transient	No data	IR+ P +	Mice	Shoji et al. (2009)
H5HA fusion with ELP	AIV A/Hatay/2004 (H5N1)	*N. benthamiana*/transient	120 mg/kg of FW (purified), 1% TSP	IR+ P +	Mice	Phan et al. (2013)
H5HA fusion with ELP	AIV A/Hatay/2004 (H5N1)	*N. tabacum*/ transgenic	90 mg/kg of FW (purified), 0.5% TSP	IR+ P +	Mice	Phan et al. (2013)
H5HA (HFBI, ELP)	AIV A/Hatay/2004 (H5N1)	*N. tabacum*/transgenic seed	0.03–0.5% TSP	No data	No data	Phan et al. (2014)
H5HA	AIV A/swan/Poland/305-135V08/2006 (H5N1)	*N. benthamiana*/transient	3.33 mg/kg of FW	No data	No data	Redkiewicz et al. (2014)

(continued)

Table 2 (continued)

Antigen	Strain	Expression system	Expression levels	Immune response	Animal	References
H5HA HA-zein	AIV A/Hatay/2004 (H5N1)	*N. benthamiana*/transient	400 mg/kg of FW (purified)	IR+ P (No data)	Mice	Hofbauer et al. (2016)
H5HA	AIV A/Hong Kong/ 213/03 (H5N1)	Arabidopsis/transgenic	140 mg/kg of FW	IR+ P +	Mice	Lee et al. (2015)
H7HA	AIV A/Hangzhou/1/ 2013. (H7N9)	*N. benthamiana*/transient	No data	IR+ R +	Mice and Ferrets	Pillet et al. (2015)
M2e	AIV A/chicken/ Kurgan/5/2005 (H5N1)	Duckweed/transgenic	40 mg/kg of FW	No data	No data	Firsov et al. (2015)
M2e (Flg-4M)	AIV A/Chicken/ Kurgan/05/2005 (H5N1	*N.benthamiana*/transient	1 mg/g of FW	IR+ P +	Mice	Mardanova et al. (2015)
M2eHBc	AIV A/Duck/Potsdam/ 1402_6/1986 (H5N2)	*N. benthamiana*/transient	1–2% of FW	IR+ P ±	Mice	Ravin et al. (2012)
H7HA	AIV A/mallard/ Sweden/7206/2004 (H7N7)	*N. benthamiana*/transient	210 mg/kg FW, (purified)	No data	No data	Kanagarajan et al. (2012)

IR: immune response; P: protection; +: positive result; −: negative result separate 5′ and 3′ provector modules that are delivered into *N. benthamiana* plant cells via *A. tumefaciens*. This method has been used to obtain efficient production of the recombinant HA1 subunit of HA (17-338 aa) from H5N1 (A/swan/Poland/305-135V08/2006) virus in *N. benthamiana* (Redkiewicz et al. 2014). The highest amount of purified protein was obtained for HA1 targeted to endoplasmatic reticulum (ER) (3.3 mg/kg of fresh weight of tissue (FW)). Comparable results were reported by another group (Spitsin et al. 2009) who obtained various versions of HA from H5N1 (A/Vietnam/1203/2004), including the full-length HA, and shorter versions containing only major antigenic domains either alone or in a fusion with Fc (human or mouse). The production level of HA1-Fc using the MagnICON® system was about 1 mg/kg FW. Although significant humoral immune responses were observed in mice immunized with various HA1 variants, none of them induced virus-neutralizing antibodies. The MagnICON system was also used to produce the full-length HA from the virus strain NIBRG-14 (reassortant between A/Vietnam/1194/04 (H5N1) and A/Puerto Rico (PR)/8/23 viruses) which produced high levels of HA protein (300 mg/kg of FW) in the cell apoplast. Recombinant HA was highly immunogenic and protected immunized chickens against a lethal challenge infection with H5N1 (A/whooper swan/Germany/R65/2006) virus (Kalthoff et al. 2010).

Another non-replicating viral vector system, named pEAQ (Sainsbury et al. 2009), contains a modified 5′-untranslated region (UTR) and the 3′-UTR from Cowpea mosaic virus (CPMV) RNA-2, within the T-DNA region of binary vector. This system was used for production of H5HA virus like particles (VLP) in *N. benthamiana* with higher yields of VLP than the alfalfa plastocyanin gene-based expression system which was previously used (D'Aoust et al. 2010). The same system was employed to transiently produce biologically active recombinant full-length HA of an H7N7, isolated from a wild Swedish mallard duck, in *N. benthamiana* as a subunit vaccine candidate against influenza in poultry and humans. The yield of purified HA was 200 mg/kg of FW, and the protein maintained its native antigenicity and specificity, providing a good source of vaccine antigen to induce immune response in poultry species (Kanagarajan et al. 2012).

The pEff vector combines the advantages of pEAQ vectors, potato virus X (PVX)-based vectors, and the P24 silencing suppressor to create a new vector for quick and efficient transient expression of recombinant proteins, up to 30% of total soluble protein (TSP) for green fluorescent protein. This vector was used for expression of the influenza vaccine candidate, M2eHBc, consisting of an extracellular domain of IV M2 protein (M2e) fused to hepatitis B core antigen (HBc). The M2eHBc was expressed to 5–10% of TSP (Mardanova et al. 2017).

The promoter and terminator from the alfalfa plastocyanin gene were used for transient expression of influenza VLP in plant. D'Aoust and colleagues (D'Aoust et al. 2008) showed that *N. benthamiana* are capable of producing VLP containing HA which elicit a protective immune response in mice (D'Aoust et al. 2008; Landry et al. 2010). The AIV antigens yields obtained using this platform were in the range of 50–1300 mg/kg of FWT (D'Aoust et al. 2010; Shoji et al. 2009, 2015). Medicago Inc. adapted the plant-based influenza VLP expression to large-scale production allowing the purified end product to be obtained only 3 weeks after receiving the HA sequence (D'Aoust et al. 2010). Recently, Medicago Inc. concluded a phase II clinical trial with their avian flu H5HA pandemic vaccine candidate and reported that the candidate vaccine was found to be safe and well tolerated. (http://www.medigaco.com). Further, the rapid development of a plant-derived VLP vaccine based on the HA sequence of influenza H7N9 A/Hangzhou/1/2013 has been described (Pillet et al. 2015). The VLP vaccine was capable of eliciting a protective immune response following a single dose in mice and ferrets.

Another example of transient expression was generated with pTRA binary vectors containing 35S CaMV promoter. HA from the A/Hatay/2004/(H5N1) influenza strain fused to γ-zein protein (Zein is a water-insoluble polymer from maize seeds that has been widely used to produce carrier particles for the delivery of therapeutic molecules) was produced in leaves of *N. tabacum*. A total of 120 µg HA-γ-zein per 300 g of FW was recovered. An immune response was achieved in all mice treated with HA-zein, even at low doses (Hofbauer et al. 2016).

7 Stable Expression in Leaves

The major advantage of stably transformed plants is that the heterologous protein production is heritable, which allows simple and rapid scaling-up. However, a relatively long time (1–2 years) is required to obtain stabilized transformants.

Stable transformation of plant cells were used for the expression of the recombinant AIV antigens (Table 2). Transgenic *Nicotiana tabacum* producing different version of HA1 (from strain A/Vietnam/1203/04 (H5N1) and HA1-Fc fused proteins, up to 4 mg/kg of FW were obtained. However, while the recombinant HA1 variants induced a significant serum humoral immune response in mice, none of the HA1 preparations induced virus-neutralizing antibodies (Spitsin et al. 2009).

Two recombinant variant forms of HA from H5N1 (A/Vietnam/1194/2004) were produced in different cellular compartments under the control of a 35S CaMV promoter (cytoplasm, chloroplast, ER, and apoplastic spaces). The expression levels of HA in the stable transgenic plants ranged from 640–1440 mg/kg of FW, higher than in the transient expression assay. The HA protein was used in an immunization experiment, resulting in a serum from vaccinated chickens and mice containing specific anti-H5HA antibody (Mortimer et al. 2012).

Elastin-like polypeptide (ELP) was used to increase the yield and improve the purification process of HA (Phan et al. 2013). The HA sequence encoding the ectodomain (aa 17–520) of the A/Hatay/2004/(H5N1) IV strain was cloned into a plant expression vector fused to ELP (H5HA and H5HA-ELP). The effect of ELPylation was a notable increase in the accumulation of H5N1 HA in the leaves of transformed *N. tabacum* plants. Stable transformants of H5HA-ELP contained 10 times more recombinant HA than extracts from H5 without ELP. Plant-derived AIV HA trimers elicited neutralizing antibodies in mice.

HA from AI H5N1 expressed in the ER of *Arabidopsis thaliana* resulted in high levels of accumulation (140 mg/kg of FW). The immunogenicity of recombinant antigen was confirmed from mice immunized orally by the presence of high levels of humoral and mucosal antibodies specific for HA. This strong immune response provided protective immunity to mice against the AIV (Lee et al. 2015).

8 Stable Expression in Seeds

Seeds are designed for the synthesis and storage of proteins, which may account for 8–40% of their weight. This endows them with a considerable advantage over green tissues whose protein content is lower (Arcalis et al. 2015). Dehydration, another feature of seeds stemming from their role as a storage compartment, also reduces the levels of non-enzymatic hydrolysis and degradation of proteins (Rossi et al. 2014). It has been shown that antibodies, vaccine antigens and other heterologous proteins can be highly accumulated in seeds, where they remain stable and functional for several years even upon storage at room temperature (Hernández-Velázquez et al. 2015; De Wilde et al. 2013; Van Droogenbroeck et al. 2007; Vamvaka et al. 2016; Hudson et al. 2014).

The influenza virus antigens have been produced in plant seed. Recently, the NP (H3N2 swine) was produced in T3 maize endosperm with the endosperm-specific promoter of the maize γ-zein gene at accumulation levels up to 70 mg/kg of seed (Nahampun et al. 2015). Oral administration showed that a humoral immune response was elicited in the mice treated with maize-derived rNP. Phan (Phan et al. 2014) reported the HA (H5N1) accumulation up 0.5% of TSP in *N. tabacum* seeds, using a combination of a seed-specific promoter (USP promoter) with the fusion at C-terminus of ELP, although immunization assays with this antigen wasn't reported.

We have genetically engineered *N. tabacum* cv. BHmN seed to express HA from the A/Vietnam/1203/04(H5N1) AIV. The extracellular domain of HA (1.6 kb) gene was inserted under the regulation of a seed specific promoter (β-phaseolin) and flanked by the arcelin 5'UTR, 2S2 signal peptide and arcelin terminator. The HA in T1 tobacco seeds (sHA) accumulated at levels of up 3000 mg/kg of seed (10% of TSP), being the highest reported for this protein in plant cells (Table 2). This high production is in accordance with other antigens and antibodies produced in seeds under the β-phaseolin promoter (De Jaeger et al. 2002; Hernández-Velázquez et al. 2015; De Wilde et al. 2013; Hernández et al. 2013).

Immunogenicity of sHA was determined by hemagglutination inhibition (HI) assay from sera of immunized chickens. The chickens were immunized with two doses of aqueous extracted proteins from tobacco seed containing 20 μg of sHA mixed with M888 adjuvant. All animals vaccinated with sHA extract generated an immune response by day 35 (T35), seven days after the boosting. No sera of chicks immunized with extracts of non-transgenic seeds, taken as a negative control group, showed any humoral response (Fig. 3). Analysis of the sera collected at day

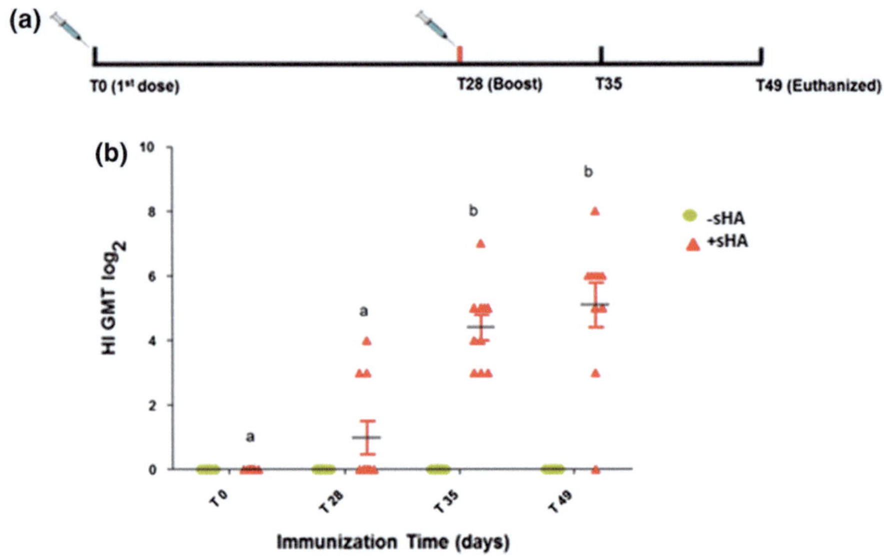

Fig. 3 Immunogenicity of seed -produced recombinant HA. **a** Immunization scheme. **b** Kinetics of the hemagglutination inhibition (HI) response in chicks immunized with sHA. Values represent the geometric mean titers (GMTs) ($\log_2$). −sHA: experimental group immunized with wild-type TSP from tobacco seeds. +sHA: experimental group immunized with 20 μg sHA extracted from seeds of the I-13 line. **a**, **b** Different letters represent significant difference ($p < 0.05$)

49 (T49) showed that 50% of chicks had HI titers of 1:64. These titers obtained in our experiments are higher than those required to protect against erythrocyte agglutination (Hannoun et al. 2004), suggesting that our approach to prevent infection with AI may be effective. Our results are similar to the HI titers obtained by D'Aoust and colleagues (D'Aoust et al. 2008), with mice immunized with 12 μg HA produced in leaves of *N. benthamiana*. In another study, (Shoji et al. 2009) obtained similar titers in mice that received 15 μg of antigen.

Development of safe and effective vaccines to reduce the economic impact on agriculture and a potential pandemic for humans of AI is significant. However, there are still no commercial plant-made vaccines against avian influenza. Medicago Inc has completed a Phase II human clinical trial with VLP H5 vaccine candidate and has reported that the vaccine was found to be safe, well tolerated and highly effective. The transient expression platform allows for rapid production of large amounts of protein and has advantages over stable expression because it needs less time to obtain the vaccine candidate. However, seeds provide an environment for stable, long-term HA storage and this potential candidate should be evaluated against the circulating strain to be used immediately in an emergency case.

References

Acha PN, Szyfres B (2003) Zoonoses and communicable diseases common to man and animals: parasitic zoonoses: Pan American Health Org

Arcalis E, Ibl V, Peters J, Melnik S, Stoger E (2015) The dynamic behavior of storage organelles in developing cereal seeds and its impact on the production of recombinant proteins. Adv Seed Biol 118

Azhakanandam K, Weissinger SM, Nicholson JS, Qu R, Weissinger AK (2007) Amplicon-plus targeting technology (APTT) for rapid production of a highly unstable vaccine protein in tobacco plants. Plant Mol Biol 63(3):393–404

Barbazan P, Thitithanyanont A, Misse D, Dubot A, Bosc P, Luangsri N et al (2008) Detection of H5N1 avian influenza virus from mosquitoes collected in an infected poultry farm in Thailand. Vector-Borne Zoonotic Dis 8(1):105–110

Ben Embarek P, Briand S, Brown I, Bruscke C, Domenech J, Formenty P et al (2009) FAO-OIE-WHO joint technical consultation on avian influenza at the human-animal interface. Influenza Other Respir Viruses 4:1–29

Buyel JF (2015) Process development strategies in plant molecular farming. Curr Pharm Biotechnol 16(11):966–982

Ceballo Y, Tiel K, López A, Cabrera G, Pérez M, Ramos O et al (2017) High accumulation in tobacco seeds of hemagglutinin antigen from avian (H5N1) influenza. Transgenic Res 26 (6):775–789

Chan HT, Daniell H (2015) Plant-made oral vaccines against human infectious diseases-are we there yet? Plant Biotechnol J 13(8):1056–1070

Chichester JA, Jones RM, Green BJ, Stow M, Miao F, Moonsammy G et al (2012) Safety and immunogenicity of a plant-produced recombinant hemagglutinin-based influenza vaccine (HAI-05) derived from A/Indonesia/05/2005 (H5N1) influenza virus: a phase 1 randomized, double-blind, placebo-controlled, dose-escalation study in healthy adults. Viruses 4(11):3227–3244

Clarke JL, Daniell H, Nugent JM (2011) Chloroplast biotechnology, genomics and evolution: current status, challenges and future directions. Plant Mol Biol 76(3):207–209

Cummings JF, Guerrero ML, Moon JE, Waterman P, Nielsen RK, Jefferson S et al (2014) Safety and immunogenicity of a plant-produced recombinant monomer hemagglutinin-based influenza vaccine derived from influenza A (H1N1) pdm09 virus: a Phase 1 dose-escalation study in healthy adults. Vaccine 32(19):2251–2259

D'Aoust MA, Lavoie PO, Couture MMJ, Trépanier S, Guay JM, Dargis M et al (2008) Influenza virus-like particles produced by transient expression in Nicotiana benthamiana induce a protective immune response against a lethal viral challenge in mice. Plant Biotechnol J 6 (9):930–940

D'Aoust MA, Couture MM, Charland N, Trepanier S, Landry N, Ors F et al (2010) The production of hemagglutinin-based virus-like particles in plants: a rapid, efficient and safe response to pandemic influenza. Plant Biotechnol J 8(5):607–619

De Jaeger G, Scheffer S, Jacobs A, Zambre M, Zobell O, Goossens A et al (2002) Boosting heterologous protein production in transgenic dicotyledonous seeds using Phaseolus vulgaris regulatory sequences. Nat Biotechnol 20(12):1265–1268

De Wilde K, De Buck S, Vanneste K, Depicker A (2013) Recombinant antibody production in Arabidopsis seeds triggers an unfolded protein response. Plant Physiol 161(2):1021–1033

Du Y, Lou B, Wu Z, Zhao P, Cui Z (2012) Influence of antibody-mediated immune pressure on neuraminidase gene mutations of avian influenza virus H9N2. Bing du xue bao = Chin J Virol 28(1):1–6

Fahad S, Khan FA, Pandupuspitasari NS, Ahmed MM, Liao YC, Waheed MT et al (2015) Recent developments in therapeutic protein expression technologies in plants. Biotechnol Lett 37 (2):265–279

FAO U (2016) Rational use of vaccination for prevention and control of H5 highly pathogenic avian influenza

Farsad A, Malekzadeh-Shafaroudi S, Moshtaghi N, Fotouhi F, Zibaee S (2017) Transient Expression of HA1 Antigen of H5N1 Influenza Virus in Tobacco (*Nicotiana tabacum* L.) via Agro-infiltration. J Agric Sci Technol 19(2):439–451

Ferguson NM, Cummings DA, Fraser C, Cajka JC, Cooley PC, Burke DS (2006) Strategies for mitigating an influenza pandemic. Nature 442(7101):448–452

Firsov A, Tarasenko I, Mitiouchkina T, Ismailova N, Shaloiko L, Vainstein A et al (2015) High-yield expression of M2e peptide of avian influenza virus H5N1 in transgenic duckweed plants. Mol Biotechnol 57(7):653–661

Floyd DL, Ragains JR, Skehel JJ, Harrison SC, van Oijen AM (2008) Single-particle kinetics of influenza virus membrane fusion. Proc National Acad Sci 105(40):15382–15387

Fujiuchi N, Matoba N, Matsuda R (2016) Environment control to improve recombinant protein yields in plants based on agrobacterium-mediated transient gene expression. Front Bioeng Biotechnol 4:23

Goodin MM, Zaitlin D, Naidu RA, Lommel SA (2008) Nicotiana benthamiana: its history and future as a model for plant–pathogen interactions. Mol Plant-Microbe Interact 21(8):1015–1026

Hannoun C, Megas F, Piercy J (2004) Immunogenicity and protective efficacy of influenza vaccination. Virus Res 103(1):133–138

Harfoot R, Webby RJ (2017) H5 influenza, a global update. J Microbiol 55(3):196–203

Hernández A, López A, Ceballo Y, Rosabal L, Rosabal Y, Tiel K et al (2013) High-level production and aggregation of hepatitis B surface antigen in transgenic tobacco seeds. Biotecnol Apl 30:97–100

Hernández-Velázquez A, López-Quesada A, Ceballo-Cámara Y, Cabrera-Herrera G, Tiel-González K, Mirabal-Ortega L et al (2015) Tobacco seeds as efficient production platform for a biologically active anti-HBsAg monoclonal antibody. Transgenic Res 24 (5):897–909

Hickling J, D'Hondt E (2006) A review of production technologies for influenza virus vaccines, and their suitability for deployment in developing countries for influenza pandemic preparedness. World Health Organization Initiative for Vaccine Research 1–34

Hofbauer A, Melnik S, Tschofen M, Arcalis E, Phan HT, Gresch U et al (2016) The encapsulation of hemagglutinin in protein bodies achieves a stronger immune response in mice than the soluble antigen. Front Plant Sci 7:142

Hu CJ, Chien CY, Liu MT, Fang ZS, Chang SY, Juang RH et al (2017) Multi-antigen avian influenza a (H7N9) virus-like particles: particulate characterizations and immunogenicity evaluation in murine and avian models. BMC Biotechnol 17(1):2

Hudson LC, Garg R, Bost KL, Piller KJ (2014) Soybean seeds: a practical host for the production of functional subunit vaccines. Biomed Res Int 2014:340804

Iowa State University Center for Food Security and Public Health, "Avian Influenza" (2016) Center for Food Security and Public Health. Technical Factsheets. 10. http://lib.dr.iastate.edu/cfsph_factsheets/10

Iyer V, Liyanage MR, Shoji Y, Chichester JA, Jones RM, Yusibov V et al (2012) Formulation development of a plant-derived h1n1 influenza vaccine containing purified recombinant hemagglutinin antigen. Human Vaccines Immunotherapeutics 8(4):453–464

Joseph T, McAuliffe J, Lu B, Vogel L, Swayne D, Jin H et al (2008) A live attenuated cold-adapted influenza A H7N3 virus vaccine provides protection against homologous and heterologous H7 viruses in mice and ferrets. Virology 378(1):123–132

Jul-Larsen Å, Madhun AS, Brokstad KA, Montomoli E, Yusibov V, Cox RJ (2012) The human potential of a recombinant pandemic influenza vaccine produced in tobacco plants. Human Vaccines immunotherapeutics 8(5):653–661

Kalthoff D, Giritch A, Geisler K, Bettmann U, Klimyuk V, Hehnen H-R et al (2010a) Immunization with plant-expressed hemagglutinin protects chickens from lethal highly pathogenic avian influenza virus H5N1 challenge infection. J Virol 84(22):12002–12010

Kalthoff D, Globig A, Beer M (2010b) (Highly pathogenic) avian influenza as a zoonotic agent. Vet Microbiol 140(3):237–245

Kanagarajan S, Tolf C, Lundgren A, Waldenstrom J, Brodelius PE (2012) Transient expression of hemagglutinin antigen from low pathogenic avian influenza A (H7N7) in Nicotiana benthamiana. PLoS ONE 7(3):e33010

Kolpe A, Schepens B, Fiers W, Saelens X (2017) M2-based influenza vaccines: recent advances and clinical potential. Expert Rev Vaccines 16(2):123–136

Landry N, Ward BJ, Trepanier S, Montomoli E, Dargis M, Lapini G et al (2010) Preclinical and clinical development of plant-made virus-like particle vaccine against avian H5N1 influenza. PLoS ONE 5(12):e15559

Landry N, Pillet S, Favre D, Poulin J-F, Trépanier S, Yassine-Diab B et al (2014) Influenza virus-like particle vaccines made in Nicotiana benthamiana elicit durable, poly-functional and cross-reactive T cell responses to influenza HA antigens. Clin Immunol 154(2):164–177

Le Mauff F, Mercier G, Chan P, Burel C, Vaudry D, Bardor M et al (2015) Biochemical composition of haemagglutinin-based influenza virus-like particle vaccine produced by transient expression in tobacco plants. Plant Biotechnol J 13(5):717–725

Lee G, Na YJ, Yang BG, Choi JP, Seo YB, Hong CP et al (2015) Oral immunization of haemaggulutinin H5 expressed in plant endoplasmic reticulum with adjuvant saponin protects mice against highly pathogenic avian influenza A virus infection. Plant Biotechnol J 13(1):62–72

Ling H-Y, Edwards AM, Gantier MP, DeBoer KD, Neale AD, Hamill JD et al (2012) An interspecific Nicotiana hybrid as a useful and cost-effective platform for production of animal vaccines. PLoS ONE 7(4):e35688

Major D, Chichester JA, Pathirana RD, Guilfoyle K, Shoji Y, Guzman CA et al (2015) Intranasal vaccination with a plant-derived H5 HA vaccine protects mice and ferrets against highly pathogenic avian influenza virus challenge. Hum Vaccin Immunother 11(5):1235–1243

Mao H, Liu Y, Sia SF, Peiris JM, Lau Y-L, Tu W (2017) Avian influenza virus directly infects human natural killer cells and inhibits cell activity. Virologica Sinica 1–8

Marangon S, Capua I (eds) (2005) Control of AI in Italy: from "Stamping-out"-strategy to emergency and prophylactic vaccination. Proc Internat Conf on Avian Influenza, Paris

Mardanova ES, Kotlyarov RY, Kuprianov VV, Stepanova LA, Tsybalova LM, Lomonosoff GP et al (2015) Rapid high-yield expression of a candidate influenza vaccine based on the ectodomain of M2 protein linked to flagellin in plants using viral vectors. BMC Biotechnol 15 (1):42

Mardanova ES, Blokhina EA, Tsybalova LM, Peyret H, Lomonossoff GP, Ravin NV (2017) Efficient transient expression of recombinant proteins in plants by the novel pEff vector based on the genome of potato virus X. Front Plant Sci 8:247

Marsian J, Lomonossoff GP (2016) Molecular pharming—VLPs made in plants. Curr Opin Biotechnol 37:201–206

Matrosovich MN, Matrosovich TY, Gray T, Roberts NA, Klenk H-D (2004) Neuraminidase is important for the initiation of influenza virus infection in human airway epithelium. J Virol 78 (22):12665–12667

Medina RA, García-Sastre A (2011) Influenza A viruses: new research developments. Nat Rev Microbiol 9(8):590–603

Mett V, Musiychuk K, Bi H, Farrance CE, Horsey A, Ugulava N (2008) A plant-produced influenza subunit vaccine protects ferrets against virus challenge. Viruses 2

Mortimer E, Maclean JM, Mbewana S, Buys A, Williamson A-L, Hitzeroth II et al (2012) Setting up a platform for plant-based influenza virus vaccine production in South Africa. BMC Biotechnol 12(1):14

Nahampun HN, Bosworth B, Cunnick J, Mogler M, Wang K (2015) Expression of H3N2 nucleoprotein in maize seeds and immunogenicity in mice. Plant Cell Rep 34(6):969–980

Nemchinov LG, Natilla A (2007) Transient expression of the ectodomain of matrix protein 2 (M2e) of avian influenza A virus in plants. Protein Expr Purif 56(2):153–159

Neuhaus V, Chichester JA, Ebensen T, Schwarz K, Hartman CE, Shoji Y et al (2014) A new adjuvanted nanoparticle-based H1N1 influenza vaccine induced antigen-specific local mucosal and systemic immune responses after administration into the lung. Vaccine 32(26):3216–3222

Nidom CA, Takano R, Yamada S, Sakai-Tagawa Y, Daulay S, Aswadi D et al (2010) Influenza A (H5N1) viruses from pigs, Indonesia. Emerg Infect Dis 16(10):1515

Olsen B, Munster VJ, Wallensten A, Waldenström J, Osterhaus AD, Fouchier RA (2006) Global patterns of influenza A virus in wild birds. Science 312(5772):384–388

Peyret H, Lomonossoff GP (2015) When plant virology met Agrobacterium: the rise of the deconstructed clones. Plant Biotechnol J 13(8):1121–1135

Phan HT, Pohl J, Floss DM, Rabenstein F, Veits J, Le BT et al (2013) ELPylated haemagglutinins produced in tobacco plants induce potentially neutralizing antibodies against H5N1 viruses in mice. Plant Biotechnol J 11(5):582–593

Phan HT, Hause B, Hause G, Arcalis E, Stoger E, Maresch D et al (2014) Influence of elastin-like polypeptide and hydrophobin on recombinant hemagglutinin accumulations in transgenic tobacco plants. PLoS ONE 9(6):e99347

Pietrzak M, Macioła A, Zdanowski K, Protas-Klukowska AM, Olszewska M, Śmietanka K et al (2016) An avian influenza H5N1 virus vaccine candidate based on the extracellular domain produced in yeast system as subviral particles protects chickens from lethal challenge. Antiviral Res 133:242–249

Pillet S, Racine T, Nfon C, Di Lenardo TZ, Babiuk S, Ward BJ et al (2015) Plant-derived H7 VLP vaccine elicits protective immune response against H7N9 influenza virus in mice and ferrets. Vaccine. 33(46):6282–6289

Pillet S, Aubin E, Trepanier S, Bussiere D, Dargis M, Poulin JF et al (2016) A plant-derived quadrivalent virus like particle influenza vaccine induces cross-reactive antibody and T cell response in healthy adults. Clin Immunol 168:72–87

Pose AG, Gomez JN, Sanchez AV, Redondo AV, Rodriguez ER, Segui RM et al (2011) Subunit influenza vaccine candidate based on CD154 fused to HAH5 increases the antibody titers and cellular immune response in chickens. Vet Microbiol 152(3–4):328–337

Pua T-L, Chan XY, Loh H-S, Omar AR, Yusibov V, Musiychuk K et al (2017) Purification and immunogenicity of hemagglutinin from highly pathogenic avian influenza virus H5N1 expressed in Nicotiana benthamiana. Human Vaccines Immunotherapeutics 13:306–313

Pushko P, Tretyakova I, Hidajat R, Zsak A, Chrzastek K, Tumpey TM et al (2017) Virus-like particles displaying H5, H7, H9 hemagglutinins and N1 neuraminidase elicit protective immunity to heterologous avian influenza viruses in chickens. Virology 501:176–182

Ravin NV, Kotlyarov RY, Mardanova ES, Kuprianov VV, Migunov AI, Stepanova LA (2012) Plant-produced recombinant influenza vaccine based on virus-like HBc particles carrying an extracellular domain of M2 protein. Biochem (Mosc) 77(1):33–44

Redkiewicz P, Sirko A, Kamel KA, Góra-Sochacka A (2014) Plant expression systems for production of hemagglutinin as a vaccine against influenza virus. Acta Biochim Pol 61(3):551–560

Rossi L, Pinotti L, Agazzi A, Dell'Orto V, Baldi A (2014) Plant bioreactors for the antigenic hook-associated flgK protein expression. Ital J Anim Sci 13(1)

Sack M, Hofbauer A, Fischer R, Stoger E (2015) The increasing value of plant-made proteins. Curr Opin Biotechnol 32:163–170

Sainsbury F, Thuenemann EC, Lomonossoff GP (2009) pEAQ: versatile expression vectors for easy and quick transient expression of heterologous proteins in plants. Plant Biotechnol J 7:682–693

Shoji Y, Chichester JA, Bi H, Musiychuk K, de la Rosa P, Goldschmidt L et al (2008) Plant-expressed HA as a seasonal influenza vaccine candidate. Vaccine 26(23):2930–2934

Shoji Y, Farrance CE, Bi H, Shamloul M, Green B, Manceva S et al (2009a) Immunogenicity of hemagglutinin from A/Bar-headed Goose/Qinghai/1A/05 and A/Anhui/1/05 strains of H5N1 influenza viruses produced in Nicotiana benthamiana plants. Vaccine 27(25–26):3467–3470

Shoji Y, Bi H, Musiychuk K, Rhee A, Horsey A, Roy G et al (2009b) Plant-derived hemagglutinin protects ferrets against challenge infection with the A/Indonesia/05/05 strain of avian influenza. Vaccine 27(7):1087–1092

Shoji Y, Chichester JA, Jones M, Manceva SD, Damon E, Mett V (2011) Plant-based rapid production of recombinant subunit hemagglutinin vaccines targeting H1N1 and H5N1 influenza. Hum Vaccines 7:41–50

Shoji Y, Jones RM, Mett V, Chichester JA, Musiychuk K, Sun X et al (2013) A plant-produced H1N1 trimeric hemagglutinin protects mice from a lethal influenza virus challenge. Human Vaccines Immunotherapeutics 9(3):553–560

Shoji Y, Prokhnevsky A, Leffet B, Vetter N, Tottey S, Satinover S et al (2015) Immunogenicity of H1N1 influenza virus-like particles produced in Nicotiana benthamiana. Hum Vaccin Immunother 11(1):118–123

Sievert K, Avarez R, Cortada R, Valks M (2006) House flies carrying avian influenza virus (AIV). Int Pest Control 48(3):114–116

Sims L, Domenech J, Benigno C, Kahn S, Kamata A, Lubroth J et al (2005) Origin and evolution of highly pathogenic H5N1 avian influenza in Asia. Vet Rec 157(6):159

Spackman E (2008) A brief introduction to the avian influenza virus. Avian Influenza Virus 1–6

Spitsin S, Andrianov V, Pogrebnyak N, Smirnov Y, Borisjuk N, Portocarrero C et al (2009) Immunological assessment of plant-derived avian flu H5/HA1 variants. Vaccine 27(9):1289–1292

Stachyra A, Pietrzak M, Maciola A, Protasiuk A, Olszewska M, Smietanka K et al (2017) A prime/boost vaccination with HA DNA and Pichia-produced HA protein elicits a strong humoral response in chickens against H5N1. Virus Res 232:41–47

Stephenson I, Wood J, Nicholson K, Zambon M (2003) Sialic acid receptor specificity on erythrocytes affects detection of antibody to avian influenza haemagglutinin. J Med Virol 70 (3):391–398

Suguitan AL Jr, McAuliffe J, Mills KL, Jin H, Duke G, Lu B et al (2006) Live, attenuated influenza A H5N1 candidate vaccines provide broad cross-protection in mice and ferrets. PLoS Med 3(9):e360

Topp E, Irwin R, McAllister T, Lessard M, Joensuu JJ, Kolotilin I et al (2016) The case for plant-made veterinary immunotherapeutics. Biotechnol Adv 34:597–604

Treanor JJ, Chu L, Essink B, Muse D, El Sahly HM, Izikson R et al (2017) Stable emulsion (SE) alone is an effective adjuvant for a recombinant, baculovirus-expressed H5 influenza vaccine in healthy adults: a Phase 2 trial. Vaccine 35(6):923–928

Vamvaka E, Twyman RM, Murad AM, Melnik S, Teh AYH, Arcalis E et al (2016) Rice endosperm produces an underglycosylated and potent form of the HIV-neutralizing monoclonal antibody 2G12. Plant Biotechnol J 14(1):97–108

Van Droogenbroeck B, Cao J, Stadlmann J, Altmann F, Colanesi S, Hillmer S et al (2007) Aberrant localization and underglycosylation of highly accumulating single-chain Fv-Fc antibodies in transgenic Arabidopsis seeds. Proc National Acad Sci 104(4):1430–1435

Veits J, Wiesner D, Fuchs W, Hoffmann B, Granzow H, Starick E et al (2006) Newcastle disease virus expressing H5 hemagglutinin gene protects chickens against Newcastle disease and avian influenza. Proc National Acad Sci 103(21):8197–8202

Velkers FC, Blokhuis SJ, Veldhuis Kroeze EJ, Burt SA (2017) The role of rodents in Avian Influenza outbreaks in poultry farms: a review. Vet Q 37(1):182–194

Vittecoq M, Gauduin H, Oudart T, Bertrand O, Roche B, Guillemain M et al (2017) Modeling the spread of avian influenza viruses in aquatic reservoirs: a novel hydrodynamic approach applied to the Rhône delta (southern France). Sci Total Environ 595:787–800

Wanaratana S, Panyim S, Pakpinyo S (2011) The potential of house flies to act as a vector of avian influenza subtype H5N1 under experimental conditions. Med Vet Entomol 25(1):58–63

Ward BJ, Landry N, Trépanier S, Mercier G, Dargis M, Couture M et al (2014) Human antibody response to N-glycans present on plant-made influenza virus-like particle (VLP) vaccines. Vaccine 32(46):6098–6106

Yang W-T, Yang G-L, Wang Q, Huang H-B, Jiang Y-L, Shi C-W et al (2017) Protective efficacy of Fc targeting conserved influenza virus M2e antigen expressed by Lactobacillus plantarum. Antiviral Res 138:9–21

Zheng M, Luo J, Chen Z (2014) Development of universal influenza vaccines based on influenza virus M and NP genes. Infection 42(2):251–262

Plant-Made Vaccines Against Avian Reovirus

Ching-Chun Chang and Hung-Jen Liu

Abstract Viral arthritis, chronic respiratory disease, malabsorption syndrome, hepatitis and myocarditis cause considerable economic losses in the poultry industry and are directly or indirectly linked to avian reovirus (ARV). The outer capsid proteins of the ARV, σC and σB, encoded by the S1 and S3 genome segments, respectively, are the prime candidates for development of a subunit vaccine against ARV infection. To develop a plant-based vaccine for immunizing poultry against ARV infection, transgenic alfalfa, Arabidopsis and tobacco expressing recombinant σC protein have been generated. Transgenic Arabidopsis with the highest expression of σC were used to subcutaneously or orally vaccinate specific-pathogen free (SPF) chickens. The Arabidopsis-expressed recombinant σC protein could induce immune response in chickens and conferred significant protection against viral challenge. Future investigations to improve the σC and/or σB protein levels in edible plants will pave the way for commercially large-scale vaccination against ARVs in the poultry industry.

Keywords Avian reovirus · σC protein · σB protein · Transgenic plants
Plant-based vaccine

C.-C. Chang (✉)
Department of Biotechnology and Bioindustry Sciences, National Cheng Kung University, Tainan 701, Taiwan
e-mail: chingcc@mail.ncku.edu.tw

H.-J. Liu
Institute of Molecular Biology, National Chung Hsing University, Taichung 402, Taiwan

H.-J. Liu
Agricultural Biotechnology Center, National Chung Hsing University, Taichung 402, Taiwan

H.-J. Liu
Rong Hsing Research Center for Translational Medicine, National Chung Hsing University, Taichung 402, Taiwan

J. MacDonald (ed.), *Prospects of Plant-Based Vaccines in Veterinary Medicine*,
https://doi.org/10.1007/978-3-319-90137-4_9

1 Genome, Protein Coding Assignment, Disease Symptoms and Occurrence

Avian reovirus (ARV) belongs to the genus *Orthoreovirus*, one of the 12 genera of the Reoviridae family (Attoui et al. 2000; Mertens 2004). This genus includes the non-fusogenic mammalian reoviruses; the two fusogenic mammalian reoviruses, Nelson Bay virus and baboon reovirus; as well as various fusogenic viruses isolated from reptilian and avian species (Mertens 2004). Avian and mammalian reoviruses are the two major groups of the Orthoreovirus genus. Although they share many morphological and physicochemical properties, they differ in coding assignment, pathogenicity, host range, and various biological and serological characteristics (Spandidos and Graham 1976; Schnitzer 1985; Rosenberger et al. 1989; Varela and Benavente 1994; Duncan and Sullivan 1998; Bodelón et al. 2001). ARVs differ from their mammalian counterparts particularly in lack of hemagglutination activity (Glass et al. 1973), the ability to induce cell fusion (Bodelón et al. 2001) and an association with naturally occurring pathological conditions (Robertson and Wilcox 1986).

ARVs are important pathogens of birds that can cause considerable economic losses in the poultry industry (Glass et al. 1973; Rosenberger et al. 1989). They were initially found to induce tenosynovitis in young chickens (Olson 1978) and were later found ubiquitous among poultry flocks, although most ARVs cause asymptomatic infections in poultry. Several diseases caused by ARVs include viral arthritis (tenosynovitis), chronic respiratory diseases, malabsorption syndrome, hepatitis, and myocarditis (Glass et al. 1973; Jones and Onunkwo 1978; Jones and Kibenge 1984; Jones and Georgiou 1984; Rosenberger et al. 1989; Jones 2000), but a direct link between the virus and disease has only been conclusively demonstrated for viral arthritis syndrome, which is characterized by swelling of the hock joints and lesions in the gastrocnemius tendons. Van de Zande and Kuhn demonstrated central nervous system signs on infecting specific-pathogen free (SPF) White Leghorn chicks with a new ARV strain belonging to the so-called enteric reovirus strains (Van de Zande and Kuhn 2007). Previous reports suggest that viral arthritis occurs occasionally in breeds that lay light-colored eggs, and that broiler chicks are more susceptible to reovirus arthritis than SPF light hybrids or commercial White Leghorns (Jones and Kibenge 1984).

Chicken susceptibility to ARV infection is clearly age-dependent, with older birds being more resistant than younger birds to infection and virus-induced lesions (Jones and Georgiou 1984). Chicks infected at 1 day old were more susceptible to experimentally induced tenosynovitis than those infected at age 2 weeks or older (Jones and Georgiou 1984). ARVs can persist in tissues of chickens for many weeks (Kerr and Olson 1969; Jones and Onunkwo 1978). Both vertical and horizontal transmission of ARVs are recognized (Menendez et al. 1975; van der Heide and Kalbac 1975; Jones and Onunkwo 1978; Al-Muffarej et al. 1996), although vertical transmission of ARVs usually occurs at a low rate (Menendez et al. 1975; Al-Muffarej et al. 1996). Most chicks appear to be infected at an early age through the fecal-oral route (Jones and Onunkwo 1978). In addition, Muscovy ducks can be affected by Muscovy duck reovirus (MDRV), which causes high morbidity and mortality. Apart from the differences in coding assignment, clinical virulence is greater with MDRV than ARV. Newly evolved types of MDRVs cause hemorrhagic-necrotic lesions in the liver and spleen of sick birds and increase morbidity and mortality (Chen et al. 2009; Yuan et al. 2013; Yun et al. 2014).

ARVs are non-enveloped viruses that replicate in the cytoplasm of infected cells. They have icosahedral symmetry and contain a double-shelled arrangement of surface protein. Virus particles can range from 70 to 80 nm. The viruses are stable between pH 3 and 9, and ambient temperatures are suitable for the survival of these ARVs, which become inactive at 56 °C in less than 1 h. Common areas where ARVs can survive include feathers, wood shavings, metal, glass, and rubber; they can survive for up to 10 days in these areas and up to 10 weeks in water (Jones 2000).

The genome consists of 10 double-stranded RNA (dsRNA) genome segments divided into 3 size classes, L (large), M (medium), and S (small), on the basis of their electrophoretic mobility (Spandidos and Graham 1976). There are three segments in the L-class (L1, L2, L3), three in the M-class (M1, M2, M3) and four in the S-class (S1, S2, S3, S4) (Spandidos and Graham 1976). The electrophoretic migration patterns of the genomic segments of ARV isolates show considerable polymorphism (Rekik et al. 1990). Among the 12 proteins encoded by the genome, four are non-structural (μNS, σNS, p17, and p10), and the remaining eight are structural (λA, λB, λC, μA, μB, σC, σA, σB) (Schnitzer 1985; Varela and Benavente 1994; Martínez-Costas et al. 1997). The lengths of these structural and non-structural proteins are as follows: λA (1293 amino acids), λB (1259 aa), λC (1285 aa), μA (732 aa), μB (676 aa), μNS (635 aa), σC (326 aa), σA (416 aa), σB (367 aa), and σNS (367 aa), p17 (146 aa), and p10 (98 aa). A phylogenetic analysis based on the sequences of the S-class genome segments grouped ARVs into six genotyping clusters (Liu et al. 2003). However, no correlation could be established between a particular genotype and the disease condition. Liu et al. (2003) suggested that genetic variability among circulating ARVs is via a combination of evolutionary mechanisms involving genetic reassortment and multiple co-circulating lineages (Liu et al. 2003).

2 Mechanism of Infection and Virus-Host Interaction

To initiate a productive infection, all viruses must cross cell membranes to express their genome inside the cells. Virus penetration can occur by fusing with the plasma membrane directly to release viral capsids into the cytosol or by endocytosis (Marsh and Helenius 2006). Enveloped viruses are able to enter host cells by fusing their lipidic envelopes with the cytoplasmic membrane (Kielian and Jungerwirth 1990), whereas non-enveloped viruses use alternative entry mechanisms. To infect host cells, viruses need to bind the cell surface, which is followed by signaling induction to facilitate their entry into host cells. Changes in viral particles, including proteolysis and/or conformational modifications of specific capsid proteins, are required for ARVs to acquire membrane-crossing ability (Marsh and Helenius 1989). ARVs initially attach to host cell receptors via the attachment protein σC (Shapouri et al. 1996), triggering receptor-mediated endocytosis (Huang et al. 2011). Recently, Huang and colleagues suggested that cell entry of ARVs follows caveolin-1-mediated and dynamin-2-dependent endocytosis pathways, which need to activate the multiple signaling pathways of Ras, p38 mitogen-activated protein kinase (MAPK), and Src to facilitate virus entry (Huang et al. 2011). After entry, ARVs require the host-encoded small GTPase Rab5 and microtubules to regulate transport to the early endosome for infection (Huang et al. 2011). Acidification of virus-containing endosomes is required for uncoating and replicating within the host cells (Duncan 1996; Huang et al. 2011). ARV uncoating mainly depends on endosome-mediated proteolytic processing of the major outer capsid protein μBC into two specific products (Duncan 1996), and lysosomal proteases may play a key role in this event (Duncan 1996; Huang et al. 2011). After intraendosomal reovirion uncoating, viral cores are released into the cytoplasm to initiate transcription of the dsRNA genome segments, which produce 10 viral mRNAs. These viral mRNAs are translated at the ribosomes into viral proteins which are recruited into newly formed core particles and used as templates for the synthesis of the viral minus strands to form progeny genomes. After completion of viral morphogenesis, the formation of mature reovirions exits the infected host cells, thus causing cell lysis. Lysis is mediated at late infection by the ARV viroporin p10 protein, which can permeabilize membranes (Bodelón et al. 2002), thereby facilitating the release of viral particles and the dissemination of the viral infection.

The ARV non-structural p10 protein appears to be a multifunctional protein that plays key roles in virus–host interaction. In addition to cell lysis, several reports suggest that p10 displays syncytium-inducing activity when expressed in virus-infected or transfected cells (Bodelón et al. 2001; Shmulevitz and Duncan 2000). Such cell-cell fusion allows ARV to spread to neighboring uninfected without exposure to the immune defenses of the host (Bodelón et al. 2001; Shmulevitz and Duncan 2000). Liu et al. demonstrated that ARVs encode this non-structural protein p10, which mediates cell syncytium formation via activation of small GTPase, RhoA, and Rac1 signaling (Liu et al. 2008). Syncytium formation in ARV-infected cells appears to be regulated by inefficient synthesis and rapid degradation of the ARV p10 protein (Bodelón et al. 2001; Shmulevitz et al. 2004). Because of their small size, unusual arrangement of structural motifs and absence of any similarity with other membrane fusion proteins, reovirus fusion proteins have been classified into a new family of fusion-associated small transmembrane (FAST) proteins (Shmulevitz and Duncan 2000). Cell–cell fusion still is important for ARV replication in infected animals, by permitting virus spread to neighboring uninfected cells without exposure to the immune defenses of the host. In addition, the ARV p10 protein is a viroporin that plays a key role in the modification of late membrane permeability (Bodelón et al. 2002). In addition to causing cell–cell fusion (Shmulevitz et al. 2004), the ARV p10 protein can permeabilize membranes

(Bodelón et al. 2002). The membrane-destabilizing activity of p10 protein causes cell lysis at late infection, thereby facilitating the release of viral particles and the dissemination of the viral infection. The ARV p10 protein appears to be a multi-functional protein that plays key roles in virus–host interaction.

ARV can also lead to cell death through apoptosis. Understanding the molecular basis of the ARV-host interaction and ARV-induced changes can shed light on normal cellular events and how specifically ARVs gain control over the host. Apoptosis plays an important role in the pathogenesis of many viral infections. It is a major mechanism of cell death induced by ARVs (Labrada et al. 2002; Shih et al. 2004; Lin et al. 2006, 2007, 2009; Chulu et al. 2007). ARVs have provided insight into the roles that specific viral genes and the proteins they encode play in virus-induced cell death and tissue injury (Labrada et al. 2002; Shih et al. 2004; Lin et al. 2007, 2009, 2011). ARVs and the ARV σC protein induce apoptosis by modulating the DAN damage signaling, Src, p53, mitogen-activated protein kinase (MAPK), and protein kinase C delta and also elicit cytochrome c release from mitochondria to the cytosol (Lin et al. 2006, 2009, 2011; Chulu et al. 2007). Recently, it has been demonstrated that ARV S1133-induced apoptosis is associated with Bip/GRP79-mediated Bim translocation to the endoplasmic reticulum (Lin et al. 2015a). In the early stages of ARV S1133 infection, activation of cell survival signals contributes to virus-induced inflammation and anti-apoptotic response (Lin et al. 2010). ARV appears to trigger the PI3-kinase/Akt/NF-kappaB and signal transducer and activator of transcription (STAT) 3 signaling pathways in the early stages of infection to result in an inflammatory response and delayed apoptosis (Lin et al. 2010). At late infection, large syncytia show syncytium formation in ARV-infected cells undergoing apoptosis (Chulu et al. 2007). The pathogenesis of ARV-induced apoptosis has been investigated in ARV-infected chicken tissues, including the tendon, liver, intestine, and bursa (Lin et al. 2007). Lin et al. demonstrated that infecting chickens with ARV S1133 caused tissue injury related to virus-induced apoptosis (Lin et al. 2007). Syncytium formation in ARV-infected chicken tissues and cultured cells undergoes apoptosis, which suggests a correlation between virus replication and apoptosis (Chulu et al. 2007; Lin et al. 2007). Furthermore, ARV-induced apoptosis could enhance both virus spreading and the processing of the viral non-structural muNS protein (Rodríguez-Grille et al. 2014).

Recently, the ARV nonstructural protein p17 was found to function as an activator of autophagy, a process that targets cytoplasmic constituents to the lysosome for degradation. Autophagy is an essential process in the control of cellular homeostasis, enabling cells to survive under certain stress conditions by removing toxic cellular components. It may also protect cells against apoptosis. However, autophagy can be beneficial for virus replication via activation of the protein phosphatase and tensin deleted on chromosome 10 (PTEN), AMP-activated protein kinase (AMPK), and dsRNA-dependent protein kinase (PKR)/eIF2 signaling pathways (Chi et al. 2013). ARV S1133 infection caused autophagy in early to middle infection stages in Vero and DF1 cell lines and apoptosis in the middle to late stages (Shih et al. 2004; Lin et al. 2010; Chi et al. 2013). Lin and colleagues further demonstrated that ARV S1133 strain regulates the switch from autophagy to apoptosis via activation of the RhoA/ROCK1 signaling pathway (Lin et al. 2015b). Recently, several reports have shown that p17-mediated Tpr suppression positively regulates p53, PTEN, and p21 and negatively regulates the PI3 K/AKT/mammalian target of rapamycin (mTOR) and extracellular signal-regulated kinase (ERK) signaling pathways, both causing host translation shutoff and cell cycle arrest, inducing autophagy and benefiting virus replication (Chiu et al. 2016; Huang et al. 2015). The ARV p17 non-structural protein appears to facilitate virus replication via induction of autophagy and cell cycle arrest as well as cellular translation shutoff (Liu et al. 2005; Ji et al. 2009; Chi et al. 2013; Huang et al. 2015; Chiu et al. 2016), thus diverting the cellular machinery required for normal cell-cycling processes for virus replication.

ARVs are highly resistant to the antiviral action chicken cytokine interferon (IFN), so these viruses use strategies to counteract the antiviral action of this cytokine (Ellis et al. 1983). IFNs binding to specific cell-surface receptors trigger the activation of signal transduction pathways that lead to increased expression of more than 30 proteins, some thought to play a critical role in combating viral infection (Goodbourn et al. 2000; Sen 2001; Haller et al. 2006). Furthermore, González-López and colleagues showed that intracellular expression of the ARV σA protein interferes with PKR function by preventing its activation. This protein may disrupt the IFN-induced cellular response against ARV by blocking enzyme pathways that depend on dsRNA. These findings suggest that σA is a virus-encoded dsRNA-binding protein that confers IFN resistance to viruses by sequestering dsRNA activators (González-López et al. 2003).

3 Plant-Made Vaccine Candidates: Rationale, Efficacy, and Yield

The σC protein is one of the outer capsid proteins encoded by the *S1* genome segment, with molecular weight ranging from 35 to 39 kD, depending on the virus strain. In addition, σC forms a homotrimer in its native state, and its C-terminal receptor-binding domain is responsible for host cell attachment (Grande et al. 2002; Guardado Calvo et al. 2005; Benavente and Martínez-Costas 2007). Furthermore, σC is involved in the induction of apoptosis of host cells (Shih et al. 2004; Lin et al. 2006, 2007, 2009; Chulu et al. 2007). Although σC is a minor capsid protein, it is one of the major antigens responsible for inducing neutralizing antibodies in host cells against the viral infection (Wickramasinghe et al. 1993; Benavente and Martínez-Costas 2007). Therefore, σC has become one of the prime candidates for producing a subunit vaccine against ARV infection. The expression of σC protein has been reported in different platforms such as *Escherichia coli* (Shapouri et al. 1995; Liu et al. 2002), lactic acid bacteria *Enterococcus faecium* (Lin et al. 2012), baculovirus/insect cells (Hu et al. 2002; Lin et al. 2008), yeast (Wu et al. 2005), and plants (Huang et al. 2006; Wu et al. 2009; Lu et al. 2011). In addition, the use of σC DNA vaccine mediated by attenuated *Salmonella typhimurium* induces a protective antibody in chickens (Wan et al. 2011, 2012).

Previously, recombinant σC protein of ARV from yeast was found to be resistant to gut degradation and induced greater immune response than a commercial vaccine when incorporated into feed at a high dose (250 μg) (Wu et al. 2005). This study supports the feasibility of a plant-based oral ARV vaccine for the poultry industry, since the cell walls of plants might similarly protect recombinant proteins from degradation within the gut.

Plant-based vaccines have several potential advantages over traditional vaccines because they are cheap, safe, and easy to store and scale up. In addition, they could be orally delivered, thus reducing the cost of syringes and needles and eliminating injection anxiety (Streatfield and Howard 2003; Aswathi et al. 2014). To investigate the possibility of developing a plant-based vaccine against ARV infection for the poultry industry, the ARV σC protein has been expressed in transgenic alfalfa (Huang et al. 2006), Arabidopsis (Wu et al. 2009) and tobacco (Lu et al. 2011) via *Agrobacteria*-mediated transformation (Table 1). In alfalfa, two different constitutive promoters, one from CaMV 35S and the other from rice actin gene, were used to drive the expression of the σC protein, but the protein level was relatively low, only up to 0.008 and 0.007%, respectively, of total soluble protein (TSP) (Huang et al. 2006). To improve the expression of σC protein in plants, codon optimization of the ARV *S1* gene was attempted by increasing the more frequently used codons in plants and eliminating the putative cryptic introns as well as the mRNA instability elements (Lu et al. 2011). In total, 61 nucleotides (6.2%) in 45 codons were optimized to tobacco codon usage (Lu et al. 2011). To investigate the effects after *S1* segment codon modification, four distinct plant expression vectors were constructed to generate transgenic tobacco. Transgenic tobacco was created by the expression of *S1* and codon-optimized *S1* (*mS1*) genes driven by CaMV 35S and/or rice actin gene promoters (Table 1). In addition, the expressed σC proteins from the *mS1* gene were accumulated in cytosol or chloroplasts. The highest σC protein expression in the transgenic *S1* tobacco lines driven by the CaMV 35S or actin gene promoter was 0.013 and 0.021% of TSP, respectively. However, the highest σC protein expression accumulated in the cytoplasm of transgenic *mS1* tobacco lines was 0.0013% of TSP, but σC protein was barely detectable in the transgenic *mS1* tobacco lines when the σC protein was accumulated in chloroplasts (Table 1). Surprisingly, codon modification of the *S1* gene did not significantly increase σC protein expression in transgenic tobacco lines. In addition, subcellular targeting of σC protein to chloroplasts did not increase the expression, although a significant amount of mRNA was detected (Lu et al. 2011). Although strong constitutive promoters (CaMV 35S and rice actin) were used to drive *S1* and/or m*S1* gene expression, neither promoter could significantly increase σC protein level in transgenic tobacco and alfalfa (Huang et al. 2006; Lu et al. 2011). These results suggest that factors other than promoter strength and codon usage might be the major obstacles to expressing σC protein in both alfalfa and tobacco (Lu et al. 2011). Ribosome stuttering resulting from the secondary structure of *S1* or *mS1*

Table 1 Development of plant-based vaccines against ARVs

Expression constructs	Plants	Protein/ gene[a]	Promoter/ subcellular localization	Expression level (% TSP)	Protection efficiency	References
1	Alfalfa	σC/*S1*	CaMV 35S/ cytosol	0.008	NA	Huang et al. (2006)
2	Alfalfa	σC/*S1*	Rice actin/ cytosol	0.007	NA	Huang et al. (2006)
3	Arabidopsis	σC/*S1*	Agrobacteria octopine synthetase/ cytosol	4.9	Subcutaneously (90%) or orally (70%)	Wu et al. (2009)
4	Tobacco	σC/*S1*	CaMV 35S/ cytosol	0.013	NA	Lu et al. (2011)
5	Tobacco	σC/*S1*	Rice actin/ cytosol	0.021	NA	Lu et al. (2011)
6	Tobacco	σC/ *mS1*[a]	CaMV 35S/ cytosol	0.0013	NA	Lu et al. (2011)
7	Tobacco	σC/ *mS1*[a]	CaMV 35S/ chloroplast	Not detectable	NA	Lu et al. (2011)

[a]*mS1*, codon-modified *S1* gene. NA, not available

mRNA might cause a major problem in the low translational efficiency of σC protein in plants (Lu et al. 2011). Alternatively, the instability of σC protein or insufficient assembly of the σC trimer might result in the low protein level in transgenic tobacco and alfalfa (Lu et al. 2011).

In Arabidopsis, the expression of σC protein could reach up to 4.9% of TSP in transgenic plants with *S1* gene driven by a strong octopine synthetase gene promoter from *Agrobacterium tumefaciens* and fused with the translational leader sequence from tobacco etch virus (Table 1; Wu et al. 2009). In addition, the higher level of σC protein expression was consistent with increased abundance of the *S1* transcripts in plants (Wu et al. 2009). This study showed that a combinatorial design on the basis of promoter strength and translational leader sequence to drive a gene expression cassette might facilitate the accumulation of σC protein in transgenic plants.

4 Comparison to Established Vaccines

No animal immunized experiments have tested the efficacy of σC proteins expressed from alfalfa or tobacco due to the low expression in plants (Huang et al. 2006; Lu et al. 2011). However, immunization and the following virus challenge have been worked out with σC proteins expressed from Arabidopsis. One-week-old

SPF chickens receiving Arabidopsis-expressed σC protein (125 μg) orally at weekly intervals for 3 consecutive weeks showed 70% protection against virus challenge, whereas broilers that subcutaneously received the plant-expressed recombinant σC protein three times at weekly intervals showed 90% protection (Wu et al. 2009). In contrast, 1-week-old chickens that received the commercial vaccine VaVac (Fort Dodge Animal Health Inc., Overland Park, KS) by eye and nose drop once showed 80% protection but with a booster with plant-expressed recombinant protein twice at weekly intervals showed up to 90% protection, with the lowest lesion scores (Wu et al. 2009). In addition, a significant T helper 2 cell-mediated immune response was conferred by broilers after oral delivery of recombinant σC protein (Wu et al. 2009). These results demonstrate that the plant-based σC protein has great potential for large-scale vaccination against ARV in commercial poultry.

5 Pathway to Commercialization

One of the major hurdles in developing a plant-based vaccine is the low antigen expression in transgenic plants (Streatfield and Howard 2003). The low foreign protein expression in alfalfa and tobacco may be due to factors affecting the transcriptional level such as promoter strength and/or post-transcriptional level—mRNA stability, translational efficiency, and protein stability. Previously, with the combination of the strong octopine synthetase gene superpromoter and a translational leader sequence from tobacco etch virus, the σC expression could reach up to 4.9% of TSP in Arabidopsis (Wu et al. 2009); however, whether the expression cassette could confer the same high expression in high-yield edible plants still needs to be demonstrated.

Expression of ARV antigen in transplastomic chloroplasts of edible plants is an alternative strategy. The chloroplast expression system offers several advantages for vaccine production such as high expression, polycistronic expression for multivalent antigens and low transgene escape to the environment. Many successful examples of vaccination via chloroplasts have been reported (Daniell et al. 2016). However, chloroplast transformation technology requires highly skilled labor and is more time-consuming and costly than nuclear transformation. This technology has been successfully applied in only about two dozen species of higher plants (Bock 2015), and only a few reports exist for most plant species except tobacco.

While immunization with *Arabidopsis*-expressed σC led to some protection against viral challenge, further improvements in the antigenicity of plant-made ARV vaccines may be possible. Previously, the structure of the $\sigma C^{151-326}$ C-terminal fragment, which contains the receptor-binding globular domain, has been solved (Guardado Calvo et al. 2005) and might facilitate vaccine development. Recently, bacterially expressed truncated $\sigma C^{122-326}$ fragment consisting of the globular head, shaft and hinge domains, excluding the intra-capsular region, could significantly induce higher levels of anti-ARV antibodies than the shorter $\sigma C^{192-326}$

fragment or full-length σC. Furthermore, the antibodies induced by $\sigma C^{122-326}$ were the most successful at neutralizing ARV infection in embryos (Goldenberg et al. 2016). The $\sigma C^{122-326}$ fragment was hypothesized to exclude the repressor region and fold correctly and to expose linear as well as conformational epitopes identical to those of the native protein (Goldenberg et al. 2016). Therefore, the future expression of the $\sigma C^{122-326}$ fragment in edible plants might be better used for developing a plant-based ARV vaccine in the commercial poultry industry.

ARV outer capsids consist of multiple copies of at least four distinct polypeptides, σB, σC, μB and μBC (Benavente and Martínez-Costas 2007). Development of plant-based multivalent subunit vaccines by co-expressing σB and σC might be feasible to increase the immune response. The σB protein encoded by the *S3* genome segment is a major component of the ARV outer capsid (Benavente and Martínez-Costas 2007). A previous report showed that an equal mixture of σC and σB proteins expressed from the surface of baculovirus in insect cells could induce higher levels of virus neutralization antibody than either alone in mice (Lin et al. 2008). In addition, the combination of σC and σB DNA vaccine mediated by attenuated *Salmonella typhimurium* could confer higher antibody production in chicken than either alone (Wan et al. 2012). The studies demonstrated the advantages of multivalent subunit vaccine in conferring an immune response against ARV infection.

Alternatively, noninfectious virus-like particles (VLPs) are very stable and confer a strong immune response (Scotti and Rybicki 2013). Plant-based expression of complex VLPs has been demonstrated in rotavirus and bluetongue virus, whose capsids consist of multiple distinct polypeptides (Scotti and Rybicki 2013). Therefore, the self-assembly of ARV-like particles in plants might be feasible by stoichiometrically co-expressing σB, σC and other capsid proteins together. Although we lack the detailed assembly mechanism of ARV particles, the combinatorial and transient expression of distinct capsid proteins in *N. benthamiana* via agroinfiltration could be rapidly used to investigate the formation of virus-like particles in plants.

In conclusion, although transgenic plants provide several advantages for the development of subunit vaccines, the low expression of antigens in commercial plants currently limits their practical application. Future investigations could involve improving σC or/and σB protein expression by chloroplast transformation or fusing the antigenic gene to a strong translational enhancer. Alternatively, enhancing protein stability and the immune response by expressing the C-terminal receptor binding domain of σC ($\sigma C^{122-326}$) or fusing antigens with an adjuvant protein such as cholera toxin B subunit in edible crops will be an important issue. In addition, plant-based multivalent vaccines that simultaneously express the stoichiometrical combination of σB, σC, and other ARV antigens to form stable virus-like particles in edible crops is an alternative strategy. These studies will pave the way to commercialize plant-based ARV oral vaccines for controlling infectious diseases in the poultry industry.

References

Al-Muffarej SI, Savage CE, Jones RC (1996) Egg transmission of avian reoviruses in chickens: comparison of a trypsin-sensitive and a trypsin-resistant strain. Avian Pathol 25(3):469–480. https://doi.org/10.1080/03079459608419156

Aswathi PB, Bhanja SK, Yadav AS, Rekha V, John JK, Gopinath D, Sadanandan GV, Shinde A, Jacob A (2014) Plant based edible vaccines against poultry diseases: a review. Adv Anim Vet Sci 2(5):305–311. https://doi.org/10.14737/journal.aavs/2014/2.5.305.311

Attoui H, Billoir F, Biagini P, de Micco P, de Lamballerie X (2000) Complete sequence determination and genetic analysis of Banna virus and Kadipiro virus: proposal for assignment to a new genus (Seadornavirus) within the family Reoviridae. J Gen Virol 81(Pt 6):1507–1515. https://doi.org/10.1099/0022-1317-81-6-1507

Benavente J, Martínez-Costas J (2007) Avian reovirus: structure and biology. Virus Res 123 (2):105–119. https://doi.org/10.1016/j.virusres.2006.09.005

Bock R (2015) Engineering plastid genomes: methods, tools, and applications in basic research and biotechnology. Ann Rev Plant Biol 66:211–241. https://doi.org/10.1146/annurev-arplant-050213-040212

Bodelón G, Labrada L, Martinez-Costas J, Benavente J (2001) The avian reovirus genome segment S1 is a functionally tricistronic gene that expresses one structural and two nonstructural proteins in infected cells. Virology 290(2):181–191. https://doi.org/10.1006/viro.2001.1159

Bodelón G, Labrada L, Martinez-Costas J, Benavente J (2002) Modification of late membrane permeability in avian reovirus-infected cells—viroporin activity of the S1-encoded nonstructural p10 protein. J Biol Chem 277(20):17789–17796. https://doi.org/10.1074/jbc.M202018200

Chen SY, Chen SL, Lin FQ, Jiang B, Wang S, Cheng XX, Zhu XL, Zhang SZ, Li ZL, Cheng YQ (2009) The primary study of pathogen of duck hemorrhagic-necrotic hepatitis. Chin Agric Sci Bull 25:28–31

Chi PI, Huang WR, Lai IH, Cheng CY, Liu HJ (2013) The p17 nonstructural protein of avian reovirus triggers autophagy enhancing virus replication via activation of phosphatase and tensin deleted on chromosome 10 (PTEN) and AMP-activated protein kinase (AMPK), as well as dsRNA-dependent protein kinase (PKR)/eIF2alpha signaling pathways. J Biol Chem 288 (5):3571–3584. https://doi.org/10.1074/jbc.M112.390245

Chiu HC, Huang WR, Liao TL, Wu HY, Munir M, Shih WL, Liu HJ (2016) Suppression of vimentin phosphorylation by the avian reovirus p17 through inhibition of CDK1 and Plk1 impacting the G2/M phase of the cell cycle. PLoS ONE 11(9):e0162356. https://doi.org/10.1371/journal.pone.0162356

Chulu JL, Lee LH, Lee YC, Liao SH, Lin FL, Shih WL, Liu HJ (2007) Apoptosis induction by avian reovirus through p53 and mitochondria-mediated pathway. Biochem Biophys Res Commun 356(3):529–535. https://doi.org/10.1016/j.bbrc.2007.02.164

Daniell H, Chan HT, Pasoreck EK (2016) Vaccination via chloroplast genetics: affordable protein drugs for the prevention and treatment of inherited or infectious human diseases. Ann Rev Genet 50:595–618. https://doi.org/10.1146/annurev-genet-120215-035349

Duncan R (1996) The low pH-dependent entry of avian reovirus is accompanied by two specific cleavages of the major outer capsid protein mu 2C. Virology 219(1):179–189. https://doi.org/10.1006/viro.1996.0235

Duncan R, Sullivan K (1998) Characterization of two avian reoviruses that exhibit strain-specific quantitative differences in their syncytium-inducing and pathogenic capabilities. Virology 250 (2):263–272. https://doi.org/10.1006/viro.1998.9371

Ellis MN, Eidson CS, Brown J, Kleven SH (1983) Studies on interferon induction and interferon sensitivity of avian reoviruses. Avian Dis 27(4):927–936. https://doi.org/10.2307/1590194

Glass SE, Naqi SA, Hall CF, Kerr KM (1973) Isolation and characterization of a virus associated with arthritis of chickens. Avian Dis 17(2):415–424. https://doi.org/10.2307/1589226

Goldenberg D, Lublin A, Rosenbluth E, Heller ED, Pitcovski J (2016) Optimized polypeptide for a subunit vaccine against avian reovirus. Vaccine 34(27):3178–3183. https://doi.org/10.1016/j.vaccine.2016.04.036

González-López C, Martinez-Costas J, Esteban M, Benavente J (2003) Evidence that avian reovirus sigma A protein is an inhibitor of the double-stranded RNA-dependent protein kinase. J Gen Virol 84:1629–1639. https://doi.org/10.1099/vir.0.19004-0

Goodbourn S, Didcock L, Randall RE (2000) Interferons: cell signalling, immune modulation, antiviral responses and virus countermeasures. J Gen Virol 81:2341–2364

Grande A, Costas C, Benavente J (2002) Subunit composition and conformational stability of the oligomeric form of the avian reovirus cell-attachment protein sigmaC. J Gen Virol 83(Pt 1):131–139. https://doi.org/10.1099/0022-1317-83-1-131

Guardado Calvo P, Fox GC, Hermo Parrado XL, Llamas-Saiz AL, Costas C, Martinez-Costas J, Benavente J, van Raaij MJ (2005) Structure of the carboxy-terminal receptor-binding domain of avian reovirus fibre sigmaC. J Mol Biol 354(1):137–149. https://doi.org/10.1016/j.jmb.2005.09.034

Haller O, Kochs G, Weber F (2006) The interferon response circuit: induction and suppression by pathogenic viruses. Virology 344(1):119–130. https://doi.org/10.1016/j.virol.2005.09.024

Hu YC, Liu HJ, Chung YC (2002) High level expression of the key antigenic protein, sigma C, from avian reovirus into insect cells and its purification by immobilized metal affinity chromatography. Biotechnol Lett 24(12):1017–1022. https://doi.org/10.1023/A:1015689313560

Huang LK, Liao SC, Chang CC, Liu HJ (2006) Expression of avian reovirus sigmaC protein in transgenic plants. J Virol Methods 134(1–2):217–222. https://doi.org/10.1016/j.jviromet.2006.01.013

Huang WR, Chiu HC, Liao TL, Chuang KP, Shih WL, Liu HJ (2015) Avian reovirus protein p 17 functions as a nucleoporin Tpr suppressor leading to activation of p53, p21 and PTEN and inactivation of PI3 K/AKT/mTOR and ERK signaling pathways. Plos One 10 (8). https://doi.org/10.1371/journal.pone.0133699

Huang WR, Wang YC, Chi PI, Wang L, Wang CY, Lin CH, Liu HJ (2011) Cell entry of avian reovirus follows a caveolin-1-mediated and dynamin-2-dependent endocytic pathway that requires activation of p38 mitogen-activated protein kinase (MAPK) and Src signaling pathways as well as microtubules and small GTPase Rab5 protein. J Biol Chem 286 (35):30780–30794. https://doi.org/10.1074/jbc.M111.257154

Ji WT, Wang L, Lin RC, Huang WR, Liu HJ (2009) Avian reovirus influences phosphorylation of several factors involved in host protein translation including eukaryotic translation elongation factor 2 (eEF2) in Vero cells. Biochem Biophys Res Commun 384(3):301–305. https://doi.org/10.1016/j.bbrc.2009.04.116

Jones RC (2000) Avian reovirus infections. Rev Sci Tech 19(2):614–625

Jones RC, Georgiou K (1984) Reovirus-induced tenosynovitis in chickens the influence of age at infection. Avian Pathol 13(3):441–457. https://doi.org/10.1080/03079458408418546

Jones RC, Kibenge FSB (1984) Reovirus-induced tenosynovitis in chickens: the effect of breed. Avian Pathol 13(3):511–528. https://doi.org/10.1080/03079458408418552

Jones RC, Onunkwo O (1978) Studies on experimental tenosynovitis in light hybrid chickens. Avian Pathol 7(1):171–181. https://doi.org/10.1080/03079457808418268

Kerr KM, Olson NO (1969) Pathology of chickens experimentally inoculated or contact-infected with an arthritis-producing virus. Avian Dis 13(4):729–745

Kielian M, Jungerwirth S (1990) Mechanisms of enveloped virus entry into cells. Mol Biol Med 7(1):17–31

Labrada L, Bodelón G, Vinuela J, Benavente J (2002) Avian reoviruses cause apoptosis in cultured cells: viral uncoating, but not viral gene expression, is required for apoptosis induction. J Virol 76(16):7932–7941

Lin HY, Chuang ST, Chen YT, Shih WL, Chang CD, Liu HJ (2007) Avian reovirus-induced apoptosis related to tissue injury. Avian Pathol 36(2):155–159. https://doi.org/10.1080/03079450701261262

Lin KH, Hsu AP, Shien JH, Chang TJ, Liao JW, Chen JR, Lin CF, Hsu WL (2012) Avian reovirus sigma C enhances the mucosal and systemic immune responses elicited by antigen-conjugated lactic acid bacteria. Vaccine 30(33):5019–5029. https://doi.org/10.1016/j.vaccine.2012.04.043

Lin PY, Liu HJ, Chang CD, Chen YC, Chang CI, Shih WL (2015a) Avian reovirus S1133-induced apoptosis is associated with Bip/GRP79-mediated Bim translocation to the endoplasmic reticulum. Apoptosis 20(4):481–490. https://doi.org/10.1007/s10495-015-1085-5

Lin PY, Chang CD, Chen YC, Shih WL (2015b) RhoA/ROCK1 regulates Avian reovirus S1133-induced switch from autophagy to apoptosis. BMC Vet Res 11:103. https://doi.org/10.1186/s12917-015-0417-6

Lin PY, Lee JW, Liao MH, Hsu HY, Chiu SJ, Liu HJ, Shih WL (2009) Modulation of p53 by mitogen-activated protein kinase pathways and protein kinase C delta during avian reovirus S1133-induced apoptosis. Virology 385(2):323–334. https://doi.org/10.1016/j.virol.2008.12.028

Lin PY, Liu HJ, Chang CD, Chang CI, Hsu JL, Liao MH, Lee JW, Shih WL (2011) Avian reovirus S1133-induced DNA damage signaling and subsequent apoptosis in cultured cells and in chickens. Arch Virol 156(11):1917–1929. https://doi.org/10.1007/s00705-011-1063-3

Lin PY, Liu HJ, Lai MJ, Yu FL, Hsu HY, Lee JW, Shih WL (2006) Avian reovirus activates a novel proapoptotic signal by linking Src to p53. Apoptosis 11(12):2179–2193. https://doi.org/10.1007/s10495-006-0291-6

Lin PY, Liu HJ, Liao MH, Chang CD, Chang CI, Cheng HL, Lee JW, Shih WL (2010) Activation of PI 3-kinase/Akt/NF-kappaB and Stat3 signaling by avian reovirus S1133 in the early stages of infection results in an inflammatory response and delayed apoptosis. Virology 400(1):104–114. https://doi.org/10.1016/j.virol.2010.01.024

Lin YH, Lee LH, Shih WL, Hu YC, Liu HJ (2008) Baculovirus surface display of sigmaC and sigmaB proteins of avian reovirus and immunogenicity of the displayed proteins in a mouse model. Vaccine 26(50):6361–6367. https://doi.org/10.1016/j.vaccine.2008.09.008

Liu HJ, Kuo LC, Hu YC, Liao MH, Lien YY (2002) Development of an ELISA for detection of antibodies to avian reovirus in chickens. J Virol Methods 102(1–2):129–138

Liu HJ, Lee LH, Hsu HW, Kuo LC, Liao MH (2003) Molecular evolution of avian reovirus: evidence for genetic diversity and reassortment of the S-class genome segments and multiple cocirculating lineages. Virology 314(1):336–349

Liu HJ, Lin PY, Lee JW, Hsu HY, Shih WL (2005) Retardation of cell growth by avian reovirus p17 through the activation of p53 pathway. Biochem Biophys Res Commun 336(2):709–715. https://doi.org/10.1016/j.bbrc.2005.08.149

Liu HJ, Lin PY, Wang LR, Hsu HY, Liao MH, Shih WL (2008) Activation of small GTPases RhoA and Rac1 is required for avian reovirus p10-induced syncytium formation. Mol Cells 26(4):396–403

Lu SW, Wang KC, Liu HJ, Chang CD, Huang HJ, Chang CC (2011) Expression of avian reovirus minor capsid protein in plants. J Virol Methods 173(2):287–293. https://doi.org/10.1016/j.jviromet.2011.02.021

Marsh M, Helenius A (1989) Virus entry into animal cells. Adv Virus Res 36:107–151. https://doi.org/10.1016/S0065-3527(08)60583-7

Marsh M, Helenius A (2006) Virus entry: open sesame. Cell 124(4):729–740. https://doi.org/10.1016/j.cell.2006.02.007

Martínez-Costas J, Grande A, Varela R, Garcia-Martinez C, Benavente J (1997) Protein architecture of avian reovirus S1133 and identification of the cell attachment protein. J Virol 71(1):59–64

Menendez NA, Calnek BW, Cowen BS (1975) Experimental egg-transmission of avian reovirus. Avian Dis 19(1):104–111

Mertens P (2004) The dsRNA viruses. Virus Res 101(1):3–13. https://doi.org/10.1016/j.virusres.2003.12.002

Olson NO (1978) Reovirus infections. In: Hofstad MS (ed) Diseases of poultry. Iowa State University Press, Ames, Iowa, USA, pp 641–647

Rekik MR, Silim A, Elazhary MA (1990) Characteristics and analysis of electropherotypes of avian reovirus field isolates. Vet Microbiol 23(1–4):273–281. https://doi.org/10.1016/0378-1135(90)90158-R

Robertson MD, Wilcox GE (1986) Avian reovirus. Vet Bull 56:759–766

Rodríguez-Grille J, Busch LK, Martinez-Costas J, Benavente J (2014) Avian reovirus-triggered apoptosis enhances both virus spread and the processing of the viral nonstructural muNS protein. Virology 462–463:49–59. https://doi.org/10.1016/j.virol.2014.04.039

Rosenberger JK, Sterner FJ, Botts S, Lee KP, Margolin A (1989) In vitro and in vivo characterization of avian reoviruses. I. Pathogenicity and antigenic relatedness of several avian reovirus isolates. Avian Dis 33(3):535–544

Schnitzer TJ (1985) Protein coding assignment of the S genes of the avian reovirus S1133. Virology 141(1):167–170

Scotti N, Rybicki EP (2013) Virus-like particles produced in plants as potential vaccines. Expert Rev Vaccines 12(2):211–224. https://doi.org/10.1586/erv.12.147

Sen GC (2001) Viruses and interferons. Ann Rev Microbiol 55:255–281. https://doi.org/10.1146/annurev.micro.55.1.255

Shapouri MR, Arella M, Silim A (1996) Evidence for the multimeric nature and cell binding ability of avian reovirus sigma 3 protein. J Gen Virol 77(Pt 6):1203–1210. https://doi.org/10.1099/0022-1317-77-6-1203

Shapouri MR, Kane M, Letarte M, Bergeron J, Arella M, Silim A (1995) Cloning, sequencing and expression of the S1 gene of avian reovirus. J Gen Virol 76(Pt 6):1515–1520. https://doi.org/10.1099/0022-1317-76-6-1515

Shih WL, Hsu HW, Liao MH, Lee LH, Liu HJ (2004) Avian reovirus sigmaC protein induces apoptosis in cultured cells. Virology 321(1):65–74. https://doi.org/10.1016/j.virol.2003.12.004

Shmulevitz M, Corcoran J, Salsman J, Duncan R (2004) Cell-cell fusion induced by the avian reovirus membrane fusion protein is regulated by protein degradation. J Virol 78(11):5996–6004. https://doi.org/10.1128/JVI.78.11.5996-6004.2004

Shmulevitz M, Duncan R (2000) A new class of fusion-associated small transmembrane (FAST) proteins encoded by the non-enveloped fusogenic reoviruses. EMBO J 19(5):902–912. https://doi.org/10.1093/emboj/19.5.902

Spandidos DA, Graham AF (1976) Physical and chemical characterization of an avian reovirus. J Virol 19(3):968–976

Streatfield SJ, Howard JA (2003) Plant-based vaccines. Int J Parasitol 33(5–6):479–493

Van de Zande S, Kuhn EM (2007) Central nervous system signs in chickens caused by a new avian reovirus strain: a pathogenesis study. Vet Microbiol 120(1–2):42–49. https://doi.org/10.1016/j.vetmic.2006.10.024

van der Heide L, Kalbac M (1975) Infectious tenosynovitis (viral arthritis): characterization of a Connecticut viral isolant as a reovirus and evidence of viral egg transmission by reovirus-infected broiler breeders. Avian Dis 19(4):683–688

Varela R, Benavente J (1994) Protein coding assignment of avian reovirus strain S1133. J Virol 68(10):6775–6777

Wan J, Wang C, Wen X, Huang X, Ling S, Huang Y, Cao S (2011) Immunogenicity of a DNA vaccine of Avian reovirus orally delivered by attenuated Salmonella typhimurium. Res Vet Sci 91(3):382–383. https://doi.org/10.1016/j.rvsc.2010.09.007

Wan J, Wen X, Huang X, Tang Y, Huang Y, Yan Q, Zhao Q, Cao S (2012) Immunogenic analysis of two DNA vaccines of avian reovirus mediated by attenuated Salmonella typhimurium in chickens. Vet Immunol Immunopathol 147(3–4):154–160. https://doi.org/10.1016/j.vetimm.2012.04.009

Wickramasinghe R, Meanger J, Enriquez CE, Wilcox GE (1993) Avian reovirus proteins associated with neutralization of virus infectivity. Virology 194(2):688–696. https://doi.org/10.1006/viro.1993.1309

Wu H, Scissum-Gunn K, Singh NK, Giambrone JJ (2009) Toward the development of a plant-based vaccine against reovirus. Avian Dis 53(3):376–381. https://doi.org/10.1637/8589-011309-Reg.1

Wu H, Williams Y, Gunn KS, Singh NK, Locy RD, Giambrone JJ (2005) Yeast-derived sigma C protein-induced immunity against avian reovirus. Avian Dis 49(2):281–284. https://doi.org/10.1637/7284-092904R1

Yuan YH, Wang JF, Wu ZX, Huang XG, He DS, Huang SJ (2013) Biological identification of a new-type duck reovirus (QY strains) isolated from Muscovy duck. Chinese J Vet Sci 33:40–44

Yun T, Yu B, Ni Z, Ye W, Chen L, Hua J, Zhang C (2014) Genomic characteristics of a novel reovirus from Muscovy duckling in China. Vet Microbiol 168(2–4):261–271. https://doi.org/10.1016/j.vetmic.2013.11.005

Part III
Vaccines for Swine

Toward the Optimization of a Plant-Based Oral Vaccine Against Cysticercosis

Edda Sciutto, Marisela Hernández, Jacquelynne Cervantes-Torres, Elizabeth Monreal-Escalante, Omayra Bolaños-Martínez, Juan Francisco Rodríguez, Gladis Fragoso and Sergio Rosales-Mendoza

Abstract It is recognized that an effective, low-cost oral vaccine may effectively contribute to prevent *Taenia solium* cysticercosis; plant-based vaccines, on the other hand, can make this goal feasible. Plants are optimal platforms for the massive production of oral vaccines. In this chapter, advances toward the development of oral plant-based vaccine against cysticercosis are reviewed.

Keywords Oral vaccine · Transgenic plant · Transplastomic plant
Carica papaya · *Daucus carota* · *Nicotiana tabacum* · *Taenia solium*

1 Introduction

1.1 Relevance of Cysticercosis

Taeniasis/cysticercosis is a parasitic infection caused by *Taenia solium*. It is prevalent in areas with low socioeconomic and sanitary standards in developing

E. Sciutto · M. Hernández · J. Cervantes-Torres · E. Monreal-Escalante · O. Bolaños-Martínez
J. F. Rodríguez · G. Fragoso
Dpto. Inmunología. Instituto de Investigaciones Biomédicas, Universidad Nacional Autónoma de México. Circuito Escolar. Ciudad Universitaria, 04510 México, CDMX, Mexico
e-mail: edda@servidor.unam.mx

S. Rosales-Mendoza (✉)
Laboratorio de Biofarmacéuticos Recombinantes, Facultad de Ciencias Químicas, Universidad Autónoma de San Luis Potosí, Av. Dr. Manuel Nava 6, 78210 San Luis Potosí, SLP, Mexico
e-mail: rosales.s@fcq.uaslp.mx

S. Rosales-Mendoza
Sección de Biotecnología, Centro de Investigación en Ciencias de la Salud y Biomedicina, Universidad Autónoma de San Luis Potosí, Av. Sierra Leona 550, Lomas 2ª. Sección, 78210 San Luis Potosí, Mexico

J. MacDonald (ed.), *Prospects of Plant-Based Vaccines in Veterinary Medicine*,
https://doi.org/10.1007/978-3-319-90137-4_10

countries. *T. solium* cysticercosis is still endemic in most countries of Asia, Africa, and Latin America despite significant progress in developing effective tools for its prevention, diagnosis, and treatment (Fleury et al. 2012, 2013).

T. solium cysticercosis can be acquired by humans (the definitive host) and pigs (the intermediate host) after ingesting parasite eggs in food or water contaminated with feces from human *T. solium* tapeworm carriers (Sciutto et al. 2000). A single tapeworm carrier may produce thousands of *T. solium* eggs per day, which are shed to the environment, contaminating vegetables, running waters, and soils, upon open-air defecation (De Aluja 2008). Humans also acquire intestinal tapeworms by eating insufficiently cooked meat from cysticercus-infected pigs.

In humans, cysticerci frequently establishes in the central nervous system (CNS), causing neurocysticercosis (NC), the most severe form of the disease. NC is a clinically and radiologically heterogeneous disease, ranging from an asymptomatic infection to a severe and eventually fatal disease. NC severity mainly depends on the location of the parasite. The most severe clinical forms occur when parasites are located in the subarachnoid space at the base of the brain. This form of the disease is also less susceptible to cysticidal drugs and more difficult to diagnose based on neuroradiological studies (Marcin Sierra et al. 2017).

Pig vaccination may curtail *T. solium* transmission by reducing the number of cysticerci, and thence the incidence of adult intestinal tapeworms in humans. Although several vaccines have been developed and successfully field-trial tested, no cysticercosis vaccines for pigs have been approved for commercialization. All of them induced high level of protection, but in all cases they are injectable vaccines (Huerta et al. 2001; Morales et al. 2008; Assana et al. 2010). Injectable vaccines are costly and their administration at a massive scale implies logistic difficulties, since pigs roaming in rural areas must be captured by trained personal for vaccination (Huerta et al. 2001; Morales et al. 2008; Assana et al. 2010). These difficulties limit the use of parenteral vaccines in nationwide programs and may underlie the lack of interest in companies to produce the vaccine commercially. An orally administered vaccine, which could be delivered by pig owners, would elude these difficulties.

1.2 Context of Veterinary Vaccines

Veterinary vaccines are aimed to reduce morbidity in animals for human consumption (chickens, cows, fish, and pigs), pets, and in wildlife species, to prevent the loss or contamination of animal derivatives (improve productivity), and to reduce the risk of disease transmission from animals to humans (zoonosis) (Meeusen et al. 2007).

According to the World Organization for Animal Health (OIE), 116 animal infections are included in the 2017 list of diseases, infections, and infestations. Most of these diseases are caused by bacteria such as *Salmonella* spp., *Listeria monocytogenes,* and *Escherichia coli;* viruses such as avian influenza virus, norovirus, and rabies virus, as well as parasites such as *Taenia saginata*, *T. solium*, *Toxoplasma gondii,* and *Giardia duodenalis* (www.oie.int/es).

Several conventional commercial pig vaccines are currently being used to prevent infections in pigs caused by pathogenic *E. coli*, *Salmonella*, and *Lawsonia intracellularis* strains (Table 1) (Meeusen et al. 2007).

While most veterinary vaccines have been formulated for parenteral administration, the number of oral vaccines has increased in recent years, considering the advantages of this alternative route. Oral administration of vaccines can be performed by animal owners themselves. In addition, oral administration is less-costly than the parenteral route since neither trained personal for administration nor the use of needles are required. Moreover, the oral route is non-invasive and particularly attractive for the prevention of orally acquired infections like cysticercosis (Wang and Coppel 2008). Oral vaccines efficiently stimulate the mucosal system, improving protection by emulating the entry path of most pathogens. In addition, oral immunization can induce a concomitant systemic immune response (Muir et al. 2000; Mutoloki et al. 2015).

The mentioned advantages are especially relevant when designing an economic vaccine to prevent these neglected tropical diseases (NTDs) (WHO 2010) that affect rural pigs in marginalized populations of poor countries. In this context, an oral recombinant vaccine could overcome these limitations.

Several avirulent and live-attenuated oral vaccines are applied to pigs to prevent diseases like salmonellosis and erysipeloid. As in the case of poultry, a lyophilizate can be mixed with sterile water and administered as a beverage to 3–8 weeks-old pigs.

On the other hand, subunit vaccines based in immunoprotective proteins provide a safer approach for vaccination. An example is the rabies vaccine known as Raboral, in which glycoprotein G is used; it is administered as bait to wild animals in France, Belgium, Germany, and the USA (Bano et al. 2017).

Table 1 Orally-administered, commercial pig vaccines against bacterial pathogens

Pathogen	Disease	Brand name	Distributor	Composition	References
Lawsonia intracellularis	Porcine proliferative enteropathy	Enterisol® Ileitis	Boehringer Ingelheim Vetmedica, Inc.	LAT	Park et al. (2013)
Salmonella choleraesuis and *typhimurium*	Salmonellosis	Enterisol® Salmonella T/C		LAV	Wales and Davies (2017)
Salmonella choleraesuis		Enterisol® SC-54		LAV	
Erysipelothrix rhusiopathiae	Erysipelas	Ingelvac® ERY-ALC		LAV	http://www.bivetmedica.com/
Escherichia coli strain K-88	Enteritis	Entero Vac	Arko labs	LAV	Cox et al. (2014)

LAT live attenuated; *LAV* live avirulent

In addition to enhanced safety, an advantage of subunit vaccines is their ability to address multiple genetic variants of a pathogen in a single chimeric protein. Several pathogens, such as RNA viruses, exhibit a high mutation rate, resulting in a great variability within a short period; in addition, multiple serotypes are reported for several virus strains. Since many existing viral vaccines cannot recognize new viral strains, novel strategies to produce vaccines against these new infectious variants are much needed (Meeusen et al. 2007). In the case of parasitic diseases, an immune response against several antigens is desired to achieve an efficient immunoprotection. Subunit vaccines constitute an alternative to address these challenges, since the development of multivalent vaccines based in mixtures of several antigen variants or even in multiepitopic proteins carrying a set of relevant epitopes can provide broad immune responses.

1.3 Advances in the Development of Plant-Based Oral Vaccines

The expression of recombinant vaccine antigens in plants to elicit and maintain protective immunity is an attractive option that has been explored for the last two decades. Edible vaccines currently under development are based on fruits, leaves, and seeds of transgenic plants. Such vaccines are prepared without expensive antigen purification steps, commonly required for parenterally administered vaccines (Lugade et al. 2010) (Table 2).

Besides protecting against pathogen viruses and bacteria infecting domestic animal species, this strategy is appropriate to delivery parasite-derived antigens to gut-associated lymphoid tissues (e.g. for fasciolosis, schistosomiasis, coccidiosis, cysticercosis, and ascariasis) (Walmsley and Arntzen 2000; Chambers et al. 2016).

1.4 Transgenic Plants as Alternative Platforms to Produce an Anti-cysticercosis Vaccine

Significant progress has been attained in the development of oral plant-based vaccine candidates against porcine cysticercosis (Table 3). An oral vaccine against cysticercosis was developed expressing the three peptides (KETc1, KETc12, and KETc7) that constitute the injectable vaccine against pig cysticercosis named S3Pvac. When parenterally applied, synthetically and recombinantly expressed S3Pvac reduced by 50% the number of infected pigs, and by 80–90% the number of established cysticerci under natural conditions of transmission (Huerta et al. 2001; Morales et al. 2008). To develop the oral version of the vaccine, the three peptides were expressed in three independent papaya embryogenic cell lines, obtained by bioballistics at the nuclear level under the CaMV35S promoter (Hernández et al.

Table 2 Experiences in immune response induced by oral vaccination with transgenic plants

Disease	Antigen	Plant	Dose	No. doses (interval)	Immune response	References
Gastroenteritis	LT-B	Potato tubers	20–50 μg	3 (wk)	Specific IgG and IgA Abs; partial protection	Mason et al. (1998)
Hepatitis	HBsAg	Potato tubers	20 μg	3 (wk)	Specific IgG Abs on day 36–50 after first feeding	Rukavtsova et al. (2015)
Dengue	cEDIII-CTB	Rice calli	100 μg	4 (0, 2 4, 9 wk)	cEDIII specific IgG and IgA Abs; splenic T cell proliferation	Kim et al. (2016)
Fasciolosis	CPFhW-HBcAg	Lettuce leaves	10 μg	2 (0, 4 wk)	65.4% protection; specific IgG1 and IgM Abs; increased blood CD4 + and CD8+	Kesik-Brodacka et al. (2017)
Cysticercosis	KETc1, KETc7, KETc12	Papaya calli	0.1–10 μg	2 (1, 10 days)	55–89% protection; specific IgG Abs; CD4 and CD8 proliferation	Fragoso et al. (2017)
Rheumatoid arthritis	APL-CII	Rice seeds	100–120 μg	14 (daily)	IL-10 production in spleen; reduced joint inflammation	Lizuka et al. (2014)

LT-B: *E. coli* heat-labile enterotoxin B subunit; HBsAg: surface hepatitis B antigen; cEDIII-CTB: envelope protein domain III (cEDIII) fused to cholera toxin B subunit; CPFhW-HBcAg: Cysteine proteinases from *F. hepatica;* APL-CII: Altered peptide ligands fused of type II collagen; wk: weekly; Abs: antibodies

Table 3 Transgenic plants and antigens evaluated to design an anti cysticercosis vaccine

Plant	Antigen	Expression Vector	Transformation	Specie	/Pathogen/ #immunizations/ route	% Protection	References
Carica papaya L.	KETc1, KETc12, KETc7	pUI235–5.1	Bioballistics	Mouse	*T. crassiceps*/ Two/sc		Hernández et al. (2007), Rosales-Mendoza et al. (2012)
				Rabbit	*T. pisiformis* Two/oral		Betancourt et al. (2012)
				Mouse	*T. crassiceps*/ Two/oral	55–66	Fragoso et al. (2017)
				Pig	*T. solium*/Two/ oral	ND	
Nicotiana tabacum	KETc1 KETc12 KETc7, GK1 HP6/TSOL18	pBI-Helios2A polyprotein system	*A. tumefaciens GV3101* strain	Mouse	Three/sc	ND	Monreal-Escalante et al. (2015)
Daucus carota L.	HP6/TSOL18	pBin	*A. tumefaciens* GV3101 strain	Mouse	*T. crassiceps*/ Two/oral	80%	Monreal-Escalante et al. (2016)

sc subcutaneous

2007). The combination of three embryogenic transgenic papaya callus lines was designated as S3Pvac-papaya. The expression of the respective peptide in each clone was confirmed at the transcriptional level by RT-PCR. Soluble extracts from the transgenic papaya clones were found to be immunogenic when subcutaneously administered to mice. Indeed, all three clones expressing the vaccine peptides induced high levels of protection against murine cysticercosis when injected to mice (Hernández et al. 2007).

Furthermore, orally administered S3Pvac-papaya was found to be protective against murine and rabbit cysticercosis caused by *T. crassiceps* and *T. pisiformis*, respectively (Betancourt et al. 2012; Fragoso et al. 2017). The protective properties of the vaccine were maintained when formulated with different excipients that could eventually be attractive for pigs (Fragoso et al. 2017).

The effectiveness of the vaccine against these highly predictive experimental models let us consider its usefulness to be employed for pig cysticercosis prevention. To further examine this possibility, the immunity of pigs orally vaccinated with S3Pvac-papaya was explored. Oral vaccination with S3Pvac-papaya elicited an exacerbated humoral and cellular response in pigs (Fragoso et al. 2017).

Given the promising potential of papaya-made *T. solium* antigens, their expression in plant systems has been expanded to add new advantages to the plant-made vaccine candidates. The simultaneous expression of *T. solium* antigens in a single plant line would facilitate vaccine formulation; thus, innovative approaches have been recently explored to address this objective. An alternative Helios2A polyprotein system was developed, which relies on the use of the 2A sequence (LLNFDLLKLAGDVESNPG-P) of the foot and mouth disease virus that is placed between each of the antigens in a translational fusion arrangement. During the translation process of the polyprotein-encoding mRNA coding for the target antigens, the 2A sequence induces self-cleavage events by modifying the activity of the ribosome to allow hydrolysis of the ester linkage 2A-tRNAgly to be released, while the translation of the downstream product continues (Ryan and Drew 1994). Thus, this approach would allow the production of a multicomponent vaccine through the insertion of a single expression cassette coding for the polyprotein arrangement (Liu et al. 2007; Minskaia et al. 2013; Minskaia and Ryan 2013). Following a 2A-based polyprotein expression approach, a new multicomponent vaccine called Helios-2A, comprising the KETc1, KETc12, and KETc7 peptides from the S3Pvac along with the TSOL18/HP6-TSOL protective antigen was generated. The latter was included to assess whether vaccine efficacy is improved, since it has been reported as a highly protective antigen against porcine cysticercosis (Lightowlers et al. 2016), The Helios-2A system allowed the successful expression of the KETc1, KETc12, KETc7, GK1 (a short protective sequence inserted in the KETc7 peptide), and Tsol18/HP6 individual antigens in tobacco plants transformed with *Agrobacterium tumefaciens* at the nuclear level using the CaMV35S promoter. Interestingly, plant-made Helios-2A antigens were recognized by cerebral spinal fluid of neurocysticercosis patients and induced humoral responses in mice upon subcutaneous immunization (Monreal-Escalante et al. 2015). Although the efficacy of the Helios-2A is still under assessment, it is proposed as a highly convenient vaccine that could be produced by propagating and characterizing a single transformed line (instead of the three lines required to formulate S3Pvac-papaya), and possibly conferring higher protection than the original S3Pvac vaccine.

Another innovation developed by our group consisted in the use of carrot cells as expression host to produce anti-cysticercosis vaccines. Carrot cell lines constitute a pioneering case in the molecular pharming arena, since the first commercialized plant-made biopharmaceutical was produced in this system. Thus, carrot was adopted to produce a candidate vaccine against cysticercosis based in cell lines expressing the TSOL18/HP6-Tsol antigen. Carrot lines were obtained by *A. tumefaciens* transformation at the nuclear level to express the TSOL18/HP6-Tsol antigen under the control of the CaMV35S promoter. Upon oral immunization with carrot-made TSOL18/HP6-Tsol, mice developed humoral responses and were protected against *T. crassiceps* challenge (Monreal-Escalante et al. 2016). Immunization trials to compare the efficacy of this vaccine with that of S3Pvac are ongoing.

Looking to enhance antigen yields, transplastomic approaches have been implemented to produce the target *T. solium* antigens. The S3Pvac-papaya components were produced along with the TSOL18/HP6-Tsol antigen in tobacco plastids.

Synthetic operons under the control of the Prnn promoter led to the expression of individual target antigens through a single transformation event. Chloroplast-made antigens retained their immunogenic properties, as revealed by immunization experiments in mice. The immunoprotective properties of this transplastomic vaccine are currently being assessed (Rosales-Mendoza et al., unpublished). As an additional advantage, this vaccine offers enhanced biosafety with respect to the nuclear transformed plants, since plastomes are maternally inherited, and thus transgene transmission via pollen is avoided. Thus, the transplastomic approach is likely to yield an optimized anti-cysticercosis vaccine; however, its detailed characterization and the assessment of its protective efficacy are still in progress.

2 Conclusions and Perspectives

Cysticercosis control is theoretically possible, and the disease was declared to be eradicable by the International Task Force for Disease Eradication in 1993. However, *T. solium* cysticercosis persists to date, and new cases are continually reported in non-developed countries, where the parasite life cycle is well established, and also in developed regions due to immigration of infected individuals. Control strategies based on mass-treatment for human taeniasis in identified transmission foci have been proposed by WHO (2010) and the Pan American Health Organization. The inclusion of an effective oral, low-cost vaccine that could be administered directly by pig owners may significantly improve the effectiveness of a control program. The production of anti-cysticercosis vaccines using plants can accomplish this goal. Substantial advances have been achieved over the last

10 years in this area. Both nuclear and transplastomic approaches have been assessed to test the biosynthetic capacity of plants to produce immunoprotective *T. solium* antigens. S3Pvac-papaya vaccine was a pioneering case for a vaccine tested in the field (Hernández et al. 2007); this first experience demonstrated that plants are promising biofactories for anti-cysticercosis vaccines, and justify the projection to generate other vaccine candidates, facilitate vaccine formulation, and maximize antigen productivity. The promising results reported in pigs prompt us to start the scale-up process to produce an oral vaccine in airlift bioreactors and obtain enough material for conducting field trials. On the other hand, a vaccine based in carrot cell lines expressing the TSOL18/HP6-Tsol provided the first evidence on the production of the functional antigen at appropriate levels to immunize mice. Since TSOL18/HP6-Tsol also confers immunoprotection against *T. saginata* and the S3Pvac peptides are highly conserved in this parasite, this vaccine candidate will allow us not only to perform studies on its role as a supplementary antigen for the S3Pvac vaccine, but also to develop a new anti-*T. saginata* vaccine for cows and cattle (Parkhouse et al. 2008).

Vaccines produced in tobacco, either by nuclear or plastid expression, exemplify the potential of synthetic polycistrons and viral sequences to engineer plant cells as efficient biofactories to produce multicomponent vaccines in a single transformed line. This expression modality will greatly facilitate vaccine formulation, since the upstream process will deal with a single seed stock, and during downstream processing a single line will be used for antigen quantification and encapsulation prior to dosage. Preclinical evaluation of these 'single line' vaccine candidates will be completed soon, and will provide the basis for field evaluations in pigs.

In conclusion, plants have proven to be suitable platforms to produce anti-cysticercosis vaccines, and promising prospects are being projected in terms of field evaluations and the development of innovative candidates, based on alternative expression approaches. Such plant-based vaccines will be valuable tools to control cyticercosis especially in poor countries, since formulations based on freeze-dried plant biomass have very low production costs and lower logistic costs, since they are easy to apply and do not require purification, cold-chain, sterile devices nor trained personnel to be applied. Altogether, these features would make a more robust and easier to handle vaccine (Hirlekar and Bhairy 2017).

References

Assana E, Kyngdon CT, Gauci CG et al (2010) Elimination of *Taenia solium* transmission to pigs in a field trial of the TSOL18 vaccine in Cameroon. Int J Parasitol 40:515–519. https://doi.org/10.1016/j.ijpara.2010.01.006

Bano I, Sajjad H, Shah AM et al (2017) A review of rabies disease, its transmission and treatment. J Anim Health Prod 4:140–144

Betancourt MA, de Aluja AS, Sciutto E et al (2012) Effective protection induced by three different versions of the porcine S3Pvac anticysticercosis vaccine against rabbit experimental *Taenia pisiformis* cysticercosis. Vaccine 30:2760–2767. https://doi.org/10.1016/j.vaccine.2012.02.022

Chambers MA, Graham SP, La Ragione RM (2016) Challenges in veterinary vaccine development and immunization. Chapter 1. Vaccine design: methods and protocols. Edited by Sunil Thomas. Human Press, USA

Cox E, Melkebeek V, Devriendt B et al (2014) Vaccines against enteric *E. coli* infections in animals. In: Pathogenic *Escherichia coli:* molecular and cellular microbiology. Caister Academic Press, pp 255–270

De Aluja AS (2008) Cysticercosis in the pig. Curr Top Med Chem 8:368–374

Fleury A, Sciutto E, Larralde C (2012) Neurocysticercosis is still prevalent in Mexico. Salud Publica Mex 54:632–636

Fleury A, Sciutto E, de Aluja AS et al (2013) Control of *Taenia solium* transmission of taeniosis and cysticercosis in endemic countries: the roles of continental networks of specialists and of local health authorities. In: Fleury A (ed) Novel aspects on cysticercosis and neurocysticercosis. InTech, Austria

Fragoso G, Hernández M, Cervantes-Torres J et al (2017) Transgenic papaya: a useful platform for oral vaccines. Planta 245:1037–1048. https://doi.org/10.1007/s00425-017-2658-z

Hernández M, Cabrera-Ponce JL, Fragoso G et al (2007) A new highly effective anticysticercosis vaccine expressed in transgenic papaya. Vaccine 25:4252–4260

Hirlekar R, Bhairy S (2017) Edible vaccines: an advancement in oral immunization. Asian J Pharm Clin Res 10:82–88

Huerta M, De Aluja AS, Fragoso G et al (2001) Synthetic peptide vaccine against *Taenia solium* pig cysticercosis: successful vaccination in a controlled field trial in rural Mexico. Vaccine 20:262–266

Kesik-Brodacka M, Lipiec A, Kozak Ljunggren M et al (2017) Immune response of rats vaccinated orally with various plant-expressed recombinant cysteine proteinase constructs when challenged with *Fasciola hepatica metacercariae*. PLoS Negl Trop Dis 11:e0005451. https://doi.org/10.1371/journal.pntd.0005451

Kim MY, Kim BY, Oh SM et al (2016) Oral immunisation of mice with transgenic rice calli expressing cholera toxin B subunit fused to consensus dengue cEDIII antigen induces antibodies to all four dengue serotypes. Plant Mol Biol 92:347–356. https://doi.org/10.1007/s11103-016-0517-0

Lightowlers MW, Donadeu M, Elaiyaraja M et al (2016) Anamnestic responses in pigs to the *Taenia solium* TSOL18 vaccine and implications for control strategies. Parasitology 143:416–420

Liu BS, Liu XY, Qian C (2007) An efficient tool for the construction of multiple-cistronic vectors: FMDV 2A. Sheng Wu Gong Cheng Xue Bao 23:765–769

Lizuka M, Wakasa Y, Tsuboi H et al (2014) Prophylactic effect of the oral administration of transgenic rice seeds containing altered peptide ligands of type II collagen on rheumatoid arthritis. Biosci Biotechnol Biochem 78:1662–1668. https://doi.org/10.1080/09168451.2014.936349

Lugade AA, Kalathil S, Heald JL et al (2010) Transgenic plant-based oral vaccines. Immunol Invest 39:468–482

Marcin Sierra M, Arroyo M, Cadena Torres M et al (2017) Extraparenchymal neurocysticercosis: demographic, clinicoradiological, and inflammatory features. PLoS Negl Trop Dis 9:e0005646

Mason HS, Haq TA, Clements JD et al (1998) Edible vaccine protects mice against Escherichia coli heatlabile enterotoxin (LT): potatoes expressing a synthetic LT-B gene. Vaccine 16:1336–1343

Meeusen Els NT, Walker J, Peters A et al (2007) Current status of veterinary vaccines. Clin Microbiol Rev 20:489–510

Minskaia E, Nicholson J, Ryan MD (2013) Optimisation of the foot-and-mouth disease virus 2A co-expression system for biomedical applications. BMC Biotechnol 13:67. https://doi.org/10.1186/1472-6750-13-67

Minskaia E, Ryan MD (2013) Protein coexpression using FMDV 2A: effect of "linker" residues. Biomed Res Int 291730. https://doi.org/10.1155/2013/291730

Monreal-Escalante E, Bañuelos-Hernández B, Hernández M et al (2015) Expression of multiple *Taenia solium* Immunogens in plant cells through a ribosomal skip mechanism. Mol Biotechnol 57:635–643
Monreal-Escalante E, Govea-Alonso DO, Hernández M et al (2016) Towards the development of an oral vaccine against porcine cysticercosis: expression of the protective TSOL18/HP6-TSOL antigen in transgenic carrots cells. Planta 243:675–685
Morales J, Martínez JJ, Manoutcharian K et al (2008) Inexpensive anti-cysticercosis vaccine: S3Pvac expressed in heat inactivated M13 filamentous phage proves effective against naturally acquired *Taenia solium* porcine cysticercosis. Vaccine 26:2899–2905. https://doi.org/10.1016/j.vaccine.2008.03.042
Muir WI, Bryden WL, Husband AJ (2000) Immunity, vaccination and the avian intestinal tract. Dev Comp Immunol 24:325–342
Mutoloki S, Munang'andu HM, Evensen Ø (2015) Oral vaccination of fish-antigen preparations, uptake, and immune induction. Front Immunol 6:519. https://doi.org/10.3389/fimmu.2015.00519.eCollection
Park S, Lee JB, Kim KJ et al (2013) Efficacy of a commercial live attenuated *Lawsonia intracellularis* vaccine in a large scale field trial in Korea. Clin Exp Vaccine Res 2:135–139
Parkhouse RM, Bonay P, González LM et al (2008) TSOL18/HP6-Tsol, an immunogenic *Taenia solium* oncospheral adhesion protein and potential protective antigen. Parasitol Res 102:921–926
Rosales-Mendoza S, Govea-Alonso DO, Monreal-Escalante E et al (2012) Developing plant-based vaccines against neglected tropical diseases: where are we? Vaccine 31:40–48
Rukavtsova EB, Rudenko NV, Puchko EN et al (2015) Study of the immunogenicity of hepatitis B surface antigen synthesized in transgenic potato plants with increased biosafety. J Biotechnol 203:84–88
Ryan MD, Drew J (1994) Foot-and-mouth disease virus 2A oligopeptide mediated cleavage of an artificial polyprotein. EMBO J 13:928–933
Sciutto E, Fragoso G, Fleury A et al (2000) *Taenia solium* disease in humans and pigs: an ancient parasitosis disease rooted in developing countries and emerging as a major health problem of global dimensions. Microbes Infect 2:1875–1890
Wales AD, Davies RH (2017) *Salmonella* vaccination in pigs: a review. Zoonoses Public Health 64:1–13
Walmsley AM, Arntzen CJ (2000) Plants for delivery of edible vaccines. Curr Opin Biotechnol 11:126–129
Wang L, Coppel RL (2008) Oral vaccine delivery: can it protect against non-mucosal pathogens? Expert Rev Vaccines 7:729–738. https://doi.org/10.1586/14760584.7.6.729
World Health Organization (2010) Working to overcome the global impact of neglected tropical diseases: first WHO report on neglected tropical diseases. Geneva, 172 p

Classical Swine Fever Virus

Han Sang Yoo

Abstract Classical swine fever (CSF) caused by classical swine fever virus (CSFV), *Pestivirus* genus of the *Flavivirdae* family, is a highly contagious, multisystemic, and hemorrhagic disease of pigs and one of the most important disease in pig industry, economically and clinically. Domestic and wild pigs are the only natural reservoirs of CSFV. Since the first report of the outbreak of CSF in the early 19th century, a lot of attempts have been made to prevent and/or eradicate the disease. Many countries have successfully eradicated CSF due to the attempts and trials. However, some countries are still sporadic outbreak and/or endemic. Therefore, World Organization for Animal Health is still designated as a notifiable disease. As a step to prevent and/or eradicate the disease, vaccines have been developed. After reviewing etiology, clinical symptoms and diagnosis of CSF, development of vaccines, especially recent vaccines were intensively reviewed. Researches on plant-made vaccine candidates against CSF were introduced in this chapter.

Keywords Classical swine fever virus · Vaccines · Plant · Prevention

1 Introduction

Classical swine fever (CSF) is caused by the classical swine fever virus (CSFV), a small, enveloped, single-stranded RNA virus that belongs to the *Pestivirus* genus of the *Flaviviridae* family. CSF is one of the most economically important pig diseases worldwide, affecting profitability in large farms and livelihood of small backyard farms. Formerly known as "Hog Cholera", is a highly contagious, multi-systemic disease of worldwide importance that is notifiable to the World Organisation for Animal Health (OIE) (Song et al. 2013).

H. S. Yoo (✉)
Department of Infectious Diseases, College of Veterinary Medicine, Seoul National University, Seoul 08826, Korea
e-mail: yoohs@snu.ac.kr

J. MacDonald (ed.), *Prospects of Plant-Based Vaccines in Veterinary Medicine*,
https://doi.org/10.1007/978-3-319-90137-4_11

Clinical outbreaks suggestive of CSF were reported in the early 19th century, and CSF was recognized as viral in nature by 1903. The first reports of a CSF outbreak in South Korea were in 1908, and the Buri strain was first isolated from the Buri pig farm in Seoul in October of 1947. Since then, CSF had been recognized as one of the most devastating diseases in Korea, with sporatic outbreaks that have continuously threatened the swine industry (Song et al. 2013).

Wild and domestic pigs are the only natural reservoirs of CSFV, which is endemic in parts of Eastern Europe, Southeast Asia, Central America, and South America. Many countries, including the United States, Australia, Canada, Ireland and New Zealand, have successfully eradicated CSF and obtained disease-free recognition from the OIE (Song et al. 2013). Now, all countries in North America, Australia and Western Europe are free of CSF. Although eradicated from domestic pigs in Western Europe, CSFV remains endemic in some populations of wild boars; therefore, farms in these areas are at risk of reinfection. Japan was recently added to the list of countries with CSF-free status, which they obtained upon successful completion of their 10-year eradication program in 2006 (Song et al. 2013). However, the reemergence of CSF is always a risk, and several areas previously free of CSF have had incursions in recent years.

2 The Etiology of CSF

The etiological agent responsible for CSF is an enveloped, single-stranded RNA virus known as classical swine fever virus (CSFV) that belongs to the *Pestivirus* genus of the *Flaviviridae* family. *Bovine viral diarrhea virus* (BVDV) 1 and 2 and *Border disease virus* (BDV) are related viruses of the same genus (Blome et al. 2017a, b). The genome of CSFV consists of a positive-stranded RNA molecule of about 12.3 kb encoding a single open reading frame that is translated into a 3898-amino-acid polyprotein, giving rise to 13 CSFV mature proteins by co-and post-translational processing (Meyers et al. 1989; Meyers and Thiel 1996, Blome et al. 2017a, b). These 13 mature proteins comprise of four structural proteins, namely the core (C) protein, and enveloped glycoproteins Erns, E1 and E2; and the non-structural proteins N^{pro}, p7, NS2-3, NS2, NS3, NS4A, NS4B, NS5A, and NS5B (Blome et al. 2017a, b). The coding region is flanked by non-translated regions (NTR) at 5' and 3' end (Blome et al. 2017a, b). Of these, the E2 structural protein encompasses major antigenic domains and cytotoxic T lymphocyte (CTL) epitopes, suggesting it is a promising candidate for use as an immunogen with the capacity to induce neutralizing antibodies and CTL activities against CSFV (Rumennapf et al. 1991; Tarradas et al. 2010; Lin et al. 2012; Xu et al. 2015). Therefore, genetic typing is most commonly based on the E2 glycoprotein because abundant sequence data are available.

Classical swine fever virus is relatively stable with very slow evolution for an RNA virus, but is antigenically and genetically diverse. Based on sequencing of the 5' non-translated region, of 190 nucleotides of the E2 envelope glycoprotein gene and of 409 nucleotides of the NS5B polymerase gene, CSFV was classified into three major genetic groups (groups 1, 2 and 3), each with three or four subgroups: 1.1, 1.2, 1.3; 2.1, 2.2, 2.3; and 3.1, 3.2, 3.3 and 3.4 (Paton et al. 2000; Kirkland et al. 2012). A link between genotype and geographic origin was revealed by phylogenetic analyses of the CSFV isolates during the last decade. Group 1 isolates are present in South America and Russia, most group 2 isolates were from outbreaks in Western, Central, or Eastern Europe, and some Asian countries and group 3 viruses are apparently confined to Asia (Kirkland et al. 2012).

Although outbreaks of the disease could be caused by transmission of the etiological agent via several routes, oronasal transmission by direct or indirect contact with infected wild or domestic pigs is the primary route under natural circumstances, followed by ingestion of contaminated foodstuff (Edwards 2000; Fritzemeier et al. 2000; Kirkland et al. 2012). Transport and/or introduction of infected pigs might be the major source of outbreaks and spread of the disease in finishing units and areas with small pig farms (Ribbens et al. 2004). Analysis of 1997/98 CSF epidemic data in the Netherlands suggested that airborne transmission is also possible within a holding pen or within a radius of less than 500 meters, even though the transmission did not occur over long distances (Elbers et al. 2001). Rats and dogs have been shown experimentally to be mechanical vectors (Ribbens et al. 2004) indicating that they have the potential for transmission of CSF. As a result, veterinarians, inseminators, pig farmers, screening teams and other people who come in contact with infected animals might act as indirect transmission vehicles when biosecurity is deficient (Ribbens et al. 2004).

Typical of enveloped viruses, CSFV is inactivated by organic solvents (ether or chloroform) and by detergents. Sodium hydroxide (2%) is still considered the most suitable for disinfection of contaminated premises. However, CSFV survives for prolonged period under certain conditions (e.g., cool, moist, protein-rich conditions such as occur in meat) (Ribbens et al. 2004; Kirkland et al. 2012), and can even survive for 2 weeks at 20 °C and more than 6 weeks at 4 °C in liquid manure (Edwards 2000; Kirkland et al. 2012).

3 Disease Symptoms and Diagnosis

It is well known that the most common transmission route of CSFV is oronasal, with primary virus replication occurring in the tonsils. The replicated virus can spreads to regional lymph nodes, then via peripheral blood to the bone marrow, visceral lymph nodes, and lymphoid structures associated with the small intestine and spleen. Complete spread of the virus in pigs usually takes less than 6 days (Kirkland et al. 2012).

CSF is a highly contagious, multisystemic, and hemorrhagic viral disease of pigs that can manifest as acute, subacute, chronic, or late onset disease based on its progression (Lim et al. 2016a, b). In the acute form of CSF, the initial clinical signs include anorexia, lethargy, conjunctivitis, respiratory signs and constipation followed by diarrhea (Floegel-Niesmann et al. 2009; Kirkland et al. 2012). The most common pathological feature in this form is hemorrhagic leucopoenia, thrombocytopenia, petechiae, and ecchymoses, which occur in the skin, lymph nodes, larynx, bladder, kidney, and ileocecal junction (MacLachlan and Dubovi 2011; Kirkland et al. 2012). Other characteristics of CSF include multifocal infarction of the margin of the spleen, even though lesions are not always present. Swollen or hemorrhagic lymph nodes or tonsils are also common. Death from the acute form usually occurs 2–4 weeks after CSFV infection.

In the chronic form, the same clinical signs are observed, but the pigs survive for 2–3 months before dying. Non-specific signs such intermittent hyperthermia, chronic enteritis, and wasting may also be seen (Kirkland et al. 2012). The pathological feature is extensive ulceration of the mucosa of the large intestine (button ulcers), as well as a generalized depletion of lymphoid tissues. Opportunistic bacterial pneumonia and enteritis might also be observed. Hemorrhages, inflammatory lesions and infractions are less common, or even absent, despite the degeneration of endothelial cells. General exhaustion of the lymphoid system, with atrophy of the thymus and germinal centers in the spleen, is the most prominent lesion in swine dying from chronic CSF (MacLachlan and Dubovi 2011; Kirkland et al. 2012).

Transplacental infection of CSFV is possible at any stage of pregnancy, and the infection can cause abortion and stillbirths depending on the strain and the time of gestation. Congenital malformations such as central dysmyelinogenesis, cerebellar hypoplasia, microencephaly, and pulmonary hypoplasia might be observed (Kirkland et al. 2012). Moreover, infection at 50–70 days of gestation can lead to the birth of persistently viremic piglets (Kirkland et al. 2012).

Early recognition of CSF and prompt elimination of CSFV-infected animals are key to controlling the disease; therefore, the importance of diagnosis has long been recognized. The longer CSF remains undetected, the greater the opportunity for the virus to spread. More than 75% of the recent CSF epidemics were detected based on clinical signs by farmers and veterinarians (Kirkland et al. 2012). Therefore, measures of the average daily gain, feed consumption, and body temperature before the first clinical signs can be monitored for changes (Kirkland et al. 2012). Diagnosis of CSFV infection is conducted by detection of CSFV and/or detection of antibodies to CSFV. However, it is essential to discriminate between CSFV and BVDV or BDV, because of cross-reactivity due to sharing of common antigens. Several diagnostic methods to detect components of the virion (antigens or nucleic acids) or specific antibodies against viral antigens, are available. CSFV may be isolated from tissue homogenates, serum, plasma, buffy coat, and whole blood collected in heparin or EDTA. The tissues most likely to contain the virus are the tonsils, spleen, kidney, ileocecal lymph node, and retropharyngeal lymph node. The isolation of CSFV followed by real-time, reverse transcription-polymerase chain reaction (qRT-PCR) or antigen-capture ELISA are the most popular methods to detect CSFV (MacLachlan and Dubovi 2011; Kirkland et al. 2012). Although virus identification (VI) is the reference method in most CSFV eradication programs, it is labor intensive, time consuming, and incompatible with the rapid response required to prevent further spread of the virus. Moreover, CSFV cannot induce a cytopathic effect (CPE) in cell culture; therefore, additional experiments are required to detect the virus. As a result, qRT-PCR assays have many advantages over other methods and are now preferred for the detection of viruses or RNA. These assays have both high sensitivity and specificity, particularly when they are probe-based (Hoffmann et al. 2009). A wide variety of samples are suitable for testing by qRT-PCR, but whole blood, swabs, and tissue samples are primarily used for the diagnosis of CSF. Several CSFV-specific qRT-PCR kits are commercially available (Le Dimna et al. 2008). The high sensitivity of qRT-PCR will support testing of pooled samples (Le Dimna et al. 2008), which can significantly increase throughput. Antigen-capture ELISAs may be used for early diagnosis of CSFV in live pigs. Double-antibody sandwich ELISAs are based on monoclonal and/or polyclonal antibodies directed against a variety of viral proteins. Serum, buffy coat fraction, whole blood in heparin or EDTA, or tissue homogenates can be used in these assays. However, ELISA is only recommended for samples from animals with clinical signs or pathological lesions compatible with CSF and for screening herds suspected to have been recently infected (Kirkland et al. 2012). Direct fluorescent antibody (FA) on frozen sections is commonly used to detect viral antigens with high sensitivity, throughput, and rapid turnaround (Kirkland et al. 2012).

Virus neutralization (VN) and ELISA for the detection of anti-CSFV antibodies are useful methods for detection of antibodies to CSFV. Virus neutralization has been considered a reference assay for the detection of CSFV-specific antibodies; however, it requires good quality serum samples and the use of a cell culture system. The method is also time consuming, requiring 3–5 days to obtain results. Accordingly, ELISAs are useful for the detection of anti-CSF antibodies during epidemiological surveys and for monitoring of CSF-free areas (Kirkland et al. 2012).

4 Commercial Vaccines

The two main strategies to control CSF are systematic prophylactic vaccination and non-vaccination with stamping-out of infected animals in outbreaks. No effective treatment for CSF is available, nor have there been any attempts to develop one. In countries with vaccination policies, live attenuated vaccines are widely used to control CSF and demonstrate efficient protection against the virus. In endemic regions, prophylactic vaccination is often used to prevent the spread of infection. Once CSF is under control, vaccination can be stopped while surveillance is continued (Moennig and Becher 2015). In endemic CSF situations or in the crisis of a larger epidemic, where the usual control measures (stamping out, zoning, standstill, etc.) may prove to be insufficient, systemic vaccination of the susceptible population is one of the most promising options for successful control and eradication, particularly in commercial and industrialized pig production systems (Milicevic et al. 2013).

In disease-free areas, new outbreaks are controlled by early detection, stamping out, movement control, safe disposal of carcasses, and cleaning and disinfection

(Moennig and Becher 2015). Although some regions are free of CSFV, it is still present at the borders between free and endemic areas and in some wild boar populations (Laddomada 2000; Rossi et al. 2015). For purposes of international trade, free areas maintain a "no vaccination" policy against CSF in order to differentiate infected animals by antibody detection. In such areas, control is based on stamping out infected or suspected herds, with concomitant implementation of quarantine measures (Kirkland et al. 2012).

While vaccination in endemically infected areas is meant to prevent economic losses and may be a first step towards eradication using highly efficacious vaccines, emergency vaccination campaigns during epidemics are mainly intended to prevent spread of the disease; accordingly, they require fast-acting vaccines, preferably with marker potential, to allow a rapid return to regular trade conditions (van Oirschot 2003; Blome et al. 2017a, b). Because different senarios require different vaccine characteristics (van Oirschot 2003; Blome et al. 2017a, b), several commercial vaccines are available. A first generation of vaccines against CSF was developed, initially consisting of virus and porcine hyper immune serum, followed by the crystal-violet inactivated vaccine in 1936 (Saulmon 1973; Blome et al. 2017a, b). However, problems with the vaccines such as low safety and efficacy led to further investigations to develop live attenuated vaccines. The strains most commonly used for vaccines were the Lapinized Philippines Coronel (LPC) strain, the Chinese vaccine strain (C-strain), known as the Chinese hog cholera lapinized virus (HCLV), the low-temperature-adapted Japanese guinea-pig exaltation-negative (GPE^-) strain, the French cell culture-adapted Thiverval strain, and Mexican PAV strains (Dong and Chen 2007; Blome et al. 2017a, b). There are several constraints associated with the conventional, live attenuated vaccines made from these strains, especially the lack of ability to differentiate between infected and vaccinated animals (DIVA or marker strategy). To address this issue, research was directed towards the development of marker vaccines (van Oirschot 2003; Dong and Chen 2007). As the first generation of marker vaccines, E2 subunit vaccines using recombinant E2 expressed by baculoviruses in insect cells were developed (Hulst et al. 1993). Although the safety of E2 subunit vaccines was confirmed, limitations were shown with respect to its efficacy, application and production of the vaccine in comparison with live attenuated vaccines (Bounna et al. 1999; Blome et al. 2017a, b). Although several studies have investigated the development of marker vaccines, only1 one subunit vaccine is available on the market. This vaccine consists of the E2 glycoprotein of CSFV strain "Alfort/Tübingen" in a water-in-oil adjuvant (Porcilis® Pesti, MSD Animal Health) (Blome et al. 2017a, b).

Subsequently, a new next-generation marker vaccine, CP7_E2alf (Suvaxyn CSF Marker, Zoetis), was developed using a chimera based on the infectious cDNA clone of the cytopathogenic BVDV strain "CP7" (Meyers et al. 1997). This vaccine was confirmed to have better genetic stability, safety, efficacy and protective effects (Blome et al. 2014; Goller et al. 2015). For the DIVA diagnosis, a specific real-time RT-PCR and serological marker system based on the detection of CSFV-specific E^{rns} antibodies was developed (Leifer et al. 2009; Liu et al. 2009, Blome et al. 2017a, b).

However, the humoral immune response induced by these vaccinated animals cannot serologically differentiate infected from vaccinated animals (DIVA), hampering controlled eradication of CSFV (Moennig and Becher 2015; Blome et al. 2017a, b).

Therefore, there have been several other attempts to develop effective vaccines against CSF that enable differential diagnosis.

5 Experimental Vaccines Expressed in Non-plant Systems

For development of CSFV vaccines, the roles of neutralizing antibody-based humoral immune responses (van Gennip et al. 2002; Xu et al. 2015) and CTL-based cellular immune responses have been both highlighted (Rau et al. 2006; Xu et al. 2011, 2015). After oral infection of CSFV, the invasion of CSFV is often initiated at mucosal surfaces, particularly intestinal tissues. Thus, vaccination inducing IgA-based protective mucosal immunity via the mucosal approach, has been attempted to prevent the virus from entering the body via mucosa and its further spread to systemic circulation. New vaccines that can trigger protective antiviral mucosal and systemic immune responses will provide a promising strategy for the development of a vaccine against CSFV infection (Xu et al. 2015).

Of three glycoproteins in the envelope of CSFV, Erns, E1, and E2, E2 is the most immunogenic, and is a virulence determinant. E2 is the main target of the humoral immune response to CSFV infection. Post-translational modification (N-linked glycosylation) is required for the immunogenicity of E2. Recombinant CSFV E2 glycoprotein (rE2) has been produced in a variety of expression systems, including bacteria, yeast, plants, adenovirus, goat, and a baculovirus/insect system (Barrera et al. 2010; Wu et al. 2010; Lin et al. 2012; Jung et al. 2014; Sanchez et al. 2014; Xu et al. 2015).

Immunogenic CSFV peptides, DNA vaccines, viral vector vaccines, trans-complemented deletion mutants (replicons) and chimeric pestiviruses were developed and evaluated in experimental and/or target animals (Blome et al. 2017a, b).

5.1 Subunit Vaccines

Vaccines based on synthetic immunogenic CSFV peptides contained either one peptide or a mixture of different peptides covering different parts of the antigenic domains of the CSFV glycoprotein E2. The vaccines are highly safe, but require parenteral administration, adjuvants and multiple vaccination schemes (Dong et al. 2002; Dong et al. 2006; Liu et al. 2006).

Recently, additional subunit vaccines were reported consisting of E2 expressed using different expression systems such as the mammary glands of goats after adenoviral transduction, yeast, *Pichia pastories*, a transgenic mammalian cell line, and baculovirus (Toledo et al. 2008; Lin et al. 2012; Hua et al. 2014; Sanchez et al. 2014; Madera et al. 2016). These vaccines reportedly had several advantages including stability, protection, and DIVA. Moreover, genetically engineered *Lactobacillus plantarum* expressing the CSFV E2 protein in combination with Thymosin alpha-1 (Xu et al. 2015) and E2 protein expressed in transgenic rice calli from *Oriza sativa* L. *cv.* were generated and found to induce humoral and cellular immune responses upon oral administration (Jung et al. 2014).

The use of live vehicles has received a great deal of attention. Lactic acid bacteria have been investigated as delivery systems for heterologous antigens to the mucosal immune system with a number of advantages over the traditional parental vaccination, such as noninvasiveness and the possibility of eliciting both mucosal and systemic immune responses (Xu et al. 2011; Villena et al. 2011). Using *Lactobacillus plantarum*, E2, a main structural protein of CSFV, and thymosin α-1 (Tα1), a nontoxic immune-modifier peptide hormone from the thymus, were cloned, expressed and developed as vaccines in E2 alone and in combination with E2 with Tα1. Upon oral administration of each vaccine, protective immune responses were induced in pigs against CSFV infection by eliciting IgA-based mucosal, IgG-based humoral and CTL-based cellular immune responses. The results suggest that the recombinant lactobacillus microecological agent expressing CSFV E2 protein combined with Tα1 as an adjuvant might be a promising candidate for vaccine development against CSFV (Xu et al. 2015).

CSFV E2 envelope viral glycoproteins were produced in the milk of adenovirally transduced goats and formulated as a water-in oil emulsion after purification. Pigs immunized with the vaccine showed an effective response in immune responses and clinical signs in the challenge with highly pathogenic CSFV strain (Barrera et al. 2010; Sanchez et al. 2014). Early onset and long lasting protection were induced in pigs by vaccination (Barrera et al. 2010). A simpler procedure for the introduction and implementation of a commercial subunit vaccine was also demonstrated with this E2 protein by optimization of purification steps and dosage of the vaccine (Sanchez et al. 2014).

5.2 *DNA Vaccines*

All DNA vaccines were based on plasmid constructs that express CSFV glycoprotein E2 (Ganges et al. 2005; Wienhold et al. 2005; Andrew et al. 2006) and partially co-express genes for cytokines such as IL-3, IL-12, IL-18 or regulatory cell surface molecules (CD154) to enhance their immunogenic potential (Wienhold et al. 2005; Andrew et al. 2006). With those vaccines, immunogenicity, DIVA and safety were demonstrated. However, high dosages and multiple vaccinations were required to protect pigs against challenge infection with highly virulent CSFV (Blome et al. 2017a, b). Moreover, application of those vaccines to pigs was not economically viable because of the high cost of DNA preparation and inefficient delivery (Sun et al. 2011). For viral vector vaccines, vectors using vaccinia virus, pseudorabies virus, porcine and human adenovirus, swine pox virus, fowlpox virus, parapox virus and canarypox virus were used (Blome et al. 2017a, b). Although some vector vaccines are able to confer full protection, they still remain as prototypes, and licensing of a candidate is not yet in sight (Blome et al. 2017a, b).

5.3 Trans-Complemented Deletion Mutants

Trans-complemented deletion mutants (replicons) of CSFV were constructed to avoid the risk of reversion to virulent viruses (Beer et al. 2007). Trans-complemented CSFV E^{rns} or E2 deletion mutants were constructed (Maurer et al. 2005; Frey et al. 2006). To accomplish this, the RNA of mutants was transfected into porcine kidney cells, which led to autonomous replication without the production of virus progeny (replicons) in E^{rns} or E2 expressing recombinant cell lines. Complemented virions are replication-deficient during the second replication cycle and thus referred to as DISC (defective in second cycle). Vaccination efficiency was dependent on the application route, with complete protection conferred by intradermal injection of the replicon A187delErns, but partial protection by oral application of the replicon (Frey et al. 2006; van Gennip et al. 2002).

5.4 Viral Vectored Vaccines

Use of DNA-based *Semliki Forest Virus* replicons expressing CSFV E2 demonstrated induction of reliable protection by high dose applied three times (Li et al. 2007a; Li et al. 2007b). Furthermore, the adenovirus/alphavirus-replicon chimeric vector-based vaccine rAdV-SFV-E2 delivered into a *Semliki Forest Virus* replicon expressing the CSFV E2 gene induced strong immune responses in pigs and rabbits (Li et al. 2016; Sun et al. 2011). Moreover, pigs were fully protected against lethal challenges with CSFV "Shimen" seven weeks after two vaccine applications which were given three weeks apart (Sun et al. 2013). Recently, enhanced protective immunity of rAdV-SFV-E2 was demonstrated by a *Salmonella* Enteritidis bacterial ghost (cell envelope preparation) adjuvant (Xia et al. 2016a, b). Chimeric pestiviruses based on infectious cDNA clones of CSFV or BVDV are the most promising candidates for marker vaccines (Meyers et al. 1996, 1997). Several chimeric pestiviruses have been described, some of which have been extensively investigated in target species. However, problems preventing improvement in discriminatory assays still remain (Blome et al. 2017a, b).

Recombinant CSFV E2 in the bi-cistronic baculovirus/larvae expression system induced high titers of anti-CSFV E2 antibodies with neutralizing activity in mice and piglets (Wu et al. 2010). Moreover, CSFV E2 (yE2) expressed in the yeast, *Pichia pastoris*, showed a protective immune response against CSFV challenge and prevented horizontal transmission of CSFV during cohabitation of unimmunized sentinels 3 days after challenge infection (Lin et al. 2012).

The safety of classical swine fever virus vaccine (LOM strain) was investigated in pregnant sows and their offspring. One study clarified the issue by vaccination of pregnant sows with or without CSFV antibody with the CSFV vaccine (LOM strain) at early or mid-stages of pregnancy (Lim et al. 2016b). The LOM strain may induce sterile immunity and provide rapid, long-lasting, and complete protection against CSFV. The LOM strain may be capable of crossing the placenta of pregnant sows with free CSFV antibody during gestation and could be transmitted from the pregnant sow to the fetus (Plateau et al. 1980; Vannier et al. 1981). Potential adverse effects of the LOM strain vaccine in pregnant sows without CSFV antibody may result in stillbirth or fetal mummification. Therefore, vaccination of the LOM strain in pregnant sows without CSFV antibody should receive attention, although there are no potential adverse effects caused by the LOM vaccine in pregnant sows with antibodies to CSFV (Lim et al. 2016b). Moreover, the efficacy of the CSF vaccine (LOM strain) was investigated in pigs infected with immunosuppressive pathogens, porcine reproductive and respiratory syndrome virus (PRRSV) and/or porcine circovirus-2 (PCV2). One study demonstrated that LOM strain may induce an immune response without interference with the immunosuppression during the development of protective immunity (Lim et al. 2016a). The levels of CSFV antibodies in LOM-vaccinated pigs were not affected by infection of PRRSV or/and PCV2. Co-infection with both PRRSV and PCV2 may affect CSF vaccine virus replication or viral activity in pigs vaccinated with LOM strain (Lim et al. 2016a).

6 Plant-Made Vaccine Candidates

The major challenges in animal vaccine development are the cost and volume. Commercial vaccines with good results still have several disadvantages such as complex and expensive production and purification, requirement for low-temperature storage, safety issues and the need for a skilled person for administration. Plants offer an attractive and affordable platform for vaccines against animal diseases because of their low cost and production in large quantities, and are free of attenuated pathogens and cold chain requirements, especially in industries with low profit margins (Shahid and Daniell 2016). Therefore, edible, orally delivered, low-cost vaccines are urgently needed for the production of disease-free animals. Also, transgenic plant based systems have added advantages, such as antigen encapsulation and stability for extended periods at ambient temperature. Most importantly, products from transgenic plants are very unlikely to be contaminated by animal pathogens or microbial toxins (Streatfield et al. Streatfied 2006; Gomez et al. 2008). Although several attempts have been made with different expression systems to develop vaccines against CSF, only a few studies have been conducted with plants.

A ubiquitin-containing vector was used to express CSFV E2 glycoprotein at a level of 10 μg of antigen in 1 g of lyophilized transgenic alfalfa leaves and 160 μg in 1 g of dry mass of lettuce. Mice were immunized with 0.5 μg of E2 CSFV two times at one-month intervals, and the kinetics of immune response (IgG in serum and IgA in fecal pellets) were monitored. A significant increase in antibodies was demonstrated in both IgG and IgA levels after the second immunization (Eble et al. 2014; Legocki et al. 2005).

With the goal of improving purification efficiency and IgG response, CSFV E2 glycoprotein was expressed as a fusion to the coat protein of potato virus X peptide in *Nicotiana benthamiana*, using a modified PVX vector. Correct retention of the peptide encoding sequences in the PVX construct after three sequential passages in the plant was confirmed by RT-PCR. Moreover, the epitope coding sequences were replicated with high fidelity during PVX infection. In addition, they demonstrated induction of an immune response in rabbits by three subcutaneous injections of partially purified virions (Marconi et al. 2006).

Several transgenic plants, including corn, wheat, potato, tomato, and rice, have received attention for vaccine production, storage and delivery systems for oral immunization. Of these, rice has been considered an effective expression system based on storage, processing, yield, and available genomic information (Nochi et al. 2007; Yuki et al. 2009). Moreover, rice can be cultured in the callus form using a bioreactor, thereby reducing the risk of transgenic rice disturbing the ecosystem (Jung et al. 2014).

The E2 glycoprotein of CSFV SW03 strain was expressed in transgenic rice calli (TRCs) from *Oriza sativa* L. ev. Dongjin. Immune responses to the rE2-TRC in mice and pigs were investigated after oral administration through oral gavage four times in mice and three times in pigs with 5 mg of rE2 per weight (mg/kg) at 10-day intervals. The administration of rE2-TRCs to mice and pigs increased E2-specific antibody titers in IgG and IgA-levels, IgA-specific antibody-secreting cells in the Peyer's patches of mice and PBMC of pigs and IgG-specific antibody-secreting cells in the PBMC of pigs when compared to animals receiving the vector alone. In addition, mice receiving rE2-TRCs had a higher level of CD8+ lymphocytes and Th1 cytokine immune responses to purified rE2 (prE2) in vitro than controls. Pigs receiving rE2-TRCs also showed an increase in IL-8, CCL2, and the CD8+ subpopulation in response to stimulation with prE2 (Jung et al. 2014).

Development of plant-based vaccines against CSF should focus on edible vaccines because these are easier to administer to domestic pigs, and can also be practical for use in wild species. CSF remains problematic in the wild boars of several countries including Western European countries. However, progress in the genetic manipulation of some edible plants such as corn is very slow due to the complicated life cycle of the plants. Appropriate plants should therefore be selected with the ability not only to express the major epitopes of CSFV, but to be used as feed for pigs (Shao et al. 2008). In addition, many pathobiological characteristics of CSFV infection remain to be revealed, especially in immune responses. After solving these problems, large-scale immunological and challenge experiments can be conducted using mass-produced CSFV antigen(s).

Classical swine fever is an important disease in both clinically and economically in the swine industry. Although the disease has been eradicated in the developed countries, it still occurs in several countries, especially developing countries. Moreover, the importance of the disease has been recognized in wild boars. Therefore, several attempts have been made to control the disease. Among these, vaccination has been the most effective. Although live attenuated vaccines have been widely recognized and used in the field, they have several disadvantages. To overcome these disadvantages, different types of vaccine have been investigated experimentally and clinically.

Several vaccines have been investigated in efforts to develop safe and effective vaccines using plant-based delivery systems; therefore, this system has also been applied to CSF vaccine. However, few studies have investigated plant-based CSF vaccines. Accordingly, additional attempts should be made to develop safe and effective vaccines for CSF in future studies.

References

Andrew M, Morris K, Coupar B, Spoat K, Oke P, Bruce M, Broadway M, Morrissy C, Strom D (2006) Porcine interleukin-3 enhances DNA vaccination against classical swine fever. Vaccine 24:3241–3247. https://doi.org/10.1016/j.vaccine.2006.01.023

Barrera M, Sanchez O, Farnos O, Rodriguez MP, Dominguez P, Tait H, Frias M, Avila M, Vega E, Toledo JR (2010) Early onset and long lasting protection in pigs provided by a classical swine fever E2-vaccine candidate produced in the milk of goats. Vet Immunol Immunopath 133:25–32. https://doi.org/10.1016/j.vetimm.2009.06.015

Beer M, Reimann I, Hoffmann B, Depner K (2007) Novel marker vaccines against classical swine fever. Vaccine 25:5665–5670. https://doi.org/10.1016/j.vaccine.2006.12.036

Blome S, Gabriel C, Schmeiser S, Meyer D, Meindl-Bohmer A, Koenen F, Beer M (2014) Efficacy of marker vaccine candidate CP7_E2alf against challenege with classical swine fever virus isolates of different genotypes. Vet Microbiol 169:8–17. https://doi.org/10.1016/j.vetmic.2013.12.002

Blome S, Moβ C, Reimann I, Konig P, Beer M (2017a) Classical swine fever vaccines—State-of-the-art. Vet Microbiol 206:10–20. https://doi.org/10.1016/j.vetmic.2017.01.001

Blome S, Staubach C, Henke J, Carlson J, Beer M (2017b) Classical swine fever—An updated review. Viruses 9:86. https://doi.org/10.3390/v9040086

Bounna A, de Smit AJ, de Kluijver EP, Terpstra C, Moormann RJ (1999) Efficacy and stability of a subunit vaccine based on glycoprotein E2 of classical swine fever. Vet Microbiol 66:101–114

Dong XN, Chen YH (2007) Marker vaccine strategies and candidate CSFV marker vaccines. Vaccine 25:205–230. https://doi.org/10.1016/j.vaccine.2006.07.033

Dong XN, Qi Y, Ying J, Chen X, Chen YH (2006) Candidate peptide-vaccine induced potent protection against CSFV and identified a principal sequential neutralizing determinat on E2. Vaccine 24:426–434. https://doi.org/10.1016/j.vaccine.2005.08.002

Dong XN, Wei K, Liu ZQ, Chen YH (2002) Candidate peptide vaccine induced protection against classical swine fever virus. Vaccine 21:167–173

Eble PL, Quak S, Geurtus Y, Moonen-Leusen HWM, Loeffen WLA (2014) Efficacy of CSF vaccine CP7_E2alf in piglets with maternally derived antibodies. Vet Microbiol 174:27–38. https://doi.org/10.016/j.vetmic.2014.08.030

Edwards S (2000) Survival and inactivation of classical swine fever virus. Vet Microbiol 73:175–181

Edwards S, Fukusho A, Lefevre P, Lipowski A, Pejsak Z, Roehe P, Westergaard J (2000) classical swine fever: the global situation. Vet Microbiol 73:103–119

Elbers AR, Stegeman JA, de Jong MC (2001) Factors associated with the introduction of classical swine fever virus into pig herds in the central area of the 1997/98 epidemic in The Netherlands. Vet Rec 149(13):377–382

Floegel-Niesmann G, Blome S, Gerβ-Dulmer H, Bunzenthal C, Moennig V (2009) Virulence of classical swine fever virus isolates from Europe and other areas during 1996 until 2007. Vet Microbiol 139:165–169

Frizemeier J, Teufferet J, Greiser-Wilke I, Staubach Ch, Schiuter H, Moenning V (2000) Epidemiologu of classical swine fever in Germany in the 1990s. Vet Microbiol 77:29–41

Frey CF, Bauhofer O, Ruggli N, Summerfield A, Hofmann MA, Tratschin JD (2006) Classical swine fever virus replicon particles lacking the Erns gene: a potential marker vaccine for intradermal application. Vet Res 37:655–670. https://doi.org/10.1051/vetres:2006028

Ganges L, Barrera M, Nunez JI, Blanco I, Frias MT, rodriguez F, Sobrino F (2005) A DNA vaccine expressing E2 protein of classical swine fever virus elicits T cell responses that can prime for rapid antibody production and confer total protection upon viral challenge. Vaccine 24: 3741–3752. https://doi.org/10.1016/j.vaccine.2005.01.153

Goller KV, Dräger C, Höper D, Beer M, Blome S (2015) classical swine fever virus m, arker vaccine strain CP7_E2alf: genetic stability in vitro and in vivo. Arch Virol 160:3121–3125. https://doi.org/10.1007/s00705-015-2611-z

Gomez E, Zoth SC, Carrillo E, Roux ME, Berinstein A (2008) Mucosal immunity induced by orally administered transgenic plants. Immunobiology 213:671–675. https://doi.org/10.1016/j.imbio.2008.02.002

Hoffmann B, Beer M, Reid SM, Mertens P, Oura CAL, van Rijn PA, Slomka MJ, Banks J, Brown IH, Alexander DJ, King DP (2009) A review of RT-PCR technologies used in veterinary virology and disease control: sensitive and specific diagnosis of five livestock diseases notifiable to the World Organisation for Animal Health. Vet Microbiol 139:1–23. https://doi.org/10.1016/j.vetmic.2009.04.034

Hua RH, Huo H, Li YN, Xue Y, Wang XL, Guo IP, Zhou B, Song Y, Bu ZG (2014) Generation and efficacy evaluation of recombinant classical swine fever virus E2 glycoprotein expressed in stable transgenic mmalian cell line. PLoS One 9:e106891. doi;https://doi.org/10.1371/journal.pone.0106891

Hulst MM, Westra DF, Wensvoort G, Moormann RJ (1993) Glycoprotein E1 of hog cholera virus expression in insect cells protects swine from hog cholera. J Virol 67:5435–5442

Jung MH, Shin YJ, Kim J, Cha SB, Lee WJ, Shin MK, Shin SW, Yang MS, Jang YS, Kwon TH, Yoo HS (2014) Induction of immune response in mice and pigs by oral administration of classical swine fever virus E2 protein expressed in rice calli. Arch Virol 159:3219–3230. https://doi.org/10.1007/s00705-014-2182-4

Kirkland PD, Le Portier M, Vannier P, Finlaisom D (2012) Classical swine fever (Hog Cholera). In: Zimmerman J (ed) Diseases of Swine, 10th edn. Wiley-Blackwell, Ames, pp 539–546

Laddomada A (2000) Incidence and control of CSF in wild boar in Europe. Vet Microboil 73:121–130

Le Dimna M, Vrancken R, Koenen F, Bougeard S, Mesplede A, Hutet E, Kuntz-Simon G, Le Potier MF (2008) Validation of two commercial real-time RT-PCR kits for rapid and specific diagnosis of classical swine fever virus. J Vriol Methods 157:136–142. https://doi.org/10.1016/j.jvirmet.2007.08.013

Legocki AB, Miedzinska K, Czaplinska M, Plucieniczak A, Wedrychowicz H (2005) Immunoprotective properties of transgenic plants expressing E2 glycoprotein from CSFV and cysteine protease from *Fascioloa hepatica*. Vaccine 23:1844–1846. https://doi.org/10.1016/j.vaccine.2004.11.015

Leifer I, Dpner K, Blome S, Le Potier MF, Le Dimna M, Beer M, Hoffmann B (2009) Differentiation of C-strain Riems or CP7_E2alf vaccinated animals from animals infected by classical swine fever virus field strains using real-time RT-PCR. J Virol Methods 158:114–122. https://doi.org/10.1016/j.virolmet.2009.02.002

Li H, Gao R, Zhang Y (2016) A promsing trigene recombinant human adenovirus vaccine against classical swine fever virus. Viral Immunol 29(4):244–251. https://doi.org/10.1089/vim.2015.0068

Li N, Qiu HJ, Zhao JJ, Li Y, Wang MJ, Lu BW, Han CG, Hou Q, Wang ZH, Gao H, Peng WP, Li GX, Zhu QH, Tong GZ (2007a) A Semliki Forest virus replicon vectored DNA vaccine expressing the E2 glycoprotein of classical swine fever virus protects pigs from lethal challenge. Vaccine 25:2907–2912. https://doi.org/10.1016/j.vaccine.2006.08.020

Li N, Zhao JJ, Zhao HP, Sun Y, Zhu QH, Tong GZ, Qiu HJ (2007b) Protection of pigs from lethal challenge by a DNA vaccine based on an alphavirus replicon expressing the E2 glycoprotein of classical swine fever virus. J Virol Methods 144:73–78. https://doi.org/10.1016/j.jviromet.2007.03.018

Lim SI, Jeoung HY, Kim B, Song JY, Kim JJ, Kim HY, Cho IS, Kim B, Woo GH, Lee JB, An DJ (2016a) Impact of porcine reproductive and respiratory syndrome virus and porcine circovirus-2 infection on the potency of the classical swine fever vaccine (LOM strain). Vet Microbiol 193:36–41. https://doi.org/10.1016/j.vetmic.2016.07.027

Lim SI, Song JY, Kim JJ, Hyun BH, Kim HY, Cho IS, Kim B, Woo GH, Lee JB, An DJ (2016b) Safety of classical swine fever virus vaccine strain LOM in pregnant sows and their offspring. Vaccine 34:2021–2026. https://doi.org/10.1016/j.vaccine.2016.02.062

Lin GJ, Deng MC, Chen ZW, Liu TY, Wu CW, Cheng CY, Chien MS, Huang C (2012) Yeast expressed classical swine fever E2 subunit vaccine candidate provides complete protection against lethal challenge infection and prevents horizontal virus transmission. Vaccine 30:2336–2341. https://doi.org/10.1016/j.vaccine.2012.01.051

Liu I, Hoffmann B, Baule C, Beer M, Belak S, Widen F (2009) Two real-time RT-PCR assays of classical swine fever virus developed for the genetic differentiation of naturally infected from vaccinated wild boars. J Virol Methods 159:131–133. https://doi.org/10.1016/j.jviromet.2009.03.003

Liu S, Yu X, Wang C, Wu J, Kong X, Tu C (2006) Quadruple antigenic epitope peptide producing immune protection against classical swine fever virus. Vaccine 24:7175–7180. https://doi.org/10.1016/j.vaccine.2006.06.042

MacLachlan NJ, Dubovi EJ (ed) (2011) Classical swine fever virus. In: Fenner's veterinary virology, 4th ed. Elsevier, Tokyo, p 480–481

Madera R, Gong W, Wang L, Burakova Y, Lleellish K, Galliher-Beckley A, Nietfeld J, Henningson J, Jia K, Li P, Bai J, Schlup J, McVey S, Tu C, Shi J (2016) Pigs immunized with a novel E2 subunit vaccine are protected from subgenotype heterologous classical swine fever virus challenge. BMC Vet Res 12:197. https://doi.org/10.1186/s12917-016-0823-4

Marconi G, Albertini E, Barone P, De Marchis F, Lico C, Marusic C, Rutili D, Veronesi F, Porceddu A (2006) *In planta* production of two peptides of the classical swine fever virus (CSFV) E2 glycoprotein fused the coat protein of potato virus X. BMC Biotech 6:29. https://doi.org/10.1186/1472-6750-6-29

Maurer R, Stettler P, Ruggli N, Hofmann MA, Tratschin JD (2005) Oronasal vaccination with classical swine fever virus (CSFV) replicon particles with either partial or complete deletion of the E2 gene induces partial protection against lethal challenge with highly virulent CSFV. Vaccine 23:3318–3328. https://doi.org/10.1016/j.vaccine.2005.01.076

Meyers G, Rumenapf T, Thiel HJ (1989) Molecular cloning and nucleotide sequence of the genome of hog cholera virus. Virology 171:555–567. https://doi.org/10.1016/0042-6822(89)90625-9

Meyers G, Tautz N, Becher P, Thiel HJ, Kummerer BM (1997) Recovery of cytopathogenic and noncytopathogenic bovine viral diarrhea viruses from cDNA constructs. J Virol 71:1735

Meyers G, Thiel HJ, Rümenapf T (1996) Classical swine fever virus: recovery of infectious viruses from cDNA constructs and generation of recombinant cytopathogenic defective interfering particles. J Virol 70:1588–1595

Meyers G, Thiel HJ (1996) Molecular characterization of pestiviruses. Adv Virus Res 47:53–118

Milicevic V, Dietze K, Plavsic B, Tikvicki M, Pinto J, Depner K (2013) Oral vaccination of backyard pigs against classical swine fever. Vet Microbiol 163:167–171. https://doi.org/10.1016/j.vetmic.2012.12.005

Moenning V, Becher P (2015) Pestivirus control programs: how far have we come and where are we going? Anim Health Res Rev 16(1):83–87. https://doi.org/10.1017/S1466252315000092

Nochi T, Takagi H, Yuki Y, Yang L, Masumura T, Mejima M, Nakanishi U, Matsumura A, Uozumi A, Hiroi T, Morita S, Tanaka K, Takaiwa F, Kiyono H (2007) Rice-based mucosal vaccine as a global strategy for cold-chain- and needle-free vaccination. Proc Natl Acad Sci USA 104:10986–10991. https://doi.org/10.1073/pnas.0703766104

Paton DJ, McGoldrick A, Greiser-Wilke I, Parchariyanon S, Sing JY, Liou PP, Stadejek T, Lowings JP, Bjorklund H, Belak S (2000) Genetic typing of classical swine fever virus. Vet Microbiol 73:137–157

Plateau E, Vannier P, Tillon JP (1980) Atypical hog cholera infection: viral isolation and clinical study of in utero transmission. Am J Vet Res 41:2012–2015

Rau H, Revets H, Balmelli C, McCullough KC, Summerfield A (2006) Immunological proeprties of recombinant classical swine fever virus NS3 protein *in vitro* and *in vivo*. Vet Res 37:155–168. https://doi.org/10.051/vetres:2005049

Ribbens S, DEwulf J, Koenen F, Laevens H, de Kruif A (2004) Transmission of classical swine fever. A review. Vet Q 26:146–155

Rossi S, Staubach C, Blome S, Guberti V, Thulke HH, Vos A, Koenen F, Le Potier M (2015) Controlling of CSFV in European wild boar using oral vaccination: a review. Front Micorbiol 6:1141. https://doi.org/10.3389/fmicb.2015.01141

Rumennapf T, Stark R, Meyers G, Thiel HJ (1991) Structural proteins of hog cholera virus expressed by vaccinia virus: further characterization and induction of protective immunity. J Virol 65:589–597

Sanchez O, Barrera M, Farnos O, Parra NC, Salgado ER, Saavedra PA, Meza CD, Rivas CI, Cortez-San Martin M, Toledo JR (2014) Effectiveness of the E2-classcial swine fever virus recombinant vaccine produced and formulated within whey from genetically transformed goats. Clin Vac Immunol 21(12):1628–1634. https://doi.org/10.1128/CVI.00416-14

Saulmon EE (1973) Hog cholera eradication? Dream or reality. J Am Vet Med Assoc 163:1103–1105

Shao HB, He DM, Qian KX, Shen GF, Su ZL (2008) The expression of classical swine fever virus structural protein E2 gene in tobacco chloroplasts for applying chloroplasts as bioreactors. Mol Biol Gene 331:179–184. https://doi.org/10.1016/j.crvi.2007.12.007

Shahid N, Daniell H (2016) Plant-based oral vaccines against zoonotic and non-zoonotic diseases. Plant Biotech J 14:2079–2099. https://doi.org/10.1111/pbi.12604

Song JY, Lim SI, Jeoung HY, Choi EJ, Hyun BH, Kim B, Kim J, Shin YK, dela Pena RC, Kim JB, Joo H, An DJ (2013) Prevalence of classical swine fever virus in domestic pigs in South Korea: 1999–2011. Trans Emer Dis 60:546–551. https://doi.org/10.1111/j.1865-1682.2012.0137.x

Streatfied S (2006) Mucosal immunization using recombinant plant-based oral vaccines. Methods 38:150–157. https://doi.org/10.1016/j.ymeth.2005.09.013

Sun Y, Li HY, Tian DY, Han QY, Zhang X, Li N, Qiu HJ (2011) A novel alphavurys replicon-vectored vaccine delivered by adenovirus induces sterile immunity against classical swine fever. Vaccine 29:8364–8372. https://doi.org/10.1016/j.vaccine.2011.08.085

Sun Y, Tian DY, Li S, Meng QL, Zhao BB, Li Y, Li D, Ling LJ, Liao YJ, Qiu HJ (2013) Comprehensive evaluation of the adenovirus/alphavirus-replicon chimeric vector-based vaccine rAdV-SFV-E2 against classical swine fever. Vaccine 31:538–544. https://doi.org/10.1016/j.vaccine.2012.11.013

Tarradas J, Argilague JM, Rosell R, Nofrarias M, Crisci E, Cordoba L, Perez-Martin E, Diaz I, Rodriguez F, Domingo M, Montoya M, Ganges L (2010) Interferon-gamma induction correlates with protection by DNA vaccine expressing E2 glycoprotein against classical swine fever virus infection in domestic pigs. Vet Microbiol 142:51–58. https://doi.org/10.1016/j.vetmic.2009.09.043

Toledo JR, Sanchez O, Montesino R, Farnos O, Rodriguez MP, Alfonso P, Oramas N, Rodriguez E, Santana E, Vega E, Ganges L, Frias MT, Cremata J, Barrera M (2008) Highly protective E2-CSFV vaccine candidate produced in the mammary gland of adenoviral transduced goats. J Biotechnol 133:370–376. https://doi.org/10.1016/j.jbiotec.2007.09.014

van Gennip HGP, Bouma A, van Rijn PA, Widjojoatmodjo MN, Moormann RJM (2002) Experimental non-transmissible marker vaccine for classical swine fever (CSF) by transcomplementation of E^{rns} or E2 of CSFV. Vaccine 20:1544–1556. https://doi.org/10.1016/S0264-410X(01)00497-2

van Oirschot JT (2003) Vaccinology of classical swine fever: from lab to field. Vet Microbiol 96:367–384. https://doi.org/10.1016/j.vetmic.2003.09.008

Vannier P, Plateau E, Tillon JP (1981) Congenital tremor in pigs farrowed from sows-given hog cholera virus during pregnancy. Am J Vet Res 42:135–137

Villena J, Salva S, Aguero G, Alvarez S (2011) Immunomodulatory and protective effect of probiotic *Lactobacillus casei* against *Candida albicans* infection in malnourished mice. Microbiol Immunol 55:434–445. https://doi.org/10.111/j.1348-0421.2011.00334x

Wienhold D, Armengol E, Marquardt C, Voigt H, Buttner M, Saalmuller A, Pfaff E (2005) Immunomodulatory effect of plasmid co-expressing cytokines in classical swine fever virus subunit gp55/E2-DNA vaccination. Vet Res 36:571–587. https://doi.org/10.1051/vetres:2005019

Wu CM, Hsuan SL, Chen ZW, Jinn TR, Huang C, Liao JW, Chen TH, Liao CM, Lee WC, Wu TY, Chien MS (2010) Expression and immunological studies of classical swine fever virus glycoprotein E2 in the bi-cistronic baculovirus/larvae expression system. Biosci Biotechnol Biochem 74(7):1343–1349. https://doi.org/10.1271/bbb.90894

Xia SL, Xiang GT, Lei JL, Du M, Wang Y, Zhou M, Liu Y, Ji S, Wang YL, Luo Y, Sun Y, Qiu HJ (2016a) Efficacy of the marker vaccine rAdV-SFV-E2 against classical swine fever in the presence of maternally derived antibodies to rAdV-SFV-E2 or C-strain. Vet Microbiol 196:50–54. https://doi.org/10.1016/j.vetmic.2016.10.001

Xia SL, Lei JL, Du M, Wang Y, Cong X, Xiang GT, Li LF, Yu S, Du E, Liu S, Sun Y, Qiu HJ (2016b) Enhanced protective immunity of the chimeric vector-based vaccine rAdV-SFV-E2 against classical swine fever in pigs by a Salmonella bacterial ghost adjuvant. Vet Res 47:64. https://doi.org/10.1186/s13567-016-0346-9

Xu YG, Cui LC, Tian CY, Zhang GC, Huo GC, Tang LJ, Li YJ (2011) Immunogenicity of recombinant classical swine fever virus CD8+ T lymphocyte epitope and porcine parvovirus VP2 antigen coexpressed by *Lactobacillus casei* in swine via oral vaccination. Clin Vaccine Immunol 18:1979–1986. https://doi.org/10.1128/CVI.05204-11

Xu YG, Guan XT, Liu ZM, Tian CY, Cui LC (2015) Immunogenicity in swine of orally administered recombinant *Lactobacillus plantarum* expressing classical swine fever virus E2 protein in conjunction with thymosin α-1 as an adjuvant. Appl Environ Micorbiol 81 (11):3745–3752. https://doi.org/10.1128/AEM.00127-15

Yuki Y, Tokuhara D, Nochi T, Yasuda H, Mejima M, Kurokawa S, Takahashi Y, Kataoka N, Nakanishi U, Hagiwara Y, Fujihashi K, Takaiwa F, Kiyono H (2009) Oral MucoRice expressing double-mutant cholera toxin A and B subunits induces toxin-specific neutralising immunity. Vaccine 27:5982–5988. https://doi.org/10.1016/j.vaccine.2009.07.071

Porcine Epidemic Diarrhea Virus

Zayn Khamis and Rima Menassa

Abstract Porcine epidemic diarrhea virus (PEDv) causes disease and mortality to piglets worldwide. Most vaccines used to combat the disease have been ineffective live attenuated virus vaccines. Research has emerged showing both the spike (S) and membrane (M) proteins of the virus have potential for use as subunit vaccines. This research has been largely undertaken using plants as expression platforms, with some promising candidates having emerged.

Keywords Porcine epidemic diarrhea virus · Recombinant protein
Subunit vaccine · Coronavirus · Plant biotechnology

1 Disease Symptoms and Occurrence

Porcine epidemic diarrhea virus (PEDv) is a coronavirus that causes porcine epidemic diarrhea (PED) in pigs. Its high mortality rate for piglets, at 90–95% (Stevenson et al. 2013) is a key distinguishing factor between PEDv and the similar transmissible gastroenteritis virus (TGEV). In suckling piglets, the PEDv incubation period is around two days, varying from 1 to 8 days. Diarrhea and vomiting can develop within 24 h, and as a result dehydration, anorexia and severe weight loss occur. In older pigs and sows, morbidity varies, and the period between onset and end of clinical symptoms is 3–4 weeks (Lee 2015; Stevenson et al. 2013).

The first known PEDv outbreaks occurred in Europe in the 1970s and 1990s, including in Belgium (Pensaert and de Bouck 1978) and Hungary (Nagy et al. 1996). While PEDv has since posed less of a threat to Europe, it has re-emerged in Italy (Martelli et al. 2008), Germany (Hanke et al. 2015), and France (Grasland

Z. Khamis · R. Menassa (✉)
Agriculture and Agri-Food Canada, London Research and Development Centre, London, ON, Canada
e-mail: rima.menassa@agr.gc.ca

Z. Khamis · R. Menassa
Department of Biology, The University of Western Ontario, London, ON, Canada

J. MacDonald (ed.), *Prospects of Plant-Based Vaccines in Veterinary Medicine*,
https://doi.org/10.1007/978-3-319-90137-4_12

et al. 2015) affecting pigs of all ages. PEDv spread to Asia in the early 1980s, where it was first detected in Japan in 1982, and then in South Korea, China, and Thailand (Chen et al. 2014). Since October 2010 China has seen a severe outbreak of PEDv, resulting in high porcine mortality rates and economic losses (Sun et al. 2012). PEDv was first detected in the United States (U.S.) in May 2013 (Stevenson et al. 2013) and then in Canada in January 2014 (Ojkic et al. 2015). These two countries have experienced severe economic losses due to the death of millions of suckling piglets and to diarrhea-derived weight loss in fattening pigs (Chen et al. 2014). The North American serotypes are most closely related to a recently emerged Chinese serotype (Huang et al. 2013) and since the North American epidemic, they have spread to South Korea, Taiwan, and Japan (Lee 2015).

While a few vaccines are available against PEDv, those targeting the Asian strains are not effective against the North American strains, while the effectiveness of a conditionally approved vaccine in the U.S. is unknown (see below). By vaccinating sows with an effective vaccine, the suckling piglets can receive lactogenic immunity through IgA antibodies secreted into the milk (Bae et al. 2003), and may be spared from vertical transmission of PEDv from sow milk (Sun et al. 2012).

2 Mechanisms of Infection

PEDv primarily enters pigs' bodies through the fecal-oral route, although airborne transmission may play a role (Alonso et al. 2014). Diarrhea and vomiting result in the spread of PEDv through contaminated environmental sources including pigs, trailers, clothing (Lowe et al. 2014), sow's milk (Sun et al. 2012), feed, and feed-supplements such as spray-dried porcine plasma (Pasick et al. 2014), highlighting the importance of biosecurity.

PEDv causes sickness through its actions in the intestines of pigs. The virus enters porcine enterocytes, which line the inner surface of the intestines, via interaction between the viral S protein and the enterocyte aminopeptidase N which acts as a cellular receptor for PEDv. Through this receptor PEDv enters the enterocyte cells, where new virions assemble by budding through the endoplasmic reticulum and Golgi apparatus membranes (Ducatelle et al. 1981; Li et al. 2007).

Once PEDv enters enterocytes it causes them to undergo acute necrosis (Jung et al. 2014). PEDv also causes a reduction in the number of goblet cells, the cells which secrete mucins to defend against microbial infection (Jung and Saif 2015). Cytolysis additionally leads to shortening and severe atrophy of intestinal villi, and causes the tips of villi to erode or become covered with attenuated epithelial cells (Jung et al. 2014; Stevenson et al. 2013). The atrophy of intestinal villi results in microscopic lesions, which are typical of an enteritic infection, (Sueyoshi et al. 1995). PEDv also results in swollen cells, and the detachment of cells from adjacent cells and from the membranes of the basal surfaces (Stevenson et al. 2013). These factors inhibit the pig's ability to absorb water and nutrients, and result in the malabsorption and diarrhea discussed above.

3 PEDv Vaccine Design and Plant-Made Candidates

To produce an effective vaccine against PEDv, it is important to understand the structure of the virus. The Coronavirinae subfamily consists of three genera: alphacoronavirus, betacoronavirus, and gammacoronavirus. PEDv is an enveloped alphacoronavirus encoded by a 28 kilobase single-stranded, positive-sense RNA genome (Song and Park 2012). Coronaviruses have the largest known RNA genomes of all viruses (King 2011). The PEDv genome has seven open reading frames (ORFs), which code for three non-structural polyproteins, and four structural proteins [spike (S), envelope (E), membrane (M) and nucleocapsid (N)]. These open reading frames are flanked by a 5′ cap and a 3′ polyadenylated tail (Figs. 1 and 2) (Song and Park 2012). The ORFs encoding the non-structural proteins consist of two overlapping open reading frames, ORF1a and ORF1b, encoding two polyproteins. These polyproteins are processed by three virus-encoded proteases, a 3C-like proteinase (3CLpro) and two papain-like proteinases (PLP) which results in 16 non-structural proteins required for genome replication and mRNA transcription (John et al. 2016; Prentice et al. 2004). The accessory protein ORF3 is a potassium ion channel, but its role is not well defined (Wang et al. 2012). Reports on transmissible gastroenteritis coronavirus (TGEV), another alphacoronavirus, indicate that S and E are only present in the virion in small quantities, with E estimated to occur 20 times in a virion (Godet et al. 1992). N and M occur in higher numbers, at a ratio of 1N:3M (King 2011). A recent study examining mouse hepatitis virus (MHV, a betacoronavirus), severe acute respiratory syndrome-coronavirus (SARS-CoV, a betacoronavirus), and feline coronavirus (FCoV, an alphacoronavirus), has determined that coronaviruses have approximately 1100 M dimers, 90 S trimers, and N proteins in a ratio from 3M:1N to 1M:1N (Neuman et al. 2011). Thus, M is the most abundant structural protein displayed at the viral surface.

Of the four structural proteins, S and M are the most antigenic. To date, the S protein has been the primary focus of subunit vaccine design due to its antigenicity, and the role it plays in viral entry, as it regulates interactions with host cell receptor protein, aminopeptidase N (Bosch et al. 2003). S contains three antigenic regions. The first epitope that was recognized spans amino acids 1495–1913. Called the CO-26K equivalent (COE), this epitope induced a neutralizing immune response and was identified through sequence homology with TGEV, which induces similar clinical symptoms in pigs (Chang et al. 2002). Subsequently, the motif spanning

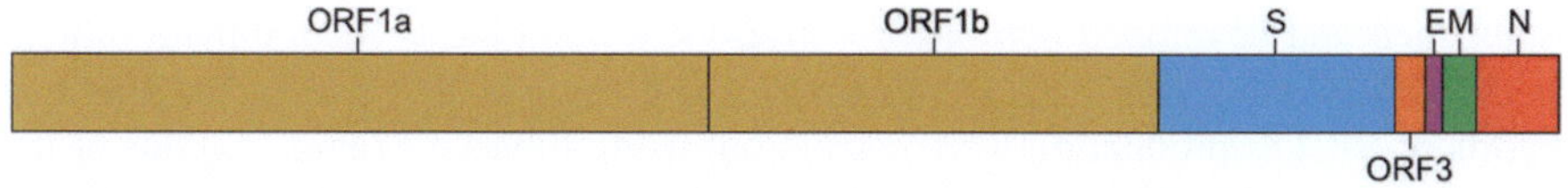

Fig. 1 A schematic of the genome of PEDv. The first two ORFs from the 5′ end of the genome cover two thirds of the genome, and code for polyprotein 1a and polyprotein 1b, respectively. The next part of the genome codes for the S protein, then ORF3, E, M, and finally the N protein (Khamis 2016)

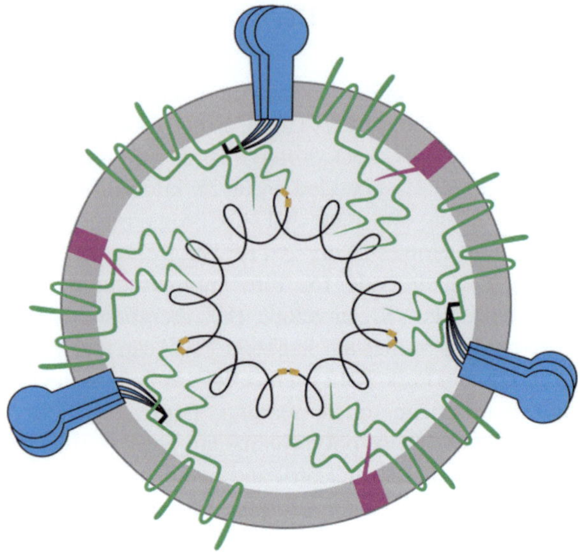

Fig. 2 Schematic of an assembled PEDv virion. The nucleocapsid protein (N, small yellow rectangles) forms a ribonucleoprotein complex with viral RNA (black line) inside the virion. The envelope protein (E, pink) is embedded in the membrane (darker grey) as is the membrane protein (M, green). The spike protein (S, blue) also embeds in the membrane, and forms surface projections, or 'spikes'. Stoichiometry not to scale. Figure modified from Khamis (2016)

amino acids 1368–1374, and the epitope region from amino acids 636–789 have also been found to induce the production of neutralizing antibodies (Cruz et al. 2008; Sun et al. 2006).

While S is antigenic, the use of S as a subunit vaccine poses challenges. *S* has been shown to be prone to mutations through serial passages (Sato et al. 2011), and has high genetic variability among PEDv strains–strains have even been found with 582 nucleotide deletions in the *S* gene (Masuda et al. 2015). Indeed, the variability of *S* makes it the gene of choice to study the genetic relatedness of different PEDv strains (Chen et al. 2014). The difference in amino acid sequence between the S protein of different strains can lead to different epitopes being presented, and may explain why previous vaccines have failed to provide effective immunity against infectious strains (Sun et al. 2012).

Nonetheless, due to the early recognition of the antigenic importance of the protein, PEDv recombinant protein production has focussed on S. Work has been done to produce S or S epitopes in *Escherichia coli* (Van Noi and Chung 2017), *Lactobacillus casei* (Ge et al. 2012), and via the Orf virus as a vaccine delivery vector (Hain et al. 2016). However, most recombinant PEDv S production has occurred in plants, with all of these plants producing S-COE. The protein yield has varied widely, as has the plant host utilized. Historically, tobacco has been the most established and developed platform for high-yield plant-based recombinant protein production (Conley et al. 2011). As such, it follows suit that five of the fourteen reports of S-COE production were carried out using tobacco (Table 1). Four of the reports expressed S-COE in a stable transgenic line, with yields between 0.1% of TSP (Kang et al. 2005b) and 2.1% of TSP, depending on codon optimization (Kang et al. 2005a), while one reported transient production at 5% of TSP (Kang et al. 2004).

Table 1 Production of S-COE in plants

Plant host	Transient or transgenic	Fusions	Yield	Promoter and/or enhancer used	References
Tobacco	Transgenic	–	10 mg/kg of protein per fresh weight[a]	2x35S, TOL	Bae et al. (2003)[b]
Tobacco	Transient	–	5% TSP	TMV RNA	Kang et al. (2004)
No-nicotine Tobacco	Transgenic	–	2.1% TSP	2x35S, TOL	Kang et al. (2005a)
Tobacco	Transgenic	–	0.1% TSP	2x35S, TOL	Kang et al. (2005b)
Potato	Transgenic	–	0.1% TSP	2x35S, TOL	Kim et al. (2005)
Duckweed	Transgenic	–	Not reported	35S	Ko et al. (2011)
Sweet potato	Transgenic	–	Not reported	35S	Yang et al. (2005)
Corn seed	Transgenic	–	0.122% TSP	2x35S, maize intron Hsp70	Kun et al. (2014)
Carrot	Transgenic	–	Not reported	2x35S, TOL	Kim et al. (2003)
Tobacco	Transgenic	LTB	1.6% TSP	Ubiquitin promoter	Kang et al. (2006)
Lettuce	Transgenic	CTB	0.0065% TSP	Ubiquitin promoter	Huy et al. (2011)
Lettuce	Transgenic	LTB	0.048% TSP	Ubiquitin promoter	Huy et al. (2009)
Rice endosperm	Transgenic	LTB	1.3% TSP	HMW-Bx17-p, Act1-i	Oszvald et al. (2007)
Rice endosperm	Transgenic	LTB	1.9% TSP	HMW-Bx17-p, Act1-i	Tamás (2010)
Rice calli	Transgenic	Co1	0.083% TSP	RAmy3D	Huy et al. (2012)[b]

2x, double-enhanced; 35S, cauliflower mosaic virus (CaMV) 35S promoter; Act1-I, rice actin first intron; Co1, M cell-targeting ligand; CTB, cholera toxin B subunit; HMW-Bx17-p, wheat high molecular weight glutenin subunit Bx17 endosperm-specific promoter; LTB, heat-labile enterotoxin B subunit of *Escherichia. coli*; RAmy3D, rice α-amylase 3D promoter; TOL, TMV Omega-prime leader, containing transcriptional and translational enhancer from the coat protein gene of TMV; TSP, total soluble protein; all yield values are highest levels reported

Table modified from Khamis (2016)

[a]Total soluble protein levels were not reported in this study

[b]Study also showed antibody production against protein

Many of the alternative plant choices resulted in yields that were either not quantified, or very low, despite often using the same promoters and enhancers as the tobacco studies. Like the tobacco studies, a double-enhanced 35S (2x35S) cauliflower mosaic virus (CaMV) promoter was used in carrot and potato expression experiments. Expression in potato reached 0.1% of tuber TSP, while expression in carrot was not reported (Kim et al. 2003, 2005). Expression of S-COE has also been reported in *Lemna minor* [35S promoter, not quantified, (Ko et al. 2011)], sweet potato [35S promoter, not quantified (Yang et al. 2005)], and corn [2x35S, 0.122% TSP (Kun et al. 2014)].

The use of fusion proteins, particularly the heat-labile enterotoxin B subunit of *Escherichia coli* (LTB), have recently gained popularity in subunit vaccine design, and have been utilized in the production of S-COE. Three studies have expressed such fusions in rice, which has shown promise in reaching equivalent accumulation levels as tobacco. Rice was used to express S-COE both in the endosperm and in calli. While accumulation levels in the calli using the rice α-amylase 3D promoter (RAmy3D) only reached 0.083% for a S-COE-M cell-targeting ligand (Co1) fusion (Huy et al. 2012), accumulation levels in the endosperm for LTB-S-COE using the wheat high molecular weight glutenin subunit Bx17 endosperm-specific promoter (HMW-Bx17-p) and rice actin first intron (Act1-I) reached 1.3 and 1.9% (Oszvald et al. 2007; Tamás 2010). LTB-COE has also been produced in tobacco, reaching comparative levels of 1.6% TSP (Kang et al. 2006).

So far, only three studies have tested the immunogenicity of plant-produced S-COE. LTB-S-COE produced in tobacco was able to bind to the GM1-ganglioside intestinal membrane receptor (Kang et al. 2006). More directly, mice orally immunized with Co1-S-COE fusion protein produced in rice calli had threefold and eightfold higher levels of IgG and IgA secreting cells in their lymphocytes, respectively, compared to unimmunized mice (Huy et al. 2012). While this is encouraging, in order for a vaccine to be effective, the antibodies produced must be virus-neutralizing. Bae et al (2003) took this next step, showing that feeding ground lyophilized transgenic tobacco containing S-COE was able to induce an immune response in mice that could inhibit PEDv plaque formation by 49.7% in comparison to controls. While full immunity was not conferred by the mounted immune response, this study marks the most significant achievement to date in plant-based PEDv vaccine research.

M is the other antigenic protein of PEDv, and the most abundant component of the viral envelope (Utiger et al. 1995). In contrast to *S*, *M* is more conserved, showing 5.5% the number of mutations as *S* after 100 serial passages (Chen et al. 2014; Sato et al. 2011). The use of a protein that remains stable is important for subunit vaccine design to ensure that a vaccine can be used for a wide variety of strains and locations. The M protein of TGEV and SARS-CoV show virus-neutralizing activity in the presence of complement, the component of the immune system that enhances the ability of antibodies to clear pathogens. This virus-neutralizing activity was demonstrated to be higher or comparable to the neutralizing capacity of 8 individual S protein fragments (Pang et al. 2004; Woods et al. 1988).

Through sequence homology with infectious bronchitis virus (IBV, a gamma-coronavirus), M was found to have a B-cell epitope on its C-terminus from amino acids 195–200 (Zhang et al. 2012). However, it is likely that with further research sequence homology studies will find more epitopes on PEDv M. For example, the M protein of SARS-CoV has B-cell epitopes on the N and C-termini (He et al. 2005), and two cytotoxic T-cell epitopes in the second and third transmembrane domain (Liu et al. 2010), while the M protein of MHV has a CD4+ T-cell epitope on its C-terminus (Xue et al. 1995).

Expression of M transiently in *N. benthamiana* as a fusion with elastin-like polypeptide (Khamis 2016) resulted in the production of virus-like particles (VLPs), an important development of subunit vaccine design. While subunit vaccines have previously failed to completely protect piglets from PEDv infection due to subpar immunogenicity, VLPs are more immunogenic because they resemble the native structure of the virion. Plant-made VLPs for influenza are shown to induce four- to six-fold higher levels of antibody response than fifty times more flu antigen not in VLPs (D'Aoust et al. 2008). The ability to produce PEDv VLPs using M may lead to the first commercial plant-produced PEDv vaccine.

4 Existing Commercial Vaccines

Much of the PEDv vaccine research has occurred in Asia, where outbreaks have been most severe, but none of the produced vaccines are completely effective against Asian PEDv strains (Song and Park 2012). Available Asian vaccines are based on strains that are genetically different from those sequenced in the U.S. that are currently causing epidemics globally (Huang et al. 2013). This has been on display in China, where vaccinated herds experienced PEDv breakouts that were found to be due to newer strains of the virus (Li et al. 2012). Two PEDv vaccines were given conditional licenses in the U.S. by the United States Department of Agriculture (USDA). The first was a vaccine originally produced by Harrisvaccines, Inc., and is based on their SirraVax^{SM} RNA platform. Using this platform, part of the RNA genome of a Venezuelan equine encephalitis alphavirus is replaced with a gene for PEDv S protein. After injection with genetic material, the pig's dendritic cells produce the S protein and an immune response is launched against the produced protein (Harrisvaccines, 2015). Merck Animal Health acquired Harrisvaccines in 2015 (Merck Animal Health 2015) but the product is still sold under a conditional license in the U.S., where safety and field trials are ongoing. The second conditionally licensed PEDv vaccine in the U.S. is an inactivated virus particle vaccine produced by Zoetis, Inc. (Zoetis 2016). Efficacy and potency studies are still in progress for the Zoetis vaccine, and duration of immunity has not been evaluated. It must be refrigerated, and used all at once when opened (Zoetis 2016). A third vaccine candidate is being developed by VIDO-Intervac (Vaccine and Infection Disease Organization—International Vaccine Centre) in Canada and is currently undergoing field testing. This vaccine candidate is a subunit S1 protein expressed in mammalian HEK-293 T cells, and although it induced production of neutralizing antibodies, it was not fully protective to suckling piglets (Makadiya et al. 2016).

None of the PEDv vaccines used commercially, at least in the U.S. and Canada, are made *in planta*. The ability to vaccinate orally through feeding plant tissue is important as (Song et al. 2007) demonstrated that oral vaccination was more effective than injection for their PEDv vaccine. When comparing oral to intramuscular administration of their attenuated virus vaccine, Song et al. found that more IgA's were produced by orally vaccinated pigs, and that the mortality rate for this group was 13% in comparison to 60% for the intramuscular group. Current vaccines on the market are also either based on S, which is prone to mutations, or are killed or live attenuated vaccines. Live attenuated vaccines present risks, as they can potentially mutate and become pathogenic again, and allow for genome segment re-assortment on farms—one Chinese PEDv strain is thought to have evolved from a live attenuated vaccine (Chen et al. 2010). Inactivated vaccines, in comparison, are safer, but have a high cost of production, and present concerns over the reliability of inactivation methods (Calvo-Pinilla et al. 2014). Both live attenuated and inactivated vaccines rule out the possibility of using "distinguish infected from vaccinated animals" (DIVA) assays, as they contain the entire virus. As such, the market has a need of an effective subunit vaccine based on the current infectious strains. Further research on current plant-made candidates could prove to fill this need.

5 Pathways to Commercialization

While proof of concept studies show that S-COE can be produced in plants, many of these studies show poor yields. Generally, even when higher yields are reported, the push forward to testing the immunogenicity of these vaccine candidates does not occur. There have been two instances where the direct immunogenicity of plant-produced S-COE has been tested. The first was incomplete, as it did not test whether the mounted immune response was virus-neutralizing (Huy et al. 2012). The other candidate that successfully took steps to test immunogenicity was the vaccine candidate produced by (Bae et al. 2003). However, while the immune response in mice inhibited PEDv plaque formation by 49.7% in comparison to controls, complete immunity was not achieved (Bae et al. 2003). Likely due to this, and due to the need for higher expression levels, no further efforts were made with this vaccine candidate.

Generally, even when higher expression levels have been obtained (Kang et al. 2004), immunogenicity was not tested, and research focused on increasing expression levels and not on producing an immunogenic commercial product. The focus on S-COE has left a gap in the research on plant production of other PEDv proteins. However, our recent results have shown that M can be produced *in planta* and that VLPs can be produced for PEDv (Khamis 2016). VLPs represent a step-forward for subunit vaccine design, demonstrating higher immunogenicity than regular subunit vaccines. In the future, advances in subunit vaccine design, such as VLPs, and a commitment to improving yields and testing immunogenicity and protective immune responses to these VLPs may lead to the commercialization of a plant-produced subunit PEDv vaccine.

References

Alonso C, Goede DP, Morrison RB et al (2014) Evidence of infectivity of airborne porcine epidemic diarrhea virus and detection of airborne viral RNA at long distances from infected herds. Vet Res 45:73

Bae JL, Lee JG, Kang TJ et al (2003) Induction of antigen-specific systemic and mucosal immune responses by feeding animals transgenic plants expressing the antigen. Vaccine 21:4052–4058

Bosch BJ, Van Der Zee R, CaM De Haan et al (2003) The coronavirus spike protein is a class I virus fusion protein: structural and functional characterization of the fusion core complex. J Virol 77:8801–8811

Calvo-Pinilla E, Castillo-Olivares J, Jabbar T et al (2014) Recombinant vaccines against bluetongue virus. Virus Res 182:78–86

Chang SH, Bae JL, Kang TJ et al (2002) Identification of the epitope region capable of inducing neutralizing antibodies against the porcine epidemic diarrhea virus. Mol Cells 14:295–299

Chen J, Wang C, Shi H et al (2010) Molecular epidemiology of porcine epidemic diarrhea virus in China. Arch Virol 155:1471–1476

Chen Q, Li G, Stasko J et al (2014) Isolation and characterization of porcine epidemic diarrhea viruses associated with the 2013 disease outbreak among swine in the United States. J Clin Microbiol 52:234–243

Conley AJ, Zhu H, Le LC et al (2011) Recombinant protein production in a variety of *Nicotiana* hosts: a comparative analysis. Plant Biotechnol J 9:434–444

Cruz DJ, Kim CJ, Shin HJ (2008) The GPRLQPY motif located at the carboxy-terminal of the spike protein induces antibodies that neutralize Porcine epidemic diarrhea virus. Virus Res 132:192–196

D'aoust MA, Lavoie PO, Couture MM et al (2008) Influenza virus-like particles produced by transient expression in Nicotiana benthamiana induce a protective immune response against a lethal viral challenge in mice. Plant Biotechnol J 6:930–940

Ducatelle R, Coussement W, Pensaert MB et al (1981) In vivo morphogenesis of a new porcine enteric coronavirus, CV 777. Arch Virol 68:35–44

Ge JW, Liu DQ, Li YJ (2012) Construction of recombinant lactobacilli expressing the core neutralizing epitope (COE) of porcine epidemic diarrhea virus and a fusion protein consisting of COE and Escherichia coli heat-labile enterotoxin B, and comparison of the immune responses by orogastric immunization. Can J Microbiol 58:1258–1267

Godet M, L'haridon R, Vautherot JF et al (1992) TGEV corona virus ORF4 encodes a membrane protein that is incorporated into virions. Virology 188:666–675

Grasland B, Bigault L, Bernard C et al (2015) Complete genome sequence of a porcine epidemic diarrhea s gene indel strain isolated in France in December 2014. Genome Announc 3

Hain KS, Joshi LR, Okda F et al (2016) Immunogenicity of a recombinant parapoxvirus expressing the spike protein of Porcine epidemic diarrhea virus. J Gen Virol 97:2719–2731

Hanke D, Jenckel M, Petrov A et al (2015) Comparison of porcine epidemic diarrhea viruses from Germany and the United States, 2014. Emerg Infect Dis 21:493–496

Harrisvaccines I (2015) SirraVax platform. http://www.merck-animal-health.com/news/2014-6-16.aspx. Accessed 4 Jan 2018

He Y, Zhou Y, Siddiqui P et al (2005) Identification of immunodominant epitopes on the membrane protein of the severe acute respiratory syndrome-associated coronavirus. J Clin Microbiol 43:3718–3726

Huang YW, Dickerman AW, Pineyro P et al (2013) Origin, evolution, and genotyping of emergent porcine epidemic diarrhea virus strains in the United States. MBio 4:e00737–00713
Huy NX, Kim YS, Jun SC et al (2009) Production of a Heat-labile Enterotoxin B Subunit-porcine Epidemic Diarrhea Virus-neutralizing Epitope Fusion Protein in Transgenic Lettuce (Lactuca sativa). Biotechnol Bioproc E 14:731–737
Huy NX, Yang MS, Kim TG (2011) Expression of a cholera toxin B subunit-neutralizing epitope of the porcine epidemic diarrhea virus fusion gene in transgenic lettuce (Lactuca sativa L.). Mol Biotechnol 48:201–209
Huy NX, Kim SH, Yang MS et al (2012) Immunogenicity of a neutralizing epitope from porcine epidemic diarrhea virus: M cell targeting ligand fusion protein expressed in transgenic rice calli. Plant Cell Rep 31:1933–1942
John SES, Anson BJ, Mesecar AD (2016) X-Ray structure and inhibition of 3C-like protease from porcine epidemic diarrhea virus. Sci Rep 6
Jung K, Saif LJ (2015) Porcine epidemic diarrhea virus infection: etiology, epidemiology, pathogenesis and immunoprophylaxis. Vet J 204:134–143
Jung K, Wang Q, Scheuer KA et al (2014) Pathology of US porcine epidemic diarrhea virus strain PC21A in gnotobiotic pigs. Emerg Infect Dis 20:662–665
Kang TJ, Kang KH, Kim JA et al (2004) High-level expression of the neutralizing epitope of porcine epidemic diarrhea virus by a tobacco mosaic virus-based vector. Protein Expr Purif 38:129–135
Kang TJ, Kim YS, Jang YS et al (2005a) Expression of the synthetic neutralizing epitope gene of porcine epidemic diarrhea virus in tobacco plants without nicotine. Vaccine 23:2294–2297
Kang TJ, Seo JE, Kim DH et al (2005b) Cloning and sequence analysis of the Korean strain of spike gene of porcine epidemic diarrhea virus and expression of its neutralizing epitope in plants. Protein Expr Purif 41:378–383
Kang TJ, Han SC, Yang MS et al (2006) Expression of synthetic neutralizing epitope of porcine epidemic diarrhea virus fused with synthetic B subunit of *Escherichia coli* heat-labile enterotoxin in tobacco plants. Protein Expr Purif 46:16–22
Khamis Z (2016) Producing a subunit vaccine for porcine epidemic diarrhea virus. Dissertation, The University of Western Ontario
Kim Y-S, Kwon T-H, Yang M-S (2003) Expression of porcine epidemic diarrhea virus spike gene in transgenic carrot plants. Korean J Plant Resour 6:108–113
Kim YS, Kang TJ, Jang YS et al (2005) Expression of neutralizing epitope of porcine epidemic diarrhea virus in potato plants. Plant Cell Tissue Organ Cult 82:125–130
King AMQ (ed) (2011) Virus taxonomy: classification and nomenclature of viruses: Ninth Report of the International Committee on Taxonomy of Viruses. Elsevier, New York
Ko SM, Sun HJ, Oh MJ et al (2011) Expression of the protective antigen for PEDV in transgenic duckweed, Lemna minor. Hortic Environ Biotech 52:511–515
Kun M, Bing Y, Jing X et al (2014) Expression of core neutralizing epitope gene of porcine epidemic diarrhea virus in maize. J Agric Sci Technol 16:28–35
Lee C (2015) Porcine epidemic diarrhea virus: an emerging and re-emerging epizootic swine virus. Virol J 12:193
Li BX, Ge JW, Li YJ (2007) Porcine aminopeptidase N is a functional receptor for the PEDV coronavirus. Virology 365:166–172
Li W, Li H, Liu Y et al (2012) New variants of porcine epidemic diarrhea virus, China, 2011. Emerg Infect Dis 18:1350–1353
Liu J, Sun Y, Qi J et al (2010) The membrane protein of severe acute respiratory syndrome coronavirus acts as a dominant immunogen revealed by a clustering region of novel functionally and structurally defined cytotoxic T-lymphocyte epitopes. J Infect Dis 202:1171–1180
Lowe J, Gauger P, Harmon K et al (2014) Role of transportation in spread of porcine epidemic diarrhea virus infection, United States. Emerg Infect Dis 20:872–874
Makadiya N, Brownlie R, Van Den Hurk J et al (2016) S1 domain of the porcine epidemic diarrhea virus spike protein as a vaccine antigen. Virol J 13:57

Martelli P, Lavazza A, Nigrelli AD et al (2008) Epidemic of diarrhoea caused by porcine epidemic diarrhoea virus in Italy. Vet Rec 162:307–310

Masuda T, Murakami S, Takahashi O et al (2015) New porcine epidemic diarrhoea virus variant with a large deletion in the spike gene identified in domestic pigs. Arch Virol 160:2565–2568

Nagy B, Nagy G, Meder M et al (1996) Enterotoxigenic *Escherichia coli*, rotavirus, porcine epidemic diarrhoea virus, adenovirus and calici-like virus in porcine postweaning diarrhoea in Hungary. Acta Vet Hung 44:9–19

Neuman BW, Kiss G, Kunding AH et al (2011) A structural analysis of M protein in coronavirus assembly and morphology. J Struct Biol 174:11–22

Ojkic D, Hazlett M, Fairles J et al (2015) The first case of porcine epidemic diarrhea in Canada. Can Vet J 56:149–152

Oszvald M, Kang TJ, Tomoskozi S et al (2007) Expression of a synthetic neutralizing epitope of porcine epidemic diarrhea virus fused with synthetic B subunit of Escherichia coli heat labile enterotoxin in rice endosperm. Mol Biotechnol 35:215–223

Pang H, Liu Y, Han X et al (2004) Protective humoral responses to severe acute respiratory syndrome-associated coronavirus: implications for the design of an effective protein-based vaccine. J Gen Virol 85:3109–3113

Pasick J, Berhane Y, Ojkic D et al (2014) Investigation into the role of potentially contaminated feed as a source of the first-detected outbreaks of porcine epidemic diarrhea in Canada. Transbound Emerg Dis 61:397–410

Pensaert MB, De Bouck P (1978) A new coronavirus-like particle associated with diarrhea in swine. Arch Virol 58:243–247

Prentice E, Mcauliffe J, Lu X et al (2004) Identification and characterization of severe acute respiratory syndrome coronavirus replicase proteins. J Virol 78:9977–9986

Sato T, Takeyama N, Katsumata A et al (2011) Mutations in the spike gene of porcine epidemic diarrhea virus associated with growth adaptation in vitro and attenuation of virulence in vivo. Virus Genes 43:72–78

Song D, Park B (2012) Porcine epidemic diarrhoea virus: a comprehensive review of molecular epidemiology, diagnosis, and vaccines. Virus Genes 44:167–175

Song DS, Oh JS, Kang BK et al (2007) Oral efficacy of Vero cell attenuated porcine epidemic diarrhea virus DR13 strain. Res Vet Sci 82:134–140

Stevenson GW, Hoang H, Schwartz KJ et al (2013) Emergence of porcine epidemic diarrhea virus in the United States: clinical signs, lesions, and viral genomic sequences. J Vet Diagn Invest 25:649–654

Sueyoshi M, Tsuda T, Yamazaki K et al (1995) An immunohistochemical investigation of porcine epidemic diarrhoea. J Comp Pathol 113:59–67

Sun D, Feng L, Shi H et al (2006) Spike protein region (aa 636–789) of porcine epidemic diarrhea virus is essential for induction of neutralizing antibodies. Acta Virol 51:149–156

Sun RQ, Cai RJ, Chen YQ et al (2012) Outbreak of porcine epidemic diarrhea in suckling piglets, China. Emerg Infect Dis 18:161–163

Tamás L (2010) Molecular farming, using the cereal endosperm as bioreactor. Acta Agron Hung 58:55–64

Utiger A, Tobler K, Bridgen A et al (1995) Identification of the membrane protein of porcine epidemic diarrhea virus. Virus Genes 10:137–148

Van Noi N, Chung Y-C (2017) Optimization of expression and purification of recombinant S1 domain of the porcine epidemic diarrhea virus spike (PEDV- S1) protein in Escherichia coli. Biotechnol Biotechnol Equip 31:619–629

Wang K, Lu W, Chen J et al (2012) PEDV ORF3 encodes an ion channel protein and regulates virus production. FEBS Lett 586:384–391

Woods RD, Wesley RD, Kapke PA (1988) Neutralization of porcine transmissible gastroenteritis virus by complement-dependent monoclonal antibodies. Am J Vet Res 49:300–304

Xue S, Jaszewski A, Perlman S (1995) Identification of a CD4+ T cell epitope within the M protein of a neurotropic coronavirus. Virology 208:173–179

Yang K-S, Lim S, Kwon S-Y et al (2005) Transgenic sweet potato (*Ipomoea batatas*) expressing spike gene of porcine epidemic diarrhea virus. J Plant Biotech 32:263–268
Zhang Z, Chen J, Shi H et al (2012) Identification of a conserved linear B-cell epitope in the M protein of porcine epidemic diarrhea virus. Virol J 9:225
Zoetis (2016) Porcine epidemic diarrhea vaccine. https://www.zoetisus.com/products/pork/porcine-epidemic-diarrhea-vaccine.aspx. Accessed July 2017

Porcine Reproductive and Respiratory Syndrome (PRRS)

Elizabeth Loza-Rubio and Edith Rojas-Anaya

Abstract Porcine reproductive and respiratory syndrome (PRRS) is an economically frustrating viral disease of pigs, characterized by severe reproductive failure in pregnant sows and respiratory disorders in piglets and growing pigs. Several research groups around the world have developed PRRSV vaccines. Some of these have been effective; however, owing to the complications that the syndrome presents and the viral evasion of the immune system, vaccines have not always been 100% effective. Biotechnological tools, such as the generation of plant-derived vaccines, offer alternatives to obtain more stable biologics, free of fermentation and cold chains. According to the literature, these vaccines are cost effective. In this document, we present some of the vaccines that have been developed against PRRSV, both traditional and new, and describe some alternatives developed in plants.

Keywords PRRS · Traditional vaccines · New generation vaccines

1 History or Antecedents

Porcine reproductive and respiratory syndrome (PPRS) has become one of the most important diseases of intensive pig production worldwide.

The late 1980s marked the emergence of a 'mystery swine disease' in the USA causing reproductive complications in sows and respiratory disease in growing pigs (Hill 1990). In near synchrony, reports began trickling in from swine-rearing countries across Europe about a swine reproductive disease matching the clinical presentations seen in North America (Plana et al. 1992). The mystery disease was

E. Loza-Rubio (✉) · E. Rojas-Anaya
Centro Nacional de Investigación Disciplinaria en Microbiología Animal
(CENID-Microbiología, INIFAP), Carretera México-Toluca Km 15.5, 05110 Palo Alto, Cuajimalpa, Mexico
e-mail: loza.elizabeth@inifap.gob.mx

J. MacDonald (ed.), *Prospects of Plant-Based Vaccines in Veterinary Medicine*,
https://doi.org/10.1007/978-3-319-90137-4_13

formally recognized as a novel swine illness named PRRS, with an aetiological agent called PRRS virus (PRRSV).

PRRSV was isolated for the first time in 1991 in Lelystad, the Netherlands, from animals with severe reproductive disorders, and the virus was called “Lelystad virus” (Type 1) (Wensvoort et al. 1992). Later, in 1992 in Minnesota, USA, another isolation from pigs with respiratory and reproductive problems was carried out, and the virus was called VR-2332 (Type 2). These two isolates induced a similar respiratory and reproductive disease when inoculated in susceptible pigs (Collins et al. 1992). Each genotype spread rapidly in its respective continent, and PRRS has gained increased attention because of its large-scale outbreaks and tremendous losses in the global swine industry (Luping et al. 2013; Renukaradhya et al. 2015).

Today, far from being eradicated, PRRSV is endemic to virtually all swine-rearing regions and presents substantial challenges for management (Brar et al. 2015).

In Mexico, the first antibody detection study was carried out from 1992–1993 in imported and local pigs. Currently, studies conducted at CENID-Microbiology have shown frequencies of more than 70% in the central area of the country (Diosdado et al. 2015).

2 Economic Impact

The economic impact of PRRS in breeding and farrowing units is caused mostly by a reduction in the number of weaned pigs and by a decrease in farrowing rates. Infection in growing-finishing pigs may increase secondary infections and mortality rates, as well as result in retarded growth, high dispersion of weights at slaughter age, and increased antimicrobial usage. In 2005, the total annual cost of PRRS outbreaks in the USA was estimated to be USD 560 million, which included USD 67 million for the breeding-farrowing phase, USD 201 million for the nursery phase and USD 292 million for the grower-finisher phase of production (Neumann et al. 2005).

Nieuwenhuis et al. (2012) in Europe estimated the economic impact of a PRRS outbreak in 9 sow herds during the first 18 weeks after the outbreak by comparing the overall costs between pre- and post-outbreak periods using different factors (production data, medication, diagnostics, labour, etc.). An outbreak of PRRSV resulted in a reduction in the number of sold pigs per sow by 1.7. The economic loss varied between €59 and €379 for one sow per 18-week period outbreak, with a mean loss per sow per outbreak of €126. The costs after the outbreak varied significantly from €3 to €160 per sow, due to the different methods used by farmers to control PRRSV outbreaks. More recently, Holtkamp et al. (2013) calculated a cost of USD 664 million/year for the United States, representing a 10% increase compared to Neumann et al. (2005). Therefore, the negative impact of PRRS on the economic margin of pig production has stimulated efforts to control and eventually eradicate the disease (Pileri and Mateu 2016).

These significant production losses and resultant increases in production costs have encouraged the use of various strategies to control PRRSV.

3 Clinical Signs

The acute presentation of the disease is characterized by lack of appetite, weakness, fever and, in some cases, difficulty breathing. Two to three weeks after infection, it manifests as a systemic disease that can affect 5–75% of animals of all ages. In general, it provokes respiratory distress in pigs of all ages, but it is especially problematic when infecting pregnant sows, leading to late abortion, early farrowing and birth of dead or weakened piglets. Although virus produce mortality in piglets, lately, more virulent strains with an increased incidence of pig mortality have been circulating and may also kill adult pigs (Zhou and Yang 2010).

Clinical signs of PRRS disease are variable, depending on the strain, animal age, immune state, productive state and environment. In sows, there are reproductive problems such as hot flashes, mummified fetuses, abortions, premature or late births, stillbirths and weak births. The sows may stop producing milk and may have incoordination or respiratory problems. It is also possible to observe purple coloration of the ears, udders and vulva. Boars tend to have decreased libido and reductions in the quality of semen as volume, motility and spastic concentration below the standards.

In clinically healthy piglets, the first and most frequent sign is severe respiratory disease. Mortality can be 10–60%, or 100% in weak piglets. Other clinical signs include swelling of the eyelids, conjunctivitis, apathy, extreme thinning, diarrhoea, shaggy hair, purple ears, behavioural symptoms, and increased secondary infections, since generally, one pathogen acts as key agent for secondary invaders by lowering the local and sometimes also the systemic defence mechanisms of the host. The pathogens involved can vary considerably between different production sites, such as Circovirus-2, Porcine epidemic diarrhea, *Pneumocystis carinii f.* sp. *Suis*, *Haemophilus* and *Mycoplasma* (Weissenbacher-Lang et al. 2017; Schweer et al. 2016; Palzer et al. 2015).

In weaned and fattened piglets, we can observe uneven litters, decreased appetite, respiratory problems and redness of the skin in some parts of the body. Mortality at this stage ranges from 10 to 20% and is a function of the sanitation level and operation management, as mortality is closely correlated with the presence of other microorganisms in the herd (Diosdado et al. 2015).

4 Agent

PRRSV, the causative agent of PRRS, is a small, enveloped, single-stranded, positive-sense RNA virus belonging to the family *Arteriviridae*. The PRRSV genome, with a size of approximately 15 kb, contains 10 open reading frames (ORFs). ORFs 1a and 1b encode for non-structural proteins, and ORFs 2–7 encode for structural proteins (Brar et al. 2015). There are two small stretches of UTRs, one at each end of the genome (Fig. 1). Three-quarters of the genome from the 5′ end is occupied by ORF1a and ORF1b, which encode replicative enzymes. Translation of these ORFs produces polyprotein (pp) 1a and pp1b, respectively. The latter is expressed by a −1 ribosomal frameshift signal in the overlapping region of the two ORFs, which lengthens pp1a into pp1ab at the C terminus (Kappes and Faaberg 2015; Snijder and Meulenberg 1998).

PRSS has been divided into European genotype 1 and North American genotype 2, with Lelystad and VR-2332 as prototypical strains, respectively (Shi et al. 2010). The viruses in the two genotypes could be further divided into different subgenotypes according to the virus genome characteristics based on phylogenetic analysis (Brar et al. 2015).

Genetic variability has been demonstrated among different strains of PRRSV, due to mutation and recombination as intrinsic causes; while extrinsic factor such as transmission dynamics, swine management practice and vaccination practices are also involved (Brar et al. 2015). The two main genotypes (I and II) share only 50–70% nucleotides and 50–80% similarity of amino acids (Forsberg 2005). Within each genotype, nucleotide identity has been estimated to be 85–87.5% with a maximum genetic distance of 21–30% (Murtaugh et al. 2010; Shi et al. 2013; Drigo et al. 2014).

For production management and disease control strategies to keep pace with a rapidly evolving virus, access to publicly available, high-resolution phylogenetic knowledge linked to phenotypic information is necessary. Such resources could help bridge the gap between known and actual diversity as well as inform the design of studies that link variation in PRRSV genomes to variation in key phenotypic characteristics (Brar et al. 2015).

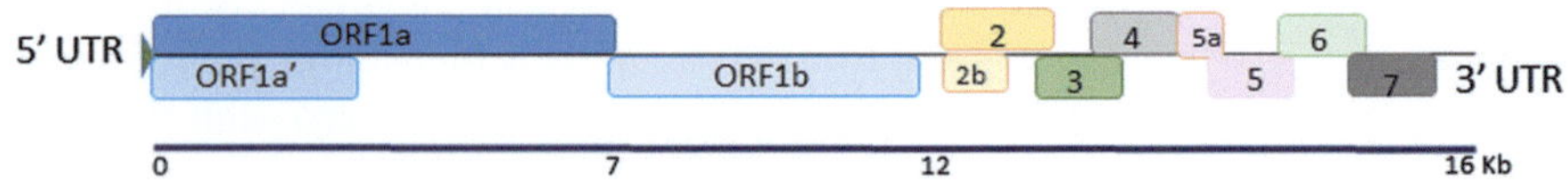

Fig. 1 Schematization of the PRRsV genome *Source* Brar et al. (2015)

5 Mechanisms of Infection

PRRSV has significant genetic diversity with important differences in pathogenicity and infection patterns (Kappes and Faaberg 2015).

PRRSV mainly infects cells of monocyte and macrophage lineage, including pulmonary alveolar macrophages. The virus has a number of complex features, such as the ability to mutate, recombine, and suppress the immune system and cause persistent disease. However, to date, there are several unknown aspects about its pathogenesis and forms of dissemination (Renson et al. 2017).

Transmission occurs mainly by close contact between pigs through respiratory secretions, saliva, urine, faeces and semen. There are also indirect routes of dissemination to susceptible populations, which include fomites, arthropods and aerosols. Once in the pig, PRRSV is distributed mainly by macrophages to various tissues. In piglets, PRRSV causes interstitial pneumonia characterized by respiratory distress, somnolence, lethargy, and fever. It has been hypothesized that infection with the virus is the main cause of death, and severe microscopic lesions were observed in the uterus and fetal placenta are not observed in infected fetus (Novakovic et al. 2016).

The integral membrane/matrix (M) protein and the primary envelope glycoprotein (GP5) seem to be responsible for the infection of macrophages, but cellular tropism has not yet been demonstrated. Once the virus enters macrophages, it replicates and kills the lymphoid tissue containing the infected macrophages. Other targets are cells with a $CD8^{+}CD3^{-}$ phenotype in the endometrium, connective tissue and uterine epithelium (Karniychuk et al. 2013).

Such infection allows the virus to persistently evade the host immune response (Meulenberg et al. 2000). The ability of the virus to infect both macrophages and dendritic cells is due to its ability to evade the response to interferon (Loving et al. 2015).

Reproductive signs due to PRRSV infection largely depend on the stage of gestation and mainly occur in late gestation. The infection starts by respiratory inoculation, with viremia occurring rapidly and resulting in reproductive failure and losses in weight and size at birth. Ladinig and coworkers (2015) concluded that the concentration of virus in systemic or lymphoid tissues has little impact on the reproductive effects of PRRS. What does have an impact comes after systemic infection, when the virus travels to the endometrium. There, it induces inflammation and vasculitis with an inflammatory cell infiltrate, apoptosis and placental separation (Karniychuk et al. 2011). The mechanism by which the virus crosses the endometrium is not well known, but transmission is tissue specific (Karniychuk and Nauwynck 2013).

The permissive cells express adequate receptors that include heparin sulphate, sialoadhesin or CD169, vimentin CD151 and intercellular adhesion molecule-3-grabbing non-integrin (DC-SIGN or CD209) (Zhang and Yoo 2015). These molecules allow viral replication by adherence, entrance, uncoating, release and budding. The expression of these molecules in macrophages and during viral replication causes apoptosis and death of cells around the site of replication and often leads to the eventual death of the individual. Moreover, inter-foetal transmission is associated with free virus or infected foetal cells of any type. Viral load and foetal death are not associated with sex or foetal position (Ladinig et al. 2015).

We previously described the main mechanisms of infection and pathology when the virus affects the reproductive system. However, as mentioned above, the virus also affects the respiratory system. The infection starts by the oronasal route, allowing colonization of the respiratory tract and provoking an immunomodulatory response that decreases the response of the host to the virus and therefore allows persistent infection (Chand et al. 2012).

The fact that the target cells of the virus are phagocytic cells allows an unavoidable evasion of the immune response. This means that the immune system plays a central role in infection by this virus.

Pro-inflammatory cytokines, such as interferon-alpha (IFN-α), tumour necrosis factor-alpha (TNF-α) and interleukin 1 (IL-1), play a key role in disease severity. PRRSV also induces the expression of other proinflammatory cytokines, including IL-6 (Van Reeth et al. 1999; Liu et al. 2010). Infected macrophages influence blood monocytes a few days after infection, promoting viral replication and exacerbating the expression of proinflammatory cytokines such as IL1β and IL8 (Han et al. 2014). Renson and coworkers (2017) have hypothesized that the IFN-γ levels detected during a post-PRRSV infection could also originate from NK cells and CD8+ T cells.

6 Traditional Vaccines

PRRSV infection induces a defective immune response with the late appearance of neutralizing antibodies and delayed cell-mediated immunity (Lopez and Osorio 2004; Piron et al. 2014a, b). This allows the infection to be chronic and makes it difficult to eliminate PRRSV from an infected herd.

Preventive vaccination against PRRSV has been considered as a strategy to minimize PRRS-associated losses when pig populations become infected with wild-type virus (Murtaugh and Genzow 2011). In fact, it has been shown that breeding herds with a recent history of PRRSV infection (i.e., prior PRRSV immunity) achieved stability more quickly and had smaller production impact than those without a history of PRRSV infection (i.e., no PRRSV herd immunity). However, immunizing breeding herds using attenuated PRRSV results in an increased cost of production due to the commercial vaccine cost and the potential negative impact of the attenuated replicating virus on productivity levels (Savard et al. 2016).

Live virus immunization (LVI) has been used mainly for introducing replacement gilts that enter into a positive herd. However, even if LVI has been beneficial and could be used for eradicating PRRS from a farm, it also presents some disadvantages, such as causing severe disease in exposed pigs. The inoculum may contain adventitious agents and may also predispose the pigs to secondary bacterial agents.

Inactivated and live attenuated vaccines are used to aid PRRS control in swine herds, but the efficacy and/or safety of current licensed vaccines are not satisfactory (Charerntantanakul 2012). Inactivated PRRSV vaccines induce only weak neutralizing antibody responses, even against homologous isolates, and weaker to no response against heterologous isolates (Geldhof et al. 2012). Live attenuated PRRSV vaccines contribute to clinical protection by unknown mechanisms without preventing infection, but a high probability of reversion to virulence is a major safety concern. To better control PRRSV infections worldwide, it is crucial to develop a safer and more efficacious vaccine that confers protective immunity against diverse PRRSV isolates.

When a sow herd becomes infected, in an attempt to hasten control and eliminate PRRSV from the breeding herd, some veterinarians have adopted a strategy called load-close-expose, which consists of interrupting replacement gilt introductions into the herd for several months (herd closure) and exposing the whole herd to replicating PRRSV. Either a modified live virus (MLV) vaccine or field-virus inoculation (FVI) is used (Corzo et al. 2010).

Vaccination with a modified live virus (MLV) vaccine is considered to be of value for decreasing losses; however, the diversity of the virus tends to interfere with the protection provided by vaccines. Herds that used MLV required 7 additional weeks to reach PRRSV-stability compared to herds that used FVI (Linhares et al. 2014). However, MLV herds recovered production levels 11 weeks sooner and had a smaller total loss of weaned pigs (advantage of 1443 pigs per 1000 sows).

Currently, the virus has 2 main genotypes, type I (European) and type II (North American), which are genetically and antigenically different (Brar et al. 2015). Furthermore, 4 subtypes of type I PRRSV, based on sequence analyses of open reading frame 5 (ORF5), have been identified. Although similarities exist between and within the type I strains, there is enough diversity that vaccines may not induce sufficient cross-protection against heterologous strains (Goldberg et al. 2003). Pileri et al. (2017) stated that the genetic diversity of PRRSV is such that all challenge situations in the field could be considered heterologous. Some studies have determined that the level of genetic similarity between the vaccine strain and the challenge strain was not necessarily an accurate predictor of vaccine efficacy, whereas others have concluded that the level of protection a vaccine provides against PRRSV infection may depend on the degree of relatedness between the vaccine and challenge strains (Renukaradhya et al. 2015).

Experimentally, vaccination with modified live PRRS virus decreases reproductive loss after infection; therefore, many veterinarians recommend preventatively vaccinating sow herds in case of infection with field virus. Vaccination has a cost, however, and not all vaccinated herds become infected. Therefore, there is a need to estimate the benefit/cost ratio of preventatively vaccinating a sow herd (Linhares et al. 2015).

Bai and collaborators (2016), evaluated five commercial PRRSV modified live vaccines that have been widely used in China for their protection against

NADC30-like PRRS infection. This is a group of viruses that show a high similarity with the NADC30 strain, which belongs to a North American type 2 PRRSV isolated in 2008. The clinical symptoms and high viremia and viral load in different tissues in vaccinated pigs after a NADC30-like PRRSV challenge suggested the inefficacy of these vaccines (Sun et al. 2016a, b).

7 New Generation Vaccine

The persistent nature of infection, coupled with the limited efficacy of vaccines, has made the control of PRRS particularly problematic (Chand et al. 2012). To better control PRRSV infections worldwide, it is crucial to develop a safer and more efficacious vaccine that confers protective immunity against diverse PRRSV isolates (Chung et al. 2016).

New generation vaccines are inherently safe because no virus is present, and by incorporating elements known to be important in immunity against PRRSV, the vaccines could be very effective. ORF5 is commonly used, which encodes for a major envelope glycoprotein (GP5). This is one of the key immunogenic proteins of PRRSV and is the leading target for the development of genetically engineered vaccines against PRRS (Kheyar et al. 2005). The modified GP5 had significantly enhanced immunogenicity, particularly in its ability to induce neutralizing antibody responses and cellular immune responses, compared to the native GP5. Consequently, this modified GP5 may be useful to facilitate the development of a new generation of vaccines, such as DNA vaccines, live attenuated chimeric virus vaccines, and live virus-vectored vaccines, against the highly pathogenic PRRSV (Kheyar et al. 2005).

7.1 *Oral Vaccination*

Edible vaccines have been viewed as the panacea of oral vaccines in terms of their use as an approach to control disease.

Oral vaccination has many distinct advantages over parenteral administration, but has proven difficult to achieve thus far, as reflected by the scarcity of licensed oral vaccines. Perhaps the most significant benefit of oral vaccination is the ability to elicit both mucosal and systemic immunity. Oral vaccines also obviate the need for trained medical personnel to administer them and reduce the risks of infection associated with needles. In veterinary medicine, oral vaccination avoids difficult management. Both of these latter aspects are important considerations for successful vaccination campaign coverage in remote or resource-limited settings (Specht and Mayfield 2014).

7.2 *Plant Derived Vaccines*

The literature indicates several potential advantages related to plant-derived vaccines, for example, heat-stable formulation for storage and transport (avoiding cold chain), which is important in tropical and subtropical areas. Other attractive benefits are that plants serve as an inexpensive means of processing and expressing proteins that can be quite complex to handle, as plants require only sunlight, water, and minerals to carry out the process. There is no risk of contamination with animal pathogens. Recombinant proteins can be expressed in non-edible plants or plant parts and administered parenterally. They can also be expressed in edible plant parts for oral administration, sometimes via incorporation into food (edible vaccine). Plant-derived edible vaccines are able to promoted mucosal and systemic immune responses (Loza-Rubio and Rojas-Anaya 2014; Saroja et al. 2011).

There are few experimental plant-derived antigen used against PRRSV. Different plant species and tissues have been employed for the production of PRRSV vaccines (Table 1).

7.3 *Plant Derived Antigen Against PRRSV*

Tobacco. Chia et al. (2010), evaluated the immunogenicity of the GP5 of PRRSV strain MD-001 expressed in tobacco plants. Six, 6-week-old pigs were fed four times orally (days 0, 14, 28, and 42) with 50 g of chopped fresh GP5 transgenic

Table 1 Plant-derived vaccines against PRRSV

Type of plant	Antigen used	Total soluble protein (TSP) %	Immunogenicity	References
Arabidopsis thaliana	Glycoproteins 3,4 and 5		Immunogenic in mice and piglets	Piron et al. (2014a, b)
Tobacco	Gp5	0.011	Immunogenic in pigs	Chia et al. (2010, 2011)
Nicotiana sylvestris	Gp5, M, N	NR	Immunogenic in mice	Uribe-Campero et al. (2015)
Potato	Gp5	NR	Immunogenic in mice	Chen and Liu (2011)
Banana	Gp5	0.021–0.037	Immunogenic in piglets	Chan et al. (2013)
Corn	M protein	NR	Immunogenic in mice	Hu et al. (2012)
Soybean	N protein	0.65	Immunogenic in mice	Vimolmangkang et al. (2012)

NR non reported

tobacco (GP5-T) leaves, corresponding to 5.5 μg of GP5-T protein. They collected samples of serum, saliva, and peripheral blood mononuclear cells (PBMCs) at different times after and leading up to the initial oral vaccination. Pigs that were fed GP5-T developed serum neutralizing antibodies to PRRSV at a titre of 1:4–1:8 after the 4th vaccination with 48 days post-initial oral vaccination. No detectable anti-PRRSV antibody response or PRRSV-specific blastogenic response were seen in non-treated pigs. This study demonstrated that pigs fed GP5-T could develop specific mucosal and systemic humoral and cellular immune responses against PRRSV.

Later, an *Escherichia coli* heat-labile enterotoxin B subunit (LTB) was used as an adjuvant to co-administer the same antigens given orally to six-week-old PRRSV-free pigs. In this study three consecutive doses, using same quantity of recombinant GP5 protein (Chia et al. 2010), administered at 2-week interval and challenged with PRRSV MD001 strain at 5×10^5 50% tissue culture infection dose ($TCID_{50}$) at 7 weeks post-initial immunization. Pigs receiving LTB-GP5-T or GP5-T developed PRRSV-specific antibody immunity and cell-mediated immunity and had significantly lower viremia and tissue viral load and milder lung lesions than those fed the wild-type tobacco plant (W-T). The LTB-GP5-T-treated group had relatively higher immune responses than the GP5-T-treated group, although the differences were not statistically significant (Chia et al. 2011). These results also suggested that transgenic plants can be an effective system for oral delivery of recombinant subunit vaccines in pigs.

Arabidopsis thaliana. Piron et al. (2014a, b), utilized GP3, GP4 and GP5 of the European prototype PRRSV, the 'Lelystad Virus' (LV), in *Arabidopsis thaliana*. In this study, the experimental vaccine was evaluated in target species. Three piglets were injected twice intramuscularly with 100 μg purified each antigen Gp3, Gp4, GP5 (300 μg antigen in total) mixed with an oil-in-water adjuvant. A fast a and strong neutralizing antibody response after challenge was founded They concluded that this platform would allow high production and correct folding and assembly of the antigens and in vivo experiments these recombinant proteins showed to be a prominent plant-derived vaccine.

Nicotiana sylvestris. To generate a plant-derived vaccine alternative against PRRSV, Uribe-Campero and coworkers (2015) used complete ORFs of genes encoding the major antigens of the virus (GP5, M, N), using genotype II. They obtained VLPs expressed in *N. sylvestris* using the agroinfiltration method. Groups of five 8-week old mice were immunized intraperitoneally with 34 μg of recombinant VLPs. They demonstrated that these purified VLPs were able to induce a humoral immune response in immunized animals.

7.4 Edible Vaccines Against PRRSV

Potato. In this study, transgenic potatoes expressing GP5 protein of PRRSV strain CH-1a were produced by *Agrobacterium*-mediated transformation. Eight mice were immunized with 0.3 μg of GP5 protein contained into transgenic potato extracts generated both serum and mucosal-specific antibodies, although low levels of neutralizing antibodies were elicited. Furthermore, IgA were detected in the intestinal wash. This research provided a new approach for the production of vaccines against PRRSV (Chen and Liu 2011).

Maize calluses. Hu et al. (2012), expressed PRRSV M protein strain VR-2385 in corn. Eight mice were immunized with 2.6 mg of protein contained in 30 mg of dried transgenic corn callus powder. This edible vaccine stimulated antigen-specific serum and mucosal antibodies and also a cellular immune response with good IFNγ production. The authors chose Matrix (M) protein because it is the most abundant viral antigen on the viral envelope and the most conserved structural protein of the virus, with 78–81% amino acid identity between type 1 and type 2 strains (Dockland 2010).

Soybean. Vimolmangkang et al. (2012), designed and evaluated a soybean-based vaccine. In this study, a construct carrying a synthesized PRRSV-ORF7 antigen of strain VR-2332, encoding the nucleocapsid N protein of PRRSV, was introduced into soybeans using *Agrobacterium*-mediated transformation. The amount of the antigenic protein accumulated in the seeds of these transgenic lines was up to 0.65% of the total soluble protein (TSP). A significant induction of a specific immune response, both humoral and mucosal, against PRRSV-ORF7 was observed following intragastric immunization of eight BALB/c female mice with 60 mg of transgenic soybean seeds powder containing 157 μg of recombinant ORF7. These findings provide a 'proof of concept' and serve as a critical step in the development of a subunit plant-based vaccine against PRRS, although the efficacy should be evaluated through simply feeding the mice, as intragastric immunization is not practical.

Banana. This fruit is considered an ideal host for producing edible vaccines because it can be eaten raw without any modification. In addition, bananas are clonally propagated through suckers and pose no risk for genetic containment. From an ecological point of view, the transgenic banana provides an added benefit of having low risk in the agricultural environment when used for the purpose of pharmaceutical protein production. Chan and collaborators (2013) expressed GP5 PRRSV strain MD001 in banana and these were used to immunized six6-week-old, PRRsV seronegative pigs using 50 g fresh leaves with 12.9 μg of recombinant antigen per dose three times in a 2-week period. The immunized animals were challenged intranasally with same strain of PRRSV. They demonstrated that pigs orally immunized with transgenic bananas showed a specific anti-GP5 response and a significant reduction in serum and tissue viral loads. Thus, they proposed that a banana vaccine can be a potent oral vaccine.

8 Conclusions and Perspectives

The advent of molecular farming has provided a cost-effective strategy for the development of transgenic plants as bioreactors to produce recombinant proteins. To better control PRRSV infections worldwide, it is crucial to develop a safer and more efficacious vaccine that confers protective immunity against diverse PRRSV isolates (Chung et al. 2016). Experimental PRRSV vaccines, including live attenuated vaccines, recombinant vectors expressing PRRSV viral proteins, DNA vaccines and plant-made subunit vaccines, have been developed. However, the genetic and antigenic heterogeneity of the virus limits the value of almost all of the PRRSV vaccines tested to date. It has been demonstrated, however, that chimeric viruses could provide protection, and a strategy using multivalent chimeric viruses could thus provide cross-protection against antigenically diverse PRRSV strains (Sun et al. 2016b). However, developing a universal vaccine that can provide broad protection against circulating PRRSV strains is still a major challenge for current vaccine development.

It has been reported that cereals produce a large quantity of transgenic protein, and in pigs, transformed corn is the most convenient and cost-effective system for the production of edible vaccines.

References

Bai X, Wang Y, Xu X, Sun Z, Xiao Y, Ji G, Li Y, Tan F, Li X, Tian K (2016) Commercial vaccines provide limited protection to NADC30-like PRRSV infection. Vaccine 34:5540–5545

Brar MS, Shi M, Murtaugh MP, Leung FC (2015) Evolutionary diversification of type 2 porcine reproductive and respiratory syndrome virus. J Gen Virol 96:1570–1580

Chan HT, Chia MY, Pang VF, Jeng CR, Do YY, Huang PL (2013) Oral immunogenicity of porcine reproductive and respiratory syndrome virus antigen expressed in transgenic banana. Plant Biotechnol J 11:315–324

Chand RJ, Trible BR, Rowland RR (2012) Pathogenesis of porcine reproductive and respiratory syndrome virus. Curr Opin Virol 2:256–263

Charerntantanakul W (2012) Porcine reproductive and respiratory syndrome virus vaccines: immunogenicity, efficacy and safety aspects. World J Virol 1:23–30

Chen X, Liu J (2011) Generation and immunogenicity of transgenic potato expressing the GP5 protein of porcine reproductive and respiratory syndrome virus. J Virol Methods 173:153–158

Chia MY, Hsiao SH, Chan HT, Do YY, Huang PL, Chang HW, Tsai YC, Lin CM, Pang VF, Jeng CR (2010) Immunogenicity of recombinant GP5 protein of porcine reproductive and respiratory syndrome virus expressed in tobacco plant. Vet Immunol Immunopathol 135:234–242

Chia MY, Hsiao SH, Chan HT, Do YY, Huang PL, Chang HW, Tsai YC, Lin CM, Pang VF, Jeng CR (2011) Evaluation of the immunogenicity of a transgenic tobacco plant expressing the recombinant fusion protein of GP5 of porcine reproductive and respiratory syndrome virus and B subunit of Escherichia coli heat-labile enterotoxin in pigs. Vet Immunol Immunopathol 140:215–225

Chung CJ, Cha S-H, Grimm AL, ChungG Gibson KA, Yoon K-J et al (2016) Recognition of highly diverse Type-1 and -2 porcine reproductive and respiratory syndrome viruses (PRRSVs)

by T-Lymphocytes induced in pigs after experimental infection with a Type-2 PRRSV strain. PLoS ONE 11:e0165450
Collins JE, Banfield DA, Christianson WT, Harris L, Hennings JC, Shaw DP et al (1992) Isolation of swine infertility and respiratory syndrome virus (isolate ATCC VR-2332) in North America and experimental reproduction of the disease in gnotobiotic pigs. J Vet Diagn Invest 4:117–126
Corzo CA, Mondaca E, Wayne S, Torremorell M, Dee S, Davies P, Morrison RB (2010) Control and elimination of porcine reproductive and respiratory syndrome virus. Virus Res 154:185–192
Diosdado VF, Socci EG, Martínez LCA, Carrera SME (2015) Síndrome reproductivo y respiratorio porcino (PRRS). Folleto para productores N° 1. INIFAP, Mexico. ISBN 978-607-37-0412-0. Junio, 2015, pp 1–15
Dockland T (2010) The structural biology of PRRSV. Virus Res 154:86–97
Drigo M, Franzo G, Gigli H, Martini M, Mondin A, Gracieux P, Ceglie L (2014) The impact of porcine reproductive and respiratory syndrome virus genetic heterogeneity on molecular assay performances. J Virol Meth 202:79–86
Forsberg R (2005) Divergence time of porcine reproductive and respiratory syndrome virus subtypes. Mol Biol Evol 22:2131–2134
Geldhof MF, Vanhee M, Van BW, Van DJ, Karniychuk UU, Nauwynck HJ (2012) Comparison of the efficacy of autogenous inactivated Porcine reproductive and respiratory syndrome virus (PRRSV) vaccines with that of commercial vaccines against homologous and heterologous challenges. BMC Vet Res 8:182
Goldberg TL, Lowe JF, Milburn SM, Firkins LD (2003) Quasispecies variation of Porcine reproductive and respiratory syndrome virus during natural infection. Virology 317:4069–4080
Han D, Hu Y, Li L, Tian H, Chen Z, Wang L, Ma H, Yang H, Teng K (2014) Highly pathogenic porcine reproductive and respiratory syndrome virus infection results in acute lung injury of the infected pigs. Vet Microbiol 169:135–146
Hill H (1990) Overview and history of mystery swine disease (swine infertility/respiratory syndrome). In Proceedings of the mystery swine disease committee meeting, 1990. Denver, CO, pp 29–30. Public Livestock Conservation Institute, Madison, WI
Holtkamp DJ, Kliebenstein JB, Neumann EJ, Zimmerman JJ, Rotto HF, Yoder TK, Wang C, Yeske PE, Mowrer CL, Haley CA (2013) Assessment of the economic impact of porcine reproductive and respiratory syndrome virus on United States pork producers. J Swine Health Prod 21:72–84
Hu J, Ni Y, Dryman BA, Meng XJ, Zhang C (2012) Immunogenicity study of plant-made oral subunit vaccine against porcine reproductive and respiratory syndrome virus (PRRSV). Vaccine 30:2068–2074
Kheyar A, Jabrane A, Zhu C, Cléroux P, Massie B, Dea S, Gagnon CA (2005) Alternative codon usage of PRRS virus ORF5 gene increases eucaryotic expression of GP(5) glycoprotein and improves immune response in challenged pigs. Vaccine 23:4016–4022
Kappes MA, Faaberg KS (2015) PRRSV structure, replication and recombination: origin of phenotype and genotype diversity. Virology 479–480:475–486
Karniychuk UU, Nauwynck HJ (2013) Pathogenesis and prevention of placental and transplacental porcine reproductive and respiratory syndrome virus infection. Vet Res 44:95
Karniychuk UU, Saha D, Geldhof M, Vanhee M, Cornillie P, Broeck W, Nauwynck HJ (2011) Porcine reproductive and respiratory syndrome virus (PRRSV) causes apoptosis during its replication in fetal implantation sites. Microb Pathog 51:194–202
Karniychuk UU, Spiegelaere WD, Nauwynck HJ (2013) Porcine reproductive and respiratory syndrome virus infection is associated with an increased number of Sn- positive and CD8-positive cells in the maternal-fetal interface. Virus Res 176:285–291
Ladinig A, Ashley C, Detmer SE, Wilkinson JM, Lunney JK, Plastow G, Harding JCS (2015) Maternal and fetal predictors of fetal viral load and death in third trimester, type 2 porcine reproductive and respiratory syndrome virus infected pregnant gilts. Vet Res 46:107

Linhares DC, Cano JP, Torremorell M, Morrison RB (2014) Comparison of time to PRRSv-stability and production losses between two exposure programs to control PRRSv in sow herds. Prev Vet Med. 116:111–119

Linhares DCL, Johnson C, Morrison RB (2015) Economic analysis of vaccination strategies for PRRS control. PLoS ONE 10:e0144265

Liu Y, Shi W, Zhou E, Wang S, Hu S, Cai X, Rong F, Wu J, Xu M, Xu M, Li L (2010) Dynamic changes in inflammatory cytokines in pigs infected with highly pathogenic porcine reproductive and respiratory syndrome virus. Clin Vaccine Immunol 17:1439–1445

Lopez OJ, Osorio FA (2004) Role of neutralizing antibodies in PRRSV protective immunity. Vet Immunol Immunopathol 102:155–163

Loving CL, Osorio FA, Murtaugh MP, Zuckermann FA (2015) Innate and adaptive immunity against porcine reproductive and respiratory syndrome virus. Vet Immunol Immunopathol 167:1–14

Loza-Rubio E, Rojas-Anaya E (2014) Case study 1: rabies. In: Hefferon KL (ed) Plant–derived pharmaceuticals. Principles and applications for developing countries. CABI Biotechnologies series, Ithaca NY, USA, pp 117–136

Luping D, Bin L, Kongwang H, Haibin Z, Kehe H, Shaobo X (2013) Construction and immunogenicity of DNA vaccines encoding fusion protein of porcine IFN-g1 and GP5 gene of porcine reproductive and respiratory syndrome virus. Biomed Res Int 2013:318698

Meulenberg J (2000) PRRSV, the virus. Vet Res 31:11–21

Murtaugh MP, Genzow M (2011) Immunological solutions for treatment and prevention of porcine reproductive and respiratory syndrome (PRRS). Vaccine 29:8192–8204

Murtaugh MP, Stadejek T, Abrahante JE, Lam TTY, Leung FC (2010) The ever expanding diversity of porcine reproductive and respiratory syndrome virus. Virus Res 154:18–30

Neumann EJ, Kliebenstein JB, Johnson CD, Mabry JW, Bush EJ, Seitzinger AH, Green AL, Zimmerman JJ (2005) Assessment of the economic impact of porcine reproductive and respiratory syndrome on swine production in the United States. J Am Vet Med Assoc 227:385–392

Nieuwenhuis N, Duinhof TF, van Nes A (2012) A economic analysis of outbreaks of porcine reproductive and respiratory syndrome virus in nine sow herds. Vet Rec 170:225

Novakovic P, Harding JCS, Al-Dissi AN, Ladinig A, Detmer SE (2016) Pathologic evaluation of Type-2 porcine reproductive and respiratory syndrome virus infection at the maternal-fetal interface of Late gestation pregnant gilts. PLoS ONE 11:e0151198

Palzer A, Haedke K, Heinritzi K, Zoels S, Ladinig A, Ritzmann M (2015) Associations among Haemophilus parasuis, Mycoplasma hyorhinis, and porcine reproductive and respiratory syndrome virus infections in pigs with polyserositis. Can Vet J 56:285–287

Pileri E, Mateu E (2016) Review on the transmission porcine reproductive and respiratory syndrome virus between pigs and farms and impact in vaccination. Vet Res 47:108

Pileri E, Gibert E, Martín-Valls GE, Nofrarias M, López-Soria S, Martín M, Díaz I, Darwich L, Mateu E (2017) Transmission of porcine reproductive and respiratory syndrome virus 1 to and from vaccinated pigs in a one-to-one model. Vet Microbiol 201:18–25

Plana J, Vayreda M, Vilarrasa J, Bastons M, Rosell R, Martinez M, San Gabriel A, Pujols J, Badiola JL et al (1992) Porcine epidemic abortion and respiratory syndrome (mystery swine disease). Isolation in Spain of the causative agent and experimental reproduction of the disease. Vet Microbiol 33:203–211

Piron R, De Koker S, De Paepe A, Goossens J, Grooten J, Nauwynck H, Depicker A (2014a) Boosting in planta production antigens derived from the porcine reproductive and respiratory syndrome virus (PRRSV) and subsequent evaluation of their immunogenicity. PLoS ONE 10: e91386

Piron R, De Koker S, De Paepe A, Goossens J, Grooten J, Nauwynck H, Depicker A (2014b) Boosting *in planta* production of antigens derived from the porcine reproductive and respiratory syndrome virus (PRRSV) and subsequent evaluation of their immunogenicity. PLoS One 9:e91386

Renukaradhya GJ, Meng XJ, Calvert JG, Roof M, Lager KM (2015) Live porcine reproductive and respiratory syndrome virus vaccine: current status and future direction. Vaccine 33:4069–4080

Renson P, Rose N, Le Dimma M, Mahé S, Keranflec'h A, Paboeuf F, Belloc C, Le Potier M-F, Bourry O (2017) Dynamic changes in bronchoalveolar macrophages and cytokines during infection of pigs with a highly or low pathogenic genotype 1 PRRSV strain. Vet Res 48:15

Saroja CH, Lakshmi PK, Bhaskaran S (2011) Recent trends in vaccine delivery systems: a review. Int J of Pharm Invest 1:64–74

Savard C, Alvarez F, Provot C, Chorfi Y, D'Allaire S, Benoit-Biancamano MO, Gagnon CA (2016) Efficacy of Fostera PRRS modified live virus vaccine against a Canadian heterologous virulent field strain of porcine reproductive and respiratory syndrome virus. Can J Vet Res 80:1–11

Schweer WP, Schwartz K, Burrough ER, Yoon KJ, Sparks JC, Gabler NK (2016) The effect of porcine reproductive and respiratory syndrome virus and porcine epidemic diarrhea virus challenge on growing pigs I: growth performance and digest. J Anim Sci 94:514–522

Shi M, Lam TT, Hon CC, Hui RK, Faaberg KS, Wennblom T, Murtaugh MP, Stadejek T, Leung FC (2010) Molecular epidemiology of PRRSV: a phylogenetic perspective. Virus Res 154:7–17

Shi M, Holmes EC, Brar MS, Leung FC (2013) Recombination is associated with an outbreak of novel highly pathogenic porcine reproductive and respiratory syndrome virus in China. J Virol 87:10904–10907

Snijder EJ, Meulenberg JJ (1998) The molecular biology of arteriviruses. J Gen Virol 79:961 979

Specht EA, Mayfield SP (2014) Algae-based oral recombinant vaccines. Front Microbiol 5:60

Sun Z, Wang J, Bai X, Ji G, Yan H, Li Y, Wang Y, Tan F, Xiao Y, Li X, Tian K (2016a) Pathogenicity comparison between highly pathogenic and NADC30-like porcine reproductive and respiratory syndrome virus. Arch Virol 161:2257–2261

Sun D, Khatun A, Kim WI, Cooper V, Cho YI, Wang C, Choi EJ, Yoon KJ (2016b) Attempts to enhance cross-protection against porcine reproductive and respiratory syndrome viruses using chimeric viruses containing structural genes from two antigenically distinct strains. Vaccine 34:4335–4342

Uribe-Campero L, Monrroy-García A, Durán-Meza AL, Villagrana-Escarreñõ MV, Ruiz-García J et al (2015) Plant-based porcine reproductive and respiratory syndrome virus VLPs induce an immune response in mice. Res Vet Sci 102:59–66

Van Reeth K, Labarque G, Nauwynck H, Pensaert M (1999) Differential production of proinflammatory cytokines in the pig lung during different respiratory virus infections: correlations with pathogenicity. Res Vet Sci 67:47–52

Vimolmangkang S, Gasic K, Soria-Guerra R, Rosales-Mendoza S, Moreno-Fierros L, Korban SS (2012) Expression of the nucleocapsid protein of porcine reproductive and respiratory syndrome virus in soybean seed yields an immunogenic antigenic protein. Planta 235:513–522

Weissenbacher-Lang C, Nedorost N, Knecht C, Hennig-Pauka I, Huber M, Voglmayr T, Weissenböck H (2017) Comparison with other respiratory pathogens in two Austrian pig herds. PLoS ONE 12:e0185387

Wensvoort G, de Kluyver EP, Pol JM, Wagenaar F, Moormann RJ, Hulst MM, Bloemraad R, den Besten A, Zetstra T, Terpstra C (1992) Lelystad virus, the cause of porcine epidemic abortion and respiratory syndrome: a review of mystery swine disease research at Lelystad. Vet Microbiol 33:185–193

Zhang Q, Yoo D (2015) PRRS virus receptors and their role for pathogenesis. Vet Microbiol 177:229–241

Zhou L, Yang H (2010) Porcine reproductive and respiratory syndrome in China. Virus Res 154:31–37

Part IV
Vaccines for Ruminants

The Benefit of a Plant-Based Cattle Vaccine for Reducing Enterohemorrhagic *Escherichia Coli* Shedding and Improving Food Safety

Adam Chin-Fatt, Ed Topp and Rima Menassa

Abstract Upon ingestion, enterohemorrhagic *Escherichia coli* (EHEC) can colonize intestinal mucosa and cause hemorrhaging of nearby tissue. The failure to adequately control its contamination of food and water can consequently compromise the health of a population and incur economic losses to all stages of the food supply chain. EHEC is currently one of the foremost foodborne pathogenic threats worldwide because of its virulence across all age groups and demographics, a low infective dose, a relatively high resilience in diverse environments and its widespread prevalence across cattle herds. EHEC primarily colonizes the bovine digestive tract from which it can be transmitted via fecal shedding or during slaughter. Considering its threat to food security and in accord with the 'One Health' framework, the development of a bovine vaccine as a pre-harvest intervention strategy to curtail the transmission of EHEC is of great interest. Although two EHEC vaccines have already been developed using bacterial production platforms, their market penetrance has been markedly low. As an alternative, production in a plant platform may have the potential to redress the reasons for this low penetrance by providing a better economy of scale and a more convenient mode of delivery. This chapter summarizes the scope of the threat posed by EHEC and discusses the prospects for developing a commercial plant-based vaccine for EHEC within the framework of the North American beef industry.

Keywords EHEC · O157 · VTEC · STEC · Shiga · Cattle · Vaccine

A. Chin-Fatt · E. Topp · R. Menassa (✉)
Agriculture and Agri-Food Canada, London, ON, Canada
e-mail: rima.menassa@agr.gc.ca

A. Chin-Fatt · E. Topp · R. Menassa
Biology Department, University of Western Ontario, London, ON, Canada

J. MacDonald (ed.), *Prospects of Plant-Based Vaccines in Veterinary Medicine*,
https://doi.org/10.1007/978-3-319-90137-4_14

1 Occurrence and Disease Symptoms

1.1 Problem and Context

Diarrhea is the second leading cause of death among toddlers under the age of five globally, with an estimated occurrence of 2.5 billion cases overall, and an estimated mortality of 1.5 million annually (Unicef 2010). While diarrhea may be a common symptom of a broad spectrum of gastrointestinal upsets, a relatively small handful of micro-organisms are the primary causes for most acute diarrheal cases, including *Escherichia coli*. The pathogenic *E. coli* strains that cause diarrheal disease in humans, collectively known as diarrheagenic *E. coli,* are broadly categorized based on clinical symptoms and virulence attributes into: enteropathogenic *E. coli* (EPEC), enterotoxigenic *E. coli* (ETEC), enteroinvasive *E. coli* (EIEC), enteroaggregative *E. coli* (EAEC), diffusely adherent *E. coli* (DAEC) and Vero toxin-producing/Shiga toxin-producing *E. coli* (VTEC/STEC). The latter category is further divided into enterohemorrhagic *E. coli* (EHEC) and non-enterohemorrhagic subgroups though in practice, the terms EHEC, STEC and VTEC are often used interchangeably. As the name suggests, the EHEC subgroup is typified by hemorrhaging of the intestines but constitutes more than 100 different serotypes that are identifiable based on variations of their O (somatic lipopolysaccharide), H (flagellar) and K (capsular) antigens. While lipopolysaccharides are found in all Enterobacteriaceae, flagellar and capsular antigens are not always present in some strains. Therefore, routine epidemiological surveillance has conventionally screened primarily for O serogroups as the primary biomarker, with subsequent H-subtyping if presumptive pathogenic O strains are detected. Subtyping for the K antigen is not part of routine surveillance since few labs are equipped for the requisite assay. The most prevalent and virulent EHEC serotype in North America is O157:H7 and has been classified as a major food adulterant by the United States Department of Agriculture (USDA) and Canadian Food Inspection Agency (CFIA) for almost 20 years. Although non-O157 strains are individually less prevalent, the collective contribution of non-O157 strains to gastrointestinal illness has as of late been of growing concern, particularly since recent surveillance indicates a 41% increase in the average annual incidence of infection of non-O157 strains over the last five years across the US (Gill and Gill 2010). Six additional EHEC serogroups O26, O45, O103, O111, O121 and O145, known as the "Big Six", generally comprise >90% of non-O157 infections of any given year and have been traced to at least 22 human disease outbreaks in the US since 1990. In the US, national surveillance was only recently enabled in 2012 by the USDA to individually track non-O157 serotypes in human illness (Mathusa et al. 2010). In 2011, Canadian national surveillance by the Public Health Agency of Canada expanded their monitoring of O157 to include all VTEC strains in agricultural, water, retail and human health components (Public Health Agency of Canada 2015).

1.2 Epidemiology of Human Infections

The recognition of EHEC as a discrete and important class of diarrheagenic *E. coli* originally stems from two reports in 1983. The first was a clinical report detailing two separate outbreak events in the United States of a distinctive gastrointestinal illness, subsequently called hemorrhagic colitis (HC), characterized by severe abdominal pain and acute watery diarrhea that later developed into bloody diarrhea (Riley et al. 1983). In both cases, the illness was associated with consuming undercooked hamburger meat from two fast food chains and dubbed by news media as the "hamburger disease". Also, stool cultures sampled from the patients both yielded a previously unidentified *E. coli* strain. The second report provided strong association between fecal cytotoxin producing *E. coli* and the occurrence of hemolytic uremic syndrome (HUS) (Karmali et al. 1983). HUS is characterized by the triad combination of acute renal failure, thrombocytopenia and microangiopathic hemolytic anemia, and was already known to be preceded by a bloody diarrhea that was symptomatically similar to that observed in the two fast food chain outbreak events. This discovered link between EHEC, its enteric disease causing ability and its route of transmission via undercooked beef products has subsequently prompted a series of surveillance efforts in the food industry to curtail the outbreak potential of EHEC (Doyle et al. 2006).

Since then, EHEC, particularly the O157:H7 strain has been detected worldwide. Based on a data mining approach of incidence studies covering 21 countries, a review has conservatively estimated that each year VTEC causes 2,801,000 acute illnesses, 3890 cases of HUS and 230 deaths (Majowicz et al. 2014). Based on these estimates, on a global ranking, VTEC places behind typhoid fever, foodborne trematodes and nontyphoidal salmonellosis in importance. EHEC is estimated to affect approximately 230,000 people in the United States each year, with ~73,000 of these being caused by O157:H7 (Hale et al. 2012). In terms of most frequently isolated overall food-borne pathogen ranking in North America, it places fourth after *Campylobacter, Salmonella* spp and *Shigella* spp based on stool samples collected from patients (Griffin 1995). However, if restricted to only stool samples with visible blood then EHEC, particularly O157:H7, is the most frequently isolated (Slutsker et al. 1997).

In the US, the national surveillance program for foodborne pathogens, FoodNet, reported that in 2015 (most recent available report) the average incidence rate for that year for O157 was 0.95 per 100,000 persons and for non-O157 strains was 1.65 per 100,000 persons (CDC 2017). Among the approximately 1200 EHEC infections (out of a total sample of ~49 million), the most common serogroups were O157 (39.8%), O26 (17.6%) and O103 (14.3%) (CDC 2017). Although surveillance for non-O157 strains is still fairly recent, the growth in incidence over the past five years is stark. Compared with the average annual incidence rate 2012–2014, non-O157 incidence has increased by 41% (CDC 2017). For that same period, there has been no significant change for O157 incidence (CDC 2017). This is possibly because most EHEC diagnostic and control measures have historically been specific for O157, despite the clinical relevance of non-O157 strains. Since discovering O157 in the 1980's, the trend of infection has progressively shown a decreasing incidence in North America, mirrored by a decrease in HUS (CDC 2011). Between 1996 and 2010, the incidence of infection for O157 has decreased by 44% and the number of HUS cases has decreased by 90% (CDC 2017). There are many likely contributing factors such as improved regulatory and biosecurity control, cleaner slaughter methods, better microbial testing and improved food awareness by consumers. In Canada, the national surveillance system for foodborne pathogens, FoodNet Canada, reported an average incidence rate for VTEC to be 3.00 per 100,000 persons (Public Health Agency of Canada 2015). Targeted surveillance on retail ground beef products across Ontario for 2015 indicated VTEC in 2.3% of samples, with a similar prevalence to Salmonella (1.5%), and placing second behind the consistent frontrunner, *Listeria monocytogenes* (25%) (Public Health Agency of Canada 2015). The 10-year trend for VTEC in contamination in retail ground beef reveals that VTEC consistently hovers around 2% positive with the exception of 2010–2011 in Ontario when incidence spiked to ~8% due to large scale outbreak (Public Health Agency of Canada 2015).

Both incidence rates and occurrence of HUS are consistently highest in toddlers <5 years compared with all other age groups. In FoodNet's latest report (2015), toddlers <5 had incidence rates of 3.72 and 6.76 per 100,000 for O157 and non-O157 strains respectively (Gill and Gill 2010). In comparison, all other age groups ranged between 0.33–2.39 and 0.62–2.04 per 100,000 for O157 and non-O157 strains respectively (Gill and Gill 2010). Approximately 1 in 5 toddlers <5 years with an O157 infection will develop HUS. Out of all HUS patients, more than 90% are due to O157, followed by O121 (4.8%) and then O111 (2.4%) (Gill and Gill 2010). Compared with 2006–2008, the incidence of pediatric HUS has decreased by 32%, which likely corresponds to the 30% decrease in O157 infections (Gill and Gill 2010).

Large-scale outbreaks are rare but can affect large numbers of people and may be transmitted from a variety of sources, though most commonly from raw foodstuff or untreated water. For example, the five largest EHEC outbreaks worldwide were from: radish sprouts in Japan (12,680 cases) (Fukushima et al. 1999), drinking water in Canada (2300 cases) (Hrudey et al. 2003), well water in the US (>1000 cases) (Charatan 1999), raw beef in the US (788 cases) (Wendel et al. 2009) and undercooked hamburger meat in the US (>700 cases) (Bell et al. 1994). In comparison, sporadic EHEC infections are more frequent and comprise the major disease burden in a population. The average frequency of sporadic cases has slightly risen over the past five years of surveillance (CDC 2017; Public Health Agency of Canada 2015). Of these sporadic cases, the incidence is distributed unevenly across North America, being more common in Canada and the northern US states than the southern US states and more common in western Canada than eastern Canada (Griffin 1995).

EHEC primarily occupies a bovine intestinal reservoir and correspondingly, its main route of transmission is via cattle's excretion of fecal matter carrying the bacterium, a process known as 'shedding'. Sporadic EHEC incidence can be affected by seasonality with the most common reports of EHEC shedding occurring during the summer through fall seasons. An investigation by the USDA on the seasonal occurrence of O157 suggests that the increased shedding of *E. coli* O157 during the summer season is strongly associated with an increased likelihood of product contamination and a corresponding increase of enterohemorrhagic cases in humans (Williams et al. 2010). Both O157 and non-O157 serogroups exhibit this trend.

1.3 Disease Symptoms in Humans

Milder forms of EHEC infection are typically associated with watery diarrhea while more aggressive forms may develop into HC or HUS, and in uncommon cases, accompanied by cardiovascular or nervous system abnormalities (Griffin and Tauxe 1991). In humans, the incubation period for EHEC O157:H7 ranges from 1 to 16 days. Symptoms usually become apparent after 3–4 days, typically manifesting as moderate to severe diarrhea. Most resolve without treatment whereas others can progress to HC after a few days, characterized by severe, bloody diarrhea with abdominal tenderness and cramping. Mild fevers, nausea, vomiting and dehydration are also possible accompanying symptoms (Cleary 2004). Although this will typically resolve in approximately 1 week, 16% will develop into HUS, characterized by the triad combination of kidney failure, hemolytic anemia and thrombocytopenia. In more severe cases, paresis, stroke, cerebral edema or coma are accompanying symptoms. Although 65–85% of patients recover from HUS without permanent injury, long term complications including hypertension, renal insufficiency and end-stage renal failure are possible. Certain demographics of patients seem to be more susceptible to the development of the infection into more serious symptoms. Patients who are younger than five, older than 60 or who are immunocompromised are significantly more likely to develop HC or HUS (Gould et al. 2009; Karmali 2004; Tuttle et al. 1999). In the elderly, a form of HUS, known as thrombocytopenia purpura, is more common, characterized by less kidney damage but more severe occurrence of neurologic symptoms such as stroke, seizure and central nervous system deterioration.

1.4 Histopathology

The typical histopathology characteristic of EHEC infection includes hemorrhaging and edema of the lamina propria (Griffin et al. 1990). Biopsy samples taken from the colon of infected patients also show focal necrosis and neutrophil infiltration. One of the hallmarks of EHEC infection is the attaching-and-effacing (A/E) lesion. This histopathology is apparent by microscopy in a variety of animal models and can also be reproduced in in vitro cell cultures (Donnenberg et al. 1993; Ismaili et al. 1995;

Pai et al. 1986). In vitro organ culture of human endoscopic biopsy samples suggests EHEC adhere and form lesions on the terminal ileum (Chong et al. 2007). This distinct phenotype is caused during the EHEC colonisation phase when microvilli become effaced and various secreted proteins enable the intimate adherence between the EHEC pathogen and the outer membrane of the intestinal epithelium. Following attachment, the accumulation and rearrangement of polymerized actin leads to an altered cytoskeleton in which a pedestal-like structure protruding from the epithelium emerges. These structures can extend up to 10 μm in a pod-like formation upon which the bacterium is ensconced (Moon et al. 1983).

2 Transmission

2.1 *Route of Transmission*

The intestines of ruminants, especially cattle, are considered the primary reservoirs of EHEC and can transmit EHEC via excreted fecal matter or after slaughter during processing (Beutin et al. 1993; Montenegro et al. 1990). High levels of EHEC colonization have been reported in cattle herds from various countries, ranging typically between 10 and 25%, but can be as high as 60%. Healthy cattle transiently host EHEC in their gastrointestinal tract and can directly or indirectly transmit this pathogen to humans (Rangel et al. 2005a, b). EHEC can persist in various environments that range extensively from soil, to water to the ruminant GI tract. In North America, most cases are caused by ingestion of contaminated food or water (Rangel et al. 2005a, b). When shed in bovine feces, the pathogen can remain viable in the farm environment and may contaminate nearby agricultural crops, other holding pens and ground water (Sanderson et al. 2006). Aside from undercooked or unpasteurized animal products and contaminated fruits and vegetables, exposure may come from contaminated soil, such as at campgrounds or other sites grazed by cattle, or from open water sources, such as swimming lakes or private wells that are drainage sinks from agricultural run-off. O157:H7 has been reported to persist for up to a year in manure-treated agricultural soil and for 21 months in non-composted raw manure (Jiang et al. 2002). Its resilience in water especially is a major factor for its dissemination and persistence across various transmission routes. Culturable O157 has been demonstrated to be able to survive for at least 8 months in contaminated water troughs (Lejeune et al. 2001). Furthermore, O157 strains that survived longer than 6 months still retained the capacity to colonize cattle (Lejeune et al. 2001). EHEC's robustness has implications for crop contamination considering that bovine manure often is used as fertilizer as well as after irrigation when surface water containing EHEC collects in sumps. Even if the use of bovine fertilizer were to be avoided, a recent report indicated that airborne transport of O157: H7 could contaminate leafy greens that were up to 180 m away from a cattle feedlot, particularly when pen surfaces were under arid conditions (Berry et al. 2015). A safe set-back distance between feedlots and crops has not yet been determined. Additionally, EHEC requires a much lower infectious dose than other foodborne pathogens when ingested, with fewer than 40 bacterial cells being sufficient to cause illness (Strachan et al. 2005). To a lesser degree than contaminated food and water, EHEC can also be transmitted from direct contact between humans as well as from animal to human contact, likely via fecal residues (Heuvelink et al. 2002).

Although infected cattle remain asymptomatic, cattle that have been exposed to EHEC develop a local immune response, an associated inflammatory response and attaching-effacing (A/E) lesions suggesting not only that EHEC is an active bovine pathogen but also that there is a limit to which the bovine host will tolerate pathogen load and after which host resistance mechanisms may actively function to reduce pathogen burden (Baines et al. 2008; Nart et al. 2008).

2.2 Super Shedders

Generally, there are three distinct patterns observed for EHEC carriage in cattle that are characterized in terms of increasing severity of intestinal colonization, duration of shedding and magnitude of shedding. First, some cattle, known as passive shedders, lack colonization, transiently shed for only a few days and in small numbers. Second, cattle that are colonized, shed for approximately 1–2 months (Besser et al. 1997). Third, a small subset of cattle populations, known as "super shedders", are colonized for extended periods, shed EHEC for longer periods at 3–12 months and at significantly higher levels (between 10^4 and 10^4 colony forming units/g of faeces) (Omisakin et al. 2003; Stephens et al. 2009). These super shedders are suggested to be important hubs in a cattle population for maintaining the penetrance of EHEC infection that perhaps would otherwise be transient and short-lived. While there is as of yet no definitive explanation of the causes of the super shedding phenomenon, it is thought to collectively be mediated by factors from the EHEC pathogen, the bovine host and the environment. Hide contamination associated with super shedders rapidly resulted in the transmission of *E. coli* O157:H7 among cattle housed in a common pen (Stanford et al. 2011). An assessment of the link between shedding density and human risk suggested that even though super shedding events were relatively rare, they dominated as the environmental contamination source as well as the relative human risk of acquiring illness (Matthews et al. 2013). Almost half of all EHEC shed from cattle in an Alberta feedlot was due to super-shedders, even though these animals represent less than a tenth of the cattle population (Stephens et al. 2009). While super-shedders are increasingly considered to have a significant role in population-level persistence of EHEC, this small proportion of super-shedding cattle is not a stable, consistent subset of the population but rather varies transiently and dynamically making quarantining of the super-shedding animal an unviable option. Consequently, targeting them for interventions such as vaccination is difficult, unless applied to the entire herd for herd immunity. However, the exception to this is if immediately prior to slaughter, there were tools available to quickly diagnose and identify these super-shedders, these could be targeted for intervention to reduce the likelihood of meat product contamination.

3 Mechanism of Infection

The ability of EHEC to successfully colonize the gastrointestinal tracts of both humans and cattle despite peristaltic movements and resource competition with neighboring microflora is one of the most defining features across all strains. In particular, although all *E. coli* strains have some form of fimbrial structure to enable surface adherence, EHEC strains express specific fimbrial antigens that seem to specialize in adherence to the gut mucosa, enhancement of colonization of the intestinal epithelium, and defining of host specificity (Vial et al. 1988). In cattle, EHEC principally adheres to and colonizes the lymphoid follicle dense mucosa at the terminal rectum known as the rectoanal junction, whereas in humans, it adheres to and colonizes the follicle-associated epithelium of ileal Peyer's patches (Lim et al. 2007; Naylor et al. 2003; Phillips et al. 2000). Successful colonisation in both humans and cattle will typically be marked by a canonical A/E lesion.

The mechanism of colonization by EHEC of a mucosal site in either cattle or humans is a conserved process requiring the expression of at least 59 genes (Büttner 2012; Dziva et al. 2004). The main virulence genes cluster together on a chromosomal 43-kb pathogenicity island known as the locus of enterocyte effacement (LEE), the presence of which is both necessary and sufficient for showing the A/E phenotype (Perna et al. 1998). The LEE contains 41 open reading frames including genes encoding various subunit proteins that assemble to form a type III secretion system (T3SS), the major adhesin protein known as intimin (Eae) and its cognate Translocated intimin receptor (Tir), a lytic transglycosylase EtgA to remove glycans near to the site of colonisation (Burkinshaw et al. 2015), various effector proteins that are secreted through this system and various chaperones to stabilize the folding and assembly of these proteins (Wong et al. 2011). The T3SS consists of a syringe-like structure that permits the secretion of multiple effector proteins stored within the bacterial cell and into the host cytosol (Jarvis and Kaper 1996).

The first step of colonization is likely through contact to an intestinal epithelial membrane by an extended hollow, filamentous structure consisting of multiple polymerized subunits of *E. coli* secreted protein A (EspA) (Delahay et al. 1999; Knutton et al. 1989). Upon initial contact, two other LEE-encoded proteins, EspB and EspD, are translocated via the EspA filament into the host cell where they will assemble along with EspA to form a translocon pore stabilizing the entry point (Fivaz and Van Der Goot 1999; Kenny and Finlay 1995; Lai et al. 1997; Warawa et al. 1999). At least 39 other effector proteins are then secreted into the host cell, altering a variety of host cell processes that ultimately improve the likelihood of the bacterium's survival and replication (Tobe et al. 2006; Wong et al. 2011). Several of these effectors along with components of the T3SS are potential vaccine candidates because of their efficacy in engaging the host's active immune response. One of these effector proteins known as the non-Lee encoded effector A (NleA) protein is also secreted into the host cell where it may have a role in disruption of intestinal tight junctions and inhibition of intercellular protein trafficking (Gruenheid et al. 2004; Kim et al. 2007). Another effector known as Tir integrates into the host cell membrane where it allows docking of the adhesin protein, intimin (Kenny et al. 1997). Docking enables intimate attachment of the bacterium to the host cell and signals the recruitment and polymerization of actin at the pore resulting in a protrusion of the membrane toward the bacterium forming the canonical A/E lesion (Garmendia et al. 2004).

Subsequent to colonization, EHEC will produce a variety of virulence factors including verocytotoxins, also called Shiga-like toxins (Stx) because of their similarity to toxins produced by *Shigella dysenteriae.* In humans, the production of Stx is the primary cause of the microvascular endothelial damage associated with HUS and HC. There are two major immunologically distinct types of Shiga-like toxins, Stx1 and Stx2, that are encoded by separate phage-derived *stx* genes on the bacterial chromosome (Wagner and Waldor 2002). Although Stx1 tends to be highly conserved across serotypes, there are many variants for Stx2. Nonetheless, all Shiga toxins form a basic A-B5 subunit structure. Typically, the 32-kDa A subunit is cleaved to yield an enzymatically active 28-kDa A1 peptide that is bridged via a 4-kDa A2 peptide to a pentamer consisting of five 7.7-kDa B subunits. The B subunit pentamer is able to bind to a specific glycolipid receptor, globotriaosylceramide (Gb3) that is found on the cell membrane surface of intestinal epithelial cells. A Gb4 receptor may also be targeted by some Stx2 variants. Upon successful binding to a receptor, the toxin is endocytosed via clathrin coated pits. The internalized toxin is then delivered to endosomes where they are primarily targeted to lysosomes for degradation. However, a fraction can be delivered to the trans-Golgi network, followed by retrograde transport via Golgi cisterns into the ER. Similar to the effects of ricin, the A1 peptide of the cytotoxin is an N-glycosidase that catalytically removes a single adenine residue from the 28S RNA of 60S ribosomal subunits to effectively suppress protein synthesis by preventing binding of tRNAs to the ribosome and consequently triggering apoptosis in affected cells (Endo et al. 1988). The presence of the Gb3 receptor on the cell surface is required for Stx toxicity (Jacewicz et al. 1995). Although Stx production occurs in both humans and cattle, the former exhibit Stx-related pathophysiology primarily because of vascular expression of the Gb3 receptor in intestinal epithelial cells while the latter lack vascular Gb3 receptor expression in their GI tracts (Pruimboom-Brees et al. 2000). Although the Gb3 receptor is expressed in the bovine brain and kidney, cells in the recto-anal junction do not permit Stx to be endocytosed and transported across the GI tract vasculature and consequently, the toxin is isolated from susceptible cells (Pruimboom-Brees et al. 2000). In contrast, EHEC's colonisation of human ileal tissue is proximal to the intestinal epithelial cells that express Gb3. The selective apoptosis of absorptive villus tip intestinal epithelial cells, carrying the Gb3 receptor, and the preservation of Gb3-absent secretory crypt cells may then lead to the osmotic dysregulation that manifests as diarrhea (Kandel et al. 1989).

The development to HUS is assumed to be based on the translocation of Stx across the epithelial cell layer and into the bloodstream. The Gb3 receptor is abundant in human renal tissue (Boyd and Lingwood 1989). Upon contact, Stx is cytotoxic to the glomerular endothelial cells leading to blocking of the glomerular microvasculature with platelets and fibrin (Louise et al. 1997). This disrupted ability to filter fluid through the glomerulus may lead to the acute renal failure characteristic of HUS.

The significance of Stx in intestinal pathology can vary depending on the animal model used. In cattle, which lack the Gb3 receptor, the occurrence of the diarrhea is independent of the presence or absence of Stx but is rather determined by the extent and distribution of the A/E lesions. This pattern is similar across cattle, sheep, goats, chickens and rabbits that do not display clinical symptoms despite the formation of A/E lesions in their GI tracts, presumably due to a lack of Gb3 receptors (Best et al. 2005; La Ragione et al. 2005, 2006; Tzipori et al. 1989; Tzipori et al. 1995; Woodward et al. 2003). Overall, reports from various animal models suggest that the occurrence of the A/E lesions is sufficient to cause non-bloody diarrhea but the cellular entry of the Stx is essential for inducing clinically relevant symptoms such as bloody diarrhea, HUS and HC.

4 Interventions

4.1 Pre-harvest and Post-harvest Interventions Against EHEC

EHEC be transmitted to humans via multiple routes such as crops, water and meat products. Towards the implementation of strategies to prevent EHEC infection of humans, the prevailing train of thought is to curtail its colonization of cattle and to minimize its spread from fecal shedding and at harvest. These strategies are broadly grouped into pre-harvest and post-harvest interventions with the former typically being adopted by beef producers and the latter by meat processors. Intervention strategies that are most commonly used or are most promising have been summarized in Table 1.

Post-harvest interventions involve removing contamination from the hide and/or carcass with various antimicrobial agents such as organic acids, oxidizing agents, heat exposure, irradiation or high pressure systems. Hide contamination can occur during skinning of the animal and to a lesser degree rupturing of the intestines. As an initial step, the carcass is often rinsed or steamed and visibly contaminated parts removed by knife trimming. Subsequently, a combination of treatments is typically used to reduce the contamination. Acid treatment is the most commonly employed method in North America likely due to its cost effectiveness. Promising newer methods such as high pressure and electron beam irradiation are twice as effective as acid treatment and have the highest efficacy amongst known interventions,

Table 1 A summary of intervention strategies that have been investigated in mitigating EHEC carriage in cattle

Strategy	Description
Pre-harvest interventions	
(1) Exposure reduction	Modulates rearing conditions to minimize transmission to cattle
Treatment of drinking water	Destroys bacteria residing in drinking water, typically by chlorination, electrolysis or ozonation
Feed strategies	Reduces ingested bacteria by change of standard grain-based feed a few days before slaughter, usually by fasting or replacement with forage or hay
Maintaining closed herds	Prevents cross-contamination across herds by quarantining of cattle herds and facilities
Pest and wildlife management	Prevents transmission from various pests and wildlife which can act as EHEC transmission vectors
Sanitation practices	Ensures clean pens, bedding and transport to prevent EHEC growth in immediate environment
(2) Exclusion strategies	Alters the mucosal site of colonisation within the GI tract to either interrupt or displace attachment and colonisation
Vaccination	Engages host active mucosal immunity by immunization with an EHEC specific antigen
Probiotics	Alters the gut microbiota by a viable preparation of microorganisms that outcompete EHEC at the ecological niche needed for colonization
Prebiotics	Enriches native competitive microbiota species by providing selectively digestible organic compounds
Competitive exclusion	Competes for EHEC binding to sterically block EHEC access
(3) Direct anti-pathogen strategies	Live animal treatments that specifically target and kill EHEC
Sodium chlorate	metabolized by an EHEC-specific nitrate reductase to chlorite, a bactericidal metabolite
Antibiotics (Neomycin sulfate)	A broad spectrum compound that binds 30S ribosomal subunit and inhibits protein translation
Bacteriophages	Viruses specific for a narrow bacterial host range that infect and lyse the EHEC bacteria
Colicins	Antimicrobial proteins that bind EHEC outer membrane receptors and subsequently translocate to the cytoplasm where they exert various cytotoxic effects
Post-harvest interventions	
Physical removal	Removes visibly contaminated parts and rinses excess unattached EHEC off carcass, usually by knife trimming, steam-vacuuming and ambient temperature water washing
Acid antimicrobials	Disrupts proton motive force and substrate transport mechanisms leading to bacteriostasis, usually acetic, citric and lactic acids
Oxidizer antimicrobials	Generates oxidative damage to a broad array of cellular structures leading to cell death, usually by peracetic acid, acidified sodium chlorite, ozone or hypobromous acid

(continued)

Table 1 (continued)

Strategy	Description
Heat exposure	Uses heat treatment to denature bacterial enzymes and nucleic acid degradation, usually by hot water sprays or steam pasteurization
Irradiation	Uses a stream of high energy electrons or UV light to damage bacterial genetic material leading to cell death
High pressure	Uses hydrostatic pressure to damage bacterial cell membranes causing lysis

This list is not intended to be exhaustive but describes the most commonly used or most promising strategies currently used

though they require specialized equipment for implementation (Wheeler et al. 2014).

Pre-harvest interventions are further sub-grouped into 3 categories: (1) exposure reduction, (2) exclusion, and (3) direct anti-pathogen strategies. Exposure reduction strategies involve management of the rearing conditions of the herd to minimize EHEC exposure such as by water and feed hygiene, by limiting exposure to pests, wildlife, and other cattle herds and by sanitation of living and transport conditions. Exclusion strategies seek to interrupt or displace attachment and colonisation of EHEC to the GI tract by altering the site of colonisation such as by engaging active immunity with vaccination, outcompeting niches with prebiotics and/or probiotics or sterically hindering access with competitive exclusion. Direct anti-pathogen strategies are live animal treatments that directly kill EHEC such as by sodium chlorate, antibiotics, bacteriophages and colicins. Based on systematic reviews of published reports, only three methods of pre-harvest interventions for EHEC have been validated to be reliably efficacious in reducing colonisation in cattle—the feeding of the probiotic combination *Lactobacillus acidophilus* NP51 (NPC 747) and *Propinionibacterium freudenreichii*, feeding of sodium chlorate and vaccination with T3SS proteins or Siderophore Receptor and Porin proteins (SRPs) (Sargeant et al. 2007; Snedeker et al. 2012). Meta-analysis also indicated no consistent association of antimicrobials with degree of shedding, and indicated that there are still an insufficient number of studies to confirm efficacy of other promising interventions such as bacteriophages and colicins.

4.2 Vaccine Products that Have Reached Market

Only two vaccine products have successfully transitioned from research to market: a T3SS formulation known as Econiche® (Bioniche Life Sciences Inc., Belleville, Ontario, Canada) and a SRP formulation known as Epitopix® (Epitopix LLC, Willmar Poultry Company (WPC), Minnesota, USA). EHEC secrete T3SS proteins during colonisation and when injected directly through a host cell wall, these secreted proteins enable a receptor-mediated bacterial adhesion event to firmly anchor the bacterium to the site of the A/E lesion. The plausibility of using T3SS proteins as a vaccine was first reported on by the Finlay lab which demonstrated the secretion of extracellular proteins via a putative T3SS in both EHEC and EPEC (Jarvis et al. 1995; Jarvis and Kaper 1996). After partnering with the Vaccine and Infectious Diseases Organization (VIDO) in Saskatchewan, they demonstrated in a pilot study using a bacterial production platform that these attachment proteins reduced shedding of O157:H7 in cattle. With the intent of moving this product to market, Bioniche Life Sciences Inc. was contacted for scale-up and commercial manufacture of the vaccine. The product, called Econiche™, obtained full licensure by the CFIA in 2008 after clearing safety and efficacy requirements but has since been discontinued due to poor market penetration, likely because of the cost and the frequency of animal handling that fell outside of regular handling schedules. The vaccine required three doses and in Phase II and Phase III studies using about 30,000 cattle, the vaccine efficacy was demonstrated to reduce duration (by 64%) and magnitude of shedding (2.3 $\log_{10}$ reduction), reduce mucosal colonization (by 98%) and reduce hide contamination (by 54%) (Smith et al. 2009a, b).

A SRP vaccine developed by Pfizer and marketed by Zoetis, known as Epitopix™, was granted a conditional marketing license by the USDA in 2009 and is currently the only licensed vaccine available on the market. Siderophore receptor proteins are highly conserved outer membrane proteins that use high affinity ferric iron chelators, known as siderophores, to transport iron inside the bacterial cell. The vaccine consisted of multiple types of purified SRPs, of molecular weights of about 72–96 kDa, extracted from the outer bacterial membrane. By engaging immunity against cell-surface SRP proteins, the vaccine was suggested to possibly restrict iron acquisition and thus competitively disadvantage the bacterium from finding a foothold in the gut. In the initial field study using three doses, efficacy was demonstrated to reduce fecal shedding (by 39% magnitude), reduce mucosal colonization (by 48%) and reduce hide contamination (by 70%). Like Econiche™, recommended usage is for three doses applied subcutaneously over the course of 8–10 weeks with an annual revaccination.

4.3 *Plant-Based Vaccines for EHEC*

Both Econiche™ and Epitopix™ vaccines when placed on the market required three injections to the animals. This required skilled labor and handling of the animals outside of their normal handling and vaccination schedules, which usually are only twice per individual cow. With additional injections, the risk of infection is also increased and the area around the injection site can sometimes become adulterated. A valuable advantage of plant-based vaccines is the utility of oral delivery with edible plant tissue containing the bioactive therapeutic. The plant matrix has been shown to confer protection against low gastric pH to recombinant proteins stored within the cell's interior (Kolotilin et al. 2012; Kwon et al. 2013; Pelosi et al. 2012). However, while oral immunization offers more convenience, a larger dose is usually required to effectively generate an active immune response, requiring milligram to gram quantities versus the microgram quantities needed for injectable delivery (Rybicki 2010). A viable plant-based EHEC vaccine therefore needs to be of high yield and stability to meet these requirements. On a general level, a plant-based method of vaccine production may be uniquely advantageous in offering a safer and easier mode of administration, and a better cost-benefit ratio for scaling up production. Table 2 summarizes all reports of plant-based subunit vaccines for EHEC to date.

While production of a SRP vaccine in plants has not yet been reported, a number of T3SS antigens have successfully been produced in plants. Perhaps the greatest technical hurdle at the moment for developing a plant-based T3SS vaccine is improving accumulation. Because many of the T3SS proteins are membrane proteins and partially intrinsically disordered, aggregation and solubilisation are technical problems that need remedying. The choice of subcellular localization in the plant cell can drastically affect the folding and accumulation of T3SS antigens and screening is often needed to select the most optimal compartment. Recently, it was demonstrated that co-expressing the native *E. coli* chaperone for recombinant Tir improved its accumulation and its in vivo and ex vivo stability when both were targeted to the chloroplast (Table 2) (Macdonald et al. 2017). This is of great value because most T3SS proteins require chaperone-mediated folding inside EHEC and suggests the possibility that post-translational regulation may be significant in causing low accumulation of T3SS proteins in heterologous hosts. Another viable strategy is to fuse the vaccine to another protein such as green fluorescent protein (GFP), elastin-like polypeptide (ELP) or hydrophobin (HFBI) which can impart added solubility, stability or accumulation and has been used effectively for EspA, NleA and Tir (Table 2) (Macdonald et al. 2017; Miletic et al. 2017).

While Econiche™ has focused on producing a cocktail of various T3SS proteins for immunization, higher yields in plants may be possible if production is focused on a few individual antigens. Among the T3SS proteins, a select few have been demonstrated to induce higher immune responses than others, namely the 24-kDa EspA and the 37-kDa EspB proteins, and to a lesser degree, intimin. Sera taken from HUS patients contain antibodies that react strongly to these proteins, compared to control patient sera which had no reactivity (Jarvis and Kaper 1996). In addition to reactivity from O157:H7 strains, antigens prepared from O26 strains also show strong reactivity. Therefore, these proteins are great candidates for the possibility of engendering multi-strain protection (Mckee and O'brien 1996).

When lyophilized plant tissue containing a 5 mg dose of a chimeric EspA vaccine (expressed transplastomically in *Nicotiana benthamiana*) was administered to sheep three times over a six week period, five of the six animals inoculated stopped shedding O157:H7 after 48 days with about a 95% reduction in magnitude compared to control animals which persisted in shedding. Of the plant-based EHEC vaccines developed thus far, this chimeric EspA seems to be the most promising candidate based on highest efficacy, yield and and has been the only candidate tested on ruminants. Due to the recent increase in non-O157 EHEC infections, market value of EHEC vaccines could be increased by either incorporating multivalency in vaccine design such as by epitope fusions as well as testing vaccine

Table 2 Accumulation profiles of all plant-based subunit vaccines against EHEC that have been reported to date

Antigen	Serotype	Transformation type	Production system	Subcellular localization	Fusion partner	Accumulation (%TSP)	Accumulation (mg/kg)	Efficacy	References
Chimeric EspA fusion	O157:H7, O26:H11	Transient	*N. benthamiana* leaves	ER	–	0.13	14.8	–	(Miletic et al. 2017)
Chimeric EspA fusion	O157:H7, O26:H11	Transient	*N. benthamiana* leaves	ER	ELP	0.8	87.8	–	(Miletic et al. 2017)
Chimeric EspA fusion	O157:H7, O26:H11	Transient	*N. benthamiana* leaves	ER	HFBI	0.5	52.7	–	(Miletic et al. 2017)
Chimeric EspA fusion	O157:H7, O26:H11	Transient	*N. benthamiana* leaves	Chloroplast	–	0.12	13	–	(Miletic et al. 2017)
Chimeric EspA fusion	O157:H7, O26:H11	Transplastomic	*N. tabacum* leaves	Chloroplast	–	–	480	sheep: ↓ shedding. 5 mg dose lyophilized tissue; parenteral	(Miletic et al. 2017)
EspA	O157:H7	Transient	*N. benthamiana* leaves	Chloroplast	± GFP	–	–	–	(Macdonald et al. 2017)
NleA	O157:H7	Transient	*N. benthamiana* leaves	ER	–	0.003	0.0002	–	(Miletic et al. 2017)
NleA	O157:H7	Transient	*N. benthamiana* leaves	ER	ELP	0.09	0.01	–	(Miletic et al. 2017)
NleA	O157:H7	Transient	*N. benthamiana* leaves	ER	HFBI	0.01	0.001	–	(Miletic et al. 2017)
NleA	O157:H7	Transient	*N. benthamiana* leaves	Chloroplast	–	0.01	0.001	–	(Miletic et al. 2017)

(continued)

Table 2 (continued)

Antigen	Serotype	Transformation type	Production system	Subcellular localization	Fusion partner	Accumulation (%TSP)	Accumulation (mg/kg)	Efficacy	References
NleA	O157:H7	Transient	*N. benthamiana* leaves	Cytoplasm	–	0.02	0.001	–	(Miletic et al. 2017)
NleA	O157:H7	Transient	*N. benthamiana* leaves	Chloroplast	± GFP	–	–	–	(Macdonald et al. 2017)
Stx2 (B subunit only)	O157:H7	Transient	*N. benthamiana* leaves	ER	ELP	0.3	28	–	(Miletic et al. 2017)
Tir	O157:H7	Transient	*N. benthamiana* leaves	Chloroplast	± GFP	–	–	–	(Macdonald et al. 2017)
Chimeric Tir fusion	O157:H7, O26:H11, O45:H2, O111:H8	Transient	*N. benthamiana* leaves	Chloroplast	± GFP	–	–	–	(Macdonald et al. 2017)
Tir	O157:H7	Transplastomic	*N. tabacum* leaves	Chloroplast	–	–	–	–	(Macdonald et al. 2017)
Chimeric EspA-Intimin-Tir fusion	O157:H7	Stable	*N. tabacum* leaves	Cytoplasm	–	0.3	–	mice: ↓ shedding. 15 μg dose purified; oral & parenteral	(Amani et al. 2011)
Chimeric EspA-Intimin-Tir fusion	O157:H7	Stable	*B. napus* seeds	Cytoplasm	–	0.3	–	mice: ↓ shedding. 15 μg dose purified; oral & parenteral	(Amani et al. 2011)

(continued)

Table 2 (continued)

Antigen	Serotype	Transformation type	Production system	Subcellular localization	Fusion partner	Accumulation (%TSP)	Accumulation (mg/kg)	Efficacy	References
Stx2 toxoid (inactivated A subunit + B subunit)	O157:H7	Stable	*N. tabacum* cell culture	Cytoplasm	–	–	7.35	mice: ↑IgA;↑ survivability against toxin challenge; cytotoxic neutralization. 5 g dose cell culture. oral	(Wen et al. 2006a, b)
Truncated Intimin (C-terminal peptide)	O157:H7	Stable	*N. tabacum* cell culture	Cytoplasm	–	–	11.5	mice: ↑IgA; ↑IgG;↓ duration of colonization. 7.5μg dose purified. oral	(Judge et al. 2004)

ER Endoplasmic reticulum, *HFBI* Hydrophobin, *GFP* (enhanced) Green Fluorescent Protein, *ELP* Elastin-like polypeptide, *Ig* Immunoglobulin candidates for cross-reactivity during animal trials. Accordingly, fusions of EspA epitopes from both O157 and non O157 strains, produced both transiently and transplastomically in leaves of *Nicotiana tabacum,* show promise as multivalent candidates (Miletic et al. 2017). Another candidate, an EspA-Intimin-Tir fusion, has been demonstrated to accumulate in leaves of *N. tabacum* and seeds of *Brassica napus* at about the same yield and reduce shedding when administered to a mouse model (Amani et al. 2011). In considering *B. napus* as a platform, there is some appeal as it is a much more familiar feed component than *N. tabacum* to producers if oral application is to be considered. However, *N. tabacum* has conventionally been the platform of choice primarily because it is neither a food- nor feed- crop and is less likely to contaminate a food supply. On the other hand, *N. tabacum* cell cultures show promise as a platform because they can be grown in a closed, sterile system isolated from the external environment. For example, this is the platform of choice for Protalix Biotherapeutics in their production of glucocerebrosidase in carrot cells. Further development of a cell-culture based EHEC vaccine towards a similar direction may be of value considering the recent trend of public attitudes and restrictive policy making with regards to containment of genetically engineered crops. An inactivated form of the stx2 toxin has been shown to accumulate in *N. tabacum* cell cultures and when administered to mice, IgA production is triggered and the mice have enhanced survivability against toxin challenge (Wen et al. 2006a, b). A truncated form of intimin has also been shown to accumulate in *N. tabacum* cell cultures and when administered to mice, triggers both IgA and IgG production as well as reduces the duration of EHEC colonization (Judge et al. 2004).

5 Pathways to Commercialization/Implementation for a Plant-Based EHEC Vaccine: Learning from Econiche's Business Model in the Canadian Beef Industry

Despite Econiche^{TM} having Canada-wide availability, marketing as a robust pre-harvest control and multiple validations of its efficacy, its adoption by the Canadian beef industry after product launch was marginal at an estimated level of adoption of only about 5% (Grier and Schmidt 2009). Beef producers, the primary target market, were reluctant to adopting the product, likely because the direct benefits are realized elsewhere along the supply chain, namely processors and consumers. Additionally, aside from the direct cost of the product, vaccination required extraneous labour and veterinary costs to implement. Was the Econiche^{TM} business model flawed? Can the barriers that hampered adoption be addressed to facilitate effective market transition for a similar product?

The economic story of the Econiche^{TM} product in the Canadian market has implications for the general prospects of any future EHEC vaccine to be considered for commercialization. Following proof of concept and efficacy studies of T3SS proteins, Bioniche Life Sciences Inc. was contracted for scale-up and commercialization (Jarvis et al. 1995; Jarvis and Kaper 1996). The project was financed via a substantial $25 million investment sourced from the Ontario government, Agriculture Canada, Industry Canada and the Business Development Bank of Canada (Bioniche 2012). In a 2012 letter to shareholders prior to the release of Econiche™, management disclosed that the company was suffering from a monthly burn rate of $1 million per month operating on a net income loss in prior years and that one of its foremost strategies for remedying this was from increased revenue anticipated from its new products to be released that year, including Econiche (Bioniche 2012). In the two years prior, revenue had stalled for the company at approximately $27 M. Despite this, a 2012 initiating report by Eresearch, a Canadian independent equity research corporation, recommended considerable upside potential for Bioniche's share price citing the release of Econiche™ as a main reason and forecasted Econiche™-specific revenue as bringing in $1.5 M and $3.75 M in 2013 and 2014 respectively, with steady growth in later years (Eresearch Corporation 2012). Collectively, this implies that (1) there was substantial financial capital available from multiple sources to develop the Econiche™ vaccine for market (2) Bioniche considered it a high priority revenue earner to be developed to counter its looming burn rate and (3) market research also corroborated the belief it would do well in the market. Considering that Econiche™ was announced as the world's first vaccine against EHEC with full licensure by the CFIA and provisional licensure by the USDA, its first commercial batch entered the Canadian market in mid-2012 without competition in its market niche. Epitopix™, the subsequent and only rival to-date in this market niche, obtained conditional USDA licensure a year later, and was restricted to US cattle markets. By 2014, Bioniche decided to refocus its efforts on solely human health, putting up its vaccine development unit up for sale, suspending operations and laying off most of its employees. In a statement released by Michael Berendt, CEO: "While the vaccine is an innovative and valuable product, (Bioniche) has been unable to convince the beef or dairy industries, or the federal and provincial governments, that vaccinating cattle to help reduce the human infection and deaths caused by *E. coli* is something they should support or pay for."

So, what went wrong? Perhaps the largest assumption could be that with full control over their market niche, Bioniche anticipated high demand—a far-removed prediction from its dismal 5% penetration (Grier and Schmidt 2009). The barriers to this demand directly relate to the requirements of its target market. In North America, vaccination needs to be done at least three months before slaughter and for most other pathogens, is the responsibility of beef producers. Therefore, the target market for the Econiche™ product comprises a potential total of 75,000 and 913,000 cattle/calf operations for Canada and the United States respectively (inclusive of beef farms, ranches, feedlots and dairy operations) (Statistics Canada 2017; United States Department of Agriculture 2017). Correspondingly, there are currently an estimated 13 million cattle in Canada and 103 million in the United States (Statistics Canada 2017; United States Department of Agriculture 2017). Of the 75,000 beef producers in Canada, 86.7% currently vaccinate their calves against some form of disease so producers are no strangers to the technology (Ochieng' and Hobbs 2017). However, although these producers are accustomed to the benefits of vaccination, it is considered in many respects an insurance policy. Most Canadian producers routinely sacrifice ~10% of their profit margin to ensure against the risk that the health and productivity of their herd be diminished from the most common bovine diseases. In most herds in Canada, this allocation usually goes toward protecting against infectious bovine rhinotracheitis (IBR), bovine respiratory disease (BRD), parainfluenza-3 virus (PI3V), clostridials, hemophilus and bovine respiratory syncytial virus (BRSV). However, since cattle are asymptomatic carriers of EHEC, whether or not they harbor the bacterium is independent of their health and productivity, or risk thereof. Since EHEC does not pose a risk to the health or viability of cattle, producers are less incentivised to purchase an EHEC vaccine since such an investment would not provide any direct returns.

Towards addressing what were the barriers for EHEC vaccine adoption, a survey of Canadian cattle producers indicated that only 15% of respondents believed they bear the primary responsibility for EHEC risk reduction and only 21% of respondents believed they benefited from an EHEC vaccine (Ochieng' and Hobbs 2017). While there were many perceived barriers to adoption that were reported by respondents, the issues that most agreed to be relevant included: (1) uncertainty over benefits, as indicated by 76.8% respondents, (2) meeting buyer needs, as indicated by 71.4% respondents and (3) efficacy of the vaccine, as indicated by 68.5% respondents (Ochieng' and Hobbs 2017). Additionally, 58% of the beef producers surveyed had not previously heard about a EHEC vaccine (Ochieng' and Hobbs 2017). Given that estimated marginal effects predict an average 16.1% increased willingness to adopt given prior awareness of an EHEC vaccine, this suggests appropriate marketing of this product is an essential component of its commercialization (Ochieng' and Hobbs 2017).

Additionally, many producers were resistant to adopting Econiche™ as a stand-alone technology which could not be easily incorporated into their routine vaccination schedules. For example, the recommended dose regimen was two doses in the initial year of life plus a subsequent annual dose. This required skilled labor and handling of the animals outside of their normal handling and vaccination schedules, which usually are only twice per individual calf. Application was also required to be by injection, a procedure that both carries safety hazards for the handler and the possibility of infection to the calf.

EHEC's prevalence in retail beef products has maintained a steady 2% over the last ten years in Canada and recalls of beef contaminated with EHEC are generally infrequent, though quite costly to processors when they do occur. Indeed, a food recall can sometimes lead to the closing of a meat processing plant if it fails subsequent safety inspections. However, incorporating a vaccination program in a market-driven economy begs the question of whether or not the cost of doing so is worth the added insurance against meat recalls. A cost-benefit analysis of industry adoption of Econiche™ indicated an overall approximate savings of CAD$68 M per year comprising an estimated benefit/cost ratio of 3:1 (Grier and Schmidt 2009). In particular, the analysis indicated a total approximate annual benefit of CAD

$103 M including: CAD$21 M in reduced medical costs, CAD$4 M in reduced recalls and industry costs and CAD$78 M from loss in demand. Conversely, the annual cost to the industry was estimated at CAD$3 M, which scaled directly with the dose regimen and herd numbers. Overall, the study suggested that with implementation, this technology could be both socially beneficial and financially prudent for the beef industry on a whole. Despite this, Canadian federal or provincial sponsored incentives for EHEC pre-harvest control are low or nonexistent at the farm level though processors are well motivated to reduce EHEC contamination in order to avoid recalls. Conversely, producers do not directly benefit from lowering the chance of a food recall, despite the expectation that they pay for vaccination.

Overall, if a similar product was to be considered for commercialization, its success would be dependent on: (1) more availability of sufficient adoption incentives for beef producers by government and supply chain (2) better awareness of EHEC and vaccine technology by producers and their veterinarians (3) development of a vaccine with a high economy of scale and that is easily accommodated into producers' typical vaccination schedule. Whereas the first two requirements will need a concerted dialogue between industry and government, the requirements of the latter can be met with technical innovation that scales well with producing large quantities with minimal investment. In this regard, a plant-based platform for an EHEC vaccine is arguably a competitive solution.

The production of an EHEC vaccine in plants that could be administered orally by incorporating it into livestock feed would bypass the extraneous time and labor that made the Econiche™ vaccines unappealing to producers. The prospect of oral immunization offers a strategic competitive advantage for a plant-based EHEC vaccine because of the increased safety and convenience for the producer. However, this mode of administration may require a much larger dose than parenteral to be effective since much of the protein is degraded during its movement through the animal's gut prior to reaching cells in the distal intestinal epithelium that can generate an immune response (Rybicki 2010). Therefore, if this selling point were to be developed, key research targets could arguably be: (1) better yield of the protein (2) improved protein stability to reduce the amount lost through degradation (3) designs or formulations geared toward adjuvancy such that the threshold required for the production of an immune response may be crossed with a lower concentration of therapeutic. Additionally, the efficient cost scaling of the technology lends itself well to a widespread vaccination program—which may likely prove necessary to enable consistent herd immunity against EHEC. This technology is still in its early development stage with major milestone requirements before commercialization being proof of efficacy across environments and improving yield to enable better scaling. Yet, we are optimistic that this technology will be of value to the Canadian beef industry and toward the control of food safety.

References

Amani J, Mousavi SL, Rafati S et al (2011) Immunogenicity of a plant-derived edible chimeric EspA, Intimin and Tir of *Escherichia coli* O157:H7 in mice. Plant science: an international journal of experimental plant biology 180:620–627

Baines D, Masson L, Mcallister T (2008) Escherichia coli O157: H7-secreted cytotoxins are toxic to enterocytes and increase Escherichia coli O157: H7 colonization of jejunum and descending colon in cattle. Can J Anim Sci 88:41–50

Bell BP, Goldoft M, Griffin PM et al (1994) A multistate outbreak of escherichia coli o157: h7—associated bloody diarrhea and hemolytic uremic syndrome from hamburgers: the washington experience. JAMA 272:1349–1353

Berry ED, Wells JE, Bono JL et al (2015) Effect of proximity to a cattle feedlot on *Escherichia coli* O157: H7 contamination of leafy greens and evaluation of the potential for airborne transmission. Appl Environ Microbiol 81:1101–1110

Besser TE, Hancock DD, Pritchett LC et al (1997) Duration of detection of fecal excretion of *Escherichia coli* 0157: H7 in cattle. J Infect Dis 175:726–729

Best A, La Ragione RM, Sayers AR et al (2005) Role for flagella but not intimin in the persistent infection of the gastrointestinal tissues of specific-pathogen-free chicks by Shiga toxin-negative Escherichia coli O157: H7. Infect Immun 73:1836–1846

Beutin L, Geier D, Steinrück H et al (1993) Prevalence and some properties of verotoxin (Shiga-like toxin)-producing *Escherichia coli* in seven different species of healthy domestic animals. J Clin Microbiol 31:2483–2488

Bioniche (2012) Letter to shareholders. https://hotcopper.com.au/threads/ann-2012-q1-shareholder-letter-mda-financials.1622733/?post_id=9436178#.WlPAl02oupp. Accessed 08 Jan 2018

Boyd B, Lingwood C (1989) Verotoxin receptor glycolipid in human renal tissue. Nephron 51:207–210

Burkinshaw BJ, Deng W, Lameignère E et al (2015) Structural analysis of a specialized type III secretion system peptidoglycan-cleaving enzyme. J Biol Chem 290:10406–10417

Büttner D (2012) Protein export according to schedule: architecture, assembly, and regulation of type III secretion systems from plant-and animal-pathogenic bacteria. Microbiol Mol Biol Rev 76:262–310

CDC (2011) Vital signs: incidence and trends of infection with pathogens transmitted commonly through food–foodborne diseases active surveillance network, 10 US sites, 1996–2010. Morb Mortal Wkly Rep 60:749

CDC (2017) Foodborne diseases active surveillance network (FoodNet): FoodNet 2015 Surveillance Report (Final Data). Atlanta, Georgia: U.S. Department of health and human services. https://www.cdc.gov/foodnet/pdfs/FoodNet-Annual-Report-2015-508c.pdf. Accessed 8 Jan 2018

Charatan F (1999) New York outbreak of *E coli* poisoning affects 1000 and kills two. BMJ. British Medical Journal 319:873

Chong Y, Fitzhenry R, Heuschkel R et al (2007) Human intestinal tissue tropism in *Escherichia coli* O157: H7–initial colonization of terminal ileum and Peyer's patches and minimal colonic adhesion ex vivo. Microbiology 153:794–802

Cleary TG (2004) The role of Shiga-toxin-producing *Escherichia coli* in hemorrhagic colitis and hemolytic uremic syndrome. In: Seminars in pediatric infectious diseases. Elsevier, p 260–265

Delahay RM, Knutton S, Shaw RK et al (1999) The Coiled-coil Domain of EspA Is Essential for the Assembly of the Type III Secretion Translocon on the Surface of Enteropathogenic *Escherichia coli*. J Biol Chem 274:35969–35974

Donnenberg MS, Tzipori S, Mckee ML et al (1993) The role of the eae gene of enterohemorrhagic *Escherichia coli* in intimate attachment in vitro and in a porcine model. J Clin Invest 92:1418

Doyle M, Archer J, Kaspar C et al (2006) FRI Briefing: human illness caused by E. coli O157: H7 from food and non-food sources. https://fri.wisc.edu/files/Briefs_File/FRIBrief_EcoliO157H7 humanillness.pdf. Accessed 08 Jan 2018

Dziva F, Van Diemen PM, Stevens MP et al (2004) Identification of *Escherichia coli* O157: H7 genes influencing colonization of the bovine gastrointestinal tract using signature-tagged mutagenesis. Microbiology 150:3631–3645

Endo Y, Tsurugi K, Lambert JM (1988) The site of action of six different ribosome-inactivating proteins from plants on eukaryotic ribosomes: the RNA N-glycosidase activity of the proteins. Biochem Biophys Res Commun 150:1032–1036

Eresearch Corporation NCWB (2012) Initiating Report for BioNiche life sciences Inc. http://www.baystreet.ca/articles/research_reports/eresearch/BNC_011712-I.pdf. Accessed 08 Jan 2018

Fivaz M, Van Der Goot FG (1999) The tip of a molecular syringe. Trends Microbiol 7:341–343

Fukushima H, Hashizume T, Morita Y et al (1999) Clinical experiences in Sakai City Hospital during the massive outbreak of enterohemorrhagic *Escherichia coli* O157 infections in Sakai City, 1996. Pediatr Int 41:213–217

Garmendia J, Phillips AD, Carlier MF et al (2004) TccP is an enterohaemorrhagic Escherichia coli O157: H7 type III effector protein that couples Tir to the actin-cytoskeleton. Cell Microbiol 6:1167–1183

Gill A, Gill CO (2010) Non-O157 verotoxigenic *Escherichia coli* and beef: a Canadian perspective. Can J Vet Res 74:161–169

Gould LH, Demma L, Jones TF et al (2009) Hemolytic uremic syndrome and death in persons with *Escherichia coli* O157: H7 infection, foodborne diseases active surveillance network sites, 2000–2006. Clin Infect Dis 49:1480–1485

Grier K, Schmidt C (2009) *E. coli* 0157 risk reduction: economic benefit to Canada. George Morris Centre

Griffin P (1995) *Escherichia coli* O157: H7 and other enterohemorrhagic Escherichia coli. Infections of the gastrointestinal tract:739–761

Griffin PM, Olmstead LC, Petras RE (1990) *Escherichia coli* O157: H7-associated colitis. Gastroenterology 99:142–149

Griffin PM, Tauxe RV (1991) The epidemiology of infections caused by Escherichia coli O157: H7, other enterohemorrhagic E. coli, and the associated hemolytic uremic syndrome. Epidemiol Rev 13:60–98

Gruenheid S, Sekirov I, Thomas NA et al (2004) Identification and characterization of NleA, a non-LEE-encoded type III translocated virulence factor of enterohaemorrhagic Escherichia coli O157: H7. Mol Microbiol 51:1233–1249

Hale CR, Scallan E, Cronquist AB et al (2012) Estimates of enteric illness attributable to contact with animals and their environments in the United States. Clin Infect Dis 54:S472–S479

Heuvelink A, Van Heerwaarden C, Zwartkruis-Nahuis J et al (2002) *Escherichia coli* O157 infection associated with a petting zoo. Epidemiol Infect 129:295–302

Hrudey S, Payment P, Huck P et al (2003) A fatal waterborne disease epidemic in Walkerton, Ontario: comparison with other waterborne outbreaks in the developed world. Water Sci Technol 47:7–14

Ismaili A, Philpott DJ, Dytoc MT et al (1995) Signal transduction responses following adhesion of verocytotoxin-producing *Escherichia coli*. Infect Immun 63:3316–3326

Jacewicz MS, Acheson D, Mobassaleh M et al (1995) Maturational regulation of globotriaosylceramide, the Shiga-like toxin 1 receptor, in cultured human gut epithelial cells. J Clin Invest 96:1328

Jarvis KG, Giron JA, Jerse AE et al (1995) Enteropathogenic *Escherichia coli* contains a putative type III secretion system necessary for the export of proteins involved in attaching and effacing lesion formation. Proc Natl Acad Sci 92:7996–8000

Jarvis KG, Kaper JB (1996) Secretion of extracellular proteins by enterohemorrhagic *Escherichia coli* via a putative type III secretion system. Infect Immun 64:4826–4829

Jiang X, Morgan J, Doyle MP (2002) Fate of *Escherichia coli* O157: H7 in manure-amended soil. Appl Environ Microbiol 68:2605–2609

Judge NA, Mason HS, O'brien AD (2004) Plant cell-based intimin vaccine given orally to mice primed with intimin reduces time of *Escherichia coli* O157:H7 shedding in feces. Infect Immun 72:168–175

Kandel G, Donohue-Rolfe A, Donowitz M et al (1989) Pathogenesis of *Shigella* diarrhea. XVI. Selective targetting of Shiga toxin to villus cells of rabbit jejunum explains the effect of the toxin on intestinal electrolyte transport. J Clin Invest 84:1509

Karmali M, Petric M, Lim C et al (1983) *Escherichia coli* cytotoxin, haemolytic-uraemic syndrome, and haemorrhagic colitis. The Lancet 322:1299–1300

Karmali MA (2004) Prospects for preventing serious systemic toxemic complications of Shiga toxin–producing *Escherichia coli* infections using Shiga toxin receptor analogues. J Infect Dis 189:355–359

Kenny B, Devinney R, Stein M et al (1997) Enteropathogenic E. coli (EPEC) transfers its receptor for intimate adherence into mammalian cells. Cell 91:511–520

Kenny B, Finlay BB (1995) Protein secretion by enteropathogenic Escherichia coli is essential for transducing signals to epithelial cells. Proc Natl Acad Sci 92:7991–7995

Kim J, Thanabalasuriar A, Chaworth-Musters T et al (2007) The bacterial virulence factor NleA inhibits cellular protein secretion by disrupting mammalian COPII function. Cell Host Microbe 2:160–171

Knutton S, Baldwin T, Williams P et al (1989) Actin accumulation at sites of bacterial adhesion to tissue culture cells: basis of a new diagnostic test for enteropathogenic and enterohemorrhagic *Escherichia coli*. Infect Immun 57:1290–1298

Kolotilin I, Kaldis A, Devriendt B et al (2012) Production of a subunit vaccine candidate against porcine post-weaning diarrhea in high-biomass transplastomic tobacco. PLoS ONE 7:e42405

Kwon K-C, Verma D, Singh ND et al (2013) Oral delivery of human biopharmaceuticals, autoantigens and vaccine antigens bioencapsulated in plant cells. Adv Drug Deliv Rev 65: 782–799

La Ragione R, Ahmed NM, Best A et al (2005) Colonization of 8-week-old conventionally reared goats by Escherichia coli O157: H7 after oral inoculation. J Med Microbiol 54:485–492

La Ragione R, Best A, Clifford D et al (2006) Influence of colostrum deprivation and concurrent Cryptosporidium parvum infection on the colonization and persistence of Escherichia coli O157: H7 in young lambs. J Med Microbiol 55:819–828

Lai L-C, Wainwright LA, Stone KD et al (1997) A third secreted protein that is encoded by the enteropathogenic Escherichia coli pathogenicity island is required for transduction of signals and for attaching and effacing activities in host cells. Infect Immun 65:2211–2217

Lejeune JT, Besser TE, Hancock DD (2001) Cattle water troughs as reservoirs of *Escherichia coli* O157. Appl Environ Microbiol 67:3053–3057

Lim JY, Li J, Sheng H et al (2007) *Escherichia coli* O157: H7 colonization at the rectoanal junction of long-duration culture-positive cattle. Appl Environ Microbiol 73:1380–1382

Louise CB, Tran MC, Obrig TG (1997) Sensitization of human umbilical vein endothelial cells to Shiga toxin: involvement of protein kinase C and NF-kappaB. Infect Immun 65:3337–3344

Macdonald J, Miletic S, Gaildry T et al (2017) Co-expression with the Type 3 Secretion Chaperone CesT from Enterohemorrhagic *E. coli* Increases Accumulation of Recombinant Tir in Plant Chloroplasts. Frontiers in Plant Science 8

Majowicz SE, Scallan E, Jones-Bitton A et al (2014) Global incidence of human Shiga toxin–producing *Escherichia coli* infections and deaths: a systematic review and knowledge synthesis. Foodborne pathogens and disease 11:447–455

Mathusa EC, Chen Y, Enache E et al (2010) Non-O157 Shiga toxin-producing Escherichia coli in foods. J Food Prot 73:1721–1736

Matthews L, Reeve R, Gally DL et al (2013) Predicting the public health benefit of vaccinating cattle against *Escherichia coli* O157. Proc Natl Acad Sci 110:16265–16270

Mckee ML, O'brien AD (1996) Truncated enterohemorrhagic *Escherichia coli* (EHEC) O157: H7 intimin (EaeA) fusion proteins promote adherence of EHEC strains to HEp-2 cells. Infect Immun 64:2225–2233

Miletic S, Hunerberg M, Kaldis A et al (2017) A Plant-produced candidate subunit vaccine reduces shedding of enterohemorrhagic Escherichia coli in Ruminants. Biotechnol J 12

Montenegro MA, Bülte M, Trumpf T et al (1990) Detection and characterization of fecal verotoxin-producing *Escherichia coli* from healthy cattle. J Clin Microbiol 28:1417–1421

Moon H, Whipp S, Argenzio R et al (1983) Attaching and effacing activities of rabbit and human enteropathogenic *Escherichia coli* in pig and rabbit intestines. Infect Immun 41:1340–1351

Nart P, Naylor SW, Huntley JF et al (2008) Responses of cattle to gastrointestinal colonization by Escherichia coli O157:H7. Infect Immun 76:5366–5372

Naylor SW, Low JC, Besser TE et al (2003) Lymphoid follicle-dense mucosa at the terminal rectum is the principal site of colonization of enterohemorrhagic *Escherichia coli* O157: H7 in the bovine host. Infect Immun 71:1505–1512

Ochieng' BJ, Hobbs JE (2017) Factors affecting cattle producers' willingness to adopt an *Escherichia coli* O157: H7 vaccine: a probit analysis. International Food and Agribusiness Management Review 20:347–363

Omisakin F, Macrae M, Ogden I et al (2003) Concentration and prevalence of *Escherichia coli* O157 in cattle feces at slaughter. Appl Environ Microbiol 69:2444–2447

Pai C, Kelly J, Meyers G (1986) Experimental infection of infant rabbits with verotoxin-producing *Escherichia coli*. Infect Immun 51:16–23

Pelosi A, Piedrafita D, De Guzman G et al (2012) The effect of plant tissue and vaccine formulation on the oral immunogenicity of a model plant-made antigen in sheep. PLoS ONE 7: e52907

Perna NT, Mayhew GF, Pósfai G et al (1998) Molecular evolution of a pathogenicity island from enterohemorrhagic *Escherichia coli* O157: H7. Infect Immun 66:3810–3817

Phillips A, Navabpour S, Hicks S et al (2000) Enterohaemorrhagic Escherichia coliO157: H7 target Peyer9 s patches in humans and cause attaching/effacing lesions in both human and bovine intestine. Gut 47:377–381

Pruimboom-Brees IM, Morgan TW, Ackermann MR et al (2000) Cattle lack vascular receptors for Escherichia coli O157: H7 Shiga toxins. Proc Natl Acad Sci 97:10325–10329

Public Health Agency of Canada (2015) Foodnet Canada Short Report 2015. http://publications.gc.ca/site/eng/461265/publication.html. Accessed 08 Jan 2018

Rangel JM, Sparling PH, Crowe C et al (2005a) Epidemiology of *Escherichia coli* O157: H7 outbreaks, United States, 1982–2002

Rangel JM, Sparling PH, Crowe C et al (2005b) Epidemiology of Escherichia coli O157:H7 outbreaks, United States, 1982-2002. Emerg Infect Dis 11:603–609

Riley LW, Remis RS, Helgerson SD et al (1983) Hemorrhagic colitis associated with a rare *Escherichia coli* serotype. N Engl J Med 308:681–685

Rybicki EP (2010) Plant-made vaccines for humans and animals. Plant Biotechnol J 8:620–637

Sanderson MW, Sargeant JM, Shi X et al (2006) Longitudinal emergence and distribution of *Escherichia coli* O157 genotypes in a beef feedlot. Appl Environ Microbiol 72:7614–7619

Sargeant J, Amezcua M, Rajic A et al (2007) Pre-harvest interventions to reduce the shedding of E. coli O157 in the faeces of weaned domestic ruminants: a systematic review. Zoonoses Public Health 54:260–277

Slutsker L, Ries AA, Greene KD et al (1997) *Escherichia coli* O157: H7 diarrhea in the United States: clinical and epidemiologic features. Ann Intern Med 126:505–513

Smith DR, Moxley RA, Klopfenstein TJ et al (2009a) A randomized longitudinal trial to test the effect of regional vaccination within a cattle feedyard on *Escherichia coli* O157: H7 rectal colonization, fecal shedding, and hide contamination. Foodborne Pathogens and Disease 6:885–892

Smith DR, Moxley RA, Peterson RE et al (2009b) A two-dose regimen of a vaccine against type III secreted proteins reduced *Escherichia coli* O157: H7 colonization of the terminal rectum in beef cattle in commercial feedlots. Foodborne pathogens and disease 6:155–161

Snedeker K, Campbell M, Sargeant J (2012) A systematic review of vaccinations to reduce the shedding of *Escherichia coli* O157 in the faeces of domestic ruminants. Zoonoses Public Health 59:126–138

Stanford K, Stephens TP, Mcallister TA (2011) Use of model super-shedders to define the role of pen floor and hide contamination in the transmission of Escherichia coli O157:H7. J Anim Sci 89:237–244

Statistics Canada (2017) Livestock estimates. http://www.statcan.gc.ca/daily-quotidien/170818/dq170818e-eng.htm. Accessed 08 Jan 2018

Stephens TP, Mcallister TA, Stanford K (2009) Perineal swabs reveal effect of super shedders on the transmission of Escherichia coli O157:H7 in commercial feedlots. J Anim Sci 87: 4151–4160

Strachan NJ, Doyle MP, Kasuga F et al (2005) Dose response modelling of *Escherichia coli* O157 incorporating data from foodborne and environmental outbreaks. Int J Food Microbiol 103: 35–47

Tobe T, Beatson SA, Taniguchi H et al (2006) An extensive repertoire of type III secretion effectors in Escherichia coli O157 and the role of lambdoid phages in their dissemination. Proc Natl Acad Sci 103:14941–14946

Tuttle J, Gomez T, Doyle M et al (1999) Lessons from a large outbreak of *Escherichia coli* O157: H7 infections: insights into the infectious dose and method of widespread contamination of hamburger patties. Epidemiol Infect 122:185

Tzipori S, Gibson R, Montanaro J (1989) Nature and distribution of mucosal lesions associated with enteropathogenic and enterohemorrhagic *Escherichia coli* in piglets and the role of plasmid-mediated factors. Infect Immun 57:1142–1150

Tzipori S, Gunzer F, Donnenberg MS et al (1995) The role of the eaeA gene in diarrhea and neurological complications in a gnotobiotic piglet model of enterohemorrhagic *Escherichia coli* infection. Infect Immun 63:3621–3627

Unicef (2010) Diarrhoea: why children are still dying and what can be done. https://www.unicef.org/media/files/Final_Diarrhoea_Report_October_2009_final.pdf. Accessed 08 Jan 2018

United States Department of Agriculture (2017) Cattle. http://usda.mannlib.cornell.edu/MannUsda/viewDocumentInfo.do?documentID=1017. Accessed 08 Jan 2018

Vial PA, Robins-Browne R, Lior H et al (1988) Characterization of enteroadherent-aggregative Escherichia coli, a putative agent of diarrheal disease. J Infect Dis 158:70–79

Wagner PL, Waldor MK (2002) Bacteriophage control of bacterial virulence. Infect Immun 70:3985–3993

Warawa J, Finlay BB, Kenny B (1999) Type III secretion-dependent hemolytic activity of enteropathogenic Escherichia coli. Infect Immun 67:5538–5540

Wen SX, Teel LD, Judge NA et al (2006a) Genetic toxoids of Shiga toxin types 1 and 2 protect mice against homologous but not heterologous toxin challenge. Vaccine 24:1142–1148

Wen SX, Teel LD, Judge NA et al (2006b) A plant-based oral vaccine to protect against systemic intoxication by Shiga toxin type 2. Proc Natl Acad Sci 103:7082–7087

Wendel AM, Johnson DH, Sharapov U et al (2009) Multistate outbreak of Escherichia coli O157: H7 infection associated with consumption of packaged spinach, August–September 2006: the Wisconsin investigation. Clin Infect Dis 48:1079–1086

Wheeler T, Kalchayanand N, Bosilevac JM (2014) Pre-and post-harvest interventions to reduce pathogen contamination in the US beef industry. Meat Sci 98:372–382

Williams MS, Withee JL, Ebel ED et al (2010) Determining relationships between the seasonal occurrence of Escherichia coli O157: H7 in live cattle, ground beef, and humans. Foodborne pathogens and disease 7:1247–1254

Wong AR, Pearson JS, Bright MD et al (2011) Enteropathogenic and enterohaemorrhagic Escherichia coli: even more subversive elements. Mol Microbiol 80:1420–1438

Woodward MJ, Best A, Sprigings KA et al (2003) Non-toxigenic Escherichia coli O157: H7 strain NCTC12900 causes attaching-effacing lesions and eae-dependent persistence in weaned sheep. Int J Med Microbiol 293:299–308

Foot-and-mouth Disease

Vanesa Ruiz and Andrés Wigdorovitz

Abstract Foot-and-mouth disease remains one of the most feared viral diseases affecting cloven-hoofed animals such as cattle, pigs, sheep and goats. The disease has been successfully eradicated from some regions like North America, Western Europe and Australia, but it is still endemic in most of the world. Although mortality is generally low in adult animals, outbreaks result in devastating economic consequences due to production losses and a major constraint to international trade of live animals and their products. Immunization with the current inactivated vaccine has been successfully used in many parts of the world. However, its production process requires the growth of large amounts of infectious virus in high-level bio-containment facilities, which is not only very expensive but carries the risk of escape of live virus during vaccine manufacture and/or incomplete inactivation. Because of these hazards and other limitations, such as thermal instability, short duration of immunity and lack of cross protection, intense research focused on the design of improved vaccines, has been developed. Important issues concerning foot-and-mouth disease occurrence, pathogenesis and vaccine development, are reviewed in this chapter.

Keywords Foot-and-mouth disease · pathogenesis · plant-made vaccines inactivated vaccine · alternative vaccines

V. Ruiz · A. Wigdorovitz (✉)
Instituto de Virología, Centro de Investigaciones en Ciencias Veterinarias y Agronómicas (CICVyA), Instituto Nacional de Tecnología Agropecuaria (INTA), Hurlingham, Buenos Aires, Argentina
e-mail: wigdorovitz.andres@inta.gob.ar

V. Ruiz
e-mail: ruiz.vanesa@inta.gob.ar

V. Ruiz · A. Wigdorovitz
Consejo Nacional de Investigaciones Científicas y Tecnológicas (CONICET), Buenos Aires, Argentina

J. MacDonald (ed.), *Prospects of Plant-Based Vaccines in Veterinary Medicine*,
https://doi.org/10.1007/978-3-319-90137-4_15

1 Disease symptoms and occurrence

1.1 Foot and Mouth Disease and geographical distribution

Foot-and-mouth disease (FMD) has been recognized as one of the most important contagious viral diseases of cloven-hoofed livestock. Although mortality associated with FMD is usually low, this severe disease poses a significant economic impact worldwide, due to both direct losses because of reduced milk and meat production, and indirect losses caused by costs of disease control and poor access to markets (Knight-Jones and Rushton 2013).

The disease was present in almost every livestock-containing region worldwide until the early 20th century, when it was progressively eradicated from the developed world. Since then, various major outbreaks have occurred in south-east Asia (Japan, South Korea and Taiwan), South America (Paraguay, Argentina, Uruguay and Brazil) and Western Europe (UK, The Netherlands, France and Ireland) (Knowles et al. 2001; Thomson et al. 2003; Brito et al. 2015). Nowadays, North America, most European countries, Australia, New Zealand and many island states, are recognized as FMD-free; the disease persists in South America, most African countries, the Middle East, and many parts of Asia, where the disease is endemic (Thomson et al. 2003). Updated information on the incidence and distribution of FMD as well as epidemiological reports, can be obtained from the Office International des Epizooties (OIE) website at http://www.oie.int/en/animal-health-in-the-world/official-disease-status, or the website of the World Reference Laboratory for FMD at http://www.wrlfmd.org.

Currently, the OIE recognizes countries and zones to be in one of three disease states with regard to FMD: FMD present with or without vaccination, FMD-free with vaccination and FMD-free without vaccination (Fig. 1). It has been recently estimated that outbreaks in FMD free countries and zones cause losses of more than US$1.5 billion a year (Knight-Jones and Rushton 2013). Therefore, many of the FMD-free countries, including Canada, the United States, and the UK, work hard to maintain their current status, focusing their control policy on minimizing the risk of virus incursion and the impact of an outbreak should one occur. In South America, after the regional eradication plan implemented in 1988, at present most of the countries/regions have their status recognized by the OIE as FMD-free either with or without vaccination.

The disease affects a wide variety of species, but cloven-hoofed animals (order: *Artiodactyla*) have a critical epidemiological role in maintaining the virus in the environment. Cattle, pigs, sheep, goats and water buffalo (*Bubalus bubalis*) are susceptible to viral infection and can spread the disease (Alexandersen and Mowat 2005). African buffalo (*Syncerus caffer*) play an important role as the natural maintenance host in Africa (Thomson et al. 2003). They are mainly thought to maintain the South African Territories (SAT) types of the virus, although antibodies to other serotypes have been found in buffalo populations of west and central Africa (Di Nardo et al. 2015).

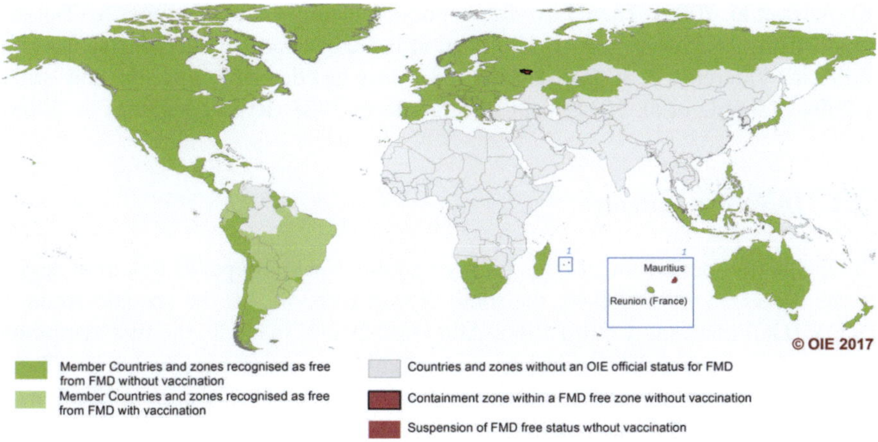

Fig. 1 FMD status recognized by the OIE. With the kind authorization of the World Organization for Animal Health (OIE), www.oie.int. Map extracted from OIE website on December 2017, available at http://www.oie.int/en/animal-health-in-the-world/official-disease-status/fmd/en-fmd-carte

Other animals may contribute to viral transmission under certain conditions including Indian elephants and any animal of the order *Artiodactyla*, like deer, camels, llamas, and alpaca. These species do not appear to play an important role in the wild, but they have to be considered as a potential risk when they are kept under farmed or crowded conditions (Thomson et al. 2003; Alexandersen and Mowat 2005). Although laboratory mice, guinea pigs and rabbits are susceptible to infection with foot-and-mouth disease virus (FMDV) under experimental conditions, there is no evidence of such animals being involved in the spread of FMD in the field (Alexandersen and Mowat 2005).

Foot-and-mouth disease virus is the aetiological agent of this infectious disease and it was the first virus of vertebrates to be discovered (Loeffler and Frosch 1897). Seven major viral serotypes have been described, termed O, A, C, Asia 1 and South African Territories (SAT) 1, 2 and 3 (Grubman and Baxt 2004), and each contains several, constantly-evolving subtype strains (Bachrach 1968). These serological types were assigned on the basis of lack of cross-protection after infection or vaccination. Viruses showing partial cross-protection were assigned to the same serotype but to a different subtype (Domingo et al. 2002).

FMDV serotypes are not distributed uniformly across the world. Six of these serotypes (O, A, C, SAT-1, SAT-2, SAT-3) have occurred in Africa, four in Asia (O, A, C, Asia 1), and only three in South America (O, A, C) (Rweyemamu et al. 2008). A particular strain of serotype O, named PanAsia, was responsible for the explosive pandemic across Asia, Africa and Europe from 1998 to 2001. The PanAsia strain caused outbreaks in the Republic of Korea, Japan, Russia, Mongolia, South Africa, the United Kingdom, Republic of Ireland, France, and the Netherlands, countries which had last experienced FMD outbreaks decades before

(Knowles et al. 2005). There have been sporadic incursions of serotypes SAT-1 and SAT-2 from Africa into the Middle East, at the crossroad between Africa, Europe and Asia (Valarcher et al. 2008). Serotype C was last detected in Kenya and Brazil in 2004 (Sangula et al. 2011), and in Ethiopia in 2005 (Rweyemamu et al. 2008).

1.1.1 Disease symptoms

The incubation period for FMD is highly variable since it depends to a high degree on the dose of virus received, the route of transmission and the specific strain of FMDV (Gailiunas and Cottral 1966; Arzt et al. 2011). In cattle, the first symptoms are high fever, anorexia, and lesions that initially present as blanched areas. These lesions progress into vesicles, which increase in size and rupture, generating areas of epithelial damage. They are typically located on the tongue, hard palate, dental pad, lips, gums, muzzle, coronary band and interdigital spaces (Fig. 2), and may also be present on teats, particularly of lactating cows (Kitching 2002). According to the location of these vesicles, animals can present abundant secretion of foamy saliva, drooling and lameness (Kitching 2002). Consequently, feeding, milking and suckling become difficult, resulting in a rapid weight loss and a marked decrease in milk production (Shahan 1962).

The disease in sheep and goats is in general clinically milder than in cattle, with lameness being the first indication of FMD, since pain may be detected in the feet for 1–2 days before vesicular lesions appear. Vesicles may develop in the interdigital cleft, on the heel bulbs, on the coronary band and in the mouth (Kitching and Hughes 2002; Hughes et al. 2002). It has been reported that up to 25% of infected sheep may fail to develop lesions, and an additional 20% may form only one lesion (Hughes et al. 2002).

In pigs, clinical disease is usually severe, and the early signs include acute lameness, reluctance to stand, depression, loss of appetite and fever (Alexandersen et al. 2003). Vesicles around the coronary bands are the most consistent finding in pigs, while lesions at other sites like the bulb of the heel, the interdigital space, the snout, lower jaw and tongue, may be found less regularly depending on environmental and other factors (Fig. 3) (Kitching and Alexandersen 2002; Stenfeldt et al. 2014b; Stenfeldt et al. 2016a).

Although most animals eventually recover from FMD, the disease can lead to myocarditis and death, especially in young calves, piglets and lambs (Kitching and Hughes 2002; Kitching 2002; Alexandersen et al. 2003; Grubman and Baxt 2004).

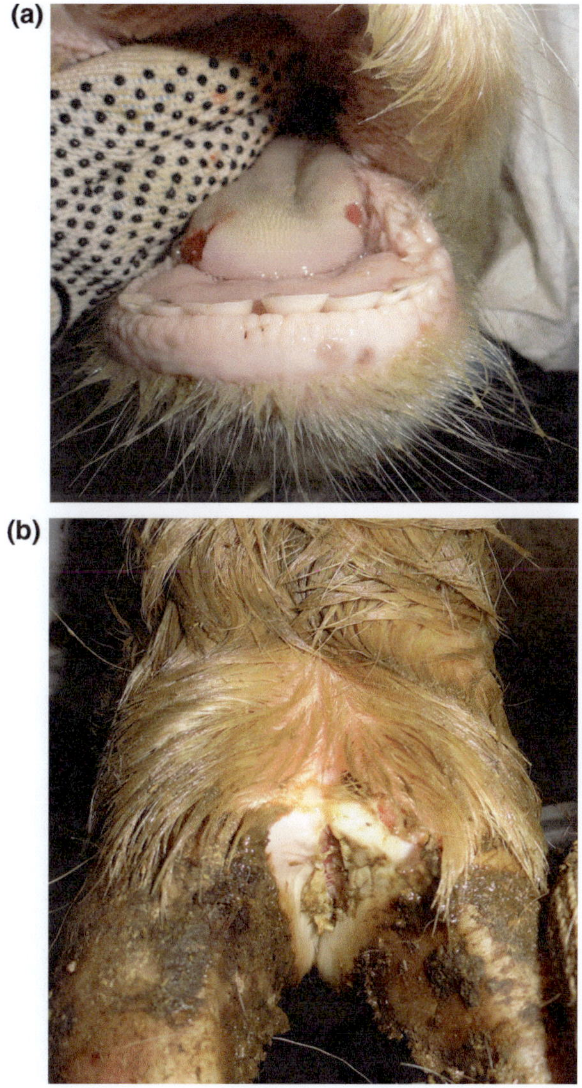

Fig. 2 FMD in cattle. **a** ruptured vesicle on the tongue, **b** lesion in the interdigital space of the feet. The pictures are courtesy of Dr. Pérez Filgueira, Instituto de Virología, CICVyA, Instituto Nacional de Tecnología Agropecuaria (INTA), Buenos Aires, Argentina

2 Mechanism of infection

2.1 The virus

FMDV is a single-stranded positive-sense RNA virus belonging to the genus Aphthovirus in the family *Picornaviridae* (Bachrach 1968). The viral genome of about 8,500 nucleotides contains a single long open reading frame (ORF) flanked by heavily structured 5′- and 3′-untraslated regions (UTR) (Fig. 4). The 5′UTR

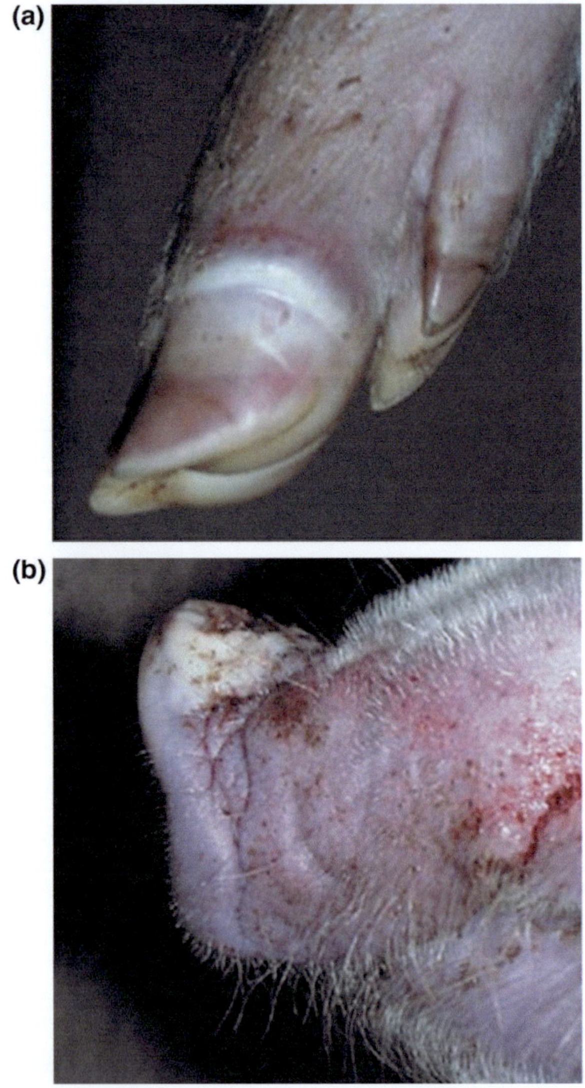

Fig. 3 FMD in pigs. **a** unruptured vesicles in the coronary bands, **b** lesion on the snout (reproduced from Alexandersen and Mowat 2005)

consists of a short (S) fragment, a poly (C) tract, repeated pseudoknots (PKs), a cis-acting replication element (cre), and an internal ribosome entry site (IRES), which is required for viral replication and translation (Belsham 2005). Viral mRNA translation begins at two in-frame AUG codons, rendering a polyprotein precursor which is co-and post-translationally proteolytically processed to generate both the intermediate and mature structural and non-structural proteins. The first mature product of viral translation is L^{pro}, a papain-like protease of fundamental role in virulence (Kleina and Grubman 1992). The capsid precursor P1 is cleaved by the 3C protease to produce the structural proteins VP0, VP3, and VP1. The most

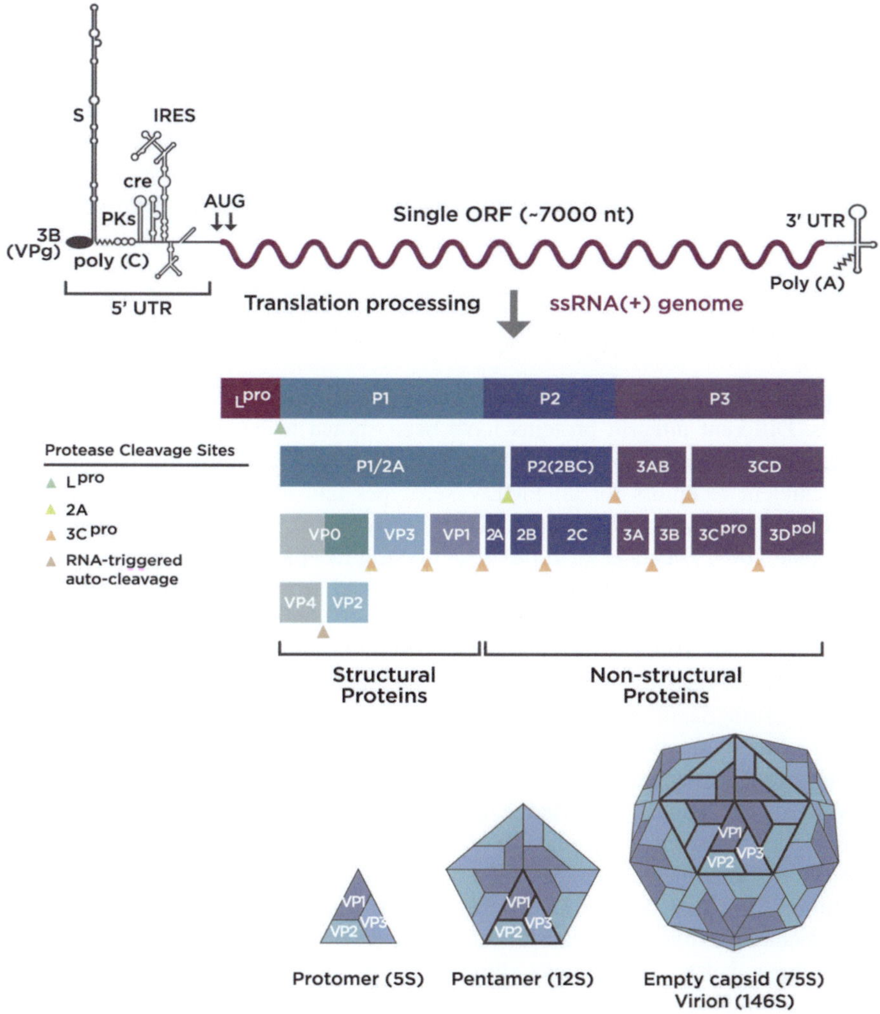

Fig. 4 Schematic diagram of FMDV genome, processing of viral polyprotein and capsid assembly. The 5'UTR, including the S-fragment, poly(C), pseudoknot structures (PKs), cis-acting replication element (cre), and IRES, is adapted from Mason et al. (2003). The intermediate and mature structural and non-structural proteins are named according to the nomenclature of Rueckert and Wimmer (Rueckert and Wimmer 1984). The sites of the primary cleavages and the proteases responsible are indicated

immunogenic epitope in all FMDV serotypes is located within a highly variable G-H loop (residues 141-160) in the VP1 protein (Borrego et al. 1995; Mateu et al. 1995). The P2 and P3 regions encode non-structural proteins (NSP), which play critical roles in RNA replication, translation, viral assembly and maturation, and virulence (Grubman and Baxt 2004; Belsham 2005). At the end of the ORF there is a relatively short 3′UTR composed of two stem–loops and a poly (A) tail, both required for viral infectivity and known to stimulate IRES activity (Rohll et al. 1995; Serrano et al. 2006). One copy of the structural proteins VP0, VP3 and VP1 spontaneously assemble and form the 5S protomer, which subsequently assembles into the 12S pentameric subunit. Twelve of these pentameric subunits associate to form the icosahedral empty capsid-like particle (75S) (Belsham 1993). Encapsidation of viral RNA to produce mature virion (146S) is accompanied by the cleavage of VP0 into VP2 and VP4. This non-enveloped icosahedral particle is about 25-30 nm in diameter (Racaniello 2001).

2.1.1 Pathogenesis

Receptors are the major determinant factors for the tropism and pathogenesis of viruses. Different integrin proteins have been shown to be receptors for FMDV: $\alpha v\beta 1$, $\alpha v\beta 3$, $\alpha v\beta 5$, $\alpha v\beta 6$, and $\alpha v\beta 8$; yet, $\alpha v\beta 6$ has been identified as one of the most efficient receptors for all FMDV serotypes and its expression is associated with FMDV tropism (Jackson et al. 2000; Jackson et al. 2002; Jackson et al. 2004; Monaghan 2005; Dicara et al. 2008). The integrin receptor recognition site involves a highly conserve Arg-Gly-Asp (RGD) amino acid sequence motif present on the G-H loop of the VP1 capsid protein (Fox et al. 1989; Baxt and Becker 1990; Verdaguer et al. 1995; McKenna et al. 1995). In addition to cellular integrins, FMDV is also able to utilize alternative membrane molecules as receptors (reviewed in Ruiz-Sáenz et al. 2009 and Wang et al. 2015). It has been demonstrated that in vitro cultivation of FMDV selects viruses that bind to heparan sulfate, resulting in the attenuation of virulence in host species (Sa-Carvalho et al. 1997).

Most of the current knowledge about the pathogenesis of FMDV has been achieved by experimental studies in cattle and pigs. Although there are numerous reports on the subject, there is still no consensus regarding the early stages of infection. This may be due to the fact that the efficiency of transmission under experimental conditions varies between different serotypes and subtypes of FMDV, which may have considerable differences in tissue tropism and virulence. Furthermore, other factors, such as the inoculation system, the duration of exposure, housing density and differences in sensitivity and/or specificity of virus detection methods, may directly influence the results of experimental transmission studies (Quan et al. 2004; Quan et al. 2009; Pacheco et al. 2012; Stenfeldt et al. 2014b; Pacheco et al. 2016).

As has been recently reported by Stenfeldt and co-workers (2016a), although cattle and pigs may be similarly susceptible to FMDV infection under most circumstances, there are critical differences between these two animals in FMD pathogenesis and infection dynamics. These include variations in susceptibility of infection by different routes of virus exposure, which lead to differences in the mechanisms of virus transmission; variations in the amount of virus shed by aerogenous routes, and variations in the capability of clearing infectious virus from all tissues. Additionally, species-specific variation has been also demonstrated, in both systemic and cellular responses to infection (Toka and Golde 2013). Therefore, the pathogenesis of FMDV will be analyzed in each of these species separately.

FMD in cattle

Susceptible cattle in contact with an infected animal are most likely to be infected via the respiratory route by aerosolized virus (Sellers and Parker 1969). Infection can also occur through abrasions on the skin or mucous membranes, although it is less efficient requiring almost 10,000 times more virus (Donaldson 1987). Indirect contact with FMDV may also occur via contaminated persons, vehicles, fomites, and wild animals. Since the virus is excreted into the milk of dairy cattle, calves can become infected by insufflation of milk droplets as they drink (Donaldson 1997). Under appropriate environmental conditions, spread by airborne carriage on the wind is likely to occur, with some isolates potentially spreading up to about 300 km by the wind (Gloster et al. 1982; Sørensen et al. 2000).

Numerous studies have investigated the events occurring during and immediately following the host's initial exposure to the virus. In cattle, FMD has been experimentally reproduced by simulated natural methods (direct or indirect contact with infected animals or virus aerosols from such animals), or by artificial methods like parenteral inoculation of virus (intradermal, intravenous, subcutaneous, or intramuscular), intranasal instillation, pulmonary inoculation or exposure to artificially created aerosols. Of these, direct introduction of virus to the respiratory tract has been widely used since it has the advantage of simulating the natural route of transmission, allowing the study of early events in pathogenesis (Alexandersen et al. 2003; Sellers and Gloster 2008).

By using a controlled aerosol inoculation system, recent reports demonstrated that after aerolization of cattle with FMDV O1-Manisa and A24-Cruzeiro, there is a primary replication in epithelial crypts of the nasopharynx, followed by extensive replication in pneumocytes in the lungs and the establishment of viremia and generalized disease (Arzt et al. 2010; Pacheco et al. 2010; Arzt et al. 2014). These reports have changed the original notion that after initial replication in the pharynx, FMDV is spread through the regional lymph nodes (Henderson 1948; Burrows et al. 1981).

Once viremia is established, the virus spreads to different tissues and organs, especially the skin and the epithelia of the tongue and mouth, where the main viral amplification occurs. This generates the characteristic vesicles that are also present in feet, teats and prepuce (Alexandersen and Mowat 2005; Arzt et al. 2009). During the viraemic and clinical phase of the disease, all excretions and secretions (saliva, nasal and lachrymal fluid, milk and expired breath) become infectious and can contain significant titres of virus. Urine and faeces also contain virus but to a lesser extent (Alexandersen et al. 2003).

Following the acute phase of FMDV infection, some ruminants may experience a long asymptomatic persistent infection. This persistent infection can occur either after a clinical or a subclinical FMD infection, and occurs in vaccinated and in non-vaccinated animals (Doel et al. 1994). Carrier animals are defined as those from which infectious FMDV can be recovered from oropharyngeal fluid for more than 28 days after infection (Van Bekkum et al. 1959). The prevalence rate of carriers depends on the species, the incidence of disease (or infection) and the immune status of the population (Alexandersen et al. 2003). In cattle, approximately 50% of infected animals become carriers, irrespective of their vaccination status (Moonen and Schrijver 2000). Studies on persistently infected animals have reported that FMDV may persist in the dorsal surface of the soft palate and the pharynx (Burrows 1966), more specifically in epithelial cells of the pharyngeal area (Zhang and Kitching 2001), and also in follicular dendritic cells within the germinal centres of lymphoid follicles (Juleff et al. 2008). The role of carrier animals in the epidemiology of the virus is still controversial, though there is evidence of transmission from buffalo to cattle both experimentally (Dawe et al. 1994b) and under natural conditions (Dawe et al. 1994a) in Africa.

FMD in pigs

Of the domestic species, pigs liberate the largest quantities of aerosolized FMDV, reaching values up to $10^{5.6}$ to $10^{8.6}$ $TCID_{50}$ per pig per day, whereas ruminants excrete about 10^4–10^5 $TCID_{50}$ per day. However, pigs are less susceptible to airborne infection when compared to ruminants, since it has been demonstrated that the latter can be infected by airborne exposure with only 10 $TCID_{50}$, while more than 10^3 $TCID_{50}$ are needed to infect pigs by this same route (Sellers and Parker 1969; Donaldson et al. 1970; Donaldson and Alexandersen 2001; Alexandersen et al. 2002a; Alexandersen and Donaldson 2002).

Recent investigations have confirmed that pigs are more susceptible to FMDV infection via exposure of the upper gastrointestinal tract (oropharynx) than through inhalation of virus (Stenfeldt et al. 2014a; Fukai et al. 2015). By using a simulated natural experimental system that has been previously demonstrated to efficiently infect pigs, Stenfeldt and co-workers have identified as the primary site of FMDV replication, segments of reticular epithelium within the paraepiglottic tonsils (Stenfeldt et al. 2014b).

After 24 h of natural or experimental exposure, pigs become viraemic and this is associated with increased shedding of virus via the oropharyngeal route (Murphy et al. 2010; Stenfeldt et al. 2014b; Stenfeldt et al. 2014c; Stenfeldt et al. 2016b). As has been recently demonstrated, the tonsil of the soft palate is the only tissue in the respiratory or gastrointestinal tract that supports substantial levels of FMDV replication (Alexandersen et al. 2001; Stenfeldt et al. 2014b), suggesting that this would be the best candidate as the source of aerosolized virus derived from pigs. Since these clinically infected pigs constitute an important source of contagion for exposed ruminants, movement of live pigs or associated products can have substantial impact on disease spread. Therefore, FMD control policies should be based on species-specific data and should be adapted to account for the animal population in each region.

Recent studies performed in domestic pigs indicate that there is a high prevalence of FMDV RNA and capsid antigen in lymphoid tissues for 30–60 days after infection (Stenfeldt et al. 2014c), indicating transient persistence of FMDV in these animals. However, the virus could not be isolated from a large number of porcine tissues processed, confirming that domestic pigs are not competent long-term carriers of FMDV (Arzt et al. 2011; Stenfeldt et al. 2014c).

3 Plant-made vaccines candidates

Vaccines play a vital role in FMD control, since they are used both to limit the spread of the virus during an outbreak in FMD-free countries and to control disease in endemic regions. Current FMD vaccine consists of binary ethyleneimine-inactivated whole virus preparation, formulated with an oil-based adjuvant (Doel 2003). Systematic vaccination programs with this inactivated vaccine have successfully reduced the number of disease outbreaks in enzootic areas, including Western Europe and parts of South America where FMDV has been eliminated with the use of this vaccine (Brown 2003). However, since there are a number of concerns about its use (see "Inactivated FMD vaccine"), there are several strategies that have been explored with the aim of developing recombinant FMDV vaccines that are safer, more effective, more stable and more economical. One of these strategies is the use of plants as bioreactors.

Most of the attempts to develop plant-based anti-FMDV vaccines have concentrated on the expression of capsid protein VP1 either in transgenic plants or transiently, since this protein contains the immunodominant epitope from the virus (summarized in Table 1) (Ruiz et al. 2015). First reports were based on the use of chimeric plant viruses engineered to display short viral sequences of FMDV on its surface as a safe way to produce vaccines. Cowpea mosaic virus (CPMV) was the first plant virus used to express a VP1 epitope of FMDV on the surface of the virus and used to infect cowpea plants, raising the possibility of producing vaccines in plants (Usha et al. 1993). However, subsequent studies revealed that the chimera did not spread systematically in cowpea, could not be purified in significant quantities, and the inserted sequence was rapidly lost during serial passages in plants, probably by a process of homologous recombination (Porta et al. 1994). Further investigation showed that both length and isoelectric point (pI) of the inserted sequence have profound effects on the growth of chimeras, indicating that chimeras with inserts that are not significantly greater than 30 amino acids and have a pI below 8.0 should give virus yields similar to those obtained with wild-type virus (up to 1 mg per gram of infected leave tissue) (Porta et al. 2003; Montague et al. 2011). When the pI of the FMDV insert was adjusted to be close to 7 by the addition of acidic residues at either or both sides of the insert, the infection could be passaged to further plants, and sequence analysis of RNA isolated from purified virus particles indicated that no reversion or mutations had taken place following serial virus passaging (Montague et al. 2011). A few years later, Wigdorovitz and

Table 1 Plant-based vaccines for FMD

Target genes	Plant host	Expression system	Yield		Animal model	References
			% TSP	mg/g FLT		
VP1	*Vigna unguiculata* (Cowpea)	Transient expression (*Cowpea mosaic virus*)	n/r	n/r	n/a	Usha et al. 1993
VP1	*N. benthamiana*	Transient expression (*Tobacco mosaic virus*)		0.05-0.15*	Mice	Wigdorovitz et al. 1999a, b
VP1	*N. tabacum*	Transient expression (*Tobacco mosaic virus*)		5*	Guinea pig Swine	Wu et al. 2003a, b
VP1	*N. tabacum*	Transient expression (*Tobacco mosaic virus*)		10-20*	Guinea pig Swine	Jiang et al. 2006
VP1	*C. quinoa, N. benthamiana*	Transient expression (*Bamboo mosaic virus*)		0.2-0.5*	Swine	Yang et al. 2007
VP1	*C. amaranticolor, N. benthamiana*	Transient expression (*Tobacco necrosis virus-A*)		0.03-0.3*	Mice	Zhang et al. 2010
VP1	*C. quinoa, N. benthamiana*	Transient expression (*Apple latent spherical virus*)	n/r	n/r	n/a	Li et al. 2014a, b
VP1	*N. benthamiana*	Transgenic plant cell-suspension culture (*Bamboo mosaic virus*)	4.7-5.0	0.06-0.1*[a]	Guinea pigs	Muthamilselvan et al. 2016
VP1	*A. thaliana*	Transgenic plants	n/r	n/r	Mice	Carrillo et al. 1998
VP1	Alfalfa	Transgenic plants	n/r	n/r	Mice	Wigdorovitz et al. 1999a
VP1	Potato	Transgenic plants	0.005-0.01		Mice	Carrillo et al. 2001
VP1	Alfalfa	Transgenic plants	0.05-0.1		Mice	Dus Santos et al. 2002
VP1	*S. guianensis*	Transgenic plants	0.1-0.5		Mice	Wang et al. 2008
VP1	*C. juncea*	Transgenic plants	0.0001-0.0012		Guinea pigs	Rao et al. 2012
VP1	*N. tabacum*	Transplantomic plants	2-3		n/a	Li et al. 2006a, b

(continued)

Table 1 (continued)

Target genes	Plant host	Expression system	Yield		Animal model	References
			% TSP	mg/g FLT		
VP1	*N. tabacum*	Transplantomic plants	51		Mice	Lentz et al. 2010
P1	*Nipponbare* (Japonica rice)	Transgenic plants	0.06-0.13		Mice	Wang et al. 2012
P12A-2B-3C	Alfalfa	Transgenic plants	0.005-0.01		Mice	Dus Santos et al. 2005
P1-2A3C	Tomato	Transgenic plants	n/r	n/r	Guinea pigs	Pan et al. 2008
VP1, VP4, 2C, 3D	*N. benthamiana*	Transient expression (*Potato Virus X*-based vector)	0.7-1		Guinea pigs	Andrianova et al. 2011

[*]Yield of purified chimeric virus is provided, [a]Yield is expressed as μg/g fresh weight of suspended biomass, *FLT* Fresh leaf tissue, *TSP* Total soluble protein, *n/r* not reported, *n/a* not assayed.

co-workers reported the expression of the complete open reading frame of VP1 using a recombinant *Tobacco mosaic virus* (TMV). Crude extracts of *Nicotiana benthamiana* plants transiently infected with the recombinant virus were injected intraperitoneally into mice. Animals received 5 doses of leaf extract containing approximately 0.5–1 μg of recombinant VP1per dose. All immunized animals developed a significant and specific immune response and were protected from the experimental challenge with virulent FMDV. Additionally, the recombinant VP1 was synthesized in comparable quantities in both inoculated and systemic leaves and its concentration was maintained at detectable levels for an extensive period during the replication of the TMV vector (Wigdorovitz et al. 1999b). It is worth noting that both guinea pigs and mice have been used extensively to study FMDV pathogenesis and vaccine efficacy trials, and that the mouse model has successfully predicted immune responses to FMDV in cattle and pigs (Habiela et al. 2014).

Using the same strategy, *Tobacco mosaic virus* (TMV)-based expression of short VP1 antigenic peptides of 11 or 14 amino acids, fused to the open reading frame of TMV coat protein (CP), has also been described (Wu et al. 2003a). Both recombinant viruses (TMVF11 and TMVF14) systemically infected tobacco plants, producing stable progeny particles as efficiently as wtTMV (about 5 g of purified particles per kilo of infected leaves). Moreover, TMVF11 and TMVF14 were quite stable and no loss of the peptide insert or degradation of the viruses was detected even after three sequential passages in tobacco plants by mechanical transmission. When the mixture of recombinant viruses was parenterally injected in guinea pigs (0.6 mg total in one dose), all the animals were protected against challenge. Preliminary tests in swine, showed that 9 pigs immunized once by parenteral injection of 3 mg of TMVF11 were protected against FMDV challenge (Wu et al. 2003a). The same group also developed an improved TMV-based vector in which up to six C-terminal amino acid residues could be deleted from the coat protein subunit, allowing the expression of a longer peptide of 25 amino acids, containing two fused epitopes (F14 and F11). Although yields of the recombinant viruses were higher, immunity in guinea pigs and swine appears to be less efficient as previously reported, possibly due to lack of a spacer arm between the two fused epitopes (Jiang et al. 2006).

Chimeric Bamboo mosaic virus (BaMV) expressing T and B-cell epitopes of VP1 was produced in *C. quinoa* and *N. benthamiana* plants, and proved to induce not only strong humoral (as indicated by neutralizing antibodies) and cell-mediated immune responses (as indicated by VP1-specific IFN-γ production), but also full protection against FMDV in swine even after just one inoculation of 1 mg of the chimeric virus (Yang et al. 2007). Recently, the same group reported the innovative development of transgenic cell-suspension cultures from *N. benthamiana* leaves carrying wild-type or chimeric BaMV expression constructs encoding VP1 epitopes (Muthamilselvan et al. 2016), providing a cost-effective and efficacious means of producing vaccine candidates. These purified chimeric virus particles triggered the production of specific antibodies in guinea pigs.

Tobacco necrosis virus-A (TNV-A) was engineered as a vector to express different peptides from FMDV VP1. Most of the obtained chimaeras contained unmodified foreign peptides even after six successive passages in *Chenopodium amaranticolor* and three passages in *N. benthamiana*, suggesting long-lasting stability. The purified chimeric virus particles could induce a strong immune response against VP1 in mice immunized intramuscularly with three doses of 0.2 mg. Mice immunized intranasally with 5 doses of 0.1 mg of chimeric virus particles developed both systemic and mucosal immune responses against FMDV VP1(Zhang et al. 2010). More recently, Li et al. showed that a nine amino acid sequence from VP1 could be expressed on the surface of apple latent spherical virus (ALSV) and that this chimeric virus can infect *N. benthamiana* and *Chenopodium quinoa* plants. However, the immunogenicity of this virus particles was not evaluated (Li et al. 2014a).

The expression of VP1 in transgenic plants has also been used in the development of experimental vaccines, some of them oriented to the development of edible vaccines. First attempts have successfully demonstrated that VP1 of FMDV can be expressed in *Arabidopsis thaliana*, alfalfa and potato plants (Carrillo et al. 1998; Wigdorovitz et al. 1999a; Carrillo et al. 2001). In these studies a binary vector containing the Cauliflower Mosaic Virus (CaMV35S) promoter was used to direct constitutive expression of the foreign gene. Mice parenterally immunized with 4 or 3 doses of 15-20 mg of fresh leaf tissue presented a FMDV VP1 specific antibody response that provided 77-100% protection against challenge (Carrillo et al. 1998; Wigdorovitz et al. 1999a; Carrillo et al. 2001). Moreover, 0.3 g of fresh leaves from alfalfa plants were used to feed mice three times a week for 2 months. All orally immunized mice developed a specific antibody response, and 66 to 75% of the animals showed protection after the experimental challenge (Wigdorovitz et al. 1999a). However, since the concentration of the expressed protein was low in all those cases (approximately, 0.005–0.01% of the total soluble protein) (Carrillo et al. 2001), the same group developed an experimental immunogen based on the expression of VP1 fused to the glucuronidase (gus A) reporter gene, which allowed the selection of transgenic plants expressing high levels of the recombinant protein by the β-glucuronidase (βGUS) enzymatic activity (Dus Santos et al. 2002). By using this alternative, authors were able to select transgenic plants of alfalfa with expression levels 10 times higher than those observed in transgenic plants previously developed (Carrillo et al. 2001). Mice immunized intraperitoneally with the transgenic selected plants (100 mg of fresh leaf tissue) developed a strong and protective antibody response against virulent FMDV.

The use of transgenic plants expressing the antigen protein of FMDV as feedstuff additives was first reported by Wang and co-workers (Wang et al. 2008). They produced transgenic *Stylosanthes guianensis* cv. Reyan II plants expressing FMDV VP1 protein, to levels of 0.1–0.5% total soluble protein (TSP). Mice that were orally immunized using the transgenic hay meal developed a specific immune response. A few years later, Rao et al. developed transgenic forage crop *Crotalaria juncea* expressing FMDV VP1 proteins of serotypes O and A linked in tandem, and tested them as bivalent FMD vaccine in guinea pigs (Rao et al. 2012). Animals were immunized twice with leaf-extracted proteins corresponding to 1 g of the leaf material or oral fed with 1 g of leaves of the transgenic plants. Guinea pigs reacted to orally or parenterally applied vaccine, by humoral as well as cell-mediated immune responses. Two of three animals (66%) were protected against challenge with the virus of both serotypes. Guinea pigs immunized with the conventional inactivated vaccine were fully protected against challenge, and the authors suggest that this could be due to the presence of conformational epitopes of the capsid proteins which can induce better immune response, than only 2 sequential epitopes of VP1 protein.

In another approach, transgenic rice expressing the capsid precursor polypeptide (P1) of FMDV was generated by *Agrobacterium*-mediated transformation. Expression levels of the recombinant P1 protein ranged from 0.6 to 1.3 mg/g of TSP in transgenic rice leaves, which was sufficient to induce a protective immune response in mice after four intraperitoneal immunizations with transformed plant extracts (containing approximately 10 μg of P1 protein). In addition, when mice were fed with 100 μl of transformed plant extract (containing approximately 10 μg P1 protein) five times per week for one month, FMDV-specific systemic and mucosal immune responses were detected, as well as partial virus clearance after challenge (Wang et al. 2012).

In order to achieve high accumulation of recombinant proteins in plants, transplastomic technology has also been attempted. In this regard, VP1 recombinant protein accumulated in tobacco chloroplasts to 3% of TSP (Li et al. 2006b). Similarly, when a highly immunogenic epitope of VP1 was fused to the ßGUS protein, high accumulation of the recombinant protein was produced in tobacco transplastomic plants, representing 51% of the soluble proteins in mature leaves (Lentz et al. 2010). This protein was also found to be immunogenic in mice.

Generally, vaccines based on individual viral proteins rarely present epitopes in their native conformation, making them less effective than whole virus preparations. Therefore, this kind of subunit vaccines often require more doses of antigen as well as adjuvanted delivery systems for immune response stimulation. In order to improve immunogenicity of subunit vaccines, alternative strategies have been explored, including the production of polyepitope and empty capsids. In this regard, an anti-FMD vaccine based on a recombinant protein formed by a set of well-studied B- and T-cell viral epitopes was developed (Andrianova et al. 2011). For this purpose, codon-optimized genes encoding B-cell epitopes of the structural proteins VP4 and VP1, and T-cell epitopes of nonstructural proteins 2C and 3D were generated and cloned together divided by 'flexible' glycine-rich linkers G4S2 to avoid potential problems of protein folding. This recombinant polyepitope protein (H-PE) was expressed in *N. benthamiana* plants using a phytovirus expression system. A single intramuscular injection of guinea pigs with an oil emulsion containing 120 μg of the purified protein induced the formation of virus-neutralizing antibodies to the homologous FMD virus and no evidence of infection was seen after challenge.

The production of recombinant FMDV empty capsids or virus-like particles (VLPs) requires the simultaneous expression of the capsid precursor P12A and the protease 3C. This viral protease processes the P12A precursor to generate structural proteins (VP0, VP3 and VP1), which then self-assemble to form the viral capsid.

The first attempt towards the production of plant-based FMDV VLPs has been the expression of the precursor P12A and the viral protease 3C in alfalfa plants (Dus Santos et al. 2005). Although the formation of FMD VLPs could not be reliably demonstrated, the administration of 4 doses of 15–20 mg of fresh leaf tissue from transgenic plants was able to evoke a strong antibody response in mice. In addition, all animals were protected from FMDV challenge. Likewise, transgenic tomato plants expressing P12A and 3C were produced (Pan et al. 2008). Although the expression and processing of the capsid precursor was demonstrated, the authors could not conclusively determine whether the capsid proteins assembled into VLPs since an electron microscopy analysis was not performed. However, all guinea pigs intramuscularly immunized with 3 doses of 50 mg fresh leaf tissue from transgenic tomato plants developed a virus-specific antibody response and were protected against challenge infection. More recently, it has been reported the expression of the P1 precursor modified to contain a CPMV 24K protease recognition site instead of the native 3C protease, as this is less toxic than 3C when expressed in plants (Saunders et al. 2009). When the modified P1 precursor was co-expressed with a CPMV 24K-containig construct in *N. benthamiana* leaves, the individual capsid proteins were identified, suggesting the correct cleavage by the CPMV 24k protease. However, no VLPs were observed by electron microscopy, indicating that either assembly or stability of FMDV capsids may not be optimal in plant cells (Thuenemann et al. 2013). Although these reports could not conclusively demonstrate the formation of VLPs, it is possible that other subviral structures were present. As mentioned in sect. 2.1, capsid proteins spontaneously assemble to form the protomer (5S), five of which subsequently form the pentamer (12S), and finally twelve pentamers assemble into the empty capsid (75S). A critical threshold concentration of pentamers is required before capsid formation can occur (Zlotnick and Stray 2003; Goodwin et al. 2009). Therefore, it is likely that recombinant pentamer protein yield was below this threshold, or that the VLPs were actually assembled, but due to instability problems would subsequent dissociate into more stable 12S pentameric capsid subunits.

4 Established vaccine and alternatives

4.1 Inactivated FMD vaccine

Foot-and-mouth disease vaccines represent the largest share of sales in the veterinary vaccine market worldwide, meaning 26.4% of the entire livestock biological business (Gay et al. 2003). Routine vaccination against FMD is often applied in countries or zones recognized as "FMD-free with vaccination" to maintain FMD-free status, and in endemic countries to control the virus.

Currently, all FMD vaccines are produced by infecting baby hamster kidney-21 (BHK-21) cells with virulent FMDV in roller bottles or in suspension under biosecure conditions in large volumes. The virus is harvested, chemically inactivated with binary ethyleneimine (BEI), concentrated by polyethylene glycol (PEG) precipitation, and purified by ultrafiltration. Some vaccine manufacturers use industrial scale chromatography to purify the antigens previously concentrated by ultrafiltration (Doel 2003). For vaccine formulation, the purified antigen is diluted with buffers and mixed with either oil or aluminum hydroxide/saponin adjuvants. The oil-emulsion vaccines are preferred for FMD prevention as they can be used to protect all susceptible species and are ideal for emergency vaccination, whereas aluminum hydroxide/saponin-adjuvanted vaccines are not recommended in pigs due to low efficacy in this species (Cao 2014).

Considering that vaccination with one serotype does not protect against other serotypes, the antigenic composition of FMD vaccines must represent the epidemiological situation of the particular country or region. This is especially important for endemic areas where several FMD viral subtypes may be circulating and where vaccination with trivalent vaccines is needed (Parida 2009). This highlights the importance of surveillance plans in order to know the local strains of FMD virus circulating in a region.

As mentioned before, the use of this inactivated vaccine has proven to be a critical component of control and eradication programs worldwide. However, there are still some concerns and shortcomings, most of them related with the vaccine production process. Since this process requires the growth of large volumes of virulent virus, expensive facilities for high biological containment are required for production. Additionally, there is a constant risk of live virus release from the manufacturing sites or inadequate inactivation of the virus. A further problem is that vaccines may contain traces of contaminating viral NSP; these vaccines will therefore induce the production of antibodies against NSP in the same way as natural infection, interfering with the NSP-based serological differentiation between infected and vaccinated animals (DIVA). Furthermore, while inactivated FMD vaccines are effective in preventing clinical disease, they do not necessarily prevent viral replication in the epithelial surface of vaccinated animals, which can result in persistent infection (Alexandersen et al. 2002b), a situation which can be very costly in lost trade if vaccination is included in the control policy of a country or a region normally free of FMD. Therefore, although there is no evidence that these vaccinated carrier animals can transmit virus, their occurrence is one of the main barriers to implement vaccination in control and eradication of disease outbreaks in FMD-free countries(Rodriguez and Gay 2011).

Additionally, the FMD vaccine does not induce long-term protective immunity, requiring multiple vaccinations to control the disease (Rodriguez and Gay 2011). Other important shortcomings of the inactivated vaccine include the short shelf life, the need of cold chain from production to delivery, and the difficulty of some serotypes or subtypes to adapt to cell culture, hindering their production.

4.1.1 Alternative vaccines

Intensive research has been carried out to achieve the production of alternative and improved FMD vaccines (Robinson et al. 2016). Different approaches have been successfully developed, based on attenuated and/or marker (DIVA) inactivated vaccines, recombinant protein vaccines, synthetic peptide vaccines and empty capsid vaccines. For a more in-depth review of these novel technologies, see Cao et al. (2016) and Diaz-San Segundo et al. (2016). The most relevant advances are summarized in this section.

In general, vaccines based in synthetic peptides or recombinant proteins with a limited number of antigenic sites have proven poor protection against challenge in host animals (DiMarchi et al. 1986; Taboga et al. 1997; Sobrino et al. 1999; Rodriguez et al. 2003). However, promising results were obtained when using dendrimeric peptides or multi-epitope proteins. A synthetic dendrimeric peptide containing one copy of a T-cell epitope and branching out into four copies of a B-cell epitope, elicited B- and T- cell specific immune responses and solid protection in pigs after two inoculations with 1.4 mg. Interestingly, despite the parental administration of the peptide, immunized pigs also developed a potent anti-FMDV immunoglobulin A response (Cubillos et al. 2008). Recently, Blanco and co-workers reported that a reduced version of the dendrimeric peptide, bearing two copies of a B-cell epitope from a type O isolate, induced in swine similar or higher B- and T-cell specific responses than the tetravalent peptide (Blanco et al. 2016). The bivalent version conferred full protection and entirely prevented virus shedding.

The commercial company UBI has developed a commercial peptide vaccine (UBITh® vaccine) for the prevention FMD in pigs, and licensed for use in Taiwan (www.unitedbiomedical.com). The UBI peptide spans the entire G-H loop and flanking sequences (amino acid positions 129–169) of VP1, has a unique consensus sequence to confront the hypervariability of serotype O viruses, and includes an artificial T helper (Th) site. This design, intended to improve and broaden VP1 G-H loop peptide immunogenicity, included disulphide bonds between the cysteine residues at positions 134 and 158 that stabilised the presentation of a flexible G–H loop-like structure providing a better immunogen than the linear equivalent. This peptide induced protective immunity against a Taiwanese isolate of FMDV O1 (Wang et al. 2002). Although this vaccine is now widely used on pig farms in China, its composition must be adjusted to extend the antigenic spectrum. Moreover, a subsequent study in cattle revealed that neutralizing antibodies titres induced by the UBI peptide were relatively low and failed to protect cattle following challenge with a serotype-O strain of FMDV at 3 weeks post-vaccination (Rodriguez et al. 2003).

Multi-epitope vaccines have also been evaluated. Cao and co-workers developed a series of multi-epitope proteins containing the G–H loops of three topotypes of FMDV serotype O and promiscuous artificial Th sites. One of these proteins (B4) showed optimal immunogenicity and cross-reactivity in a mouse model. When this protein was co-administered with polyriboinosinic-polyribocytidylic acid [poly(I:C)], this vaccine elicited FMDV-specific neutralizing antibodies, IFN-α/β as well as IFN-γ, and offered a cross-protection against three topotypes of FMDV serotype O in pigs (Cao et al. 2013; Cao et al. 2014). Similar results were recently reported for a multi-epitope vaccine containing epitopes of currently circulating strains of FMDV serotype A, using cytidine-phosphate-guanosine (CpG) DNA as adjuvant (Cao et al. 2017). In cattle, both peptide and multi-epitope protein vaccines have shown limited efficacy, providing 60 % protection after a single immunization (Zhang et al. 2015), and 80% protection when CpG DNA and Montanide ISA-206 were used as adjuvant (Ren et al. 2011).

Protein/peptide vaccines pose no risk as they do not involve infectious virus, are highly stable at room temperature, can be easily stored and may function as DIVA vaccines. However, in most cases they are poorly immunogenic, especially for T cells, requiring additional adjuvants to induce protective immune response based both on antibodies and effector T cells.

The use of genetically engineered attenuated FMDV viruses to prepare inactivated vaccines was also considered as a safer alternative. Taking into account that viruses lacking the leader protease coding region (leaderless) are attenuated *in vivo*, and that the use of these viruses after chemical inactivation proved to be as effective as the *wild-type* inactivated antigen (Mason et al. 1997; Chinsangaram et al. 1998), Uddowla et al. have reported the design of a vaccine candidate that included the deletion of the L^{pro}, and one of the three 3B coding sequences, rendering a virus that was attenuated in both cattle and swine. Moreover, this recombinant virus harbours negative markers for potential DIVA capabilities, encoded in 3D alone or in 3B and 3D (Uddowla et al. 2012). This attenuated, antigenically marked virus was chemically inactivated and used to immunize cattle, providing 100% protection from challenge with parental wild-type virus. Similarly, a marker vaccine with a deletion within FMDV 3A was constructed using an infectious cDNA clone and proved to protect pigs against challenge with homologous *wild-type* FMDV (Li et al. 2014b). These attenuated viruses might provide a safer antigen for inactivated FMDV vaccine production, but they can still produce persistent viral infections in ruminants, and there is also the possibility of recombination of field and vaccine viruses due to virus escape or incomplete inactivation (Park 2013). This would be more likely if the attenuation is not stable and complete in all susceptible species. The use of FMDV empty capsids or VLPs is a very promising alternative that has been extensively explored. These immunogens have the entire repertoire of immunogenic epitopes displayed in the correct conformation and in a highly repetitive manner, enabling the induction of strong humoral and cellular immune responses (Noad and Roy 2003; Kushnir et al. 2012). In addition, the lack of genomic material and NSP makes them safe marker vaccines. As mentioned before, for the production of recombinant FMD VLPs, it is necessary to express the viral capsid proteins VP0, VP3 and VP1. This can be achieved by expressing these proteins individually, or co-expressing the capsid precursor (P12A) and protease 3C, which cleaves this precursor to generate the capsid proteins.

FMD VLPs can be produced *in vitro* by different expression systems and used as a subunit vaccine, or they can be produced *in vivo* following immunization with a

DNA vaccine or a viral vector vaccine with the genetic material needed for capsid formation.

Many heterologous expression systems have been used for the production of recombinant FMD VLPs. Although prokaryotic expression system would not be the best choice since the correct assembly of FMDV empty capsids requires a post-translational modification (myristoylation), the SUMO fusion system for the expression of VP0, VP3 and VP1 in *E. coli* has been reported by different authors (Guo et al. 2013; Xiao et al. 2016). Expression of recombinant proteins as fusions with SUMO (small ubiquitin-like modifier) protein has significantly increased the yield of difficult-to-express proteins in *E, coli*, since SUMO usually promotes correct folding and structural stability of the fusion proteins (Lee et al. 2008). The co-expression of VP0, VP3, and VP1 assembled successfully into VLPs *in vitro*, and provided protection to both swine and cattle (Guo et al. 2013; Xiao et al. 2016). Since the issue of myristoylation was not addressed in those reports it is unclear how VLPs have assembled. The SUMO-modified *E. coli* expression system is relatively simple and economical, but requires further steps of chromatographic purification and proteolytic cleavage of the SUMO protein tag to ensure that structural proteins assembled into particulate structures *in vitro* (Xiao et al. 2016). These additional steps significantly reduce final yield and counteract the simplicity of the bacterial expression system.

The baculovirus expression vector system has been extensively explored for the production of recombinant FMD VLPs. Li and co-workers reported the use of recombinant baculoviruses containing P12A-3C coding sequences from serotypes Asia 1 and A, which produced VLPs in silkworm larvae (Li et al. 2008b) and pupae (Li et al. 2012), respectively. Cattle vaccinated with these VLPs developed high titres of FMDV-specific antibodies and were completely protected against virulent homologous virus challenge. FMD VLPs were also expressed in recombinant baculovirus-infected insect cells, but protein expression levels and efficiency of capsid assembly have been highly variable in initial reports (Roosien et al. 1990; Grubman et al. 1993; Oem et al. 2007; Li et al. 2008b). However, subsequent studies demonstrated that the down regulation of 3C protease activity, which is known to be toxic to cells, improved capsid protein expression levels (Porta et al. 2013b; Vivek Srinivas et al. 2016). Moreover, by mutating a single histidine in VP2, Porta et al. were able to produce stable empty capsids of FMDV A22, that resisted both heat and acid treatments (Porta et al. 2013a). Cattle vaccinated with these recombinant capsids showed sustained virus neutralisation titres and protection from challenge.

Mammalian cell expression systems are the ideal choice to produce recombinant eukaryotic proteins as they are able to introduce post-translational modifications, proper protein folding and product assembly, which are essential for complete biological activity. However, initial attempts to constitutively express the capsid precursor P12A with 3C protease were unsuccessful (Abrams et al. 1995), probably due to protease 3C toxicity. Indeed, using a vaccinia virus-based transient expression system, it has been shown that optimal production of the processed capsid proteins from P12A precursor (from serotypes O and A) is achieved when expression levels of 3C protease or its activity are reduced (Polacek et al. 2013; Gullberg et al. 2013). Transient gene expression in serum-free suspension-growing cells for the production of FMDV empty capsids has been recently reported (Mignaqui et al. 2013). For this purpose, a plasmid encoding the complete cassette P12A3C from serotype A was transiently transfected into 293-6E cells. The recombinant proteins were expressed at levels similar to the ones achieved in the vaccine facilities after BHK-21 infection (approximately 3 μg/ml), and assembled into VLPs that induced protective immune response against viral challenge in mice. The absence of serum in the cell culture diminishes the cost of the whole process, which can be easily scaled up.

Recombinant FMD VLPs as subunit vaccines represent the alternative vaccine most similar to the current inactivated virus, with the advantage of safe production and DIVA characteristics.

As mentioned above, DNA vaccines and viral vector vaccines can also be used to deliver genetic information to express FMD VLPs inside the host cells and potentially induce both humoral and cell-mediated immune responses. So far, DNA vaccines have not been very efficient in inducing protective immune responses, requiring multiple immunizations, the addition of adjuvants and cytokines [e.g., granulocyte-macrophage colony-stimulating factor (GM-CSF), interleukin 18 (IL-18)] (Cedillo-Barrón et al. 2001; Li et al. 2006a; Mingxiao et al. 2007; Fowler et al. 2011; Ganges et al. 2011; Borrego et al. 2011) or a prime-boost regime, using a DNA vaccine for priming and the commercial vaccine for boosting (Li et al. 2008a). Conversely, several viral vectors have been successfully used to produce FMD VLPs, including herpesvirus, poxvirus, alphavirus and adenovirus. The most effective and extensively tested platform uses a replication-defective human adenovirus 5 (hAd5) (Mayr et al. 1999; Mayr et al. 2001; Moraes et al. 2002; Wu et al. 2003b; Moraes et al. 2003; Santos et al. 2006; Pena et al. 2008). The use of hAd5 containing the P1 coding region of FMDV A24 and the 3C coding region of A12 (Ad5-24) was evaluated in both swine and cattle. Animals vaccinated with a single dose of this vaccine were protected from clinical disease after direct inoculation challenge or contact exposure as early as 7 days post-vaccination (Moraes et al. 2002; Pacheco et al. 2005). During 2012, the Ad5-FMD A24 Cruzeiro (Ad5-24) vaccine was granted a conditional license for use in cattle in emergency situations in the United States (Grubman et al. 2012). Since the vaccine lacks the coding regions of most of NS proteins, vaccinated animals can be distinguished from infected animals. Moreover, its production does not require high-level biocontainment facilities, allowing its production in biosafety level-2 laboratories in the United States. The Ad5-based FMD vaccine has been developed for other serotypes and subtypes, and several of these candidates have successfully completed the efficacy studies (Grubman et al. 2010). The first generation of the Ad5-O1Campos (Ad5-O1C) vaccine induced lower levels of neutralizing antibodies than Ad5-A24 in swine, and did not protect animals against homologous challenge, even with the co-administration of porcine GM-CSF as an adjuvant (Caron et al. 2005). A second generation of Ad5-FMD vectors, containing the full-length 2B coding region (Ad5-FMD-2B) was constructed for O1C and A24 subtypes, in order to enhance the synthesis of viral capsid proteins and the efficiency and/or stability of capsid assembly. These vaccines have shown to improve vaccine efficacy both in cattle and swine (Moraes et al. 2011; Grubman et al. 2012).

5 Pathway to commercialization

Many efforts have been made in order to develop new FMD vaccines that meet the requirements that would be expected for an ideal vaccine, such as safety production, induction of rapid and long-lasting protective immunity after a single inoculation, compatibility with DIVA principle, and cost-effectiveness. However, it is unlikely that a single vaccine would meet all these requirements, and different strategies could be implemented for different particular cases. For example, emergency vaccines should be produced with simple and versatile technologies able to match the field strain responsible for an outbreak as fast as possible, while vaccines for endemic regions should emphasize cost-effective technologies.

In order to accelerate progress, the Global FMD Research Alliance (GFRA) was established in 2003. The core of the GFRA is a consortium of research institutions with the aim of developing a new generation of accessible and efficacious vaccines, diagnostics and antiviral agents to successfully prevent, control and eradicate FMD.

It is evident that FMD remains a severe threat to the livestock industry worldwide. Recently, it has been reported that the annual impact of this disease in terms of production losses and vaccination in endemic regions alone amount to between USD 6,5 and 21 billion (Knight-Jones and Rushton 2013). It has been estimated that 2.35 billion doses of FMD vaccine are administered annually in the world, with the main regions being China (1600 million doses), South America (500 million doses) and India (150 million doses) (Knight-Jones and Rushton 2013).

In the past few years, favourable government initiatives to control FMD outbreaks have significantly increased the growth of the global FMD vaccines market. Additionally, the rising consumption of meat and dairy products across the globe has further powered the need for FMD vaccines. According to a recent report published by the Transparency Market Research (2015), the global foot and mouth disease vaccines market is expected to rise from a valuation of USD 0.51 billion in 2013 to USD 0.95 billion by 2020 (http://www.transparencymarketresearch.com/foot-mouth-disease-vaccines.html).

There is a general interest for global eradication of FMD, and vaccine-based eradication has demonstrated to be a feasible strategy regionally that could be expanded to global application. However, in order to achieve this goal, significant challenges must be overcome, especially in developing countries. For example, the need of sustainable and competent veterinary service trained to monitor vaccination and detect outbreaks; the development of effective surveillance methods to rapidly confirm clinical suspicions, preferably performed on-site; and vaccine production technologies significantly less expensive and more flexible (Smith et al. 2014).

In addition, actual FMD vaccine production has several drawbacks, specially associated with the large amount of infectious virus that must be manipulated, the existence of different serotypes and strains that complicates the maintenance of vaccine stocks and the instability of the formulated vaccine (12-18 months). This thermal instability is also problematic when vaccines are formulated from frozen antigen concentrates in vaccine banks, since it is difficult to keep them refrigerated, especially in poorer parts of the world. The need of expensive high-level bio-containment facilities and the constant risk of viral escape are undoubtedly other important disadvantages of the current vaccine production system that have prompted different countries to prohibit the production of this vaccine in their mainland.

In this scenario, the implementation of molecular farming for FMD vaccine production could be easily introduced in the market since it may not require bio-containment facilities, or these would be less stringent or less expensive. Moreover, it favors a faster design of new immunogens in case of an eventual introduction of exotic strains, potentially conceding the optimization of the time for supplying the market with a product that enables the control of the outbreak. Additionally, like any other recombinant vaccine would allow the accurate differentiation between vaccinated and infected animals, guaranteeing the control of outbreaks through sero-epidemiological surveillance programs.

Taking in mind that the immunogenicity of recombinant FMD VLPs has already been demonstrated, this approach should be the focus in the development of a novel vaccine against FMD. However, there is still a need to improve expression levels to move the production of FMDV vaccines in plants closer to the development phase. Successful commercialization of recombinant veterinary vaccines will require cooperation with business associates, the creation of a suitable business plan, and multiple stages of financing. Since the process from research to product registration can take up to 7 years, the target market should be predicted well in advance, and the ability for economical scale-up must be assessed to determine whether the product can be produced cost-effectively at consistent quality. In the field of plant-based vaccines, intellectual property protection and freedom to operate are more complicated by the intricacies of the technology, and their requirements are continually changing. While there are variations between countries in terms of how plant-based vaccines are regulated, in general, they must be shown to be safe, efficacious, and environmentally benign in order to gain approval (Macdonald et al. 2015).

Even when considering what was just said, the achievements and ripeness of molecular farming, especially when large doses of vaccine are required, certainly makes this production platform one of the most interesting options as an alternative to the conventional FMDV vaccine.

References

Abrams CC, King AM, Belsham GJ (1995) Assembly of foot-and-mouth disease virus empty capsids synthesized by a vaccinia virus expression system. J Gen Virol 76:3089–3098

Alexandersen S, Brotherhood I, Donaldson AI (2002a) Natural aerosol transmission of foot-and-mouth disease virus to pigs: minimal infectious dose for strain O-1 Lausanne. Epidemiol Infect 128:301–312. https://doi.org/10.1017/s095026880100646x

Alexandersen S, Donaldson AI (2002) Further studies to quantify the dose of natural aerosols of foot-and-mouth disease virus for pigs. Epidemiol Infect 128:313–323. https://doi.org/10.1016/S0034-5288(02)90037-8

Alexandersen S, Mowat N (2005) Foot-and-mouth disease virus: host range and pathogenesis. Curr Top Microbiol Immunol 288:9–42. https://doi.org/10.1007/b138628

Alexandersen S, Oleksiewicz MB, Donaldson AI (2001) The early pathogenesis of foot-and-mouth disease in pigs infected by contact: A quantitative time-course study using TaqMan RT-PCR. J Gen Virol 82:747–755. https://doi.org/10.1099/0022-1317-82-4-747

Alexandersen S, Zhang Z, Donaldson AI (2002b) Aspects of the persistence of foot-and-mouth disease virus in animals—the carrier problem. Microbes Infect 4:1099–1110. https://doi.org/10.1016/S1286-4579(02)01634-9

Alexandersen S, Zhang Z, Donaldson AI, Garland AJM (2003) The Pathogenesis and Diagnosis of Foot-and-Mouth Disease. J Comp Pathol 129:1–36. https://doi.org/10.1016/S0021-9975(03)00041-0

Andrianova EP, Krementsugskaia SR, Lugovskaia NN, et al (2011) Foot and mouth disease virus polyepitope protein produced in bacteria and plants induces protective immunity in guinea pigs. Biochem Biokhimii͡a 76:339–46.

Arzt J, Baxt B, Grubman MJ et al (2011) The Pathogenesis of Foot-and-Mouth Disease II: Viral Pathways in Swine, Small Ruminants, and Wildlife; Myotropism, Chronic Syndromes, and Molecular Virus-Host Interactions. Transbound Emerg Dis 58:305–326. https://doi.org/10.1111/j.1865-1682.2011.01236.x

Arzt J, Gregg DA, Clavijo A, Rodriguez LL (2009) Optimization of immunohistochemical and fluorescent antibody techniques for localization of Foot-and-mouth disease virus in animal tissues. J Vet Diagnostic Investig 21:779–792. https://doi.org/10.1177/104063870902100604

Arzt J, Pacheco JM, Rodriguez LL (2010) The early pathogenesis of foot-and-mouth disease in cattle after aerosol inoculation. Identification of the nasopharynx as the primary site of infection. Vet Pathol 47:1048–1063. https://doi.org/10.1177/0300985810372509

Arzt J, Pacheco JM, Smoliga GR et al (2014) Foot-and-mouth disease virus virulence in cattle is co-determined by viral replication dynamics and route of infection. Virology 452:12–22. https://doi.org/10.1016/j.virol.2014.01.001

Bachrach HL (1968) Foot-And-Mouth Disease. Annu Rev Microbiol 22:201–244. https://doi.org/10.1146/annurev.mi.22.100168.001221

Baxt B, Becker Y (1990) The effect of peptides containing the arginine-glycine-aspartic acid sequence on the adsorption of foot-and-mouth disease virus to tissue culture cells. Virus Genes 4:73–83

Belsham G (1993) Distinctive features of foot-and-mouth disease virus, a member of the picornavirus family; aspects of virus protein synthesis, protein processing and structure. Prog Biophys Mol Biol 60:241–260

Belsham GJ (2005) Translation and replication of FMDV RNA. Curr Top Microbiol Immunol 288:43–70

Blanco E, Guerra B, De La Torre BG et al (2016) Full protection of swine against foot-and-mouth disease by a bivalent B-cell epitope dendrimer peptide. Antiviral Res 129:74–80. https://doi.org/10.1016/j.antiviral.2016.03.005

Borrego B, Argilaguet JM, Pérez-Martín E et al (2011) A DNA vaccine encoding foot-and-mouth disease virus B and T-cell epitopes targeted to class II swine leukocyte antigens protects pigs against viral challenge. Antiviral Res. https://doi.org/10.1016/j.antiviral.2011.07.017

Borrego B, Camarero JA, Mateu MG, Domingo E (1995) A highly divergent antigenic site of foot-and-mouth disease virus retains its immunodominance. Viral Immunol 8:11–8. https://doi.org/10.1089/vim.1995.8.11

Brito BP, Rodriguez LL, Hammond JM, et al (2015) Review of the Global Distribution of Foot-and-Mouth Disease Virus from 2007 to 2014. Transbound Emerg Dis 1–17. doi: 10.1111/tbed.12373

Brown F (2003) The history of research in foot-and-mouth disease. Virus Res 91:3–7

Burrows R (1966) Studies on the carrier state of cattle exposed to Foot-and-mouth disease virus. J Hyg (Lond) 64:81–90. https://doi.org/10.1017/S0022172400040365

Burrows R, Mann JA, Garland AJM et al (1981) The pathogenesis of natural and simulated natural Foot-and-Mouth disease infection in cattle. J Comp Pathol 91:599–609

Cao Y (2014) Adjuvants for foot-and-mouth disease virus vaccines: recent progress. Expert Rev Vaccines 13:1377–1385. https://doi.org/10.1586/14760584.2014.963562

Cao Y, Li D, Fu Y et al (2017) Rational design and efficacy of a multi-epitope recombinant protein vaccine against foot-and-mouth disease virus serotype A in pigs. Antiviral Res 140:133–141. https://doi.org/10.1016/j.antiviral.2017.01.023

Cao Y, Lu Z, Li D et al (2014) Evaluation of cross-protection against three topotypes of serotype O foot-and-mouth disease virus in pigs vaccinated with multi-epitope protein vaccine incorporated with poly(I:C). Vet Microbiol 168:294–301. https://doi.org/10.1016/j.vetmic.2013.11.023

Cao Y, Lu Z, Li Y et al (2013) Poly(I:C) combined with multi-epitope protein vaccine completely protects against virulent foot-and-mouth disease virus challenge in pigs. Antiviral Res 97:145–153. https://doi.org/10.1016/j.antiviral.2012.11.009

Cao Y, Lu Z, Liu Z (2016) Foot-and-mouth disease vaccines: progress and problems. Expert Rev Vaccines 584(14760584):1140042. https://doi.org/10.1586/14760584.2016.1140042

Caron L, Brum MCS, Moraes MP et al (2005) Granulocyte-macrophage colony-stimulating factor does not increase the potency or efficacy of a foot-and-mouth disease virus subunit vaccine. Pesqui Vet Bras 25:150–158

Carrillo C, Wigdorovitz A, Oliveros JC, et al (1998) Protective Immune Response to Foot-and-Mouth Disease Virus with VP1 Expressed in Transgenic Plants. 72:1688–1690.

Carrillo C, Wigdorovitz A, Trono K et al (2001) Induction of a virus-specific antibody response to foot and mouth disease virus using the structural protein VP1 expressed in transgenic potato plants. Viral Immunol 14:49–57

Cedillo-Barrón L, Foster-Cuevas M, Belsham GJ et al (2001) Induction of a protective response in swine vaccinated with DNA encoding foot-and-mouth disease virus empty capsid proteins and the 3D RNA polymerase. J Gen Virol 82:1713–1724

Chinsangaram J, Mason PW, Grubman MJ (1998) Protection of swine by live and inactivated vaccines prepared from a leader proteinase-deficient serotype A12 foot-and-mouth disease virus. Vaccine 16:1516–1522. https://doi.org/10.1016/S0264-410X(98)00029-2

Cubillos C, de la Torre BG, Jakab A et al (2008) Enhanced mucosal immunoglobulin A response and solid protection against foot-and-mouth disease virus challenge induced by a novel dendrimeric peptide. J Virol 82:7223–30. https://doi.org/10.1128/JVI.00401-08

Dawe PS, Flanagan FO, Madekurozwa RL et al (1994a) Natural transmission of foot-and-mouth disease virus from African buffalo (Syncerus caffer) to cattle in a wildlife area of Zimbabwe. Vet Rec 134:230–232

Dawe PS, Sorensen K, Ferris NP et al (1994b) Experimental transmission of foot-and-mouth disease virus from carrier African buffalo (Syncerus caffer) to cattle in Zimbabwe. Vet Rec 134:211–5

Di Nardo A, Libeau G, Chardonnet B et al (2015) Serological profile of foot-and-mouth disease in wildlife populations of West and Central Africa with special reference to Syncerus caffer subspecies. Vet Res 46:77. https://doi.org/10.1186/s13567-015-0213-0

Diaz-San Segundo F, Medina GN, Stenfeldt C et al (2016) Foot-and-mouth disease vaccines. Vet Microbiol. https://doi.org/10.1016/j.vetmic.2016.12.018

Dicara D, Burman A, Clark S et al (2008) Foot-and-mouth disease virus forms a highly stable, EDTA-resistant complex with its principal receptor, integrin alphavbeta6: implications for infectiousness. J Virol 82:1537–46. https://doi.org/10.1128/JVI.01480-07

DiMarchi R, Brooke G, Gale C, et al (1986) Protection of cattle against foot-and-mouth disease by a synthetic peptide.

Doel TR (2003) FMD vaccines. Virus Res 91:81–99

Doel TR, Williams L, Barnett PV (1994) Emergency vaccination against foot-and-mouth disease: Rate of development of immunity and its implications for the carrier state. Vaccine 12:592–600. https://doi.org/10.1016/0264-410X(94)90262-3

Domingo E, Baranowski E, Escarmís C, Sobrino F (2002) Foot-and-mouth disease virus. Comp Immunol Microbiol Infect Dis 25:297–308. doi: 10.1016/S0147-9571(02)00027-9

Donaldson AI (1987) Foot-and-mouth disease: the principal features. Irish Vet J 41:325–327

Donaldson AI (1997) Risks of spreading foot and mouth disease through milk and dairy products. Rev Sci Tech 16:117–24

Donaldson AI, Alexandersen S (2001) Relative resistance of pigs to infection by natural aerosols of FMD virus. Vet Rec 148:600–2

Donaldson AI, Herniman KAJ, Parker J, Sellers RF (1970) Further investigations on the airborne excretion of foot and mouth disease virus. J Hyg, Camb 68:557–564. https://doi.org/10.1016/B978-0-12-382219-2.00490-7

Dus Santos MJ, Carrillo C, Ardila F et al (2005) Development of transgenic alfalfa plants containing the foot and mouth disease virus structural polyprotein gene P1 and its utilization as an experimental immunogen. Vaccine 23:1838–43. https://doi.org/10.1016/j.vaccine.2004.11.014

Dus Santos MJ, Wigdorovitz A, Trono K et al (2002) A novel methodology to develop a foot and mouth disease virus (FMDV) peptide-based vaccine in transgenic plants. Vaccine 20:1141–1147

Fowler VL, Bashiruddin JB, Maree FF et al (2011) Foot-and-mouth disease marker vaccine: Cattle protection with a partial VP1 G-H loop deleted virus antigen. Vaccine 29:8405–8411. https://doi.org/10.1016/j.vaccine.2011.08.035

Fox G, Parry NR, Barnett PV et al (1989) The cell attachment site on foot-and-mouth disease virus includes the amino acid sequence RGD (arginine-glycine-aspartic acid). J Gen Virol 70(Pt 3):625–37. https://doi.org/10.1099/0022-1317-70-3-625

Fukai K, Yamada M, Morioka K et al (2015) Dose-dependent responses of pigs infected with foot-and-mouth disease virus O/JPN/2010 by the intranasal and intraoral routes. Arch Virol 160:129–139. https://doi.org/10.1007/s00705-014-2239-4

Gailiunas P, Cottral GE (1966) Presence and persistence of foot-and-mouth disease virus in bovine skin. J Bacteriol 91:2333–2338

Ganges L, Borrego B, Fernández-Pacheco P, et al (2011) DNA immunization of pigs with foot-and-mouth disease virus minigenes: From partial protection to disease exacerbation.

Gay CG, Salt J, Balaski C (2003) Challenges and opportunities in developing and marketing vaccines for OIE List A and emerging animal diseases. Dev Biol (Basel) 114:243–50

Gloster J, Sellers RF, Donaldson AI (1982) Long distance transport of foot-and-mouth disease virus over the sea. Vet Rec 110:47–52

Goodwin S, Tuthill TJ, Arias A et al (2009) Foot-and-mouth disease virus assembly: processing of recombinant capsid precursor by exogenous protease induces self-assembly of pentamers in vitro in a myristoylation-dependent manner. J Virol 83:11275–11282. https://doi.org/10.1128/JVI.01263-09

Grubman M, Moraes M, Schutta C et al (2010) Adenovirus serotype 5-vectored foot-and-mouth disease subunit vaccines: the first decade. Future Virol 5:51–64

Grubman MJ, Baxt B (2004) Foot-and-Mouth Disease. 17:465–493. https://doi.org/10.1128/CMR.17.2.465

Grubman MJ, Lewis S a, Morgan DO (1993) Protection of swine against foot-and-mouth disease with viral capsid proteins expressed in heterologous systems. Vaccine 11:825–9

Grubman MJ, Segundo FD, Perez-martin E, Service I (2012) Use of replication-defective adenoviruses to develop vaccines and biotherapeutics against foot-and-mouth disease. Future Virol 7:767–778

Gullberg M, Muszynski B, Organtini LJ et al (2013) Assembly and characterization of foot-and-mouth disease virus empty capsid particles expressed within mammalian cells. J Gen Virol 94:1769–1779. https://doi.org/10.1099/vir.0.054122-0

Guo H-C, Sun S-Q, Jin Y et al (2013) Foot-and-mouth disease virus-like particles produced by a SUMO fusion protein system in Escherichia coli induce potent protective immune responses in guinea pigs, swine and cattle. Vet Res 44:48. https://doi.org/10.1186/1297-9716-44-48

Habiela M, Seago J, Perez-Martin E et al (2014) Laboratory animal models to study foot-and-mouth disease: a review with emphasis on natural and vaccine induced immunity. J Gen Virol 95:2329–2345. https://doi.org/10.1099/vir.0.068270-0

Henderson WM (1948) A consideration of some of the factors concerned in intracutaneous injection of cattle. J Pathol Bacteriol 60:137–139

Hughes GJ, Mioulet V, Kitching RP et al (2002) Foot-and-mouth disease virus infection of sheep: implications for diagnosis and control. Vet Rec 150:724–7

Jackson T, Clark S, Berryman S et al (2004) Integrin alphavbeta 8 functions as a receptor for foot-and-mouth disease virus: role of the beta-chain cytodomain in integrin-mediated infection. J Virol 78:4533–40

Jackson T, Mould AP, Sheppard D, King AMQ (2002) Integrin alphavbeta1 is a receptor for foot-and-mouth disease virus. J Virol 76:935–41

Jackson T, Sheppard D, Denyer M et al (2000) The epithelial integrin alphavbeta6 is a receptor for foot-and-mouth disease virus. J Virol 74:4949–56

Jiang L, Li Q, Li M et al (2006) A modified TMV-based vector facilitates the expression of longer foreign epitopes in tobacco. Vaccine 24:109–115. https://doi.org/10.1016/j.vaccine.2005.09.060

Juleff N, Windsor M, Reid E et al (2008) Foot-and-mouth disease virus persists in the light zone of germinal centres. PLoS One 3:e3434. https://doi.org/10.1371/journal.pone.0003434

Kitching RP (2002) Clinical variation in foot and mouth disease: cattle. Rev Sci Tech 21:499–504. https://doi.org/10.20506/rst.21.3.1367

Kitching RP, Alexandersen S (2002) Clinical variation in foot and mouth disease: pigs. Rev Sci Tech 21:513–8

Kitching RP, Hughes GJ (2002) Clinical variation in foot and mouth disease: sheep and goats. Rev Sci Tech 21:505–512

Kleina LG, Grubman MJ (1992) Antiviral effects of a thiol protease inhibitor on foot-and-mouth disease virus. J Virol 66:7168–75

Knight-Jones TJD, Rushton J (2013) The economic impacts of foot and mouth disease - What are they, how big are they and where do they occur? Prev Vet Med 112:162–173. https://doi.org/10.1016/j.prevetmed.2013.07.013

Knowles NJ, Samuel AR, Davies PR et al (2001) Outbreak of foot-and-mouth disease virus serotype O in the UK caused by a pandemic strain. Vet Rec 148:258–9

Knowles NJ, Samuel AR, Davies PR et al (2005) Pandemic strain of foot-and-mouth disease virus serotype O. Emerg Infect Dis 11:1887–1893. https://doi.org/10.3201/eid1112.050908

Kushnir N, Streatfield SJ, Yusibov V (2012) Virus-like particles as a highly efficient vaccine platform: Diversity of targets and production systems and advances in clinical development. Vaccine 31:58–83. https://doi.org/10.1016/j.vaccine.2012.10.083

Lee C-D, Sun H-C, Hu S-M et al (2008) An improved SUMO fusion protein system for effective production of native proteins. Protein Sci 17:1241–8. https://doi.org/10.1110/ps.035188.108

Lentz EM, Segretin ME, Morgenfeld MM et al (2010) High expression level of a foot and mouth disease virus epitope in tobacco transplastomic plants. Planta 231:387–395. https://doi.org/10.1007/s00425-009-1058-4

Li C, Yamagishi N, Kaido M, Yoshikawa N (2014a) Presentation of epitope sequences from foreign viruses on the surface of apple latent spherical virus particles. Virus Res 190:118–126. https://doi.org/10.1016/j.virusres.2014.07.013

Li P, Lu Z, Bai X et al (2014b) Evaluation of a 3A-truncated foot-and-mouth disease virus in pigs for its potential as a marker vaccine. Vet Res 45:1–11. https://doi.org/10.1186/1297-9716-45-51

Li Y, Aggarwal N, Takamatsu H et al (2006a) Enhancing immune responses against a plasmid DNA vaccine encoding a FMDV empty capsid from serotype O. Vaccine 24:4602–4606

Li Y, Stirling CMA, Denyer MS et al (2008a) Dramatic improvement in FMD DNA vaccine efficacy and cross-serotype antibody induction in pigs following a protein boost. Vaccine 26:2647–2656. https://doi.org/10.1016/j.vaccine.2008.01.037

Li Y, Sun M, Liu J et al (2006b) High expression of foot-and-mouth disease virus structural protein VP1 in tobacco chloroplasts. Plant Cell Rep 25:329–333. https://doi.org/10.1007/s00299-005-0074-5

Li Z, Yi Y, Yin X et al (2008b) Expression of foot-and-mouth disease virus capsid proteins in silkworm-baculovirus expression system and its utilization as a subunit vaccine. PLoS One 3:1–7. https://doi.org/10.1371/journal.pone.0002273

Li Z, Yi Y, Yin X et al (2012) Development of a Foot-and-Mouth Disease Virus Serotype A Empty Capsid Subunit Vaccine Using Silkworm (Bombyx mori) Pupae. PLoS One 7:e43849. https://doi.org/10.1371/journal.pone.0043849

Loeffler F, Frosch P (1897) Summarischer bericht ueber der ergebnisse der untersuchungen zur erforschung der maul- und klauenseuche. Zentbl Bakteriol Parasitenkd Infekt Hyg 22:257–259

Macdonald J, Doshi K, Dussault M et al (2015) Bringing plant-based veterinary vaccines to market: Managing regulatory and commercial hurdles ☆. Biotechnol Adv 33:1572–1581. https://doi.org/10.1016/j.biotechadv.2015.07.007

Mason PW, Grubman MJ, Baxt B (2003) Molecular basis of pathogenesis of FMDV. Virus Res 91:9–32. https://doi.org/10.1016/S0168-1702(02)00257-5

Mason PW, Piccone ME, McKenna TSC et al (1997) Evaluation of a live-attenuated foot-and-mouth disease virus as a vaccine candidate. Virology 227:96–102. https://doi.org/10.1006/viro.1996.8309

Mateu MG, Camarero JA, Giralt E et al (1995) Direct evaluation of the immunodominance of a major antigenic site of foot-and-mouth disease virus in a natural host. Virology 206:298–306. https://doi.org/10.1016/S0042-6822(95)80045-X

Mayr GA, Chinsangaram J, Grubman MJ (1999) Development of replication-defective adenovirus serotype 5 containing the capsid and 3C protease coding regions of foot-and-mouth disease virus as a vaccine candidate. Virology 263:496–506

Mayr GA, O'Donnell V, Chinsangaram J et al (2001) Immune responses and protection against foot-and-mouth disease virus (FMDV) challenge in swine vaccinated with adenovirus-FMDV constructs. Vaccine 19:2152–62

McKenna TS, Lubroth J, Rieder E et al (1995) Receptor binding site-deleted foot-and-mouth disease (FMD) virus protects cattle from FMD. J Virol 69:5787–90

Mignaqui AC, Ruiz V, Perret S et al (2013) Transient Gene Expression in Serum-Free Suspension-Growing Mammalian Cells for the Production of Foot-and-Mouth Disease Virus Empty Capsids. PLoS One 8:e72800. https://doi.org/10.1371/journal.pone.0072800

Mingxiao M, Ningyi J, Jaun L et al (2007) Immunogenicity of plasmids encoding P12A and 3C of FMDV and swine IL-18. Antiviral Res 76:59–67. https://doi.org/10.1016/j.antiviral.2007.05.003

Monaghan P (2005) The alfa(v)beta6 integrin receptor for Foot-and-mouth disease virus is expressed constitutively on the epithelial cells targeted in cattle. J Gen Virol 86:2769–2780. https://doi.org/10.1099/vir.0.81172-0

Montague NP, Thuenemann EC, Saxena P et al (2011) Recent advances of *Cowpea mosaic virus*-based particle technology. Hum Vaccin 7:383–390. https://doi.org/10.4161/hv.7.3.14989

Moonen P, Schrijver R (2000) Carriers of foot-and-mouth disease virus: A review. Vet Q 22:193–197. https://doi.org/10.1080/01652176.2000.9695056

Moraes M, Mayr G, Mason P, Grubman M (2002) Early protection against homologous challenge after a single dose of replication-defective human adenovirus type 5 expressing capsid proteins of foot-and-mouth disease virus (FMDV) strain A24. Vaccine 20:1631–1639. https://doi.org/10.1016/S0264-410X(01)00483-2

Moraes MP, Chinsangaram J, Brum MC, Grubman MJ (2003) Immediate protection of swine from foot-and-mouth disease: a combination of adenoviruses expressing interferon alpha and a foot-and-mouth disease virus subunit vaccine. Vaccine 22:268–279. https://doi.org/10.1016/S0264-410X(03)00560-7

Moraes MP, Segundo FD-S, Dias CC et al (2011) Increased efficacy of an adenovirus-vectored foot-and-mouth disease capsid subunit vaccine expressing nonstructural protein 2B is associated with a specific T cell response. Vaccine 29:9431–40. https://doi.org/10.1016/j.vaccine.2011.10.037

Murphy C, Bashiruddin JB, Quan M et al (2010) Foot-and-mouth disease viral loads in pigs in the early, acute stage of disease. Vet Rec 166:10–4. https://doi.org/10.1136/vr.b5583

Muthamilselvan T, Lee CW, Cho YH et al (2016) A transgenic plant cell-suspension system for expression of epitopes on chimeric Bamboo mosaic virus particles. Plant Biotechnol J 14:231–239. https://doi.org/10.1111/pbi.12377

Noad R, Roy P (2003) Virus-like particles as immunogens. Trends Microbiol 11:438–444. https://doi.org/10.1016/S0966-842X(03)00208-7

Oem J-K, Park J-H, Lee K-N et al (2007) Characterization of recombinant foot-and-mouth disease virus pentamer-like structures expressed by baculovirus and their use as diagnostic antigens in a blocking ELISA. Vaccine 25:4112–4121

Pacheco J, Brum M, Moraes M et al (2005) Rapid protection of cattle from direct challenge with foot-and-mouth disease virus (FMDV) by a single inoculation with an adenovirus-vectored FMDV subunit vaccine. Virology 337:205–209

Pacheco JM, Arzt J, Rodriguez LL (2010) Early events in the pathogenesis of foot-and-mouth disease in cattle after controlled aerosol exposure. Vet J 183:46–53. https://doi.org/10.1016/j.tvjl.2008.08.023

Pacheco JM, Lee K-N, Eschbaumer M et al (2016) Evaluation of Infectivity, Virulence and Transmission of FDMV Field Strains of Serotypes O and A Isolated In 2010 from Outbreaks in the Republic of Korea. PLoS One 11:e0146445. https://doi.org/10.1371/journal.pone.0146445

Pacheco JM, Tucker M, Hartwig E et al (2012) Direct contact transmission of three different foot-and-mouth disease virus strains in swine demonstrates important strain-specific differences. Vet J 193:456–463. https://doi.org/10.1016/j.tvjl.2012.01.012

Pan L, Zhang Y, Wang Y et al (2008) Foliar extracts from transgenic tomato plants expressing the structural polyprotein, P1-2A, and protease, 3C, from foot-and-mouth disease virus elicit a protective response in guinea pigs. Vet Immunol Immunopathol 121:83–90

Parida S (2009) Vaccination against foot-and-mouth disease virus: strategies and effectiveness. Expert Rev Vaccines 8:347–365. https://doi.org/10.1586/14760584.8.3.347

Park J (2013) Requirements for improved vaccines against foot-and-mouth disease epidemics. Clin Exp Vaccine Res 2:8–18. https://doi.org/10.7774/cevr.2013.2.1.8

Pena L, Moraes MP, Koster M et al (2008) Delivery of a foot-and-mouth disease virus empty capsid subunit antigen with nonstructural protein 2B improves protection of swine. Vaccine 26:5689–5699. https://doi.org/10.1016/j.vaccine.2008.08.022

Polacek C, Gullberg M, Li J, Belsham GJ (2013) Low levels of foot-and-mouth disease virus 3C protease expression are required to achieve optimal capsid protein expression and processing in mammalian cells. J Gen Virol 94:1249–1258. https://doi.org/10.1099/vir.0.050492-0

Porta C, Kotecha A, Burman A et al (2013a) Rational Engineering of Recombinant Picornavirus Capsids to Produce Safe. Protective Vaccine Antigen. PLoS Pathog 9:e1003255. https://doi.org/10.1371/journal.ppat.1003255

Porta C, Spall VE, Findlay KC et al (2003) Cowpea mosaic virus-based chimaeras: Effects of inserted peptides on the phenotype, host range, and transmissibility of the modified viruses. Virology 310:50–63. https://doi.org/10.1016/S0042-6822(03)00140-5

Porta C, Spall VE, Loveland J et al (1994) Development of Cowpea Mosaic Virus as a High-Yielding System for the Presentation of Foreign Peptides. Virology 202:949–955. https://doi.org/10.1006/viro.1994.1417

Porta C, Xu X, Loureiro S et al (2013b) Efficient production of foot-and-mouth disease virus empty capsids in insect cells following down regulation of 3C protease activity. J Virol Methods 187:406–412. https://doi.org/10.1016/j.jviromet.2012.11.011

Quan M, Murphy CM, Zhang Z et al (2009) Influence of Exposure Intensity on the Efficiency and Speed of Transmission of Foot-and-Mouth Disease. J Comp Pathol 140:225–237. https://doi.org/10.1016/j.jcpa.2008.12.002

Quan M, Murphy CM, Zhang Z, Alexandersen S (2004) Determinants of Early Foot-and-Mouth Disease Virus Dynamics in Pigs. J Comp Pathol 131:294–307. https://doi.org/10.1016/j.jcpa.2004.05.002

Racaniello V (2001) Picornaviridae: the viruses and their replication. Fields Virology, 4th edn. Lippincott Williams & Wilkins, Philadelphia, pp 685–722

Rao JP, Agrawal P, Mohammad R et al (2012) Expression of VP1 protein of serotype A and O of foot-and-mouth disease virus in transgenic sunnhemp plants and its immunogenicity for guinea pigs. Acta Virol 56:91–99. https://doi.org/10.4149/av_2012_02_91

Ren J, Yang L, Xu H et al (2011) CpG oligodeoxynucleotide and montanide ISA 206 adjuvant combination augments the immune responses of a recombinant FMDV vaccine in cattle. Vaccine 29:7960–7965. https://doi.org/10.1016/j.vaccine.2011.08.072

Robinson L, Knight-Jones TJD, Charleston B et al (2016) Global Foot-and-Mouth Disease Research Update and Gap Analysis: 3 - Vaccines. Transbound Emerg Dis 63:30–41. https://doi.org/10.1111/tbed.12521

Rodriguez LL, Barrera J, Kramer E et al (2003) A synthetic peptide containing the consensus sequence of the G-H loop region of foot-and-mouth disease virus type-O VP1 and a promiscuous T-helper epitope induces peptide-specific antibodies but fails to protect cattle against viral challenge. Vaccine 21:3751–3756. https://doi.org/10.1016/S0264-410X(03)00364-5

Rodriguez LL, Gay CG (2011) Development of vaccines toward the global control and eradication of foot-and-mouth disease. Expert Rev Vaccines 10:377–87. https://doi.org/10.1586/erv.11.4

Rohll JB, Moon DH, Evans DJ, Almond JW (1995) The 3' untranslated region of picornavirus RNA: features required for efficient genome replication. J Virol 69:7835–44

Roosien J, Belsham GJ, Ryan MD et al (1990) Synthesis of foot-and-mouth disease virus capsid proteins in insect cells using baculovirus expression vectors. J Gen Virol 71:1703–1711

Rueckert RR, Wimmer E (1984) Systematic nomenclature of picornavirus proteins. J Virol 50:957–9

Ruiz-Sáenz J, Goez Y, Tabares W, López-Herrera A (2009) Cellular Receptors for Foot and Mouth Disease Virus. Intervirology 52:201–212. https://doi.org/10.1159/000226121

Ruiz V, Mozgovoj MV, Dus Santos MJ, Wigdorovitz A (2015) Plant-produced viral bovine vaccines: What happened during the last 10 years? Plant Biotechnol J 13:1071–1077. https://doi.org/10.1111/pbi.12440

Rweyemamu M, Roeder P, MacKay D et al (2008) Epidemiological patterns of foot-and-mouth disease worldwide. Transbound Emerg Dis 55:57–72. https://doi.org/10.1111/j.1865-1682.2007.01013.x

Sa-Carvalho D, Rieder E, Baxt B et al (1997) Tissue culture adaptation of foot-and-mouth disease virus selects viruses that bind to heparin and are attenuated in cattle. J Virol 71:5115–23

Sangula AK, Siegismund HR, BelshamGJ, et al (2011) Low diversity of foot-and-mouth disease serotype C virus in Kenya: evidence for probable vaccine strain re-introductions in the field. Epidemiol Infect 139:189–196. doi: 10.1017/S0950268810000580

Santos TDL, Botton SDA, Weiblen R, Grubman MJ (2006) The Leader Proteinase of Foot-and-Mouth Disease Virus Inhibits the Induction of Beta Interferon mRNA and Blocks the Host Innate Immune Response. 80:1906–1914. https://doi.org/10.1128/JVI.80.4.1906

Saunders K, Sainsbury F, Lomonossoff GP (2009) Efficient generation of cowpea mosaic virus empty virus-like particles by the proteolytic processing of precursors in insect cells and plants. Virology 393:329–37. https://doi.org/10.1016/j.virol.2009.08.023

Sellers R, Gloster J (2008) Foot-and-mouth disease: A review of intranasal infection of cattle, sheep and pigs. Vet J 177:159–168. https://doi.org/10.1016/j.tvjl.2007.03.009

Sellers RF, Parker J (1969) Airborne excretion of foot-and-mouth disease virus.

Serrano P, Pulido MR, Sáiz M, Martínez-Salas E (2006) The 3' end of the foot-and-mouth disease virus genome establishes two distinct long-range RNA-RNA interactions with the 5' end region. J Gen Virol 87:3013–3022. https://doi.org/10.1099/vir.0.82059-0

Shahan M (1962) Shahan : Virus of Foot-and-Mouth Disease. Ann N Y Acad Sci 101:444–454
Smith MT, Bennett AM, Grubman MJ, Bundy BC (2014) Foot-and-mouth disease: Technical and political challenges to eradication. Vaccine
Sobrino F, Blanco E, García-Briones M, Ley V (1999) Synthetic peptide vaccines: foot-and-mouth disease virus as a model. Dev Biol Stand 101:39–43
Sørensen JH, Mackay DK, Jensen CO, Donaldson AI (2000) An integrated model to predict the atmospheric spread of foot-and-mouth disease virus. Epidemiol Infect 124:577–90
Stenfeldt C, Diaz-San Segundo F, de Los Santos T et al (2016a) The Pathogenesis of Foot-and-Mouth Disease in Pigs. Front Vet Sci 3:41. https://doi.org/10.3389/fvets.2016.00041
Stenfeldt C, Pacheco JM, Brito BP et al (2016b) Transmission of Foot-and-Mouth Disease Virus during the Incubation Period in Pigs. Front Vet Sci 3:105. https://doi.org/10.3389/fvets.2016.00105
Stenfeldt C, Pacheco JM, Rodriguez LL, Arzt J (2014a) Infection dynamics of foot-and-mouth disease virus in pigs using two novel simulated-natural inoculation methods. Res Vet Sci 96:396–405. https://doi.org/10.1016/j.rvsc.2014.01.009
Stenfeldt C, Pacheco JM, Rodriguez LL, Arzt J (2014b) Early Events in the Pathogenesis of Foot-and-Mouth Disease in Pigs; Identification of Oropharyngeal Tonsils as Sites of Primary and Sustained Viral Replication. PLoS One 9:e106859. https://doi.org/10.1371/journal.pone.0106859
Stenfeldt C, Pacheco JM, Smoliga GR et al (2014c) Detection of Foot-and-mouth Disease Virus RNA and Capsid Protein in Lymphoid Tissues of Convalescent Pigs Does Not Indicate Existence of a Carrier State. Transbound Emerg Dis 63:152–164. https://doi.org/10.1111/tbed.12235
Taboga O, Tami C, Carrillo E et al (1997) A large-scale evaluation of peptide vaccines against foot-and-mouth disease: lack of solid protection in cattle and isolation of escape mutants. J Virol 71:2606–14
Thomson G, Vosloo W, Bastos AD (2003) Foot and mouth disease in wildlife. Virus Res 91:145–161. https://doi.org/10.1016/S0168-1702(02)00263-0
Thuenemann EC, Lenzi P, Love AJ, et al (2013) The Use of Transient Expression Systems for the Rapid Production of Virus-like Par- ticles in Plants.
Toka FN, Golde WT (2013) Cell mediated innate responses of cattle and swine are diverse during foot-and-mouth disease virus (FMDV) infection: a unique landscape of innate immunity. Immunol Lett 152:135–43. https://doi.org/10.1016/j.imlet.2013.05.007
Transparency Market Research (2015) In: Foot Mouth Dis. Vaccines Mark. http://www.transparencymarketresearch.com/foot-mouth-disease-vaccines.html. Accessed 1 Mar 2017
Uddowla S, Hollister J, Pacheco JM et al (2012) A safe foot-and-mouth disease vaccine platform with two negative markers for differentiating infected from vaccinated animals. J Virol 86:11675–85. https://doi.org/10.1128/JVI.01254-12
Usha R, Rohll J, Spall V et al (1993) Expression of an animal virus antigenic site on the surface of a plant virus particle.pdf. Virology 197:366–374
Valarcher J-F, Leforban Y, Rweyemamu M et al (2008) Incursions of foot-and-mouth disease virus into Europe between 1985 and 2006. Transbound Emerg Dis 55:14–34. https://doi.org/10.1111/j.1865-1682.2007.01010.x
Van Bekkum J, Frenkel H, Frederiks H, Frenkel S (1959) Observations on the carrier state of cattle exposed to foot- and-mouth disease virus. Tijdschrift voor Diergeneeskd 84:1159–1164
Verdaguer N, Mateu MG, Andreu D et al (1995) Structure of the major antigenic loop of foot-and-mouth disease virus complexed with a neutralizing antibody: direct involvement of the Arg-Gly-Asp motif in the interaction. EMBO J 14:1690–6
Vivek Srinivas VM, Basagoudanavar SH, Hosamani M (2016) IRES mediated expression of viral 3C protease for enhancing the yield of FMDV empty capsids using baculovirus system. Biologicals 4:1–5. https://doi.org/10.1016/j.biologicals.2015.12.002
Wang CY, Chang TY, Walfield AM et al (2002) Effective synthetic peptide vaccine for foot-and-mouth disease in swine. Vaccine 20:2603–2610. https://doi.org/10.1016/S0264-410X(02)00148-2

Wang DM, Zhu JB, Peng M, Zhou P (2008) Induction of a protective antibody response to FMDV in mice following oral immunization with transgenic Stylosanthes spp. as a feedstuff additive. Transgenic Res 17:1163–1170. https://doi.org/10.1007/s11248-008-9188-1

Wang G, Wang Y, Shang Y et al (2015) How foot-and-mouth disease virus receptor mediates foot-and-mouth disease virus infection. Virol J 12:9. https://doi.org/10.1186/s12985-015-0246-z

Wang Y, Shen Q, Jiang Y et al (2012) Immunogenicity of foot-and-mouth disease virus structural polyprotein P1 expressed in transgenic rice. J Virol Methods 181:12–17. https://doi.org/10.1016/j.jviromet.2012.01.004

Wigdorovitz A, Carrillo C, Dus Santos M et al (1999a) Induction of a Protective Antibody Response to Foot and Mouth Disease Virus in Mice Following Oral or Parenteral Immunization with Alfalfa Transgenic Plants Expressing the Viral Structural Protein VP1 1. Virology 255:347–353

Wigdorovitz A, Pérez Filgueira DM, Robertson N et al (1999b) Protection of Mice against Challenge with Foot and Mouth Disease Virus (FMDV) by Immunization with Foliar Extracts from Plants Infected with Recombinant Tobacco Mosaic Virus Expressing the FMDV Structural Protein VP1. Virology 264:85–91. https://doi.org/10.1006/viro.1999.9923

Wu L, Jiang L, Zhou Z et al (2003a) Expression of foot-and-mouth disease virus epitopes in tobacco by a tobacco mosaic virus-based vector. Vaccine 21:4390–4398. https://doi.org/10.1016/S0264-410X(03)00428-6

Wu Q, Moraes MP, Grubman MJ (2003b) Recombinant adenovirus co-expressing capsid proteins of two serotypes of foot-and-mouth disease virus (FMDV): in vitro characterization and induction of neutralizing antibodies against FMDV in swine. Virus Res 93:211–219. https://doi.org/10.1016/S0168-1702(03)00116-3

Xiao Y, Chen H-Y, Wang Y et al (2016) Large-scale production of foot-and-mouth disease virus (serotype Asia1) VLP vaccine in Escherichia coli and protection potency evaluation in cattle. BMC Biotechnol 16:56. https://doi.org/10.1186/s12896-016-0285-6

Yang C-D, Liao J-T, Lai C-Y et al (2007) Induction of protective immunity in swine by recombinant bamboo mosaic virus expressing foot-and-mouth disease virus epitopes. BMC Biotechnol 7:62. https://doi.org/10.1186/1472-6750-7-62

Zhang Y, Li J, Pu H et al (2010) Development of Tobacco necrosis virus A as a vector for efficient and stable expression of FMDV VP1 peptides. Plant Biotechnol J 8:506–523. https://doi.org/10.1111/j.1467-7652.2010.00500.x

Zhang Z, Pan L, Ding Y et al (2015) Efficacy of synthetic peptide candidate vaccines against serotype-A foot-and-mouth disease virus in cattle. Appl Microbiol Biotechnol 99:1389–98. https://doi.org/10.1007/s00253-014-6129-1

Zhang ZD, Kitching RP (2001) The localization of persistent foot and mouth disease virus in the epithelial cells of the soft palate and pharynx. J Comp Pathol 124:89–94. https://doi.org/10.1053/jcpa.2000.0431

Zlotnick A, Stray SJ (2003) How does your virus grow? Understanding and interfering with virus assembly. Trends Biotechnol 21:536–542. https://doi.org/10.1016/j.tibtech.2003.09.012

Part V
Vaccines with Limited Research

Diseases with Limited Research of Plant-Based Vaccines

Ann Meyers

Abstract There are a number of diseases which are important globally in terms of the effect they have on livestock but for which the development of recombinant plant-produced vaccines is preliminary. For many of these diseases such as bovine viral diarrhoea (BVD), bovine rotavirus (BRV), bovine herpes (BoH), transmissible gastroenteritis (TGE) in pigs, infectious bronchitis (IB) in chickens, bluetongue (BT) in sheep, Rift Valley fever (RVF) in sheep and coccidiosis in chickens, commercially available live-attenuated or killed vaccines are available. Although most are effective to varying degrees, there are numerous issues with manufacture and potential reassortment of the vaccine strains. For some diseases such as bovine papillomavirus (BPV) infections and Crimean-Congo haemorrhagic fever (CCHF), there are no commercially available vaccines, and limited studies have been conducted on their development. This chapter discusses some of the research developments in plant-produced vaccine candidates which have potential for further development towards commercialisation.

Keywords Preliminary · BVD · BRV · BoH · TGE · IB · BT
RVF · CCHF · BPV

1 Viral Diseases

1.1 Bovine Viral Diarrhoea (BVD) in Cattle

Bovine viral diarrhoea (BVD) is an infectious disease occurring worldwide. The causative agent Bovine viral diarrhoea virus (BVDV) belongs to the *Pestivirus* genus of the *Flaviviridae* family and affects mainly cattle. It is a spherical enveloped virus with a single-stranded RNA genome encoding at least 4 structural

A. Meyers (✉)
Biopharming Research Unit, Department of Molecular and Cell Biology,
University of Cape Town, Cape Town, South Africa
e-mail: ann.meyers@uct.ac.za

J. MacDonald (ed.), *Prospects of Plant-Based Vaccines in Veterinary Medicine*,
https://doi.org/10.1007/978-3-319-90137-4_16

proteins: the capsid protein (C), E^{rns} (envelope glycoprotein with RNAse) and two external glycoproteins (E1 and E2) (Choi et al. 2004). It is persistent in cattle, causing a range of immunosuppressive symptoms as well as affecting respiratory functions and fertility (Lanyon et al. 2014). This leads to herd morbidity, and subsequent financial impact on farmers and the livestock economy. A particular variant of BVD causes 'mucosal disease' which can be transmitted in utero to the foetus by an antibody-negative mother. Such antibody-negative animals are referred to as persistently infected (PI). PI animals do not show symptoms of BVD but they shed large amounts of virus and their young most often succumb to the disease after making it desirable to terminate any such animal (Brock et al. 2008). Transmission of BVDV occurs vertically and horizontally by aerosol or through the fecal/oral route. The virus usually enters mucosal tissue lining the mouth or nose and replicates in the epithelial tissue. It can then be spread via the blood stream and enters most other types of tissue cells using this route (Lanyon et al. 2014).

Currently used vaccines comprise modified-live viruses (MLV) and killed virus (KV). MLVs are preferred as they give a broader and longer lasting effect but there are disadvantages to using them such as the induction of post vaccinal mucosal disease and immunosuppression (Ridpath 2013). Immunity induced by recommended vaccination programmes at the herd level is effective, but problems with the heterogeneity of the virus leading to genetic drift and subsequent variation in strains have led to the constant demand for new vaccine strains. This in turn has led researchers to look into the development of recombinant vaccines (Fulton 2008). Recombinant vaccines against BVDV have focussed on E2, the major glycoprotein of the virus. E2 has been shown to be immunogenic and induce neutralising antibodies (Ridpath 2013). Recombinant E2, or modifications of it, provide cattle with up to 100% protection against BVDV when produced in animal cells (Pecora et al. 2012; Thomas et al. 2009).

There has also been some work carried out on the development of recombinant vaccine production in plants. Nelson et al. (2012) cloned a truncated version of the immunogenic E2 protein (tE2) in *N. tabacum*. The proteins accumulated up to 20 μg of tE2 per gram of fresh leaves and serum from guinea pigs immunised subcutaneously with 20 μg doses mixed with adjuvant showed the presence of neutralizing antibodies. In addition, Aguirreburualde et al. (2013) made transgenic alfalfa plants which produced a recombinant fusion protein [tE2 fused to a single chain variable fragment (scFv)] that targets proteins to antigen presenting cells. Protein yields of up to 1 μg/g fresh weight were reported. Immunogenicity evaluation of serum from guinea pigs immunised intramuscularly with leaf extracts containing 0.2–0.4 μg recombinant E2 showed the induction of a strong neutralising antibody response. In addition, cattle immunised intramuscularly with 3 μg doses of the product also showed a neutralising antibody response and subsequent complete protection against infection when challenged with live virus. Interestingly, Gao et al. (2015) have developed transgenic ginseng hairy root cultures which express BVDV glycoprotein $E2^{rns}$. Sika deer were immunised subcutaneously with transgenic hairy root extracts and serum was shown to contain E2-specific antibodies. There was also a specific cell-mediated response induced as measured by an increase in phytohaemagglutinin-induced lymphocyte proliferation.

1.2 Bovine Papillomavirus (BPV) in Cattle

Bovine papillomavirus is caused by a group of double-stranded DNA bovine papillomaviruses (BPV) belonging to the family *Papillomaviridae* of which there are 13 types described to date (Munday 2014). The DNA encodes 3 viral oncoproteins (E5, E6 and E7) as well as the two structural proteins L1 and L2 which form the virion capsid (Chen et al. 1982). Generally, BPVs have been shown to be specific to bovines, but BPV-1 BPV-2 and BTV-13 have been reported to infect equids as well as bovines (Hamad et al. 2016). The virus is endemic worldwide, causing benign or malignant tumours in infected mucosa or squamous epithelium, or in the mesenchyme of embryos. Such symptoms can lead to loss of milk production, infertility, and immunosuppression, causing a decline in general animal health (Love et al. 2012). BPV is spread between cattle by direct or indirect contact and can also be transmitted by insects. Infection is thought to occur with microtrauma and subsequent entry into various tissues including cutaneous and mucosal epithelium with subsequent disruption of the cell life cycle effected by the BPV E5 protein (Munday 2014).

In 1995, Campo et al. (1995) showed that prophylactic vaccination of animals with live BPV, stimulated the production of neutralising antibodies and provided complete protection against live virus challenge. The same group of researchers also showed that subunit vaccines comprising L1 and L2 antigens of BVP-2 provided effective prophylaxis and resulted in production of neutralising antibodies and protection. Another research group describes the vaccination of calves which induced virus neutralising antibodies and their subsequent protection from infection when challenged (Borzacchiello and Roperto 2008). The co-expression of L1 and L2 or L1 alone in mammalian cells results in the assembly of virus-like particles (VLPs) which are immunogenic, stimulating neutralising antibodies and providing protection (Kirnbauer et al. 1992). Despite this, there is no commercially available vaccine against BPV as of yet.

The production of such a recombinant vaccine in plants would be more appealing as it abrogates the possibilities of contamination with other animal derived proteins as high sterility tissue culture conditions are required. Love et al. (2012) have tested the production of BPV VLPs in plants and shown that similar L1 VLPs can be expressed transiently in *N. benthamiana* on infiltration with *Agrobacterium* harbouring BPV-1 cloned into the pEAQ-HT plant expression vector. Yields were reported to be up to 183 mg/kg fresh weight. Rabbits immunised with 150 μg of L1 VLPs mixed with Freund's incomplete adjuvant produced L1-specific IgG.

1.3 Bovine Rotavirus (BRV) Group A Gastroenteritis

Rotavirus A infection of cattle is caused by a double stranded RNA virus belonging to the *Rotavirus* genus, family *Reoviridae* and occurs worldwide. It causes acute gastroenteritis and diarrhoea, particularly in young animals which are 1–3 weeks old, which progresses to increases in morbidity with severe enteritis and mortality as well as a reduction in growth rate (Dhama et al. 2009). Together with extensive treatment costs, these factors have lead to serious economic losses for livestock owners and affected countries. Although clinical signs do not last long, permanent outbreaks during certain seasons often occur as a result of viral shedding which can last for up to 3 weeks after infection. Virions are made up of an inner core mostly of the VP2 capsid protein, a middle layer comprised of the most abundant protein VP6 and 2 other proteins VP7 and VP4 forming the outer layer. The major capsid protein VP6 is highly immunogenic and contains the common antigens for typing serogroups of which there are 7 (Papp et al. 2013). The virion is very stable in the environment and is transmitted by the faecal-oral route and spread by viral shedding through faeces and water; very low doses are required for infectivity.

Most adult cattle populations are generally BRV positive, and naturally occurring passive immunisation of newborn calves is common by oral ingestion of colostrum and milk (Saif and Fernandez 1996). However, in modern times, intensive farming methods, exposure of animals to high concentrations of virus, early weaning and supplementation of feed with calf concentrates which dilutes milk antibodies has influenced and reduced the extent of immunity of the herd, leading to the requirement for vaccines. There are currently commercially available BRV maternal vaccines on the market which allow for the parenteral immunisation of females, which subsequently facilitate passive immunity in newborn calves. These vaccines comprise of attenuated-live viral strains or inactivated virus which have been shown to stimulate increased antibody production in mammary secretions of vaccinated females, which protects feeding calves from BRV challenge. However, limitations to attenuated-live vaccines such as the presence of adventitious agents and the gradual emergence of different serotypes as well as to inactivated vaccines such as alteration of the immunogenic form, make the generation of recombinant vaccines more favourable. There has also been some development in generating recombinant BRV vaccines in the form of VLPs. In 1991, Labbé et al. (1991) were able to produce rotavirus VLPs consisting of the VP2 and VP6 capsid proteins in a baculovirus expression system.

The production of BRV VLPs in plants seems to have yielded greater success and there are several examples of research into developing recombinant plant-produced BRV vaccines (Hammond and Nemchinov 2009; Ruiz et al. 2015). In 2005 Dong et al. made recombinant VP6 in transgenic alfalfa and orally gavaged mice with 24 μg purified VP6 adjuvanted with CpG (Dong et al. 2005). Serum from treated mice was shown to have anti-VP6 IgG antibodies and anti-VP6 IgA antibodies isolated from homogenised small intestine, faeces and saliva. Interestingly, pups born from the immunised mice and subsequently challenged with live rotavirus showed reduced symptoms of diarrhoea, and reduced intensity and duration of diarrhoea indicating that immunity had been passively transferred to the pups.

In another study recombinant rotavirus capsid protein VP6 was expressed in *Chenopodium amaranticolor* as a result of infection with engineered beet black scorch virus (BBSV) (Zhou et al. 2010). Partially purified doses of plant leaf extract containing 25 μg VP6 and CpG adjuvant was used to gavage mice and analysis of their mucosal IgA and IgG titres showed high titres when compared to control mice. Pups born from these immunised mice were challenged with rotavirus and showed a significant reduction in incidence, severity and duration of diarrhoea indicating that the plant-produced VP6 provided passive protection against the rotavirus.

In 2006, Saldaña et al. described the production of VP2/VP6 VLPs in transgenic tomato (*Lycopersicon esculentum* L.) fruits albeit fairly low (Saldaña et al. 2006). Serum from mice immunised with lyophilised tomato extract containing 1 μg of rotavirus proteins and Freund's adjuvant was shown to contain antibodies against the proteins but no protection studies were carried out. Yang et al. (2011) co-expressed rotavirus VP2/VP6 or VP2/VP6/VP7 in transgenic tobacco plants. All proteins were expressed in leaves and both combinations of proteins resulted in the assembly of VLPs. Oral delivery of semi-purified preparations of VLPs administered with CT as an adjuvant showed that although both vaccine candidates induced raised titres of rotavirus-specific antibodies (IgA and IgG), those mice immunised with VP2/VP6/VP7 VLPs had higher titres than VP2/VP6 VLPs.

Lentz et al. approached the rotavirus vaccine differently by testing the expression of VP8* (Lentz et al. 2011). VP4 is one of the outermost capsid proteins and on infection this is cleaved to produce VP8*—a N-terminal non-glycosylated sialic acid-recognising domain—and VP5—a C-terminal fragment which remains associated with the virion. VP8* is a major determinant of viral infectivity and one of the neutralising antigens. VP8* was expressed in transplastomic tobacco leaves and shown to form insoluble aggregates. Both soluble and insoluble fractions of crude preparations containing 2 μg of VP8* and Marcol adjuvant were used to immunise mice intraperitoneally. VP8* IgG antibody titres from mice immunised with both samples were shown to be induced. In addition, both immunogens were shown to induce virus neutralising antibodies. They were also shown to passively immunise their offspring as demonstrated by the presence of high levels of IgG antibodies in pups. Eighty to 100% of the challenged pups were protected from rotavirus challenge as determined by the absence of diarrhoea.

Wigdorovitz et al. (2004) report the expression of an immunogenic peptide having a neutralising epitope originating from VP4—eBRV4—in transgenic alfalfa. This was fused to ß-glucuronidase serving as a carrier molecule which, when injected intraperitoneally with 3 doses into mice as a crude extract adjuvanted with Freund's incomplete adjuvant containing 0.5 μg eBRV4, was shown to induce rotavirus-specific antibodies. Moreover, when fresh leaves containing 6 μg eBRV4 were fed orally to mice once a week 8 times, rotavirus-specific antibodies were shown to be induced and mice were protected from oral virus challenge.

Matsamura et al. (2002) have reported the expression of rotavirus capsid protein VP6 in transgenic potato tubers. Intraperioneal immunisation of mice with 2 doses of adjuvanted potato tuber extract containing 750 ng each of VP6 showed that anti-VP6-specific antibodies were stimulated.

These very positive results all contribute to the feasibility of producing subunit rotavirus vaccines in plants.

1.4 Bovine Herpes (BoH) in Cattle

Bovine herpes virus 1 (BoHV-1) causes a worldwide disorder in cattle referred to as infectious bovine rhinotracheitis. BHV is a double-stranded DNA virus belonging to the *Varicellovirus* genus of the *Herpesviridae* family. It causes an extensive range of symptoms including rhinotracheitis, infectious pustular vulvovaginitis, enteritis, general respiratory disease, encephalitis, decreased milk production, weight loss and abortion in pregnant cows which is why it is of great concern with regard to causing severe economic losses (Graham 2013). It affects both adult and young animals. BoH infects animals through the mucous membranes of either the upper respiratory or genital tract (Muylkens et al. 2007). Transmission between animals requires direct contact. The BoH virion encodes 10 glycoproteins, of which glycoprotein D (gD) is responsible for permissive host cell receptor binding and this and 3 others are responsible for cell entry.

Current vaccines employed to combat this virus consist of inactivated or modified live virus (van Drunen 2006). It has been shown that these vaccines are efficacious in reducing symptoms of infection after challenge, but the challenge virus can remain latent in host tissue and be re-activated and subsequently excreted, thereby not eradicating the infection in the herd. In addition, the efficacy of these types of vaccines in neonates is not as efficacious as in adult animals due to the functional immaturity of their immune system and difficulty in particular in mounting a cellular response against BoH which is essential for immunity. This which has led to the development of recombinant vaccines. Subunit vaccines are comprised of glycoprotein antigens as these have been shown to be excellent targets for neutralising antibodies (Alves Dummer et al. 2014)—and have been expressed in baculovirus, adenovirus and mammalian cell culture systems (Muylkens et al. 2007).

Perez-Filguiera et al. (2003) have expressed glycoprotein D in *N. benthamiana* from a TMV-based expression vector. Mice immunised with approximately 4 μg crude leaf extract mixed with IFA showed specific humoral responses to glycoprotein D after a single dose. Cattle, which were immunised with approximately 100 μg crude leaf extract also showed a specific antibody response after 2 vaccinations. In addition analysis of lymphocytes from PBMCs indicated positive proliferation indices suggesting that the cellular immune response was also stimulated by this vaccine. Secreted nasal fluids from immunised animals which were subsequently challenged with live virus showed that there were similar amounts of virus in that of the mice, but that the period of virus shedding in the cattle was by up to 5 days indicating that the vaccine was protective.

1.5 Rinderpest in Cattle

Rinderpest, otherwise known as cattle plague, is caused by a single-stranded RNA *Morbillivirus* from the *Paramyxoviridae* family. The RNA encodes the phosphoprotein (P), nucleoprotein (N) and polymerase (L) which are found on the inside of the virion as well as the matrix protein (M) which forms the outer capsid and the haemagglutinin neuraminidase (HN) and fusion proteins (F) which form the outer envelope of the virion (Martin 1986). The virus is highly communicable and is spread by direct contact, contaminated drinking water and sometimes by aerosol. It affects mammals including man, but affects mostly ungulates. Symptoms are characterised by fever, ocular and nasal discharges and morbidity and mortality rates are high as these primary symptoms lead on to oral and gastrointestinal tract ulceration, dysentery, diarrhoea, dehydration, and lymphocyte depletion which causes protein loss and immunosuppression (Roeder et al. 2013). It is thought that rinderpest originated in Central or South Asia and as a result of human activity, spread to Europe and Africa. Rinderpest virus (RV) replicates very quickly and is inactivated quite easily by heat or direct sunlight. After infection, it targets the lymphatic system as well as the epithelial cells of the respiratory system and gastrointestinal tract (OIE 2008a).The virus moves very quickly and symptoms can be seen as soon as three days post infection.

The first vaccine developed against rinderpest was an attenuated virus cultured in bovine kidney cells (tissue culture rinderpest vaccine—TCRV) and was extremely successful in combatting all clades of the virus with lifelong immunity to cattle with only a single dose (Plowright and Ferris 1962). Subsequent to that, a more temperature resistant vaccine ThermoVax was developed, and this was used to finally eradicate rinderpest by 2011 (Roeder et al. 2013). Leading up to the declaration of final global eradication in 2011, there was some development of recombinant vaccines against rinderpest. The envelope proteins H (haemagglutinin) and F (fusion) have been expressed in cattle immunised with a disabled human adenovirus vector Ad 5. These vectors have been shown to result in the expression of rinderpest-specific proteins which stimulated the production of neutralising antibodies and subsequently the protection of cattle from infection when challenged with live virus (Cosby and Yamanouchi 2006).

Prior to the declaration of final eradication in 2011, Khandelwal et al. described the generation of transgenic peanut plants (*Arachis hypogea* L.) expressing the RPV H protein (Khandelwal et al. 2004). Mice immunised intraperitoneally with 10 μg H together with IFA elicited H-specific antibodies which neutralised virus in vitro. Oral immunisation of mice elicited H-specific IgG and IgA antibodies. The same group carried out oral immunisation of cattle with transgenic peanut leaves expressing H protein. Serum from immunised cattle was shown to contain H-specific antibodies as well as neutralising and cross-neutralising activity (Khandelwal et al. 2003).

Around the same time, Sathyavathi et al. (2003) published a report on the expression of RPV H protein in transgenic pigeon pea plants (*Cajanus cajan* (L.)

Millsp.), yielding levels of 0.1–0.49% of total soluble protein. However, no immunogenicity tests were carried out.

1.6 Transmissible Gastroenteritis (TGE) in Pigs

Transmissible gastroenteritis is a highly contagious disease causing severe acute diarrhoea in newborn piglets which results in very high mortality rates of less than two-week old animals (Saif et al. 1994). It is caused by transmissible gastroenteritis virus (TGEV), a single-stranded RNA virus which belongs to the *Coronavirus* genus of the family *Coronaviridae*. It is widespread, causing severe economic losses to pig farmers. The virus infects the enterocytes of the small intestine where it multiplies and causes enteritis (OIE 2008b). The M protein in an integral membrane protein and the gS (spike) glycoprotein occurs on the surface of the virion (Laude et al. 1990).

The current commercially available vaccine used to treat TGEV is a modified live virus. The gS glycoprotein is responsible for inducing neutralising antibodies. However, further development has progressed on recombinant vaccines produced in mammalian expression vectors which have been shown to promote systemic and mucosal immunity as well as passive immunity to suckling piglets (Hammond and Nemchinov 2009).

There are also several examples of plant-produced proteins which have been tested as vaccines against TGEV. Gómez et al. (1998) made transgenic *Arabidopsis* encoding the full-length gS protein or the N-terminal domain of gS shown to be neutralising (N-gS). Serum from mice inoculated with these proteins showed virus-specific neutralisation and immunoprecipitation in vitro. Gómez et al. (2000) also made transgenic potato plants expressing N-gS. Tuber extracts inoculated into mice induced TGEV-specific IgG. Oral immunisation of mice with the same antigen was also shown to stimulate serum antibodies specific to gS. Tuboly et al. (2000) similarly expressed truncated TGEV protein regions expressed in transgenic tobacco and immunised pigs which showed TGEV-specific immune responses. Lamphear et al. (2002, 2004) describe the production of an oral TGEV vaccine candidate produced in corn which when milled and fed to pigs, boosted antibody levels in their serum, colostrum and milk.

1.7 Infectious Bronchitis (IB) in Chickens

Infectious bronchitis in young chickens is an acute and highly contagious respiratory disease caused by infectious bronchitis virus (IBV), a gamma *Coronavirus* belonging to family *Coronaviridae* (Jackwood 2012). Apart from affecting the upper respiratory tract, it can also affect the reproductive tract, and some strains can cause nephritis. Infections cause weight loss, decreased egg production, general poor performance of flocks and high morbidity (Bande et al. 2015). Inhaled IBV infects the respiratory tissues and replicates mainly in the upper respiratory tract (Raj and Jones 1997). The virus is then disseminated to other tissues and is epitheliotropic, entering the epithelial cells of primarily respiratory organ tissue but can move on to reproductive regions, kidneys and the intestine, causing permanent damage and dysfunction of the infected organs. IBV consists of 3 major structural proteins, S (spike), M (membrane) and N (nucleoprotein). The S protein is cleaved into the S1 and S2 forms which associate to form spikes on the virus surface. This is the protein which has been shown to be immunogenic.

IB is typically controlled with serotype-specific vaccines which are either live-attenuated or killed (Bande et al. 2015). But the problem with IBV is that there are more than 20 different serotypes, and even more variants, and there is little cross-protection between the vaccines. The constant evolution of new variants can cause the disappearance of others, requiring a constant demand for new serotype specific vaccines. The potential risk of reversion to virulence by live-attenuated vaccine strains and the weaker immune response and subsequent requirement for multiple dosing schedules has led to some development of recombinant vaccines involving expression of the S protein as this has been shown to have stimulate virus neutralisation (Cavanagh et al. 1986). Some research being carried out on the production of recombinant vaccines (Bande et al. 2015) include S1 protein production using a baculovirus expression system (Song et al. 1998) as well as a recombinant fowl adenovirus (Johnson et al. 2003) that has been shown to stimulate virus-neutralising antibodies and provide protective immunity in chickens.

Zhou et al. (2004) have shown that 2.5–5 μg transgenic potato tuber extract expressing the full length IBV S1 protein induced the production of virus neutralising antibodies and provided up to 85% protection in chickens when challenged with live IBV. In addition, cell mediated immunity which is considered to contribute to protective immunity against IBV was also stimulated, as measured by T-cell proliferations and an increase in chIL-2 levels.

1.8 Bluetongue (BT) Disease in Sheep

Bluetongue disease is caused by the Bluetongue virus (BTV) belonging to the *Orbivirus* genus of the *Reoviridae* family (Sperlova and Zendulkova 2011). It is a double-stranded RNA virus with the virion having a double capsid composed of 3 protein shells: the inner layer composed of capsid protein VP3, the intermediate layer composed of capsid protein VP5 and the outer layer composed of capsid protein VP2 and VP7 (Mertens et al. 2004). It is a non-contagious, infectious disease affecting ruminants and camelids and is transmitted by midges of the *Culicoides* genus. Up until 2014, 26 different serotypes had been recorded (Maan et al. 2012) but a 27th serotype has more recently been identified and sequenced (Jenckel et al. 2015). The presence of BTV was first recorded in South Africa at the end of the 18th century. Traditionally it has been known to be present in a region spanning 40°N and 35°S which corresponded to the geographical distribution of the midges responsible for its transmission (Carpenter et al. 2008). It has become much more of a concern recently however, due to global warming which is thought to have favoured the survival of the midges over the colder winter periods. Thus the current distribution of BTV outbreaks has recently shown to extend beyond the previously recorded region and has been predicted to broaden in Africa, Russia and the United states with future-climate predictions (Samy and Peterson 2016). Bluetongue affects sheep the most with acute, chronic or subclinical conditions. Clinical signs are usually lameness, swelling, infertility and in severe cases, death. Goats are less frequently infected, rarely showing clinical signs of the disease although if they do, symptoms are less severe. Cattle act more as reservoir hosts and clinical signs of infection are rare, except those infected with BTV serotype 8 which has been reported to have caused elevated numbers of morbidity, fertility and mortality (Elbers et al. 2009). Virus is introduced into animals through the bite of an infected midge and directed to the lymph nodes which are the site of initial viral replication. Virus is then circulated in the blood, infecting secondary organs including additional lymph nodes, the spleen and lungs. Pathogenesis is characterised by small blood vessel injury to the target tissues which leads to increased vascular permeability and subsequent oedema and effusions (Maclachlan et al. 2009).

There are currently two types of vaccines available for prophylactic immunisation of animals: live attenuated vaccines and inactivated vaccines. The vaccines are serotype-specific so it is important to establish which serotype is responsible for circulation in a specific area. Until recently, live attenuated vaccines were the only ones available, and multivalent preparations are currently still used in regions such as South Africa. Although these are effective, they are temperature sensitive and show poor protection against heterologous BTV serotypes. More concerning however, is the fact that these vaccines can result in negative clinical signs such as bluetongue, abortion, a reduction in milk production and foetal malformation in pregnant ewes (Bhanuprakash et al. 2009). There is also the possibility of reversion to virulence or viral reassortment which could generate a more virulent strain virus (Sperlova and Zendulkova 2011). Inactivated vaccines are reliable and protective and prevent clinical disease from developing but are fairly expensive to produce and animals require re-vaccination. They have been predominantly used in various outbreaks experienced in Europe (Zientara et al. 2010).

Recombinant vaccines could help abrogate the use of live inactivated virus and provide a less expensive alternative to killed vaccines. Such vaccines include recombinant viral vector vaccines, subunit vaccines and virus-like particles (VLPs). French et al. (1990) and Roy (1992) were able to show that co-expression of the 4 BTV capsid proteins in insect cells using a baculovirus expression system resulted in the production of assembled VLPs, with the major immunogenic determinant VP2 being presented on the outer shell. Inoculation of animals with these and those representing other BTV serotypes showed that they were immunogenic and protective (Stewart et al. 2012) against BTV challenge. However, this product is not very scalable and is fairly costly to implement.

There has been some progress with developing the production of a BTV serotype 8 VLP vaccine in plants using transient expression in *N. benthamiana* (Thuenemann et al. 2013; van Zyl et al. 2016). The 4 BTV capsid proteins were cloned into a pEAQ-HT expression vector (Sainsbury et al. 2009) and co-infiltrated into *N. benthamiana* leaves using recombinant *Agrobacterium*. All four proteins were expressed and TEM analysis of purified extracts from harvested leaves showed that all 4 proteins assembled into VLPs. It was estimated that yields were approximately 70 mg VLPs per kg fresh leaf weight. Purified preparations of the VLPs at 50 μg per dose combined with IFA injected into sheep with a second booster dose at 28 days post initial injection induced a strong immune response and provided protective immunity in sheep challenged with a BTV serotype 8 strain. However, this method of purification of VLPs (centrifugation) is not very scalable.

1.9 Rift Valley Fever (RVF) in Ruminants

Rift Valley Fever (RVF) is a zoonotic infectious disease which primarily affects ruminants. It is caused by Rift Valley fever virus (RVFV) which was first identified in 1930 in the Rift Valley in Kenya when it caused an outbreak of disease in sheep. RVFV is an enveloped negative-stranded RNA virus with a tripartite genome belonging to the *Phlebovirus* genus in the *Bunyaviridae* family (Pepin et al. 2010). It has a lipid bilayer composed of glycoproteins G_n and G_c, encapsidating the RNA-associated ribonucleoproteins (RNPs) and RNA polymerase (L). The virus was restricted to sub-Saharan Africa prior to the middle 1970s but has now become endemic to parts of Africa and the Arabian Peninsula and is considered an emerging virus. (Pepin et al. 2010). Symptoms of infection in ruminants vary between different animal species but include the occurrence of near simultaneous abortions in pregnant animals, high neonatal mortality, hepatic damage and deformed young. RVFV is predominantly spread by mosquitoes although it can be harboured by a few other vectors such as ticks and sandflies. Infections in animals occur predominantly as a result of bites from mosquitoes carrying the virus but can also be spread by direct contact with infected animal tissues and fluids. This is the main route by which humans are infected, severe cases of which lead to jaundice, neurological disease or haemorrhagic complications and possible fatalities (Pepin et al. 2010).

The spread of RVFV can be prevented by vaccination of animals and humans but development of safe and effective vaccines has been difficult (Bouloy and Flick 2009; Ikegami and Makino 2009). Live attenuated vaccines such as the Smithburn strain have been used extensively for livestock vaccination programmes, but still cause undesirable symptoms in pregnant animals and could easily revert or mutate to a more virulent strain. Research has been carried out using inactivated viral strains but these are expensive to produce and require multiple doses and boosters for immunity to be maintained in animals. The formalin-inactivated vaccine TSI-GSD-200 is only used for veterinary workers in endemic regions, high containment lab workers and others at high risk for contracting RVFV. It is not commercially available, but is used for veterinarians who work in areas endemic to RVFV as well as for others who are at high risk of contracting the disease including lab workers in high containment areas. Like the other live attenuated vaccines, this vaccine is expensive, difficult to make and requires multiple doses and boosters for efficacious immunity, making routine immunisation of animals impractical (Bouloy and Flick 2009).

The RVFV glycoproteins G_c and G_n, have been shown to be the antigenic determinants for stimulating neutralising antibodies which provide immunity to the disease. This discovery has enabled several recombinant vaccine candidates to be developed. Expression vectors such as the Venezuelan equine encephalitis (VEEV) vector and the lumpy skin disease virus (LSDV) vector (Wallace et al. 2006) have been used to generate RVFV glycoproteins which have elicited RVFV-specific immune responses in and provided protection against RFVF challenge in animals.

RVFV virus like particles (VLPs) are also a possible candidate for vaccination against RVF as they are stable, they may be more immunogenic than recombinant proteins alone and they maintain their conformational epitopes which induce neutralising antibodies. By expressing RVFV nucleocapsid (N) and glycoproteins (G_n-G_c) together in a dual baculovirus expression vector, (Liu et al. 2008) showed that enveloped VLPs resembling wildtype RVFV virions could be produced. VLPs were also produced by the dual baculovirus vector system with N and G_c alone. These VLPs were more pleomorphic than the VLPs comprised of both glycoproteins. Habjan et al (2009) have produced RVFV VLPs in mammalian cells by co-expressing recombinant RVFV polymerase and nucleocapsid protein together with a minireplicon RNA and additional expression of the viral glycoproteins (Habjan et al. 2009). These have been shown to protect mice from a lethal challenge of RVFV (Näslund et al. 2009).

Interestingly, De Boer et al. (2010) have shown that RVFV VLPs can be produced in a *Drosophila* insect cell system expressing the G_n and G_c proteins alone. Furthermore, these VLPs have been shown to provide 100% protection of mice when challenged with wildtype virus. This type of vaccine lacking the N protein is desirable as this can help with distinguishing between infected and vaccinated animals (DIVA) using diagnostic kits. De Boer et al. also tested the effect of soluble G_n in vaccinated mice and were able to show that full protection was afforded from lethal challenge with RVFV.

Some preliminary studies have been carried out on the immunogenicity of plant-produced RVFV antigens. Kalbina et al. made transgenic *Arabidopsis* plants expressing RVFV G_n (deletion mutant) and RVFV N (Kalbina et al. 2016). Leaves containing these proteins were fed to mice orally and mouse serum was shown to have elevated titres of antigen-specific IgG, suggesting that they are immunogenic.

1.10 Crimean-Congo Haemorrhagic Fever (CCHFV)

CCHF is a zoonotic disease caused by Crimean-Congo haemorrhagic fever virus (CCHFV) belonging to the genus *Nairovirus,* family *Bunyaviridae* (Whitehouse 2004). Similar to RVFV, it is an enveloped RNA virus with a lipid bilayer of glycoproteins G_n and G_c encapsidating the RNA-associated ribonucleoproteins (RNPs) and polymerase (L). CCHF is a highly contagious disease which infects a large variety of vertebrates including sheep, goats, cattle, large wild herbivores, hares and hedgehogs (Bente et al. 2013). There are also numerous bird species that have been shown to have antibodies to CCHFV although they are refractory to infection, as well as one instance of CCHFV antibodies found in a tortoise (reptile). However these animals generally show no symptoms of the disease, but develop sufficient viraemia to support transmission of the virus to uninfected ticks which then bite and infect humans. The distribution of the disease tends to follow the geographical range of this vector. Humans are the only host of CCHFV in which disease is manifested except for newborn mice. Humans also become infected by handling crushed, infected ticks and by direct contact with infected blood or tissue of animals or humans harbouring the virus. The disease is endemic in more than 30 countries in Africa, Asia, southeast Europe and the Middle East and it has recently emerged in areas previously free from the disease such as Turkey. CCHF is a notifiable disease as outbreaks have epidemic potential which constitutes a public health threat, they have a high fatality ratio, nosocomial outbreaks are extremely prevalent, the virus is potentially a bioterrorism agent and treatment and prevention of the disease is difficult. Ticks harbouring CCHFV bite animals and humans and the virus is transported in the bloodstream and absorbed into permissive cells (Bente et al. 2013).

There are currently no commercial vaccines available against CCHFV. However, due to the desire for a vaccine which is universally acceptable for use, and that can be used to inoculate animals which are the main carriers of the virus and responsible for the disease outbreaks and thereby reducing the CCHFV numbers within their populations, some progress has been made in their development.

Initially, a formalin-inactivated vaccine which was developed in Russia in the 1960s from suckling mouse brains for humans (Hoogstraal 1979). It was shown to induce neutralising antibodies in humans. This vaccine was licensed in Bulgaria for prophylactic use on people who work in army units, medical workers, agricultural workers and people living in CCHF endemic regions. Use of this vaccine is reported to have reduced CCHF cases in Bulgaria (Christova et al. 2010) however, this vaccine is not used in any other country in Europe due to concerns that its production in mouse brains may induce autoimmune or allergic responses in humans. In addition, it requires maximum containment facilities to generate virions for inactivation, and this is costly. Another type of treatment used in Russia, Bulgaria and several other countries is administration of anti-CCHF immunoglobulin derived from convalescent serum by intramuscular injection or intravenously (Keshtkar-Jahromi et al. 2011). CCHF patients treated with convalescent serum recovered from their illness.

Despite these measures of vaccination and treatment, there is currently no universally accepted method of prophylaxis against CCHF and not much work done on the prevention of CCHFV in animals. Development of alternative vaccines has been hampered by the lack of information on immunogenic epitopes of CCHFV as well as the lack of a CCHF animal model to test efficacy. The use of newborn mice which are naturally affected as models are questionable, as they are very susceptible to a wide array of pathogens and may not be that useful as a disease model. However, recent research in the USA has yielded 2 animal models which have the potential for use in testing vaccines. These models are both knockout mice models which are sensitive to CCHFV infection (Keshtkar-Jahromi et al. 2011) and should encourage further development in the CCHFV vaccine research field.

Very little research on alternative vaccine platforms has been published. Zhou et al. (2011) report the production of CCHF N protein virus like particles (VLPs) in a baculovirus-insect cell expression system with the intention of testing their immunogenicity in mice. Spik et al. (2006) have shown that a DNA vaccine containing the CCHF M segment (encoding G_c and G_n) elicits neutralizing antibodies in vaccinated mice as well as antibodies that are able to react with G_n and G_c proteins.

A small amount of work has been conducted on the production of CCHFV vaccine candidates in plants. Ghiasi et al. (2012) report the production of a CCHFV glycoprotein (G_nG_c) in transgenic tobacco plants. Serum from Balb/C mice which were orally immunised with leaves or roots containing 5–20 μg of recombinant glycoprotein combined with Freund's adjuvant showed an increase in anti-glycoprotein antibodies, although these are not ideal models for the disease.

1.11 Rabbit Haemorrhagic Fever (RHF)

Rabbit haemorrhagic fever is a lethal disease of rabbits caused by the virus rabbit haemorrhagic disease virus, a member of the *Lagovirus* genus of the *Caliciviridae* family (Abrantes et al. 2012). It is a non-enveloped RNA virus having an outer capsid made up of the major structural protein VP60 which encapsidates several non-structural proteins including RND dependent polymerase, helicase and a protease. The virions are extremely resistant and stable in the environment and cause acute necrotising hepatitis, as well as haemorrhaging in other organs. Infection causes death within 48–72 h of necrotising hepatitis. Outbreaks of this disease consequently have a severe effect on the rabbit meat and fur industry. It enters through the oral, nasal or conjunctival routes in animals and usually infects host cells through the upper respiratory or digestive tracts (Abrantes et al. 2012).

Currently, rabbits can be vaccinated using formalin-inactivated liver homogenates of liver-infected individuals. RHV capsid protein VP60 has been shown to be immunogenic and injection with recombinant baculovirus-produced VP60 or

VP60 fusion proteins has been shown to protect immunised rabbits from lethal challenge with the virus (Hammond and Nemchinov 2009).

Fernandez et al. showed that the VP60 polypeptide could be produced in *Nicotiana clevelandii* using a plum pox potyvirus (PPV) based vector (Fernández-Fernández et al. 2001). Rabbits injected subcutaneously with leaf extracts containing recombinant VP60 showed an immune response and were protected when intranasally challenged with RHDV. Castanon et al. (1999) made transgenic potato plants expressing VP60 and immunised rabbits parenterally with leaf extract containing 12 μg recombinant VP60 mixed with Freund's adjuvant. High anti-VP60 titres were measured in the rabbit serum and rabbits were fully protected when challenged with live virus. In addition, Martin-Alonso et al. (2003) showed that VP60 could be expressed in transgenic potato tubers yielding up to 3.5 μg VP60 per mg total soluble protein (TSP). Oral immunisation of rabbits with lyophilised potato extract containing up to 500 μg per dose was only partially effective in inducing immunity and protection when rabbits were challenged with live virus.

2 Parasitic Diseases

2.1 Coccidiosis in Chickens

Coccidiosis is a disease of chickens caused by protozoan parasites which belong to the *Eimeria* genus (Shirley et al. 2005). It is a diarrhoeal disease, causing large poultry losses, particularly in the US. Transmission of the parasite is via the oral-faecal route. Oocysts of *Eimeria* are ingested by chickens and infect the gut where they undergo asexual and sexual reproductive phases. Developing oocysts are shed in the faeces, whereupon they undergo meiosis on contact with oxygen and moisture, and are then re-ingested by animals for sporozoite release in the intestine (Shirley et al. 2005).

Interestingly, infection with *Eimeria* has been shown to lead to lifelong immunity to the particular species of *Eimeria* infecting the host. Hence, commercially available live attenuated vaccines comprising specific *Eimeria* genera have been developed and are routinely used in chick hatcheries (Price 2012), although they are not cross-protective (Shirley et al. 2005). However, efficacy, stability issues, quality control and the cost effectiveness of making live attenuated vaccine strains have led to some developments in recombinant Eimeriid vaccines. The highly complex nature of the Eimeriid life-cycle and the difficulty in identification of specific antigens that protect against infection as well as the method of delivery of the vaccine have presented problems however with the development of recombinant vaccines (Shirley et al. 2007). To date, the most studied recombinant vaccine antigens have been those associated with the most motile and functionally important parasite cycle stages. Proteins associated with the subcellular organelles micronemes, EtMIC1 and EtMIC2 are some of the most extensively studied (Shivaramaiah et al. 2014).

Sathish et al. (2011) have successfully transiently expressed the *E. tenella* microneme (MIC) protein EtMIC2 in *N. tabacum*. Immunisation of chicks with 50 μg of adjuvanted EtMIC2 with one primary and 2 boosts 2 weeks apart showed the stimulation of high serum-specific IgG titres as well as induction of a specific IFN-Y response. Immunised birds were also challenged and showed an increase in weight, compared to control birds. The same group have also produced recombinant EtMIC1 and tested its ability to stimulate an immune response in birds as a monovalent vaccine as well as in combination with plant-produced EtMIC2 (Sathish et al. 2012).

Zimmerman et al. (2009) describe the development of a plant-made recombinant antibody to be used against coccidiosis. They identified an anti-*Eimeria* scFv from a mouse phage display library when panning with proteins extracted from various developmental stages of the parasite. The scFv-encoding gene was cloned into a vector for the generation of transgenic pea plants (*Vicia faba*—Fodder pea). Pea seeds from the transgenic plants were dried, ground up and force-fed to chickens infected with *Eimeria* containing 1 mg scFv in pea seed flour doses.

3 Conclusion

There are several vaccine candidates targeting domestic animals that could potentially be produced in plants. Most candidates discussed in this chapter were successful in not only producing neutralising antibodies in vaccinated animals but protecting animals challenged with the corresponding virus, suggesting that the plant expression route could have potential for production of vaccines, thereby circumventing the biosafety levels required for live virus handling as well as other problems such as reassortment.

References

Abrantes J, van der Loo W, Le Pendu J, Esteves PJ (2012) Rabbit haemorrhagic disease (RHD) and rabbit haemorrhagic disease virus (RHDV): a review. Vet Res 43(1):12–12. https://doi.org/10.1186/1297-9716-43-12

Aguirreburualde MSP et al (2013) Efficacy of a BVDV subunit vaccine produced in alfalfa transgenic plants. Vet Immunol Immunopathol 151(3–4):315–324. https://doi.org/10.1016/j.vetimm.2012.12.004

Alves DL, Pereira LLF, van Drunen LH (2014) Bovine herpesvirus glycoprotein D: a review of its structural characteristics and applications in vaccinology. Vet Res 45(1):111. https://doi.org/10.1186/s13567-014-0111-x

Bande F, Arshad SS, Hair Bejo M, Moeini H, Omar AR (2015) Progress and challenges toward the development of vaccines against avian infectious bronchitis. J Immunol Res 2015:12. https://doi.org/10.1155/2015/424860

Bente DA, Forrester NL, Watts DM, McAuley AJ, Whitehouse CA, Bray M (2013) Crimean-Congo hemorrhagic fever: history, epidemiology, pathogenesis, clinical syndrome and genetic diversity. Antiviral Res 100(1):159–189. https://doi.org/10.1016/j.antiviral.2013.07.006

Bhanuprakash V, Indrani BK, Hosamani M, Balamurugan V, Singh RK (2009) Bluetongue vaccines: the past, present and future. Expert Rev Vaccines 8(2):191–204. https://doi.org/10.1586/14760584.8.2.191

Borzacchiello G, Roperto F (2008) Bovine papillomaviruses, papillomas and cancer in cattle. Vet Res 39(5):45

Bouloy M, Flick R (2009) Reverse genetics technology for Rift Valley fever virus: current and future applications for the development of therapeutics and vaccines. Antiviral Res 84(2):101–118. https://doi.org/10.1016/j.antiviral.2009.08.002

Brock KV, Grooms DL, Givens MD (2008) Reproductive disease and persistent infections bovine viral diarrhea virus. Blackwell, Oxford, pp 145–156

Campo MS (1995) Infection by bovine papillomavirus and prospects for vaccination. Trends Microbiol 3(3):92–97

Carpenter S, Mellor PS, Torr SJ (2008) Control techniques for *Culicoides* biting midges and their application in the U.K. and northwestern Palearctic. Med Vet Entemol 22:175–187

Castañón S et al (1999) Immunization with potato plants expressing VP60 protein protects against rabbit hemorrhagic disease virus. J Virol 73(5):4452–4455

Cavanagh D, Davis PJ, Darbyshire JH, Peters RW (1986) Coronavirus IBV: virus retaining spike glycopolypeptide S2 but not S1 is unable to induce virus-neutralizing or haemagglutination-inhibiting antibody, or induce chicken tracheal protection. J Gen Virol 67 (7):1435–1442. https://doi.org/10.1099/0022-1317-67-7-1435

Chen EY, Howley PM, Levinson AD, Seeburg PH (1982) The primary structure and genetic organization of the bovine papillomavirus type 1 genome. Nature 299(5883):529–534

Choi KH et al (2004) The structure of the RNA-dependent RNA polymerase from bovine viral diarrhea virus establishes the role of GTP in de novo initiation. Proc Natl Acad Sci USA 101 (13):4425–4430. https://doi.org/10.1073/pnas.0400660101

Christova I, Kovacheva O, Georgieva G, Ivanova S, Argirov D (2010) Vaccine against Congo-crimean haemorrhagic fever virus—Bulgarian input in fighting the disease. ProblInfectParasitDis 37:7–8

Cosby KC, Yamanouchi K (2006) Immunology of rinderpest—an immunosuppression but a lifelong vaccine protection. In: Barrett T, Pastoret P-P, Taylor W (eds) Rinderpest and Peste des Petits Ruminants. Elsevier, Oxford, UK, pp 196–222

de Boer SM et al (2010) Rift valley fever virus subunit vaccines confer complete protection against a lethal virus challenge. Vaccine 28:2330–2339

Dhama K, Chauhan RS, Mahendran M, Malik SVS (2009) Rotavirus diarrhea in bovines and other domestic animals. Vet Res Commun 33(1):1–23. https://doi.org/10.1007/s11259-008-9070-x

Dong J-L, Liang B-G, Jin Y-S, Zhang W-J, Wang T (2005) Oral immunization with pBsVP6-transgenic alfalfa protects mice against rotavirus infection. Virology 339(2):153–163. https://doi.org/10.1016/j.virol.2005.06.004

Elbers ARW, van der Spek AN, van Rijn PA (2009) Epidemiologic characteristics of bluetongue virus serotype 8 laboratory-confirmed outbreaks in The Netherlands in 2007 and a comparison with the situation in 2006. Prev Vet Med 92(1–2):1–8. https://doi.org/10.1016/j.prevetmed.2009.08.007

Fernández-Fernández MR, Mouriño M, Rivera J, Rodríguez F, Plana-Durán J, García JA (2001) Protection of rabbits against rabbit hemorrhagic disease virus by immunization with the VP60 protein expressed in plants with a potyvirus-based vector. Virology 280(2):283–291. https://doi.org/10.1006/viro.2000.0762

Filgueira DP et al. (2003) Bovine herpes virus gD protein produced in plants using a recombinant tobacco mosaic virus (TMV) vector possesses authentic antigenicity. Vaccine 21 https://doi.org/10.1016/s0264-410x(03)00495-x

French TJ, Marshall JJ, Roy P (1990) Assembly of double-shelled, virus like particles of bluetongue virus by the simultaneous expression of four structural proteins. J Virol 64 (12):5695–5700

Fulton RW (2008) Vaccines bovine viral diarrhea virus. Blackwell, Oxford, pp 209–222

Gao Y et al (2015) A transgenic ginseng vaccine for bovine viral diarrhea. Virol J 12:73. https://doi.org/10.1186/s12985-015-0301-9

Ghiasi SM et al (2012) Molecular farming, an effective system for the production of immunogenic Crimean-Congo Hemorrhagic fever virus glycoprotein. Prog Biol Sci 2(1):12–29

Gómez N, Carrillo C, Salinas J, Parra F, Borca MV, Escribano JM (1998) Expression of immunogenic glycoprotein S polypeptides from transmissible gastroenteritis coronavirus in transgenic plants. Virology 249(2):352–358. https://doi.org/10.1006/viro.1998.9315

Gómez N et al (2000) Oral immunogenicity of the plant derived spike protein from swine-transmissible gastroenteritis coronavirus. Adv Virol 145(8):1725–1732. https://doi.org/10.1007/s007050070087

Graham DA (2013) Bovine herpes virus-1 (BoHV-1) in cattle—a review with emphasis on reproductive impacts and the emergence of infection in Ireland and the United Kingdom. Irish Vet J 66(1):15–15. https://doi.org/10.1186/2046-0481-66-15

Habjan M et al (2009) Efficient production of Rift Valley fever virus-like particles: the antiviral protein MxA can inhibit primary transcription of bunyaviruses. Virology 385(2):400–408. https://doi.org/10.1016/j.virol.2008.12.011

Hamad MA, Al-Shammari AM, Odisho SM, Yaseen NY (2016) Molecular and phylogenetic analysis of bovine papillomavirus type 1: first report in Iraqi cattle. Adv Virol 2016:7. https://doi.org/10.1155/2016/2143024

Hammond RW, Nemchinov LG (2009) Plant production of veterinary vaccines and therapeutics. In: Karasev AV (ed) Plant-produced Microbial Vaccines, vol 332. Springer, Berlin, pp 79–102

Hoogstraal H (1979) The epidemiology of tick-borne Crimean-Congo hemorrhagic fever in Asia, Europe and Africa. J Med Entomol 15:307–417

Ikegami T, Makino S (2009) Rift valley fever vaccines. Vaccine 27(Supplement 4):D69–D72. https://doi.org/10.1016/j.vaccine.2009.07.046

Jackwood MW (2012) Review of infectious bronchitis virus around the world. Avian Dis 56 (4):634–641. https://doi.org/10.1637/10227-043012-review.1

Jenckel M et al (2015) Complete coding genome sequence of putative novel bluetongue virus serotype 27. Genome Announcements 3(2) 10.1128/genomea.00016-15

Johnson MA, Pooley C, Ignjatovic J, Tyack SG (2003) A recombinant fowl adenovirus expressing the S1 gene of infectious bronchitis virus protects against challenge with infectious bronchitis virus. Vaccine 21(21–22):2730–2736. https://doi.org/10.1016/S0264-410X(03)00227-5

Kalbina I et al (2016) Arabidopsis thaliana plants expressing rift valley fever virus antigens: mice exhibit systemic immune responses as the result of oral administration of the transgenic plants. Protein Expr Purif 127:61–67. https://doi.org/10.1016/j.pep.2016.07.003

Keshtkar-Jahromi M, Kuhn JH, Christova I, Bradfute SB, Jahrling PB, Bavari S (2011) Crimean-Congo hemorrhagic fever: current and future prospects of vaccines and therapies. Antiviral Res 90(2):85–92. https://doi.org/10.1016/j.antiviral.2011.02.010

Khandelwal A, Lakshmi Sita G, Shaila MS (2003) Oral immunization of cattle with hemagglutinin protein of rinderpest virus expressed in transgenic peanut induces specific immune responses. Vaccine 21(23):3282–3289. https://doi.org/10.1016/S0264-410X(03)00192-0

Khandelwal A, Renukaradhya GJ, Rajasekhar M, Sita GL, Shaila MS (2004) Systemic and oral immunogenicity of hemagglutinin protein of rinderpest virus expressed by transgenic peanut plants in a mouse model. Virology 323(2):284–291. https://doi.org/10.1016/j.virol.2004.02.030

Kirnbauer R, Booy F, Cheng N, Lowy DR, Schiller JT (1992) Papillomavirus L1 major capsid protein self-assembles into virus-like particles that are highly immunogenic. Proc Natl Acad Sci 89(24):12180–12184

Labbé M, Charpilienne A, Crawford SE, Estes MK, Cohen J (1991) Expression of rotavirus VP2 produces empty core like particles. J Virol 65(6):2946–2952

Lamphear BJ et al (2002) Delivery of subunit vaccines in maize seed. J Control Release 85(1–3):169–180. https://doi.org/10.1016/S0168-3659(02)00282-1

Lamphear BJ, Jilka JM, Kesl L, Welter M, Howard JA, Streatfield SJ (2004) A corn-based delivery system for animal vaccines: an oral transmissible gastroenteritis virus vaccine boosts lactogenic immunity in swine. Vaccine 22(19):2420–2424. https://doi.org/10.1016/j.vaccine.2003.11.066

Lanyon SR, Hill FI, Reichel MP, Brownlie J (2014) Bovine viral diarrhoea: pathogenesis and diagnosis. Vet J 199(2):201–209. https://doi.org/10.1016/j.tvjl.2013.07.024

Laude H, Rasschaert D, Delmas B, Godet M, Gelfi J, Charley B (1990) Molecular biology of transmissible gastroenteritis virus. Vet Microbiol 23(1):147–154. https://doi.org/10.1016/0378-1135(90)90144-K

Lentz EM, Mozgovoj MV, Bellido D, Santos MJD, Wigdorovitz A, Bravo-Almonacid FF (2011) VP8* antigen produced in tobacco transplastomic plants confers protection against bovine rotavirus infection in a suckling mouse model. J Biotechnol 156(2):100–107. https://doi.org/10.1016/j.jbiotec.2011.08.023

Liu L, Celma CCP, Roy P (2008) Rift valley fever virus structural proteins: expression, characterization and assembly of recombinant proteins. Virol J 5:82–82. https://doi.org/10.1186/1743-422X-5-82

Love AJ, Chapman SN, Matic S, Noris E, Lomonossoff GP, Taliansky M (2012) In planta production of a candidate vaccine against bovine papillomavirus type 1. Planta 236(4):1305–1313. https://doi.org/10.1007/s00425-012-1692-0

Maan NS et al (2012) Identification and differentiation of the twenty six bluetongue virus serotypes by RT–PCR Amplification of the serotype-specific genome segment 2. PLoS ONE 7 (2):e32601. https://doi.org/10.1371/journal.pone.0032601

Maclachlan N, Drew C, Darpel K, Worwa G (2009) The pathology and pathogenesis of bluetongue. J Comp Pathol 141:1–16

Martin SJ (1986) The structure and composition of morbilliviruses: a brief review. Rev Sci Tech Off Int Epiz 5(9):389–393

Martín-Alonso JM, Castañón S, Alonso P, Parra F, Ordás R (2003) Oral immunization using tuber extracts from transgenic potato plants expressing rabbit hemorrhagic disease virus capsid protein. Transgenic Res 12(1):127–130. https://doi.org/10.1023/A:1022112717331

Matsumura T, Itchoda N, Tsunemitsu H (2002) Production of immunogenic VP6 protein of bovine group A rotavirus in transgenic potato plants. Adv Virol 147(6):1263–1270. https://doi.org/10.1007/s00705-002-0808-4

Mertens PP, Diprose J, Maan S, Singh KP, Attoui H, Samuel AR (2004) Bluetongue virus replication, molecular and structural biology. Vet Ital 40(4):426–37

Munday JS (2014) Bovine and human papillomaviruses. Vet Pathol 51(6):1063–1075. https://doi.org/10.1177/0300985814537837

Muylkens B, Thiry J, Kirten P, Schynts F, Thiry E (2007) Bovine herpesvirus 1 infection and infectious bovine rhinotracheitis. Vet Res 38(2):181–209

Näslund J et al (2009) Vaccination with virus-like particles protects mice from lethal infection of rift valley fever virus. Virology 385(2):409–415. https://doi.org/10.1016/j.virol.2008.12.012

Nelson G, Marconi P, Periolo O, La Torre J, Alvarez MA (2012) Immunocompetent truncated E2 glycoprotein of bovine viral diarrhea virus (BVDV) expressed in nicotiana tabacum plants: a candidate antigen for new generation of veterinary vaccines. Vaccine 30(30):4499–4504. https://doi.org/10.1016/j.vaccine.2012.04.068

OIE (2008a) Rinderpest, vol Ch 2.1.15

OIE (2008b) Transmissible gastroenteritis, vol Ch 2.8.11

Papp H et al (2013) Review of group A rotavirus strains reported in swine and cattle. Vet Microbiol 165(3–4):190–199. https://doi.org/10.1016/j.vetmic.2013.03.020

Pecora A, Aguirreburualde MSP, Aguirreburualde A, Leunda MR, Odeon A, Chiavenna S, Bochoeyer D, Spitteler M, Filippi JL, Santos MJD, Levy SM (2012) Safety and efficacy of an E2 glycoprotein subunit vaccine produced in mammalian cells to prevent experimental infection with bovine viral diarrhoea virus in cattle. Vet Res Commun 36(3):157–164.

Pepin M, Bouloy M, Bird BH, Kemp A, Paweska J (2010) Rift valley fever virus (Bunyaviridae: Phlebovirus): an update on pathogenesis, molecular epidemiology, vectors, diagnostics and prevention. Vet Res 41(6) https://doi.org/10.1051/vetres/2010033

Plowright W, Ferris RD (1962) Studies with rinderpest virus in tissue culture: the use of attenuated culture virus as a vaccine in cattle. Res Vet Sci 3:172–182

Price KR (2012) Use of live vaccines for coccidiosis control in replacement layer pullets. J Appl Poult Res 21(3):679–692. https://doi.org/10.3382/japr.2011-00486

Raj GD, Jones RC (1997) Infectious bronchitis virus: immunopathogenesis of infection in the chicken. Avian Pathol 26(4):677–706. https://doi.org/10.1080/03079459708419246

Ridpath JF (2013) Immunology of BVDV vaccines. Biologicals 41(1):14–19. https://doi.org/10.1016/j.biologicals.2012.07.003

Roeder P, Mariner J, Kock R (2013) Rinderpest: the veterinary perspective on eradication. Philos Trans R Soc Lond B Biol Sci 368:20120139

Roy P (1992) Bluetongue virus proteins. J Gen Virol 73(Pt 12):3051–64

Ruiz V, Mozgovoj MV, Dus Santos MJ, Wigdorovitz A (2015) Plant-produced viral bovine vaccines: what happened during the last 10 years? Plant Biotechnol J 13(8):1071–1077. https://doi.org/10.1111/pbi.12440

Saif LJ, Fernandez FM (1996) Group A rotavirus veterinary vaccines. J Infect Dis 174 (Supplement_1):S98-S106 https://doi.org/10.1093/infdis/174.supplement_1.s98

Saif LJ, van Cott JL, Brim TA (1994) Immunity to transmissible gastroenteritis virus and porcine respiratory coronavirus infections in swine. Vet Immunol Immunopathol 43(1):89–97. https://doi.org/10.1016/0165-2427(94)90124-4

Sainsbury F, Thuenemann EC, Lomonossoff GP (2009) pEAQ: versatile expression vectors for easy and quick transient expression of heterologous proteins in plants. Plant Biotechnol 7:682–693

Saldaña S et al (2006) Production of rotavirus-like particles in tomato (*Lycopersicon esculentum L.*) fruit by expression of capsid proteins VP2 and VP6 and immunological studies. Viral Immunol 19(1):42–53. https://doi.org/10.1089/vim.2006.19.42

Samy AM, Peterson AT (2016) Climate change influences on the global potential distribution of bluetongue virus. PLoS ONE 11(3):e0150489. https://doi.org/10.1371/journal.pone.0150489

Sathish K et al (2011) Plant expressed EtMIC2 is an effective immunogen in conferring protection against chicken coccidiosis. Vaccine 29(49):9201–9208. https://doi.org/10.1016/j.vaccine.2011.09.117

Sathish K et al (2012) Plant expressed coccidial antigens as potential vaccine candidates in protecting chicken against coccidiosis. Vaccine 30(30):4460–4464. https://doi.org/10.1016/j.vaccine.2012.04.076

Satyavathi V, Prasad V, Khandelwal A, Shaila M, Sita LG (2003) Expression of hemagglutinin protein of rinderpest virus in transgenic pigeon pea [Cajanus cajan (L.) Millsp.] plants. Plant Cell Rep 21(7):651–658 https://doi.org/10.1007/s00299-002-0540-2

Shirley MW, Smith AL, Tomley FM (2005) The biology of avian Eimeriawith an emphasis on their control by vaccination. Adv Parasitol 60 https://doi.org/10.1016/s0065-308x(05)60005-x

Shirley MW, Smith AL, Blake DP (2007) Challenges in the successful control of the avian coccidia. Vaccine 25 https://doi.org/10.1016/j.vaccine.2006.12.030

Shivaramaiah C, Barta JR, Hernandez-Velasco X, Téllez G, Hargis BM (2014) Coccidiosis: recent advancements in the immunobiology of *Eimeria* species, preventive measures, and teh importance of vaccination as a control tool against these Apicomlexan parasites. Vet Med Res Rep 5:23–34

Song CS et al (1998) Induction of protective immunity in chickens vaccinated with infectious bronchitis virus S1 glycoprotein expressed by a recombinant baculovirus. J Gen Virol 79 (4):719–723. https://doi.org/10.1099/0022-1317-79-4-719

Sperlova A, Zendulkova D (2011) Bluetongue: a review. Vet Med 56(9):430–452

Spik K, Shurtleff A, McElroy AK, Guttieri MC, Hooper JW, Schmaljohn C (2006) Immunogenicity of combination DNA vaccines for rift valley fever virus, tick-borne encephalitis virus, Hantaan virus, and Crimean Congo hemorrhagic fever virus. Vaccine 24 (21):4657–4666. https://doi.org/10.1016/j.vaccine.2005.08.034

Stewart M et al (2012) Protective efficacy of Bluetongue virus-like and subvirus-like particles in sheep: presence of the serotype-specific VP2, independent of its geographic lineage, is essential for protection. Vaccine 30(12):2131–2139. https://doi.org/10.1016/j.vaccine.2012.01.042

Thuenemann EC, Meyers AE, Verwey J, Rybicki EP, Lomonossoff GP (2013) A method for rapid production of heteromultimeric protein complexes in plants: assembly of protective bluetongue virus-like particles. Plant Biotechnol J:n/a-n/a. https://doi.org/10.1111/pbi.12076

Thomas C, Young NJ, Heaney J, Collins ME, Brownlie J (2009) Evaluation of efficacy of mammalian and baculovirus expressed E2 subunit vaccine candidates to bovine viral diarrhoea virus. Vaccine 27(17):2387–2393.

Tuboly T et al (2000) Immunogenicity of porcine transmissible gastroenteritis virus spike protein expressed in plants. Vaccine 18(19):2023–2028. https://doi.org/10.1016/S0264-410X(99)00525-3

van Drunen LH (2006) Rationale and perspectives on the success of vaccination against bovine herpesvirus-1. Vet Microbiol 113(3–4):275–282. https://doi.org/10.1016/j.vetmic.2005.11.002

van Zyl AR, Meyers AE, Rybicki EP (2016) Transient Bluetongue virus serotype 8 capsid protein expression in Nicotiana benthamiana. Biotechnol Rep 9:15–24. https://doi.org/10.1016/j.btre.2015.12.001

Wallace DB, Ellis CE, Espach A, Smith SJ, Greyling RR, Viljoen GJ (2006) Protective immune responses induced by different recombinant vaccine regimes to rift valley fever. Vaccine 24 (49–50):7181–7189. https://doi.org/10.1016/j.vaccine.2006.06.041

Whitehouse CA (2004) Crimean-Congo hemorrhagic fever. Antiviral Res 64:145–160

Wigdorovitz A et al (2004) Protective lactogenic immunity conferred by an edible peptide vaccine to bovine rotavirus produced in transgenic plants. J Gen Virol 85(7):1825–1832. https://doi.org/10.1099/vir.0.19659-0

Yang Y et al (2011) Immunogenicity and virus-like particle formation of rotavirus capsid proteins produced in transgenic plants. Sci China Life Sci 54(1):82–89. https://doi.org/10.1007/s11427-010-4104-3

Zhou J-Y et al (2004) Generation of the transgenic potato expressing full-length spike protein of infectious bronchitis virus. J Biotechnol 111(2):121–130. https://doi.org/10.1016/j.jbiotec.2004.03.012

Zhou B et al (2010) Oral administration of plant-based rotavirus VP6 induces antigen-specific IgAs, IgGs and passive protection in mice. Vaccine 28(37):6021–6027. https://doi.org/10.1016/j.vaccine.2010.06.094

Zhou ZR, Wang ML, Deng F, Li Tx HuZH, Wang HL (2011) Production of CCHF virus-like particle by a baculovirus-insect cell expression. Virologica Sinica 26:338–346

Zientara S, Maclachlan N, Calistri P, Sanchez-Vizcaino J, Savini G (2010) Bluetongue vaccination in Europe. Expert Rev Vaccines 9:989–991

Zimmermann J et al (2009) Antibody expressing pea seeds as fodder for prevention of gastrointestinal parasitic infections in chickens. BMC Biotechnol 9(1):79. https://doi.org/10.1186/1472-6750-9-79

MIX
Papier aus verantwortungsvollen Quellen
Paper from responsible sources
FSC® C105338

If you have any concerns about our products,
you can contact us on
ProductSafety@springernature.com

In case Publisher is established outside the EU,
the EU authorized representative is:
Springer Nature Customer Service Center GmbH
Europaplatz 3, 69115 Heidelberg, Germany

Printed by Libri Plureos GmbH
in Hamburg, Germany